THERMAL RECOVERY OF OIL AND BITUMEN

ROGER M. BUTLER
Department of Chemical
and Petroleum Engineering
University of Calgary
Calgary, Alberta, Canada

Prentice Hall, Englewood Cliffs, New Jersey 07632

Library of Congress Cataloging-in-Publication Data

Butler, Roger M. (date)
Thermal recovery of oil and bitumen/Roger M. Butler.
p. cm.
Includes bibliographical references and index.
ISBN 0-13-914953-8
1. Thermal oil recovery. 2. Bitumen. I. Title.
TN871.B87 1991 90-7401
622'.33827--dc20 CIP

Cover design: Bruce Kenselaar
Manufacturing buyer: Kelly Behr and Sue Brunke

Printed in the United States of America
10 9 8 7 6 5 4 3 2 1

ISBN 0-13-914953-8

Prentice-Hall International (UK) Limited, *London*
Prentice-Hall of Australia Pty. Limited, *Sydney*
Prentice-Hall Canada Inc., *Toronto*
Prentice-Hall Hispanoamericana, S.A., *Mexico*
Prentice-Hall of India Private Limited, *New Delhi*
Prentice-Hall of Japan, Inc., *Tokyo*
Simon & Schuster Asia Pte. Ltd., *Singapore*
Editora Prentice-Hall do Brasil, Ltda., *Rio de Janeiro*

Contents

Preface

This book describes the recovery of heavy oils and bitumen by in situ thermal methods. It is based on the lecture notes, which have been developed by the author for an annual thirteen-week graduate course at the University of Calgary, to classes drawn from full-time graduate students and to a greater extent from engineers whose work is directly related to the oil industry. The author has presented the course each year since 1982 and the book has been written during this period.

The first chapter is an introduction to the subject. The heavy oil and, more importantly, the bitumen deposits in Canada are an enormous resource which will become of great economic importance. Production from these sources is already equivalent to a very significant fraction of the Canadian requirement for crude oil. Other countries as well as Canada have vast deposits of these crude oils. The deposits in the Canadian province of Alberta and those in Venezuela are each approximately equal in quantity to those of conventional crudes in place in the Middle East reservoirs. The purpose of this book is to discuss the technical factors and problems which are involved in their production by those in situ methods which involve the heating of the reservoir.

Although the book discusses, in a logical development, the theory and much of the practice in this area, it is not intended to be an encyclopedia on the subject. It describes the main ideas of the subject with the purpose of providing the reader with tools which can be used to make further advances. In places, the book summarizes well established thinking whereas in others, it describes original ideas and approaches; some of these have been published previously in papers written by the author and his colleagues while others appear here for the first time.

Chapter 2 deals with the transfer of heat within the reservoir bulk and within the adjacent regions by thermal conduction. Equations are presented, and many are derived, which allow the analysis and prediction of quantities such as the heat loss from the boundaries of a heated reservoir. Numerical examples in this chapter, like those in other chapters, provide the reader with the means for the practical understanding and application of the theoretical material. The interspersing of numerical examples within the book and, in some cases, the use of the results from the examples for the further development of concepts, are intended to make this book interesting and useful to the practical engineer. The approach employed is practical and fundamental with a minimum of academic sophistication. The author's address is now in an invory tower but he came with tar on his boots.

One of the conclusions to be drawn from Chapter 2 is that simple thermal conduction is, in most instances, an inadequate means for heating substantial reservoir volumes from small diameter wells. It is too slow. The third chapter discusses convective heating achieved by the injection of hot fluids such as steam or hot water. This allows heat to be introduced much faster and over substantial volumes. Again a practical approach involving the use of illustrative numerical examples is employed. One of the conclusions which the reader will draw from this chapter is that a very substantial quantity of heat is required simply to raise a volume of reservoir to the steam temperature and that this quantity has to be augmented, frequently several-fold, in order to also supply the losses of heat from the reservoir boundaries. The material in the third chapter provides the reader with tools which allow the estimation of these quantities and with a grasp of how the heat is distributed in steam recovery processes.

Steamflooding and results from steamflooding field projects are discussed further in the fourth chapter. The chapter also extends the theoretical ideas developed previously. For example the tendency for steam to override the oil during lateral steamflooding and the contribution of steam distillation to recovery are discussed.

Chapter 5 is concerned with the mechanism by which oil is displaced by injected fluids. A factor of major importance here is that the displacing fluid is usually much less viscous than the oil. This causes interfacial instabilities and the fingering of the displacing fluid—particularly if it is water. The situation can be different with steam since it condenses when it intrudes into colder oil and it is the resulting aqueous condensate rather than the steam which fingers. Also steam tends to float above the adjacent oil and override because of its low density. One of the subjects which is discussed with practically-oriented numerical examples is the displacement of oil by steam within a steam-saturated region using the Buckley–Leverett approach. This mechanism is surprisingly effective despite the sharp contrast between the viscosity of the steam and the oil. It is shown that the reason for this is that the flow of steam in such situations is, on a volumetric basis, much higher than that of the oil. Steam contains much less heat per unit volume than does hot water and much larger volumes are required to heat a volume of reservoir. These much larger volumes are much more effective in displacing oil from the heated zone even though the dynamic viscosity of steam is less than that of water.

The cyclic steam stimulation process is described in Chapter 6. This process was discovered by accident in 1959 and it provided the main thrust for the early development of thermal recovery in California, although most of those projects have now been converted to steamflooding. Steam stimulation is still the major process for the in situ recovery of Alberta bitumen although it is likely that it too will be surpassed eventually by steam displacement processes because of their potential to achieve higher oil recoveries.

The Steam-Assisted Gravity Drainage process is described in Chapter 7. This involves steamflooding to horizontal production wells which are located near the base of the reservoir. Steam is injected from wells which are higher in the formation. Steam-saturated zones form and grow above the production wells. The growth of these steam chambers can be both vertical and sideways. The oil near the boundary of each chamber is heated and it flows by gravity downwards to the production well. An important feature of the process is that the displaced oil remains heated as it flows to the production wells. The process has been tested in field pilots, particularly in AOSTRA's Underground Test Facility in the Athabasca tar sands near McMurray. The performance of this pilot is promising and recent reviews by AOSTRA conclude that the process should be economically competitive with steam stimulation even for projects recovering the extremely viscous crude of Athabasca.

The facilities which are used for thermal recovery including steam generators, wells, lifting practices, treating, and recycling water are discussed and analyzed in Chapter 8.

Heavy oil and bitumen recovery using in situ combustion is reviewed in Chapter 9. The main attraction to this process, as compared to steaming, lies in the much lower cost of the heat for the reservoir. This advantage continues to generate interest in the process although development activity appears to be declining. The chapter includes a discussion of the principles involved and describes several successful and economic field applications.

The final chapters in the book are Appendices which contain data and correlations useful in the analysis of thermal recovery projects.

The author is grateful to many people for the help and advice they have given to him in developing this work to its present state: Chi Tak Yee, Viera Oballa, and Philip Bakes as well as many other students, made important contributions in identifying inconsistencies, and errors, both substantial and typographical, in earlier versions of the text. Riza Konak of Esso Resources Canada and Ken Porter of Gulf Resources Canada reviewed the material of Chapter 8 and suggested valuable improvements and additions. Gordon Moore and Matthew Ursenbach of the University of Calgary reviewed the material on in situ combustion and made important and useful suggestions. The author is also indebted to his former colleagues of the Heavy Oil Research Division of Esso Resources Canada who contributed ideas, advice, and enthusiasm which became embedded in his experience. He has clear, vivid memories of many stimulating and productive disussions with G. S. McNab, H.Y. Lo, D. J. Stephens, M. Weiss, F. Greebe, D. A. Best, S. Bharatha, P. N. Troffimenkoff, P. J. Griffin, R. Leaute, and many others. For him they were exciting years and exciting people. The encouragement and support which the author has received from the Alberta Oil Sands Technology and Research Authority (AOSTRA), and particularly from its first Chairman Dr. C.W. Bowman and its first Vice-Chairman Dr. M. A. Carrigy, is also acknowledged. The Authority employed the author as Director of Technical Programs during 1983 and it was in this period that its plans for the Underground Test Facility were finalized. In March 1984 the author's proposal for the Steam Assisted Gravity Drainage process as the first process to be demonstrated in the UTF was presented at a review organized by AOSTRA for potential industry participants. The success of the subsequent test and the enthusiasm which this has generated in industry has done much to bolster his confidence in presenting the material of Chapter 7. The author is also indebted to the Calgary Section of the Petroleum Society of CIM and to the industries of Calgary for their endowment of the Chair of Petroleum Engineering which he has occupied since 1983. This support has made the writing of this book possible.

I wish to thank the following for permission to use and copy material for which they hold the copyright; in each case, credit is also give to the author where the material appears:

Alberta Energy provided the data for Fig. 1.3 from their publication *Energy Heritage* (1982).

The American Petroleum Institute granted permission to print Fig. A.2.1.

The Alberta Oil Sands Research and Technology Authority (AOSTRA) for works published in the *Aostra Journal of Research,* the proceedings of the UNITAR/UNDP International Conferences of Heavy Crude and Tar Sands, and proceedings of the AOSTRA and CANPET Seminars: Figs. 1.4, 7.30 thru 7.32, 7.45 thru 7.54, 7.67, 7.70, 8.14, 8.17 thru 8.19, 8.29, 8.30, 8.33, 9.5, 9.22, 9.23, and 9.44.

Babcock & Wilcox, Barberton, Ohio, for Figures 8.1 thru 8.3 reproduced from their publication *Steam.*

Business Information Services (BIS), copyright holders for *Petroleum Engineer International* magazine, for permission to use Fig. 9.64.

The Canadian Institute of Mining and Metallurgy, publishers of the *Journal of Canadian Petroleum Technology* for the following: Figs. 4.7, 7.12, 7.15, 7.16, 7.34, 7.39, 7.44, 7.54 thru 7.56, 7.68, 7.69, and 8.27.

The Canadian Journal of Chemical Engineering (CJChE) for the following: Figs. 1.8, 7.1 thru 7.3, 7.5, and 7.7.

The Canadian Society of Petroleum Geologists for Figs. 1.2 and 1.5.

Corod Manufacturing Ltd. provided the original drawing for Fig. 8.21.

Editions Technip, Paris, France granted permission to reproduce Figs. 9.2 and 9.34.

Esso Resources Canada Limited for permission to use Figures 8.23, 8.24, 8.25, 8.35, and 8.38.

Mr. W. H. Fairfield and Mr. P. D. White for permission to publish Figures 9.60 to 9.63 and Tables 9.7 to 9.10.

Foster Wheeler Fired Heaters Ltd., Calgary, Alberta provided us with the illustrations for Figs. 8.10(a) and 8.10(b).

Dr. G.W. Govier for permission to use his data in Table 1.2.

McGraw-Hill, N.Y., publishers of the 1st UNITAR/UNDP International Conference on Heavy Crude and Tar Sands for the following: Fig. 1.1, Table 1.9, and Figs. 8.31 and 8.32.

Natco Canada, Calgary, Alberta provided the drawings for Figs. 8.7, 8.28, and 8.34.

The National Research Council, publishers of the *Canadian Journal of Earth Science,* for Fig. 6.29.

Professor C.W. Nutt, granted permission to publish Figs. 5.32 thru 5.35 and Figs. 5.40 and 5.41.

Oxford University Press, Oxford, U.K., for Fig. 2.12. Used by permission of the Oxford University Press.

The Petroleum Society of the Canadian Institute of Mining and Metallurgy (CIM), Calgary Section, for Figs. 7.58 thru 7.66 published in the preprints of the 40th Annual Technical Meeting of the Petroleum Society of CIM.

The Society of Petroleum Engineers holds the copyright for all material published in their SPE papers, the *Journal of Petroleum Engineering,* the *Society of Petroleum Engineering Journal,* and transcripts of the SPE of AIME. Permission has been received and acknowledged for the following: Figs. 1.9, 4.9, 4.11 thru 4.14, 4.17 thru 4.21, 4.29 thru 4.40, 4.43 thru 4.51, 5.1, 5.7, 5.9, 6.2 thru 6.13, 6.15, 6.21, 6.23, 6.24, 8.39, 8.43 thru 8.45, 9.6 thru 9.21, 9.25 thru 9.32, 9.35 thru 9.39, 9.45 thru 9.47, 9.50 thru 9.55, and 9.65; Tables 3.3, 4.1 thru 4.5, 4.8, 6.2, 8.4, and 9.6.

Dr. P. G. Saffman granted permission to reproduce Figs. 5.3 and 5.4.

TOTRAN Services Limited, Calgary, Alberta provided the photograph for Fig. 8.8.
Eugene F. Traverse supplied Figs. 4.15 and 4.16.

The first draft of the book was typed by Mrs. Margaret McAuslan in 1984 and the author is grateful to her for her hard work and interest. Since then the annual revisions and extensions of the lecture notes and the manuscript for this book have been typed in a world-class manner by Mrs. Patricia Stuart-Bakes. The author wishes to thank her for her perseverance, moral support, and enthusiasm.

Finally, the author wishes to recognize the encouragement and patience of his wife Joyce who has understood and supported him. Writing books is an interesting and worthwhile endeavor but it is time consuming and hard on one's family. Thank you, Joyce.

Roger M. Butler
Calgary, Alberta

1

Introduction to Thermal Recovery

The efficient and economic recovery of heavy oil and bitumen from reservoirs in Canada, Venezuela, and elsewhere is a major technical challenge and task. As will be seen later in this chapter, the quantities of heavy oil and, particularly, bitumen in place are as large as and probably far larger than those of conventional crude oil. The challenge is twofold: recovering the oils from the reservoir and converting them to useful petroleum products. Heavy oils and bitumen contain much larger proportions of nondistillable residual material than do conventional oils. The residues contain larger proportions of asphaltenes, and this makes them particularly viscous. It is their high natural viscosity that makes the recovery of heavy oils and bitumen difficult.

The same factors that determine the viscosity of these oils also greatly affect their conversion into conventional petroleum products. The high contents of asphaltic residue make them particularly suitable for asphalt manufacture but also greatly reduce their suitability for most other purposes. Their conversion to distillate boiling-range material involves residual cracking processes such as coking and/or hydrocracking. The high contents of sulphur and nitrogen in the distillates create the need for extensive hydrotreating. The aromatic content of the middle distillates obtained reduces their value as diesel and aviation jet fuels. Improvement of these properties requires further extensive hydrotreating.

The production and utilization of heavy oils and bitumens as basic raw materials for the manufacture of the conventional products of petroleum thus involves extensive technology; there are great incentives for the extension and improvement of this technology. This book concentrates on the first of the two areas described, the recovery phase. Although it may appear that this phase is the more straightforward of the two, nevertheless it abounds in interesting facets and opportunities for development and invention.

ENHANCED OIL RECOVERY METHODS

As the availability of conventional crude oil has declined, there has developed an increased incentive for the improvement of the recovery from known reservoirs, and methods for "enhanced oil recovery" have been developed. The most important of these are as follows:

THERMAL RECOVERY

- Steam stimulation
- Steamflooding
- Hot waterflooding
- In situ combustion

CHEMICAL PROCESSES

- Surfactant floods
- Polymer floods
- Alkaline floods

MISCIBLE DISPLACEMENT

- Light hydrocarbon floods
- Carbon dioxide floods

This book is concerned with the first of these, thermal recovery, a subject area that includes the techniques which have found the most extensive use. Most of the applications of thermal methods are for the recovery of heavy oils that are too viscous at the original reservoir conditions to flow with economic rates and recoveries. The effectiveness of these techniques depends largely upon the reduction in oil viscosity that accompanies heating. Although heating the oil requires energy, this is, in economic applications, considerably less than the energy that the produced oil is capable of providing. A factor which promotes the use of thermal recovery processes is that many of the deposits of heavy crudes are large, rich, and often well known. Thermal recovery projects are usually profitable and are frequently quite large.

The following introduces the salient characteristics of the common thermal recovery approaches.

Steam Stimulation

Shell discovered the process of steam stimulation by accident in Venezuela when it was producing heavy crude by steamflooding the Mene Grande field near the eastern shore of Lake Maracaibo.

During the flood, a breakthrough of steam to the surface of the ground occurred and, in order to reduce the steam pressure in the reservoir, the injection well was allowed to flow back. Copious quantities of oil were produced; from this accidental discovery in 1959 (reported by de Haan and van Lookeren 1969) came the steam stimulation process, which also goes by the name of *steam soak* and *huff and puff.* There was a very rapid growth in the use of steam stimulation in the next decade, particularly in California. By 1967 there were 408 steam generators in use in California producing about 120 kB/d of oil (Burns 1969).

In the steam stimulation process, steam is injected into the reservoir at rates of the order of 1000 B/d[1] for a period of weeks; the well is then allowed to flow back and is later pumped. In suitable applications, the production of oil is rapid and the process is efficient, at least in the early cycles. The process is used extensively in California and Venezuela; if the steam pressure is high enough to fracture the reservoir and thus allow injection, it can also be used to produce the very viscous oil of the oil sands. For this operation, a steam pressure of about 1 psi per foot of depth is required to overcome the in situ rock stresses to cause fracturing.

Imperial Oil—and later its production wing, Esso Resources Canada—has been the leading developer of the cyclic steam stimulation process for the production of bitumen from the oil sands at Cold Lake. This development started with small-scale pilot experiments in the early 1960s. With more-or-less continuous development on an ever-increasing scale, Esso's Cold Lake field is now producing over 80,000 B/d of bitumen and this, together with its proportionate share of the production from the Syncrude operation, has now converted Imperial Oil, which is Canada's largest oil producer, to one that is dependent for about half of its production on Canadian bitumen. It is reasonable to expect that these trends will continue.

The main drawback of the cyclic steam stimulation process is that it often allows only about 15% of the oil to be recovered before the oil-to-steam ratio becomes prohibitively low.

Steamflooding

In this process steam is forced continuously into specific injection wells and oil is driven to separate production wells. The zones around the injection wells become heated to the saturation temperature of the steam, and these zones expand toward the production wells.

Oil and water from the condensation of steam are removed from the producers. With viscous oil there is a considerable tendency for the steam to override the reservoir, and this tends to limit the downward penetration of the heat and hence the recovery. Steamflooding can allow higher steam injection rates than steam stimulation; this advantage often offsets the rather lower thermal efficiency. Steam stimulation usually requires less (and in favorable cases far less) steam than flooding initially but is less efficient as depletion proceeds. Often it is economic to switch to steamflooding after initial operation of a field by steam stimulation. The recovery from steamflooding can approach 50% or even more.

[1]In the oil fields steam quantities are normally measured as the volume of water at standard conditions contained in the steam; a barrel of steam is thus 350 lb and a cubic meter is 1 tonne.

Burning bitumen as fuel in a conventional oil field steam generator would produce about 14 to 15 m^3 of 70% quality steam per cubic meter of fuel burned (or 14 to 15 B/B).

It is usual and desirable to produce oil first by steam stimulation from both the injectors and producers in a steam flood project. This provides rapid initial production and better economics and also allows effective steamflooding to be achieved more rapidly.

Hot Waterflooding

Hot waterflooding is usually less effective than steamflooding because of the lower heat content of hot water compared with steam. Also, it is found that the residual oil level that can be achieved with a hot waterflood is markedly higher than that found with steam—even at the same temperature.

It is thought that steam is more effective than hot water in displacing oil because of the following:

1. The extra pressure differential resulting from the higher kinematic viscosity of steam. A comparable mass flow of steam results in much higher fluid velocities and pressure differentials.
2. A relatively low tendency for steam to finger compared with water.
3. Steam distillation effects, which allow volatile fractions of the crude oil to evaporate into the steam and be carried by it. There are, thus, some of the characteristics of a miscible flood in displacement by steam.

These factors are discussed in subsequent chapters.

There is some application of hot waterflooding as a follow-up treatment to steamflooding; this is practiced in several areas.

In a later chapter it will be shown that, during a steamflood, oil is largely, and effectively, displaced from the steam-saturated zone (the steam chamber) and transferred through the condensation front. As the oil proceeds through the condensation front, it cools rapidly and its viscosity increases. In tar sand reservoirs with high initial oil viscosity, this displaced oil can rapidly seal off any communication passages that may exist (see, for example, Sufi 1988). In this reference, Sufi shows that the injection of steam into a permeable water-saturated zone at the base of a model reservoir containing tar sands resulted in rapid blockage; bitumen carried into the fracture plugged it as it cooled. On the other hand, the injection of hot water resulted in the gradual heating of the tar sand mass without blockage; as will be seen later, the reason for this is that hot water effects relatively little transport of bitumen as compared to steam. This difference may be useful if it is desired to heat tar sands by the injection of heat-carrying fluids into a relatively thin permeable zone or fracture. Under these circumstances, hot water is superior to steam because the permeable zone does not become blocked.

In Situ Combustion

In situ combustion involves the generation of heat by combustion within the reservoir. Air or (in some recent tests) oxygen is supplied to the combustion zone by injection into wells drilled from the surface. The main attraction of these processes is that heat is produced more cheaply than by surface steam generators. Although the fuel for heating comes from the reservoir itself, there is a substantial energy requirement for driving the compressors and—if oxygen is used—for operating the oxygen separation plant.

As the combustion zone advances through the reservoir, the oil ahead of the front becomes heated. Volatile fractions are distilled from the oil and then, as the temperature rises, thermal cracking reactions occur. The residual oil eventually forms a coke residue.

In the successful applications of this process, it is, for the most part, this coke that burns and supplies the fuel; because of the distillation and cracking that occur, the produced crude tends to be lighter and somewhat more valuable than the original crude oil. Emulsions produced by in situ combustion are often very difficult to separate.

In situ combustion tends to be less stable[2] than steam processes, and premature arrival of the combustion front at the production wells is common. This often causes well failure. Problems are also created sometimes by the bypassing of oxygen containing gas around the front into cooler parts of the reservoir. This results in low temperature oxidation (LTO) reactions in which the oxygen is added chemically to the oil. The oxygenated products have higher viscosities, and this makes the oil less easily recovered. Also, valuable oxygen is consumed wastefully by LTO reactions.

The inherent thermal advantage of in situ combustion as compared with steam should be greatest where heat losses for steam processes are greatest—in thin reservoirs and in deeply buried reservoirs. In in situ combustion, only the reservoir at and beyond the fire front needs to be at high temperature, particularly if water is injected as well as air (wet combustion). Water injection generates steam behind the combustion front. This steam passes through the front and condenses ahead of it. In this way, heat that would otherwise be left behind is utilized in steamflooding the oil ahead of the front. The vertical segregation, due to gravity, of the water and air/oxygen behind the front can be a problem.

WORLD FUEL RESOURCES

Table 1.1 compares estimates of the world's reserves of oil, gas, shale oil, and heavy oil and tar sands expressed in exajoules (1 EJ $= 10^{18}$ J $\cong 169 \times 10^6$ B of oil or 0.95×10^{12} SCF of gas). The columns in the table are not comparable because the first two are for recoverable reserves, whereas the second two represent the resource in place. However it is very clear that the oil sand and shale oil resources are enormous.

Table 1.2 compares Canadian energy resources of different types using both a proven, recoverable reserve basis and also an "ultimate" (recoverable) resource basis. The production in the year 1982 is also shown for comparison.

[2]The reasons for the greater stability of steam fronts are discussed in later chapters. Here it is sufficient to note that if a finger of steam tends to advance before a broadly moving front, the steam will tend to condense, leaving only water to advance, and this will become rapidly cooled. Thus a stable advancing steam front can have in front of it fingers of cold condensate running toward the well. It is the water that fingers, not the steam.

TABLE 1.1 World Fuel Resources in Exajoules

	Established Reserves	Resource in Place
Oil[(1)]	3970	
Gas[(1)]	3189	
Shale oil[(2)]		100,000
Heavy oil and oil sands[(3)]		22,000 to 36,000 (16000 of above in Canada)

[(1)]R. J. Enright (1982)
[(2)]F. L. Hartley, J. M. Hopkins and H. C. Huffman (1980)
[(3)]J. Janisch (1979)

Comparison of the upper and lower parts of the table shows that opportunities for finding conventional oil in Canada may be considered to be relatively limited. On the other hand, there appears to be more potential for discovering gas.

The present reserves for synthetic oil are much higher than for conventional oil, and the potential is very much higher. This is because the potential includes oil from in situ recovery as well as that from open-pit mining.

TABLE 1.2 Canadian Energy Resources in Exajoules and Exajoules per Year

RESOURCE	1982 PRODUCTION	NONFRONTIER	FRONTIERS ARCTIC	FRONTIERS OFFSHORE	TOTAL
		Proved Resources			
Conventional oil	2.4	29	1	7	37
Synthetic oil from tar sands	0.3	150	0	0	150
Natural gas	3.0	82			82
Coal	1.0	430	0	0	430
Uranium (CANDU eff.)	4.8	131	8		139
Hydro (30 yr)[(1)]	2.7	89			89
Total	14.2	911	9	7	927
		Ultimate Resources			
Conventional oil	2.4	60	89	68	217
Synthetic oil from tar sands	0.3	1,170			1,170
Natural gas	3.0	147	252	76	475
Coal	1.0	16,000	270		16,270
Uranium	4.8	?			?
Hydro	2.7	201	50		251
Total	14.2	17,578	661	144	18,383

[(1)]Hydroelectric power is a renewable resource, and the reserves are, in principle, infinite. To achieve a comparison, the quantities shown here represent the quantity of energy that would be produced from 30 y of operation.

(from Govier 1983)

The potential coal resource is now seen to be enormous and much higher than that for oil sands. The data indicate that there is sufficient coal to supply Canada's present production of energy resources for over 1000 years at the present rate of consumption—assuming that the coal can be converted into the required forms.

THE OIL SAND RESOURCE

Table 1.3 lists estimates of the volume of oil in place within the major known deposits of oil sand. There is considerable uncertainty in these figures—particularly those for Venezuela and for Alberta's Carbonate Triangle. Nevertheless, it is apparent that the heavy oil resource is, for the major part, divided between Canada and Venezuela.

Canada is not endowed with much "conventional" crude oil (at least with easily accessible conventional crude oil that can be found) but it does have tremendous

TABLE 1.3 Major Heavy Oil and Oil Sands Deposits

	Volume in Place (Billion Barrels)	Geological Age
Venezuela		
Orinoco heavy oil belt	700–3000	Tertiary and Lower Cretaceous Sands
Canada		
Athabasca	869	Lower Cretaceous Sands
Cold Lake	270	Lower Cretaceous Sands
Wabasca	119	Lower Cretaceous Sands
Peace River	92	Lower Cretaceous Sands
Lloydminster	32	Lower Cretaceous Sands
Carbonate Triangle	1350	Paleozoic Carbonates
Subtotal	2732	
U.S.S.R		
Melekess	123	Permian Sands
Siligir	13	Cambrian Carbonates
Olenek	8	Permian Sands
Subtotal	144	
U.S.A		
Tar Triangle	16	Permian Sands
Circle Cliffs	1	Permian Sands
Sunnyside	4	Eocene Sands
P.R. Springs	4	Eocene Sands
Hill Creek	1	Eocene Sands
Asphalt Ridge	1	Eocene Sands
Various heavy oils	110	Tertiary, Mesozoic
Subtotal	137	
Four-country total	3713–6013	

(from Janisch 1979)

Figure 1.1 Heavy Oil and Bitumen in Venezuela and Trinidad (after Gutierrez 1979)

quantities of oil sands and very substantial amounts of conventional heavy oil. Canada contains about one-sixth of the world's discovered oil in place, but about 95% of it is bitumen. The recovery and utilization of this bitumen is a challenge for engineers and scientists.

VENEZUELAN HEAVY OIL

The Venezuelan heavy oil fields and the extensions to them lie in a band across the northern end of South America, as may be seen from Figure 1.1 (Gutierrez 1979).

The eastern end of this band lies in Trinidad (K), where asphalt has been a product for many years. To the west lies the Gulf of Paria (J) and Guanoco (I).

To the south and west lies the Orinoco tar belt (E, F, G, and H), which contains the bulk of the material shown in the previous table. Up until now it has not been developed, although there are significant plans to do so. Area D is the Barinas subbasin.

The reservoirs around Lake Maracaibo (A, B, and C) are the most highly developed. It is here that Shell first experimented with steamflooding and discovered steam stimulation. Production from the Bolivar coast is discussed in Chapter 6.

CANADIAN HEAVY OIL AND BITUMEN

Although the origin of the Alberta oil sand deposits is speculative, the following seems to be a likely description.

Figure 1.2 (Jardine 1974) shows Alberta as it is thought to have been in Cretaceous time (120 million years ago). The climate was tropical and giant rivers, fed by water from the Canadian Shield in the east and from mountains to the west,

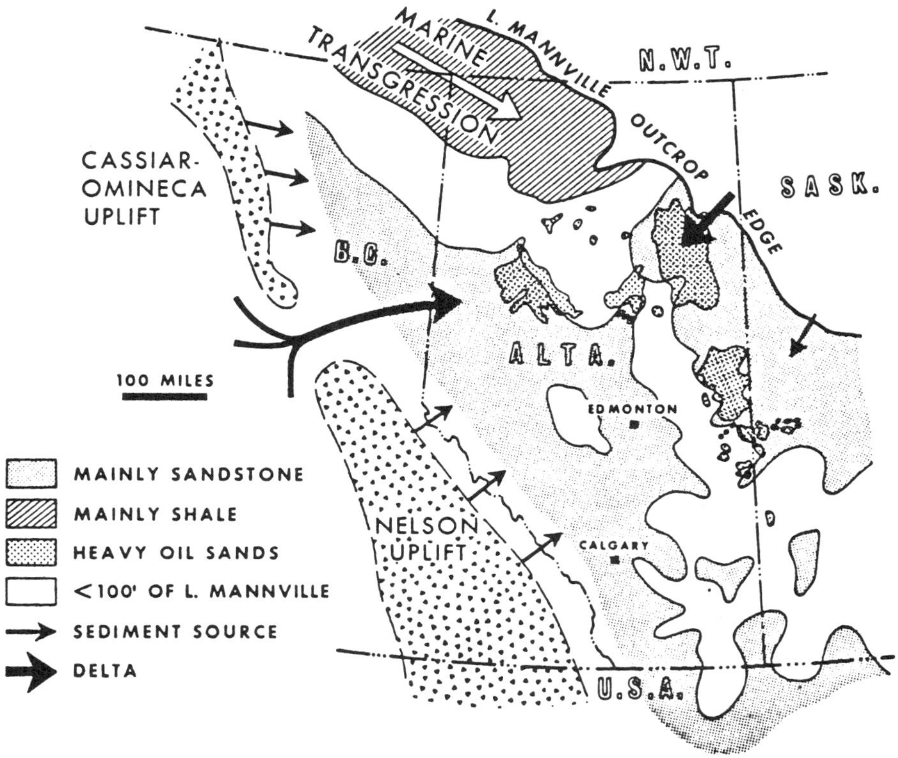

Figure 1.2 Lower Mannville Sediment Deposition in Western Canada (after Jardine 1974)

drained into what is now central Alberta and then northward to a sea. During the period this sea transgressed southward as the land sank.

One large delta formed the McMurray Sands of Athabasca, and the other sand deposits formed similarly. Eventually the sea transgressed further, and the Upper Mannville period ended with the deposition of the partly marine Grand Rapids Sands. The depositional environments under which different parts of the oil sands were laid down had a strong effect on the nature of the sand. Deposits in the relatively still water of the sea tended to be much finer than those in the channels where water motion kept the fine material suspended. Beach sand is finer than channel sand. Variations in the sea depth with time resulted in most deposits having a layered structure.

Mannville time ended with the subsidence of the entire area. The sea from the north joined the sea from the south, and the Upper Cretaceous sediments and shales were deposited. The locations of the major oil sands deposits are shown in Figure 1.3.

The preceeding paragraphs describe how the large deposits of sand were formed from the sediments carried by the rivers from the mountains and high

Figure 1.3 Heavy Oil and Bitumen in Western Canada (Courtesy Alberta Energy 1982)

ground. How the oil was deposited in these sediments and why it is so viscous are rather more obscure.

Some have argued that the tar sand oils have never been subjected to sufficiently high temperatures and pressures and that, because of this, the oils are immature. However, the view that the oil was originally a conventional crude oil that migrated to its present position and then became oxidized by bacteria that consumed the aliphatic chains of the oil while utilizing oxygen dissolved in surface water percolating through the deposit now seems more likely.

It is possible that the oil originated at a greater depth—possibly paleozoic limestone source beds beneath the Mannville Sands. High vanadium and sulphur contents have been said to indicate carbonate source rock rather than shale (Breger 1977). Others (Demaison 1977) think the source was in downdip Lower Cretaceous shales as far as 240 mi away.

It is thought that the thermally mature crude could then have migrated upwards in the Late Cretaceous period until it was trapped in the Cretaceous sand under the shale cap. Support for this idea is found in the fact that oil is found in every permeable zone within the reservoir. This oil became biodegraded by contact with oxygen-containing water percolating from the surface and through the shallow deposit. The flow of underground water driven by gravity forces is well known.

In support of this concept Jardine (1974) points out that the density of the Cretaceous oils in Alberta shows a definite correlation with the salinity of the reservoir water. Where there is high salinity, the density of the oil is low. In Athabasca, where the water is of low salinity (indicating that the original seawater has

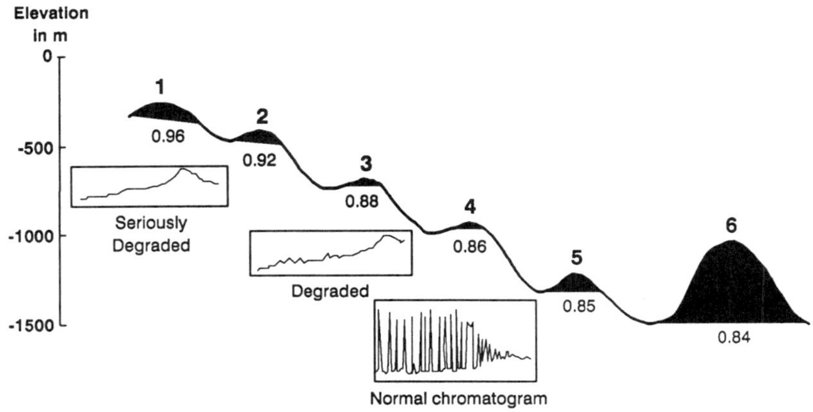

Figure 1.4 Origin of Heavy Oil from the West Slope in the Songliao Basin, China (after Hu Jianyi 1986)

been highly diluted), the density of the oil is relatively high. Paralleling this effect are the compositions of the oils, as shown by a series of chromatograms. The heaviest oils have the least paraffin chains and vice versa. This is consistent with the idea that biodegradation attacks paraffinic materials preferentially.

A very interesting series of oil pools in the Songliao Basin in northeastern China has been described by Hu Jianyi (1986). Figure 1.4, which is taken from his paper, shows a series of oil pools containing oil of increasing specific gravity and, as shown by the chromatograms, less and less aliphatic sidechains in the hydrocarbon molecules. It is thought that the oil migrated upwards through this series of traps and became more and more biodegraded as it encountered migrating surface waters containing oxygen.

Correlation of Canadian Tar Sand Deposits

Figure 1.5 is Jardine's (1974) correlation chart of the Canadian Tar Sand deposits. The names of the reservoir structures employed in the various Canadian Lower Cretaceous heavy oil deposits are compared. Thus, for example, the D unit at Cold Lake, the McMurray Sands in Athabasca and Wabasca, and the Gething and Bullhead Sands at Peace River were all deposited over the same period. However, of these, only the Athabasca and Peace River Sands contain large oil saturations; the others are water-saturated.

SIZE OF ALBERTA OIL SAND DEPOSITS

Table 1.4 (Strom and Dunbar 1979) shows an estimate of the bitumen in place in the major oil sand deposits shown in Figure 1.5; Figure 1.6 shows a comparison from another source (Allen 1979) of the volumes of oil in these deposits with those of conventional oil in Saudi Arabia.

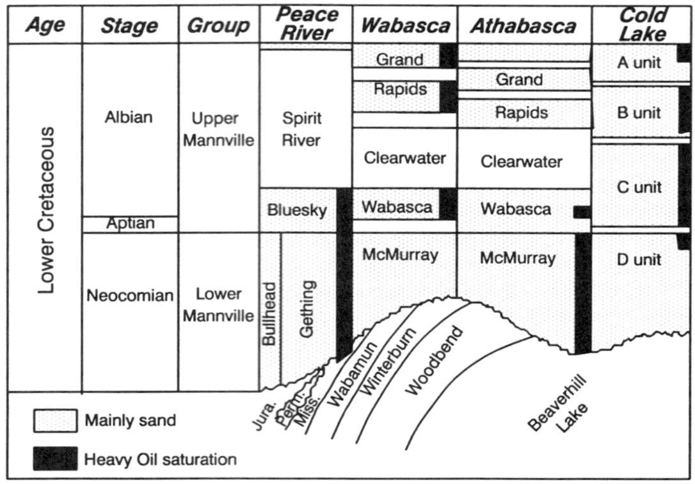

Figure 1.5 Correlation Chart of Lower Cretaceous Heavy Oil and Bitumen Deposits in Western Canada (after Jardine 1974)

COMPARISON OF HEAVY OIL AND CONVENTIONAL OIL RESOURCES

Figure 1.7, which is based upon a paper by A. Janisch (1979) compares estimates of the quantities of heavy oil plus tar sands with those for conventional oil. The quantities are shown as trillions of barrels (terabarrels). The Venezuelan and Canadian deposits are each comparable in quantity to the total of all of the known deposits of conventional crude oils.

The quantities shown in Figure 1.7 are estimates that have considerable uncertainty associated with them. However, it is believed that they are of the correct order of magnitude. It should also be noted that it is the *oil in place* that is depicted—not the recoverable oil. Little of the heavy oil and tar sands bitumen is recoverable without thermal recovery processes; even with these, much of the potential resource is uneconomic. There is a tremendous challenge to develop improved means for the production of heavy oil and bitumen.

DEPOSITS OF HEAVY OIL AND BITUMEN IN THE UNITED STATES

Although accumulations of heavy petroleum are much smaller in total than the Venezuelan and Canadian deposits, they are very substantial in the United States. These have been classified as either

1. Heavy oil: petroleum heavier than 25°API but sufficiently fluid at reservoir conditions to be produced commercially by natural flow.
2. Tar sands: sands containing bitumen, asphalt, or oil that is too viscous to flow in commercial quantities at reservoir conditions.

Table 1.5 shows estimates of the size of the U.S. deposits using these definitions.

TABLE 1.4 In-Place Bitumen Resources of Alberta—E.R.C.B. June, 1979

	Athabasca Wabiskaw-McMurray Surface Mineable	Athabasca Wabiskaw-McMurray Nonmineable	Peace River Bluesky Bullhead	Grand Rapids[1]	Cold Lake Clearwater	McMurray	Wabasca Grand Rapids[1]	Pre Cretaceous (Carbonates)[1]
Area (10^3 ha)	308	3,600	770	1,100	330	300	580	Not defined.
Initial bitumen in place (10^9 m^3)	19.2	95.2	10.2	17.3	5.7	2.2	6.0	Not determined deposits of economic value may be small.
Range of depth (m)	0–75	75–760	460–760	300–520	400–500	425–600	150–460	
Average thickness (m)	32	19	15	18	12	7	20	Bitumen carbonate thickness could be very large in local areas.
Average bitumen saturation (bulk weight fraction of bitumen sand.)	0.091	0.060	0.051	0.065	0.076	0.055	0.062	Bitumen content too low to warrant consideration in many areas.
Approximate density (kg/m^3)	1020	1010	990	980	1000	1000	1015	(1020)

[1]Composite of all units.

(from Strom and Dunbar 1979)

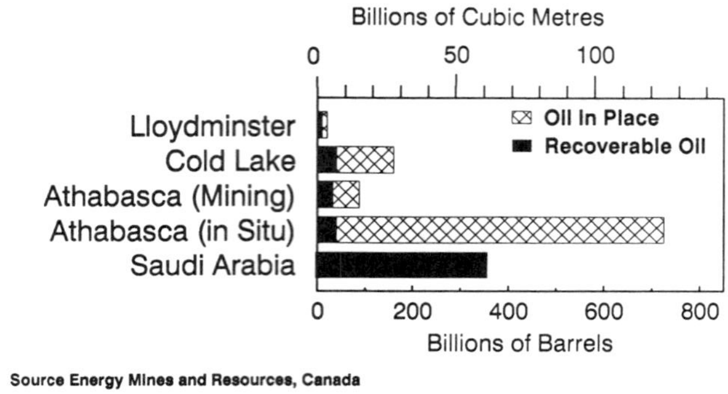

Figure 1.6 Comparison of Canadian Bitumen & Heavy Oil with Conventional Reserves in Saudi Arabia (after Allen 1979)

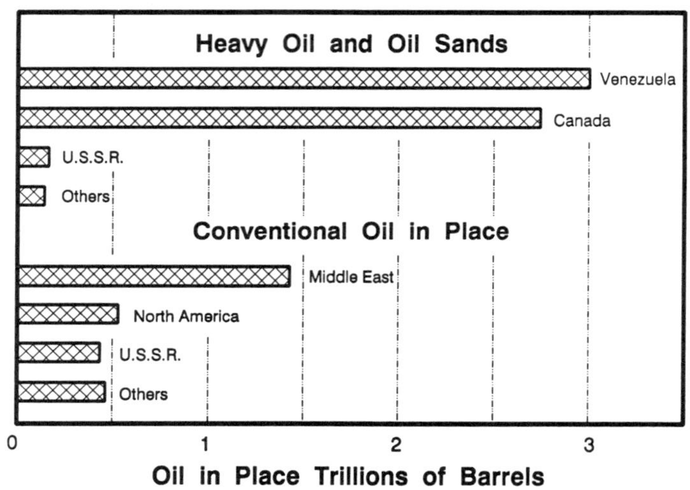

Figure 1.7 Comparison of Quantities of Heavy Oil and Tar Sands to those for Conventional Crude Oil in Place (order of magnitude estimates) (after Janisch 1979)

There is a total volume of heavy oil in place of about 107 billion barrels. About half of this occurs in California, and most of the remainder is in Texas.

Most of the tar sands in the United States are in Utah; the total volume in place is about 24–30 billion barrels.

THE NATURE OF HEAVY OIL AND BITUMEN DEPOSITS

Although this section is written with the Cretaceous deposits of heavy oil and bitumen in Alberta in mind, most of the concepts are applicable to other deposits. Oil sand as it occurs in a reservoir is a multiphase mixture with a very definite struc-

TABLE 1.5 Heavy Crudes and Tar Sands in the U.S.A.

HEAVY CRUDE OIL		Billions of Barrels in Place
Alabama and Mississippi		3
Arkansas		5
California		54
Louisiana		6
Oklahoma		2
Texas		31
Utah		1
Wyoming		5
	Total	107
TAR SANDS		
Utah		23–29
Other		1
	Total	24–30

(based on Whiting 1979)

ture. It consists largely of unconsolidated or sometimes weakly consolidated sand containing fluids—oil, water and sometimes gas within the pore space between the grains. In addition to the sand grains, other finer solids are present: silt and clays.

A characteristic feature of many of the tar sand deposits is that most or all of the solid material, is water-wet.[3] A layer of water is spread over the solid surface, which prevents the bitumen in the pore space from contracting the sand grains.

Figure 1.8 (Takamura 1982) shows the structure of typical Athabasca tar sand. Water within the sand is shown as occurring in three forms:

1. At the grain-to-grain contacts, water is present as pendular rings, which derive their shape from interfacial tension.
2. Along the surface of the solid materials, there is a thin (about 10 nm, or 0.01 μm) film of water. Although this is only a few tens of molecules thick[4], it is sufficient to protect the solid surface from contact with the bitumen.
3. Water is associated with clay and other fine material. These solids occur as clusters of fine particles within the pore space and are often distributed as a layer on the main sand particles.

[3]Some tar sands such as those in Utah are oil-wet rather than water-wet; the hot-water process is not effective in separating these. The solid matrix in some heavy oil and bitumen reservoirs is carbonate rather than sand. The Grosmont and other formations in the "Alberta carbonate triangle" contain important Canadian deposits. These formations are frequently karstic and contain fractures and voids that can provide important flow passages and reservoir volume. The Rospo Mare reservoir in the Adriatic Sea is a very large European deposit that—even though it contains a very heavy oil—can be produced by nonthermal methods. The productivity is greatly influenced by the fractures and void volumes.

[4]The average distance between molecules in liquid water is of the order of 0.3 nm.

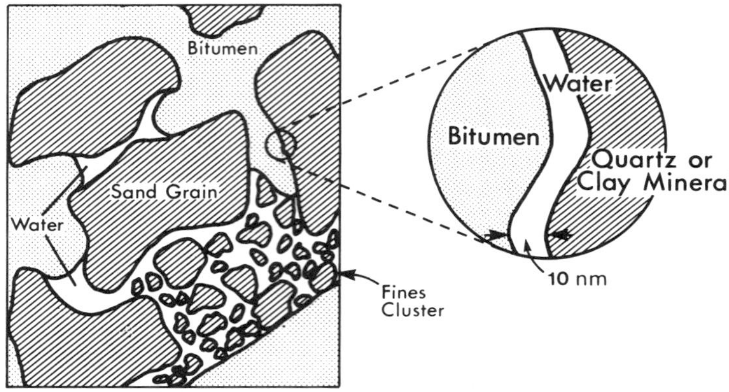

Figure 1.8 Diagram showing the Structure of Athabasca Oil Sand (Courtesy Takamura 1982)

Solid Mineral Matter

Solid mineral matter is often a very complex mixture in itself. Usually it is unconsolidated. The large grains are called sand. In some cases, particularly in the McMurray formation of Athabasca, the sand grains are almost entirely quartz. In other cases, such as the Clearwater and Grand Rapids sands of Alberta (see Figure 1.5) the grains are a complex mixture of various mineral types: quartz, chert, feldspars, and volcanic fragments.

In addition there are fine particles (less than 300 mesh), which vary from less than 5 wt% in high-grade samples to over 20 wt% in low-grade material The fines contain substantial proportions of clays (e.g., 30–60 wt%). The clays include kaolinite, smectite or montmorillonite, illite, and chlorite. The particle size of the clays is generally less than 2 μm.

Clay minerals are hydrated aluminum silicates that frequently contain other cations. They have a sheet structure similar to that of mica. On an atomic scale, there are two kinds of layers that occur in each of the preceding clay minerals.

1. *Silica tetrahedra.* These are tetrahedra of oxygen ions with central silicon ions. These units are linked together to form a hexagonal sheet of indefinite size.
2. *Alumina or aluminum hydroxide layers.* In these, oxygen ions or hydroxyl ions form two parallel sheets with aluminum ions arranged between the oxygen octahedra that constitute the structure; only two-thirds of the possible aluminum sites are occupied, and the average area occupied per aluminum ion is the same as the area per silicon ion in the silica tetrahedral layer.

The main difference between the structure of the different clay minerals arises from the relative proportions of the two types of layers.

Kaolinite

Kaolinite is based on a 1:1 combination of the two types of layer. Its composition is

$$(OH)_8Al_4Si_4O_{10}$$

Stoichiometrically, this may be looked on as follows:

Silica layer	Hydrated alumina layer
$4\ SiO_2$	$2\ Al_2O(OH)_4$

It should be realized that each silicon ion is, in fact, at the center of an oxygen tetrahedron, and the aluminum ions are each between six oxygen or hydroxyl ions (an octahedron has six corners).

When water is added to kaolinite, the lattice does not expand (i.e., the distance between adjacent layers does not increase). Another characteristic of kaolinite is that substitution of iron or magnesium for aluminum is not observed; kaolinite is usually white in color. It was named in 1867 after a hill near Jauchau Fu in China (reported in Grim 1968).

Montmorillonite

Montmorillonite (named after Montmorillon in France) is frequently used interchangeably with the mineral name smectite. Some authors imply the broad class of expanding lattice clays by the term smectite and reserve the term montmorillonite for clay minerals of this type having only a small replacement of Al by Mg (Grim 1968).

The mineral is based upon a layer structure having one layer of alumina/aluminum hydroxide sandwiched between two silica tetrahedral layers. In the idealized case with no substitution of Al it has the composition

$$(OH)_4Al_4Si_8O_{20} \cdot nH_2O$$

The Si/Al atomic ratio is now 2, as compared to 1 for kaolinite. The structure may be visualized as follows:

Silica layer	Hydrated Alumina	Silica layer	Water layer
$4\ SiO_2$	$2\ Al_2O_2(OH)_2$	$4\ SiO_2$	nH_2O

The water is present as a layer of water that penetrates the lattice, between the silica layer of one three-layer, silica-alumina-silica unit and the silica layer of the adjacent one. This quantity of water is variable. The addition of water to montmorillonite causes the lattice to expand and the clay to swell. This is an important characteristic of smectite clays.

The swelling of montmorillonite clay is greatly reduced if potassium or magnesium ions are present in the water layer, since these are able to bind the triple-layer sandwiches together and prevent the intrusion of water. This sensitivity of smectite clays to fresh water is of practical importance in petroleum engineering, since swollen clays can plug reservoirs.

Illite

Illite (the "clay mica," named after Illinois (Grim 1968)) is of a similar structure to montmorillonite except that some of the silicon ions are replaced by aluminum ions, and the resulting deficiency in charge (Al^{3+} is trivalent, whereas Si^{4+} is tetravalent) is made up by the presence of potassium atoms. These appear at the outside flat surface of the three larger units and bind them together. This prevents swelling when water is added. Other substitutions of metal ions are found commonly within illites.

Chlorites

Chlorites have a three-layer structure similar to montmorillonite but are magnesia-rich. The three-layer units are held together by a magnesium hydroxide layer. Chlorite clays are nonswelling.

For more information on clays, the reader is referred to F. J. Pettijohn (1957) and R. E. Grim (1968).

Effect of clays on permeability

Clays, and particularly swelling clays of the smectite type, can influence the permeability of a porous solid. Swelling clays can expand to block pores and particularly pore throats, and clays in general can also become detached from the surfaces to which they adhere. They can then be carried by the moving fluid and deposited so that they block the throats of pores. This is often not a serious problem with high-permeability sands.

In a recent study (M. Kwan 1988) it was shown that repacked, extracted cores from Cold Lake had a much lower permeability if they were exposed to fresh water. Presumably clays migrated and swelled and plugged the core. However, it was found that if extracted preserved cores were employed (i.e., cores that had not been broken up and repacked but were used in their original mechanical form), then fresh water had little effect on the permeability. The difference between the behavior of repacked and preserved cores was very large, and this should be a concern to experimenters wishing to test permeability and fines migration phenomena in core material.

Water

The sand grains in oil sands usually have fine clay material adhering to them; this clay is wetted with *the connate* water.[5] This water is saline and often also contains calcium and magnesium salts, which make it hard. As has been mentioned previously, the salinity of the water varies considerably from area to area.

The wetting of the mineral matter by water rather than by bitumen is a very important characteristic of some oil sands, particularly the deposits in Athabasca; this makes possible the separation of mined tar sand by the Clark hot-water process, which is used by Suncor and by Syncrude. In this process the tar sand is mixed with hot water and a little caustic soda. Most of the sand separates cleanly, leaving the liberated oil to rise to the surface as droplets adhering to bubbles of gas. The clean separation is possible because the bitumen does not wet the solid initially. It is becoming apparent that the nature of the wetting of the matrix also plays an important role in the recovery of heavy oils by steam injection. In particular, it is

[5]Connate water is related to the residual water left in the reservoir matrix after the bulk of the original water was displaced by oil during the filling of the reservoir.

found that a matrix that is wetted initially with connate water greatly reduces the water in oil emulsification that occurs on steaming.

Oil and Bitumen

The most important physical property of crude oil in recovery processes is its viscosity. Figure 1.9 shows the viscosities of several typical heavy crudes as a function of temperature (Buckles 1979).

The viscosity of heavy crude oils correlates, at least approximately, with the density of the oil. Figure 1.10 shows a correlation prepared by Farouq Ali (1983), which can be used to estimate the viscosity as a function of temperature and the density of the oil. However, because significant anomalies are found when the gravities and viscosities of heavy oils are compared, Figure 1.10 should be used only for rough estimates.

Gas

Heavy oil reservoirs often contain pockets with gas saturation and most heavy oils and bitumens contain dissolved natural gas. Also, during heating, a gas phase tends to be formed. The mechanisms involved in this include the evolution of dissolved natural gas, the decomposition of inorganic carbonates to form carbon dioxide, and the decarboxylation of organic acids.

$$RCO_2H \rightarrow RH + CO_2$$

Decarboxylation of acids

The gas produced from the steam recovery of bitumen frequently contains about 50% carbon dioxide, with the remainder being mostly methane. Most of the carbon dioxide probably comes from the chemical transformation of carbonate rocks. It has been suggested that carbon dioxide comes from the thermal decomposition of siderite (ferrous carbonate), which is less stable than other carbonates.

$$FeCO_3 \rightarrow FeO + CO_2$$

Carbonate decomposition

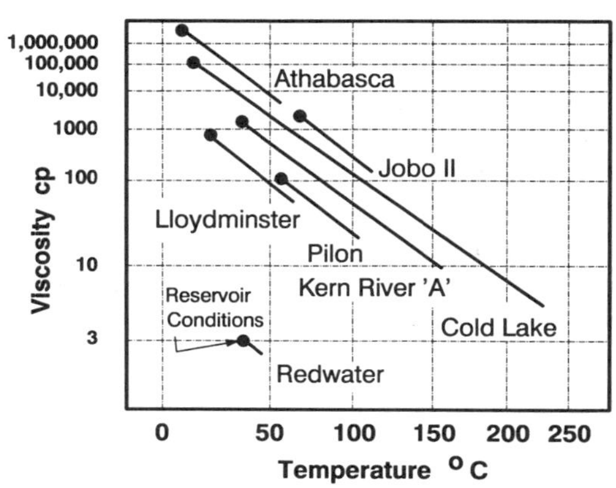

Figure 1.9 Viscosity of Heavy Crudes as a Function of Temperature (Courtesy Buckles 1979)

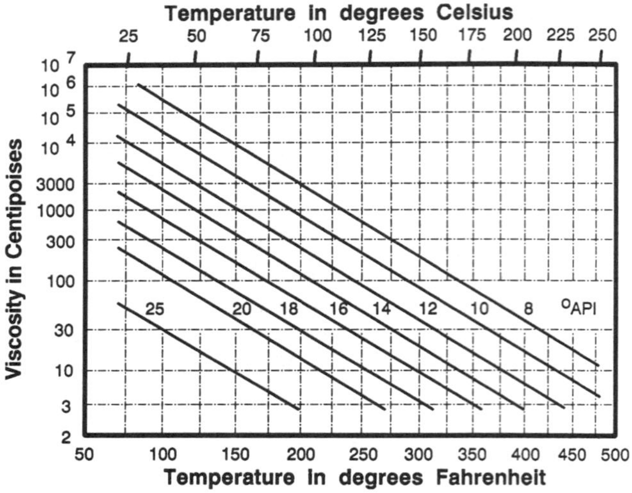

Figure 1.10 Approximate Relationship between Oil Viscosity, Gravity, and Temperature (after Farouq Ali 1983)

However, a more important source of carbon dioxide is probably the reaction of inorganic carbonates with quartz (SiO_2) to yield silicates and carbon dioxide. Gunter and Bird (1989), in a review, describe several hydrothermal reactions in which quartz reacts with carbonates to liberate carbon dioxide. For example,

$$\text{calcite} + \text{quartz} + \text{kaolinite} = \text{Ca-smectite} + CO_2 + H_2O$$

$$\text{dolomite} + \text{quartz} + \text{kaolinite} + H_2O = \text{Ca-Mg-smectite} + \text{calcite} + CO_2$$

One way of interpreting these reactions is to look on the SiO_2 as an acid which is displacing CO_2 from the carbonate.

UNITS OF MEASUREMENT

Measurements in the field of petroleum production—as in other areas—are in some confusion because many countries have switched from a hodgepodge of old, traditional units to the new, more consistent SI (Système International) units. However, the United States continues, for the most part, to use customary units; as a result, most of the literature continues to be written in these units. Even in Canada, where the SI system has been adopted, the old units still prevail in many cases. For example, the Canadian government continues to discuss the price of oil in dollars (US$) per barrel rather than per cubic meter. Even before the present trend to SI, there was confusion in the ranks of the reservoir engineers. Some authors present equations that are dimensionally consistent and into which one may substitute numerical values drawn from any dimensionally consistent set of units of measurement, whereas others write equations that involve dimensioned constants. This latter class of equation requires the use of specified units in order to provide the correct result.

For example, Darcy's law for the flow of a fluid in a porous medium may be written as the dimensionally consistent equation 1.1.[6]

[6]All symbols are listed in Appendix 1.

$$q = -\frac{kA}{\mu}\left(\frac{\Delta P}{\Delta x}\right) \quad (1.1)$$

where k is permeability L^2
A is area L^2
P is pressure $ML^{-1}T^{-2}$
μ is viscosity $ML^{-1}T^{-1}$
x is distance L
q is flow L^3T^{-1}

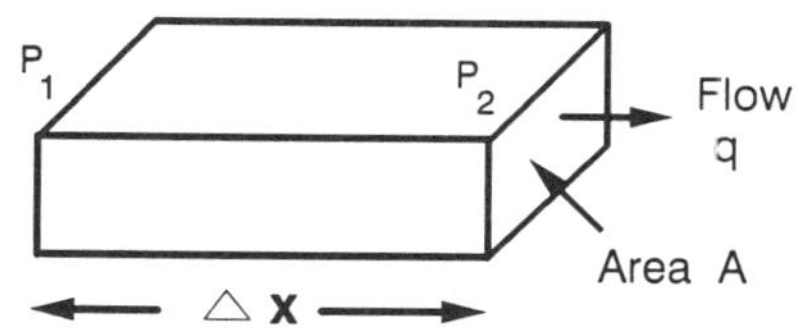

This equation will give the correct answer providing that any consistent set of units is employed. For example, it will work with SI units, with cgs. units, with fps units, and with any other set of units having a consistent basis for mass, length, and time. In reservoir engineering literature, it has been (and still is in the United States) usual to measure time in days, length in feet, viscosities in centipoise, permeabilities in millidarcys, and volumes in barrels or sometimes in acre-feet. It has also been customary to rewrite equations such as 1.1 into forms in which the so-called field units can be substituted directly.

Equation 1.2 is a frequently used dimensional form of Darcy's equation. It is correct provided that the variables are measured in the particular units shown.

$$q = -0.001127\frac{kA}{\mu}\left(\frac{\Delta P}{\Delta x}\right) \quad (1.2)$$

q, B/d; A, ft^2; k, mD; μ, cp; P, psi; x, ft

The numerical coefficient in equation 1.2 has the dimensions of (B cp ft)/(day mD psi).

Although lacking elegance and sophistication, the field-unit system has probably reduced numerical error by allowing the use of familiar and easily visualized quantities. However, the traditional field-unit system has the disadvantage of introducing awkward factors such as the 0.001127 of equation 1.2. It also requires that physical properties be converted to a rather rigid set of specific units. Conversion factors for various frequently employed quantities are shown in Tables 1.6 and 1.7. In this book, dimensionally consistent equations are normally employed. In some cases, where dimensional equations are given, specific units must be used. These are specified at the point where the equation is introduced. In descriptive material the author has employed the units that are most familiar (to the author!).

TABLE 1.6 Consistent Measurement Units

	SI	cgs	fps	Engineers
Mass	kg	gm	lb	slug
Length	m	cm	ft	ft
Time	s	s	s	s

TABLE 1.7 Oil Reservoir Units

VOLUME	
1 ac-ft	= 7757.8 B
1 B	= 5.615 ft^3 = 0.159 m^3
PRESSURE	
1 MPa	= 145 psi = 10^6 Nm^{-2} where N = Newton
1 psi	= 6.895 kPa
PERMEABILITY	
1 D	= 1(cm^3/s) (cp) (cm) $(cm^2)^{-1}(atm)^{-1}$
	= 0.9869×10^{-8} cm^2 = 0.9869×10^{-12} m^2
	= 0.9869 μm^2
DYNAMIC VISCOSITY	
1 p	= 1 g cm^{-1} s^{-1} = 0.1 kg m^{-1} s^{-1} or 0.1 Pa · s p = poise
1 cp	= 0.01 p = 1 mPa · s
KINEMATIC VISCOSITY	
1 st	= 1 cm^2 s^{-1} = 0.0001 m^2 s^{-1}
1 cst	= 1 mm^2 s^{-1}

USE OF PROGRAMMABLE CALCULATORS AND MICROCOMPUTERS

Programmable calculators and particularly microcomputers make calculations in this field much simpler. To use computers effectively, it is important to have available simple equations that allow the calculation of physical and mathematical quantities occurring in the problem at hand.

For example, a microcomputer cannot readily use a steam table, but it can easily calculate the desired value from correlation equations. Just as engineers used to use slide rules, so the modern engineer uses calculators and microcomputers. Often simple correlations are sufficiently accurate to estimate physical quantities in view of the other uncertainties involved in the problem. Slide-rule accuracy is sufficient for most engineering calculations.

The practicing engineer should search for and collect equations that are of a suitable form to be included in computer calculations. A selection of useful correlation equations is given in the appendices.

RADIAL FLOW TO A VERTICAL WELL

Figure 1.11 shows the plan view of a fully-perforated well that is producing oil in radial, steady-state flow in a reservoir of height h. It is assumed that the boundary, at radius R_e, is at a constant pressure P_e and that the well, of radius R_w, is at a constant lower pressure P_w. The effective area for flow diminishes as the fluid approaches the well; because of this, the absolute pressure gradient increases.

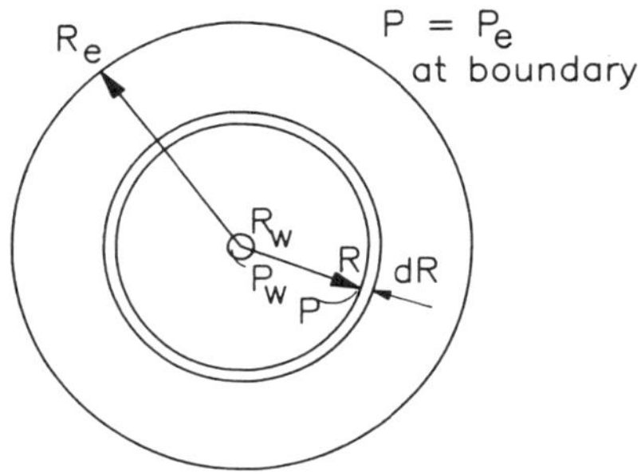

Figure 1.11

At some intermediate radius R the pressure gradient to maintain the flow q is given by substituting the area $2\pi Rh$ into Darcy's equation (1.1). In this example, q is considered positive for flow to the well (i.e., in the opposite direction to R), so the minus sign in (1.1) is omitted. The result is

$$\frac{dP}{dR} = \frac{q\mu}{k(2\pi Rh)} \tag{1.3}$$

This equation may be integrated to calculate the flow arising from the pressure difference $\Delta P = P_e - P_w$.

$$P_e - P_w = \int_{R_w}^{R_e} \frac{q\mu}{2\pi kh} \frac{dR}{R}$$

$$q = \frac{2\pi kh\,\Delta P}{\mu \ln(R_e/R_w)} \tag{1.4}$$

Equation 1.4 is written for dimensionally consistent units. If the dimensional form of Darcy's equation (equation 1.2) is employed, the result is equation 1.5. This is the form found in many texts on reservoir engineering.

$$q = 0.00708 \frac{kh\,\Delta P}{\mu \ln(R_e/R_w)}$$

$$\begin{array}{ll} q\text{, B/d;} & \mu\text{, cp} \\ k\text{, mD;} & R_e\text{, ft} \\ h\text{, ft;} & R_w\text{, ft} \\ \Delta P\text{, psi} & \end{array} \tag{1.5}$$

It is instructive to substitute numerical values into these equations to obtain an idea of the effect of viscosity on oil production rate. Table 1.8 shows values of the production rate that have been calculated for a high-quality, thick reservoir that is saturated with oils having viscosities varying from 1 cp (a low-viscosity conventional crude oil) up to 1,000,000 cp, which corresponds to a material such as Athabasca bitumen.

For a typical well bore radius of 0.3 ft, the production falls from 44,000 B/d for the light crude to only 0.4 B/d for the bitumen. The first case corresponds to a well of remarkable productivity and the latter, to a well of little value.

TABLE 1.8 Calculated Well Flow Rates

Assume k = 1000 mD (excellent sand); h = 100 ft; ΔP = 599 psi;
R_e = 1000 ft.

CALCULATED WELL FLOW RATES

Oil viscosity (cp)	1	100	10,000	100,000
		R_w = 0.3 ft		
Flow (B/d)	4.4×10^4	440	4.4	0.4
		R_w = 100 ft		
Flow (B/d)	1.5×10^5	1,500	15.0	1.5

Using a larger well bore will increase the productivity. The lower line in the table shows the effect of using an imaginary well having a radius of 100 ft. Although such a device is impractical, it may be approximated, at least in principle, by heating the reservoir around a well of normal dimensions. A similar effect might be obtained by using a horizontal well several hundred feet in length.

Such a strategy might, in the example shown, produce a useful effect for the oil having a viscosity of 10,000 cp, but the production with the bitumen would still be too meager to be effective. A flow of 15 B/d is close to the lower limit at which economic production could be anticipated for a practical well.

Comparing the results of calculations such as this with the very sharp changes of viscosity with temperature, which are shown by Figure 1.9, illustrates the importance which reservoir temperature plays in the recovery of heavy oils.

Figure 1.12 shows the oil recovery achievable for a number of Venezuelan heavy oil fields as a function of the in situ viscosity, as given by Borregales (1979). Also shown on the figure are points corresponding to the oil viscosities for the Athabasca and Cold Lake fields.

A major reason for the higher viscosity of Canadian bitumens as compared to those in Venezuela is the lower reservoir temperature. (See data for Jobo and Atha-

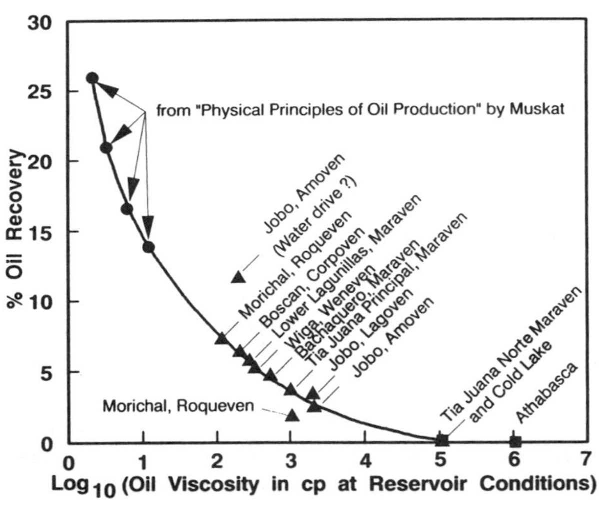

Figure 1.12 Approximate Effect of Viscosity on Oil Recovery by Solution Gas Drive (after Borregales 1979)

basca crudes in Figure 1.12.) The climate of Venezuela makes the ground surface temperature much higher than in Canada, and this difference is exaggerated by the deeper burial of the Venezuelan reservoirs. It is this difference in reservoir temperature rather than intrinsic differences between the crude oils that causes much of the difference between the productivity of the Venezuelan heavy oil wells and the Canadian ones.

Although the Canadian bitumens must be heated somewhat more than the Venezuelan ones for satisfactory production, the largest difficulty that the high initial viscosity presents is that of gaining initial access to the reservoir in order to be able to contact the material with heating media. In many respects, the problem of the production of bitumen in Canada is that of trying to heat a remote, very thick, impermeable, immobile, asphaltic concrete!

THE PROBLEM OF ECONOMIC EXPLOITATION

There are other practical problems which are encountered in the exploitation of heavy oil resources such as those in Canada. So far we have discussed the problem of recovering the crude—i.e., the problem of moving it to the surface of the ground. The concern of this book is moving it to the surface by the use of in situ heating.

Another approach to the same problem is to remove the tar sand by mining methods and then to separate it using processes such as the hot-water process. Large operations of this type are carried out in Athabasca by Suncor (formerly Great Canadian Oil Sands) and by Syncrude near Fort McMurray, Alberta. These plants are successful. However, the approach is very demanding; it depends on brute force and is suitable only for those deposits in Alberta that are relatively shallow. Ninety percent of the bitumen in Alberta and most elsewhere is too deeply buried for this to be a practical approach.

The published economics of the large Cold Lake commercial plant that was once proposed by Esso show that recovery using the cyclic steam process is competitive with mining (McMillan 1979). An Esso Cold Lake commercial plant was originally proposed in the late 1970s that involved the large-scale production of bitumen by cyclic steaming followed by the upgrading of the bitumen to synthetic crude by fluid bed coking and hydrotreating. The project was shelved because of the questionable economics and the enormous capital outlay which would have been required. However, since then, Esso has realized that the production of bitumen without upgrading can be economic. This approach involves the production of bitumen and pipeline transportation of the bitumen diluted with condensate in quantities that will soon be far above 100 kB/d.

BITUMEN TRANSPORTATION

Transportation of the product is a major problem for the bitumen producer since it cannot be pumped through a conventional pipeline. Possible solutions are shown next. All have been put into practice.

- Move the bitumen in trucks or trains
- Convert the bitumen to a more fluid material by chemical transformation

- Dilute the bitumen with a solvent such as condensate and transport it by pipeline
- Pump the bitumen with water through a pipeline under conditions that allow the water to flow as an annulus surrounding a bitumen core
- Emulsify the bitumen in water and transport the mixture by pipeline

For a number of years bitumen was moved from Cold Lake in road trucks as hot cargoes.

Suncor and Syncrude both convert the bitumen to an overhead product using coking. Hydrogenation processes provide an alternative method of conversion being developed by several groups. Syncrude's recent expansion involves the addition of an LC-fining process to their plant; Husky plans an H-oil Unit for their upgrader in Lloydminster. These processes also find use for the conversion of the residual material from the distillation of conventional crude oils.

Table 1.9 shows a comparison between the properties of bitumen and those of a typical conventional light crude and the upgraded crude product that was to have been produced by the Cold Lake commercial project. The upgrading cracks the bitumen, and the cracked products are treated with hydrogen to remove sulphur and nitrogen and to saturate some of the aromatics.

The dilution of bitumen with a solvent such as condensate to make it pumpable has been practiced for years in the Lloydminster area and more recently, and on a much larger scale, at Cold Lake. The main problem is the availability of a suitable diluent; about 30 LV% (based on bitumen volume) of a material such as condensate is required. In some cases double pipelines have been constructed, with the diluent being returned to the field from the remote refinery by a second line.

The shipment of heavy oils by pipeline has been discussed by Sloan, Ingham, and Mann (1981). They conclude that the crude oil viscosity should be less than 150 cst and that the temperature should be maintained less than 200°F in order to

TABLE 1.9 Cold Lake Project—Yield Comparisons (LV%)

	Cold Lake Bitumen	Typical Alberta Light Crude	Upgraded Crude Objective
Butane (C_4)	—	3	3–4
Naphtha (C_5–180°C)			
(C_5–350°F)	1	30	15–20
Distillate (180–345°C)			
(350–650°F)	17	30	45–50
Gas oil (345–565°C)			
(650–1050°F+)	40	30	28–30
Residuum (565°C+)			
(1050°F+)	42	7	0
Total sulphur—wt%	4.5	0.5	<0.5
Gravity—kg/m^3	992	834	885–834
—°API	11	38	34–38

(from Skrabec 1979)

TABLE 1.10 Crude Oil Shipment by Pipeline

- Needs oil viscosity less than about 150 cst
- Temperature should be less than about 200°F to prevent water boiling under insulation

Temperature (°F)[1]	Crude Viscosity (cst)	Approximate Barrels of Condensate per barrel of Crude to reduce viscosity to 150 cst
100	15,000	0.3
150	1,100	0.15
200	180	0.02

[1]Approximate temperature to give indicated viscosity for Cold Lake crude oil.

(from Sloan, Ingham, and Mann 1981)

prevent the boiling of trapped water beneath the pipe insulation. Table 1.10 shows how the viscosity of Cold Lake crude can be reduced to 150 cst by various combinations of heating and dilution with condensate.

Under some conditions it is possible to pump very viscous oils through a pipeline as a central core surrounded by an annular cylinder of water. The water acts as a lubricant, which facilitates the movement of the oil. A pipeline using this principle is being operated by Shell in the United States, but the scheme has not found extensive use.

The transportation of bitumen as an emulsion has been studied by several groups. Lagoven has emulsified Cerro Negro crude (8.5°API) in laboratory and pilot tests and by December 1986 had produced more than 3×10^6 B of emulsion that contained about 30% water. The technical work has been carried out in conjunction with British Petroleum Canada (BP).

Extensive plans have been announced by Lagoven to sell an emulsion of bitumen in water as a product ("Orimulsion") suitable for transportation in ocean tankers and for combustion as a substitute for heavy fuel oil or coal. Technology is also available to produce an emulsion that is suitable for breaking at the destination for use as a refinery feedstock.

BP and Intervep have developed a means of emulsifying bitumen in water for pipeline transportation. The emulsified bitumen is known as TRANSOIL, and a field trial involving the movement of 79 m^3/d has been reported (Hardy, Sit, and Stockwell, 1988).

BIBLIOGRAPHY

ALLEN, F. H., "The Canadian Oil Sands: A Race Against the Clock" 1st UNITAR Conference, Edmonton, Alberta (June 4–12, 1979), reported in *The Future of Heavy Oils and Tar Sands,* New York: McGraw-Hill (1981), 29–32.

BORREGALES, C. J., "Production Characteristics and Oil Recovery in the Orinoco Oil Belt", 1st UNITAR Conference, Edmonton, Alberta (June 4–12, 1979), reported in *The Future of Heavy Oils and Tar Sands,* New York: McGraw-Hill, (1981), 498–509.

BREGER, I. A., "Geochemical Considerations Regarding the Origin of Heavy Crude Oils: Suggestions for Exploration," 1st UNITAR Conference, Edmonton, Alberta (June 4–12,

1979), reported in *The Future of Heavy Crude Oils and Tar Sands,* New York: McGraw-Hill (1981), 163–167.

BUCKLES, R.S., "Steam Stimulation Heavy Oil Recovery at Cold Lake, Alberta," Preprint No. SPE 7, 994, 1979 Calif. Reg. Meeting of Soc. Pet. Eng. of AIME, Ventura, Calif. (April 18–20, 1979). © 1979 SPE.

BURNS, J., "A Review of Steam Soak Operations in California," *J. Pet. Tech.,* 25–34 (January, 1969).

GRIM, R.E., Clay Mineralogy, 2d Ed., McGraw Hill N.Y., 1968.

DE HAAN, H.J. and VAN LOOKEREN, J., "Early Results of the First Large-Scale Steam Soak Project in the Tia Juana Field, Western Venezuela," *J. Pet. Tech.* (January 1969), *Trans. AIME,* 246 (1969).

DEMAISON, G.T., "Tar Sands and Supergiant Oil Fields," *Am. Assoc. Pet. Geol.,* 61: 1950–1961 (November 1977).

DEROO, G., TISSOT, B., MCGROSSAN, R.G., AND DER, F., "Geochemistry of the Heavy Oils of Alberta," in *Oil Sands, Fuel of the Future,* Can. Soc. Pet. Geol., Memoir 3, 148–167 (1974).

FAROUQ ALI, S.M., *Improved Oil Recovery,* Chapter 7, 311–355, Oklahoma City, Okla: Interstate Oil Compact Commission, (1983).

FAROUQ ALI, S.M., *Secondary and Tertiary Oil Recovery Processes,* Chapter 6, 127–182, Oklahoma City, Okla: Interstate Oil Compact Commission, (1974); 2d printing (1978).

FAROUQ ALI, S.M., "Steam Injection Theories—A Unified Approach," *SPE* 10746 (1982).

GOVIER, G.W., "Canada's Energy Resources", presented before *The Energy Opportunities Conference*, Edmonton, Alberta, 22 March (1983).

GUNTER, W.D. and BIRD, G.W., "Inorganic Chemistry", Chapter 9 of *AOSTRA Technical Handbook on Oil Sands, Bitumen and Heavy Oils,* AOSTRA Technical Publication #6, Alberta Oil Sands Technology and Research Authority, Edmonton, Alberta (1989).

GUTIERREZ, F.J., "Occurrence of Heavy Crudes and Tar Sands in Latin America," 1st UNITAR Conference, Edmonton, Alberta (June 4–12, 1979), reported in *The Future of Heavy Crude Oils and Tar Sands,* New York: McGraw-Hill (1981), 107–117.

HARDY, W.A., SIT, S.P. and STOCKWELL, A., "Field Trials of Transoil Technology for Emulsion Pipelining of Bitumen," 4th UNITAR/UNDP Conference on Heavy Crude and Tar Sands, Edmonton, Alberta (July 1988).

JANISCH, A., "Oil Sands and Heavy Oil: Can They Ease the Energy Shortage?" 1st UNITAR Conference, Edmonton, Alberta (June 4–12, 1979), reported in *The Future of Heavy Crude Oils and Tar Sands,* New York: McGraw Hill (1981), 33–41.

JARDINE, D., "Cretaceous Oil Sands of Western Canada," in *Oil Sands, Fuel of the Future,* Can. Soc. of Pet. Geologists, Memoir 3, 50–67 (1974).

JIANYI, H., "Heavy Oil Asphalt and Oil Sand Resources and Their Distribution in China," *Advances in Petroleum Recovery and Upgrading Technology* 1986, AOSTRA (June 12–13, 1986).

KWAN, M.Y.M., CULLEN, M.P., JAMIESON, P.R. and FORTIER, R.A., "A Study of Fines Migration Related Permeability Damage in Extracted Cold Lake Heavy Oil Cores," Paper 88-39-59 presented at the 39th Annual Technical Meeting of the Petroleum Society of C.I.M., Calgary, Alberta (June 12–16, 1988).

MCMILLAN, J.C., "The Challenge of Financing Canadian Oil Sands Development," 1st UNITAR Conference, Edmonton, Alberta (June 4–12, 1979), reported in *The Future of Heavy Crude Oils and Tar Sands,* New York: McGraw-Hill (1981), 775–785.

PETTIJOHN, F. J., *Sedimentary Rocks,* 2d Ed., New York: Harper and Row, (1957).

PHIZACKERLEY, P. H. and SCOTT, L. O., "Occurrence and Prospects of Tar Sands," 7th World Pet. Cong., Mexico (1967), Vol. 3, London: Elsevier (1967), 551–571.

SKRABEC, J., "Process Selection Considerations in the Upgrading of Cold Lake Bitumen," 1st UNITAR Conference, Edmonton, Alberta (June 4–12, 1979), reported in *The Future of Heavy Crude Oils and Tar Sands,* New York: McGraw-Hill (1981), 612–617.

SLOAN, A., INGHAM, R., and MANN, W. L., "Pipeline Transportation of Heavy Oils," 1st UNITAR Conference, Edmonton, Alberta (June 4–12, 1979), reported in *The Future of Heavy Crude Oils and Tar Sands,* New York: McGraw-Hill (1981), 719–726.

STROM, N. A., and DUNBAR, R. B., "Bitumen Resources of Alberta: Converting Resources to Reserves," 1st UNITAR Conference, Edmonton, Alberta (June 4–12, 1979), reported in *The Future of Heavy Oils and Tar Sands,* New York: McGraw Hill (1981), 47–60.

SUFI, A. H., "Injectivity Enhancement in Tar Sands—A Physical Model Study," Paper No. 88-39-61 presented at the 39th Annual Technical Meeting of the Petroleum Society of C.I.M., Calgary, Alberta (June 12–16, 1988).

TAKAMURA, K., "Microscopic Structure of Athabasca Oil Sand", *Can. J. Chem. Eng.,* 60: 538–545 (1982).

WHITING, R. L., "Heavy Crude Oil and Tar Sand Resources and Reserves of the United States. Emphasis on Texas," 1st UNITAR Conference, Edmonton, Alberta (June 4–12, 1979) reported in *The Future of Heavy Crude Oils and Tar Sands,"* New York: McGraw-Hill (1981), 90–96.

GENERAL REFERENCES

BOBERG, T. C., *Thermal Methods of Oil Recovery,* New York: John Wiley (1988).

CAMP, F.W., *The Tar Sands of Alberta, Canada,* 2d Ed. Denver, Colo.: Cameron Engineers, Inc. (1974).

CARRIGY, M. A., Historical Highlights of Major Events in the History of the Athabasca Oil Sands, Alberta Research Council Contribution No. 631 (1973).

Energy Heritage — Oil Sands and Heavy Oils of Alberta, ENR I/19-1, p. 14, Alberta Energy (1982).

FITZGERALD, J. J., *Black Gold with Grit—The Alberta Oil Sands,* Sidney, B.C.: Gray's Publishing Co., (1978).

Improved Oil Recovery, Interstate Oil Compact Commission, Oklahoma City, Okla. (1983).

LATIL, M., *Enhanced Oil Recovery,* Houston, Tex.: Gulf Publishing Co. (1980).

OKANDAN, E. (Ed.), *Heavy Crude Oil Recovery,* The Hague: Martinus Nijhoff (1984).

PRATS, M., "A Current Appraisal of Thermal Recovery," *J. Pet. Tech.,* 1129–1136 (August 1978).

PRATS, M., "Thermal recovery," SPE Monograph Volume 7, Dallas, Tex: SPE (1982).

SKOLROOD, A. H., et al., "Pioneers of the Athabasca Oil Sands". Edmonton, Alberta: Syncrude (1978).

Thermal Recovery Processes, SPE Reprint Series No. 7, Dallas, Tex.: SPE (1985).

VAN POOLLEN, H. K., *Fundamentals of Enhanced Oil Recovery,* Tulsa, Okla.: PennWell Books (1980).

WHITE, P. D. and MOSS, J.T., *Thermal Recovery Methods,* Tulsa, Okla.: PennWell Books (1983).

2

Conduction of Heat within Solids

INTRODUCTION

The quantitative analysis of the transfer and movement of heat within the reservoir plays a central role in the subject of thermal recovery. There are two major mechanisms by which heat is transferred: thermal conduction through relatively stationary materials and convective transport by moving fluids. This chapter is concerned with the first of these mechanisms, thermal conduction.

The reservoir, or adjoining strata, are considered as a homogeneous solid in which the transfer of heat is by conduction. Although this process is very important in thermal recovery, it is very slow. By itself, thermal conduction is an inadequate means of transferring heat within large reservoir volumes. However, it is effective in transferring heat over relatively short distances, as, for example, in the transfer of heat from a steam-saturated region to the adjacent colder reservoir. It plays a particularly important (and undesirable) role by causing the unwanted loss of heat to the overburden and underburden during reservoir heating.

When thermal recovery processes extend over large areas, the loss of heat from the reservoir can become intolerably large. This loss is relatively more important when the reservoir is thin. For example, a given vertical heat loss from a reservoir 10 ft thick might be intolerable, whereas the same heat loss from a 150-ft reservoir could be acceptable because a larger volume of oil would be recovered.

THERMAL CONDUCTIVITY

The theory of heat conduction assumes that the heat flux is in the direction of the temperature gradient and is proportional to the magnitude of the gradient.[1] The

[1]This statement is true only if the thermal conductivity is the same in each direction. The problem of variable thermal conductivity is not considered here.

proportionality constant is defined as the thermal conductivity of the material. For the one-dimensional flow of heat by conduction, the heat flow is given by equation 2.1, where Q is the flow, A is the cross-sectional area for flow, T is the temperature, and x is the distance.

$$Q = -KA\frac{\partial T}{\partial x} \tag{2.1}$$

The negative sign arises because the heat flow is assumed to be positive in the direction of the x axis. For a positive flow of heat, a negative temperature gradient is required.

Although in this chapter the analysis is limited to one-dimensional problems, it is useful to note the form of equation 2.1, which arises in the three-dimensional case.

$$\vec{u} = -K\,\text{grad}(T) \tag{2.2}$$

$$\text{grad}(T) = \nabla T = \vec{i}\frac{\partial T}{\partial x} + \vec{j}\frac{\partial T}{\partial y} + \vec{k}\frac{\partial T}{\partial z} \tag{2.3}$$

In equation 2.2, $\vec{u}$ is the heat flux vector (i.e., the heat flow per unit area) and grad(T) is the temperature gradient vector.

FOURIER'S EQUATION

Except in the special case of steady-state heat conduction, the flow of heat is associated with a change in temperature. In a small element, such as that shown in Figure 2.1, the heat flow away from the element will usually not be equal to that flowing in.

The difference in these flows will be accommodated by a changing inventory of the heat within the element. The heat balance is represented by equation 2.4, and this may be reduced to (2.5), where ρ is the density and C is the heat capacity; the group $K/\rho C$ has been combined into a single variable, α, which is known as the thermal diffusivity with dimensions of L^2T^{-1}.

$$\frac{\partial Q}{\partial x} = -KA\left(\frac{\partial^2 T}{\partial x^2}\right) = -\rho CA\left(\frac{\partial T}{\partial t}\right) \tag{2.4}$$

$$\left(\frac{\partial^2 T}{\partial x^2}\right) = \frac{1}{\alpha}\left(\frac{\partial T}{\partial t}\right) \tag{2.5}$$

Equation 2.5 is the one-dimensional form of the more general equation 2.6, which was first derived by Fourier in 1822.

$$\left(\frac{\partial^2 T}{\partial x^2}\right) + \left(\frac{\partial^2 T}{\partial y^2}\right) + \left(\frac{\partial^2 T}{\partial z^2}\right) = \frac{1}{\alpha}\left(\frac{\partial T}{\partial t}\right) \tag{2.6}$$

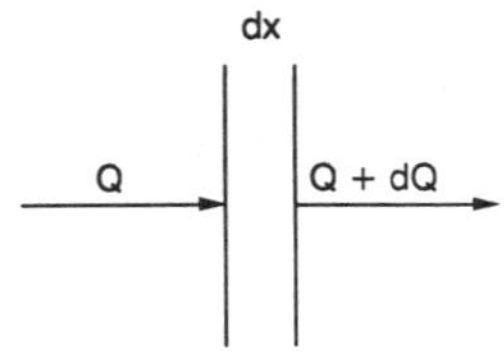

Figure 2.1

Solutions to this equation fall into two general classes:[2]

1. Steady-state solutions, where the term $(\partial T/\partial t)$ is zero
2. Unsteady-state solutions, where temperatures change with time—i.e., where the term $(\partial T/\partial t)$ is generally not zero

The first class of solutions is of interest in problems such as the steady-state loss of heat through insulation. It is the second class of solutions that is usually of most interest in thermal recovery problems. A typical case is that of the heat loss from the upper surface of a heated reservoir to the overburden, which is initially at some more-or-less uniform low temperature. This particular problem is considered in the next section.

FLOW OF HEAT INTO A SEMI-INFINITE SOLID

Consider the flow of heat into the semi-infinite solid body shown in Figure 2.2. Assume that initially the solid is at a uniform temperature T_R, that at time $t = 0$, the surface $x = 0$ is raised to a temperature T_S; and that this temperature is then maintained at the surface. The temperature within the solid is controlled by the one-dimensional Fourier heat-conduction equation 2.5.

A simple method for solving this problem is to first reduce the number of variables by means of dimensional analysis. It is useful to transform the temperature variable to a dimensionless form by making the substitution

$$T^* = \left(\frac{T - T_R}{T_S - T_R}\right)$$

[2]Solutions of equation 2.6 are also of interest in studying the transient flow around oil wells. The basic partial differential equation used to analyze transient pressure tests of oil wells is (see, for example, Matthews and Russell 1967)

$$\frac{\partial^2 P}{\partial x^2} + \frac{\partial^2 P}{\partial y^2} + \frac{\partial^2 P}{\partial z^2} = \frac{\phi\mu c}{k}\frac{\partial P}{\partial t}$$

where P is the fluid pressure at the location with coordinates x, y, and z, ϕ is the porosity, μ is the viscosity, c is the compressibility, and k is the permeability. This equation has the same form as equation 2.6 if the term $k/\phi\mu c$, which is known as the *hydraulic diffusivity,* is equated to the thermal diffusity, α. Well-testing theory is largely a study of solutions of this equation. Solutions of Fourier's equation in this chapter can thus also be applied to reservoir fluid-flow problems. For example, Figure 2.3 can be looked upon as describing the pressure distribution around a fracture maintained at a constant pressure and Figure 2.12, as describing the pressure around a vertical well that is kept at a constant pressure above or below the initial reservoir pressure.

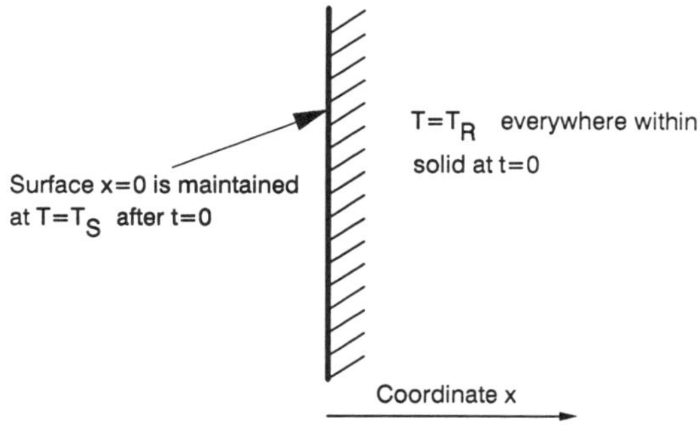

Figure 2.2 One Dimensional Conductive Heating of Semi-infinite Slab

The desired final solution will give the dimensionless temperature T^* as a function of x, t, and α. Suppose this may be written as a power series, such as

$$T^* = \cdots + B_i x^a t^b \alpha^c + \cdots \tag{2.7}$$

For such a solution to be correct, it is necessary for the dimensions of each term to be the same as for T^*; in this example, T^* is dimensionless. The dimensions of the terms on the right-hand side are as follows:

- B_i dimensionless
- x length $\mathfrak{L}$
- t time $\mathfrak{T}$
- α length²/time, i.e., $\mathsf{L}^2\mathsf{T}^{-1}$

For the right-hand side to be dimensionless, the exponents in each term must satisfy

$$c = \frac{-a}{2} \qquad \mathfrak{L} \text{ dimension}$$

and

$$b = c = \frac{-a}{2} \qquad \mathfrak{T} \text{ dimension}$$

Hence, in general, the solution must be of the form:

$$T^* \text{ is a function of } \left(\frac{x}{\sqrt{\alpha t}}\right)$$

or, without loss of generality,[3]

$$T^* = T^*(x, t, \alpha) = T^*(z), \qquad \text{where } z = \frac{x}{2\sqrt{\alpha t}} \tag{2.8}$$

The factor 2 is included for later convenience.

Equation 2.8 enables the number of independent variables to be reduced by one and the partial differential equation to be converted into an ordinary one. The substitution of z from (2.8) into equation 2.5 results in equation 2.9, with the boundary conditions that $T^* = 1$ for $z = 0$ and $T^* = 0$ for $z = \infty$.

[3]This approach is known as Boltzman's transformation; see Carslaw and Jaeger (1959), p. 89.

$$\frac{d^2T^*}{dz^2} = -2z\frac{dT^*}{dz} \tag{2.9}$$

Equation 2.9 may be integrated in two stages. First, let $p = dT^*/dz$, with the result (2.10):

$$\frac{dp}{dz} = -2zp \tag{2.10}$$

This is integrated to give equation 2.11, where C_1 is an integration constant.

$$p = C_1 e^{-z^2} \tag{2.11}$$

Substitution of dT^*/dz for p in (2.11) gives equation 2.12, which when integrated yields equation 2.13.

$$\frac{dT^*}{dz} = C_1 e^{-z^2} \tag{2.12}$$

$$T^* = C_2 + C_1 \int_0^z e^{-z^2} dz \tag{2.13}$$

The first boundary condition leads to $C_2 = 1$ and the second results in equation 2.14.

$$0 = 1 + C_1 \int_0^\infty e^{-z^2} dz \tag{2.14}$$

The value of the definite integral in equation 2.14 can be determined as follows; this derivation is based on that given by Jenson and Jeffreys (1963).

The volume of the solid of revolution that is formed by rotating the curve $z = e^{-x^2}$ about the z axis is given by

$$V = \int_0^\infty 2\pi R e^{-R^2} dR = \pi \int_0^\infty e^{-R^2} d(R^2) = \pi \left| -e^{-R^2} \right|_0^\infty = \pi \tag{2.15}$$

The same volume can also be found from the double integral

$$\begin{aligned} V &= 4\int_0^\infty \int_0^\infty e^{-R^2} dy\, dx = 4\int_0^\infty dy \int_0^\infty e^{-(x^2+y^2)} dx \\ &= 4\int_0^\infty e^{-y^2} dy \int_0^\infty e^{-x^2} dx = 4I^2 \end{aligned} \tag{2.16}$$

where I is the integral of equation 2.14. Hence

$$V = \pi = 4I^2 \quad \text{and} \quad I = \frac{\sqrt{\pi}}{2} \tag{2.17}$$

Substitution of this value into equation 2.14 leads to $C_1 = -2/\sqrt{\pi}$ and the desired solution, equation 2.18, where erfc is the complementary error function.

$$T^* = 1 - \text{erf}(z) = \text{erfc}(z) = \text{erfc}\left(\frac{x}{2\sqrt{\alpha t}}\right) \tag{2.18}$$

The error function and the complementary error function are defined by the following equations:

$$\operatorname{erf}(z) = \frac{2}{\sqrt{\pi}} \int_0^z e^{-z^2} dz \tag{2.19}$$

$$\operatorname{erfc}(z) = 1 - \operatorname{erf}(z) = \frac{2}{\sqrt{\pi}} \int_z^\infty e^{-z^2} dz \tag{2.20}$$

A graph of the complementary error function is given in Figure 2.3. Numerical values of the error function and the complementary error function are given in Table 2.1.

TABLE 2.1 Numerical Values of the Complementary Error Function

z	erfc(z)	z	erfc(z)	z	erfc(z)
0.00	1.000000	1.00	0.157299	2.00	0.004678
0.05	0.943628	1.05	0.137564	2.05	0.003742
0.10	0.887537	1.10	0.119795	2.10	0.002980
0.15	0.832004	1.15	0.103876	2.15	0.002362
0.20	0.777298	1.20	0.089686	2.20	0.001863
0.25	0.723674	1.25	0.077100	2.25	0.001463
0.30	0.671373	1.30	0.065992	2.30	0.001143
0.35	0.620618	1.35	0.056238	2.35	0.000889
0.40	0.571608	1.40	0.047715	2.40	0.000689
0.45	0.524518	1.45	0.040305	2.45	0.000531
0.50	0.479500	1.50	0.033895	2.50	0.000407
0.55	0.436677	1.55	0.028377	2.55	0.000311
0.60	0.396144	1.60	0.023652	2.60	0.000236
0.65	0.357971	1.65	0.019624	2.65	0.000179
0.70	0.322199	1.70	0.016210	2.70	0.000134
0.75	0.288844	1.75	0.013328	2.75	0.000101
0.80	0.257899	1.80	0.010910	2.80	0.000075
0.85	0.229332	1.85	0.008889	2.85	0.000056
0.90	0.203092	1.90	0.007210	2.90	0.000041
0.95	0.179109	1.95	0.005821	2.95	0.000030

$$\text{For } z > 3, \qquad \operatorname{erfc}(z) \simeq \frac{2}{\sqrt{\pi}} \cdot \frac{1}{z + \sqrt{z^2 + 2}} e^{-z^2}$$

Useful approximate formulas for calculating this function using microcomputers follow.

Rational Approximations for the Complementary Error Function

Approximation 1:

$$\operatorname{erfc}(x) = (a_1 t + a_2 t^2 + a_3 t^3) e^{-x^2} + \epsilon(x)$$

$$t = 1/(1 + px) \qquad |\epsilon(x)| \le 2.5 \times 10^{-5}$$

$$p = 0.47047 \qquad a_1 = 0.3480242 \qquad a_2 = -0.0958798 \qquad a_3 = 0.7478556$$

Approximation 2:

$$\mathrm{erfc}(x) = (a_1t + a_2t^2 + a_3t^3 + a_4t^4 + a_5t^5)e^{-x^2} + \epsilon(x)$$

$$t = 1/(1 + px) \qquad |\epsilon(x)| \le 1.5 \times 10^{-7}$$

$p = 0.3275911$ $\quad a_1 = 0.254829592$ $\quad a_2 = -0.284496736$

$a_3 = 1.421413741$ $\quad a_4 = -1.453152027$ $\quad a_5 = 1.061405429$

Approximation 3:

$$\mathrm{erfc}(x) = (1 + a_1x + a_2x^2 + a_3x^3 + a_4x^4)^{-4} + \epsilon(x)$$

$$|\epsilon(x)| \le 5 \times 10^{-4}$$

$a_1 = 0.278393$ $\quad a_2 = 0.230389$ $\quad a_3 = 0.000972$ $\quad a_4 = 0.078108$

Approximation 4:

$$\mathrm{erfc}(x) = (1 + a_1x + a_2x^2 + a_3x^3 + a_4x^4 + a_5x^5 + a_6x^6)^{-16} + \epsilon(x)$$

$$|\epsilon(x)| \le 3 \times 10^{-7}$$

$a_1 = 0.0705230784$ $\quad a_2 = 0.0422820123$ $\quad a_3 = 0.0092705272$

$a_4 = 0.0001520143$ $\quad a_5 = 0.0002765672$ $\quad a_6 = 0.0000430638$

(from Hastings 1955)

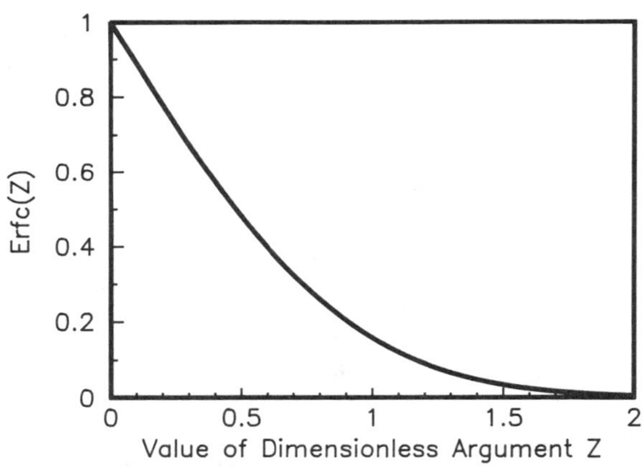

Figure 2.3 The Complementary Error Function

Significance of Solution

The single curve in Figure 2.3 contains within it all the information required to calculate temperature gradient curves within the solid for various times. It is a dimensionless plot of temperature against dimensionless distance. At later times heat penetrates farther into the solid. For a given temperature, the distance is proportional to the square root of the time. It takes four times as long for the heat to penetrate twice as far.

The heat flux into the surface of the infinite solid is given by equation 2.21.

$$\frac{Q}{A} = -K\left(\frac{\partial T}{\partial x}\right)_{x=0} \tag{2.21}$$

Substituting $(T - T_R)/(T_S - T_R)$ for T^* in equation 2.18 gives equation 2.22, where T_S is the temperature of the hot surface and T_R is the reservoir temperature.

$$T = T_R + (T_S - T_R)\,\mathrm{erfc}\left(\frac{x}{2\sqrt{\alpha t}}\right) \tag{2.22}$$

The temperature gradient at $x = 0$ may be calculated from equation 2.23.

$$\left(\frac{\partial T}{\partial x}\right)_{x=0} = \frac{T_S - T_R}{2\sqrt{\alpha t}}\left(\frac{dT^*}{dz}\right)_{z=0} \tag{2.23}$$

Substitution of dT^*/dz from equation 2.12 and using the value that was calculated previously for C_1 results in equation 2.24 for the heat flux at the surface.

$$\frac{Q}{A} = (T_S - T_R)\frac{K}{\sqrt{\pi \alpha t}} = (T_S - T_R)\sqrt{\frac{K\rho C}{\pi t}} \tag{2.24}$$

Equation 2.24 shows that the heat flux is inversely proportional to the square root of time. This decrease of heat flux with time is an important consideration in heat transfer within solids such as the overburden above reservoirs. It is interesting to note that the heat flux is equally dependent on the thermal conductivity, K, and upon the volumetric heat capacity, ρC.

The cumulative heat flux at time t may be found by integrating equation 2.24, to give equation 2.25:

$$\int_0^t \frac{Q}{A}\,dt = \frac{Q_c}{A} = 2(T_S - T_R)K\sqrt{\frac{t}{\pi \alpha}} = 2(T_S - T_R)\sqrt{K\rho C t/\pi} \tag{2.25}$$

HEAT TRANSFER FROM A SPREADING HOT ZONE

An important application of the theory developed in the previous section is to the case of a flat heated area that is increasing with time. This concept plays an important role in theories dealing with steamflooding. The heat loss to the overburden above a spreading steam zone is an example of this situation.

In the solution, it will be assumed that the problem remains one-dimensional. Heat is assumed to be transferred in a direction normal to the surface; i.e., transfer parallel to the surface is ignored. For the first case which will be considered, it is assumed that the heated area increases linearly with time.

A small element of area $dA = \dot{A}dt_0$ is heated from T_R to T_S between time t_0 and $t_0 + dt_0$; it is then maintained at T_S. No restrictions are placed on the shape of the heated area A. Also, although it is considered a constant for the present, the spreading rate $\dot{A}$ may vary with t_0.

At some later time t, the cumulative amount of heat, Q_c, that has been transferred through a particular area element $\dot{A}dt_0$ may be found using equation 2.25. The result is given by:

$$dQ_c = 2K(T_S - T_R)\sqrt{\frac{t - t_0}{\pi \alpha}}\,\dot{A}dt_0 \tag{2.26}$$

TABLE 2.2 Values of Factor ψ

n	$\psi(n)$
0	1.0000
0.1000	0.9436
0.2000	0.8955
0.3000	0.8540
0.4000	0.8176
0.5000	0.7854 = $\pi/4$
0.6000	0.7567
0.7000	0.7309
0.8000	0.7075
0.9000	0.6862
1.0000	0.6667

The total heat that has been transferred may be found by integrating 2.26 from $t_0 = 0$ to $t_0 = t$. Equation 2.27 is the result.

$$Q_c = \frac{2}{3} \cdot 2K(T_S - T_R)A\sqrt{\frac{t}{\pi\alpha}} \tag{2.27}$$

Expanding the heated area at a steady rate rather than immediately exposing the same total area to the high temperature reduces the cumulative heat flux in time t by a factor of two-thirds (compare equation 2.27 with 2.25).

The preceding derivation may be made more general by assuming that the area increases with the nth power of t_0 rather than linearly. An example for which this is of interest occurs when a steam chamber rises with the two-third power of time. The parameter n is 0 for the case where A is constant and is 1 when the area increases at a constant rate.

Using the same approach, it is found that the cumulative flow is given by equation 2.28, where the factor ψ is the function of the exponent n as defined below;[4] tabulated values of ψ are given in Table 2.2.

$$Q_c = 2K(T_S - T_R)A\sqrt{\frac{t}{\pi\alpha}}\,\psi(n) \tag{2.28}$$

where

$$\psi(n) = \frac{n\sqrt{\pi}}{2}\,\frac{\Gamma(n)}{\Gamma(n + 3/2)}$$

CONSTANT HEAT INJECTION RATE INTO A FRACTURE

A particularly interesting and useful application of equation 2.28 occurs when the exponent n is equal to 0.5. If A is proportional to the square root of time, then it is apparent from equation 2.28 that, for this special case, the rate of heat loss will be constant.

[4] $\Gamma(n)$ is the gamma function of n.

A practical case in which this occurs is where steam is injected at a constant flow rate into a fracture initially at reservoir temperature. The heat-loss rates under this condition to the material either side of the fracture will each be equal to half the heat-injection rate, and the heated area of the fracture will grow proportionally to the square root of time.[5]

Starting with equation 2.28, it is easy to show that under these conditions the heated area of the fracture will be given by

$$A = \left(\frac{Q_{ci}}{t}\right)\sqrt{\frac{\alpha t}{\pi}}\frac{1}{K(T_S - T_R)} \tag{2.29}$$

where Q_{ci} is the total cumulative heat injected into the fracture, because heat is lost from both sides of the fracture, $Q_{ci} = 2Q_c$.

This equation is derived in another manner in Chapter 3 (see equation 3.59).

CONDUCTION FROM A SPREADING CHAMBER THAT ADVANCES TO A LIMIT AND THEN STOPS

In recovery problems connected with wells confined in a repeated pattern, the heated regions of the reservoir can spread laterally until they interfere with the hot regions of neighboring wells. When this confinement takes place, the hot zones cease to expand, and heat is lost to the overburden at a decreasing rate.

During the expansion the hot zone is continually encountering fresh cold overburden, and this requires considerable heat. Once confinement takes place, this drain stops and the rate of heat loss starts to decrease.

Consider the case where the heated area is defined by

$$A = \dot{A}t \quad \text{for } 0 \le t \le t_1$$

$$A = \dot{A}t_1 \quad \text{for } t \ge t_1$$

In these equations t is looked upon as a variable and t_1, the time required to reach the outer boundary, is a constant. $\dot{A}$ is assumed constant.

Using (2.27) the cumulative heat transfer for the period $t < t_1$ may be written as

$$Q_c = \frac{4}{3}K(T_S - T_R)A\sqrt{\frac{t}{\pi\alpha}} = \frac{4}{3}\frac{K(T_S - T_R)\dot{A}}{\sqrt{\pi\alpha}}t^{3/2} \tag{2.30}$$

If t were greater than t_1, then the value of Q_c calculated from the preceding equation would be too high because the heated area stops growing at time t_1. The value of Q_c would be too high by the amount of heat that would have been supplied to the area greater than $\dot{A}t_1$ if the growth had carried on unchecked.

The heat which would have been supplied to the area beyond $\dot{A}t_1$ may be calculated by setting t in (2.27) equal to the time after t_1—i.e., to $(t - t_1)$. The result is

$$Q_c = \frac{4}{3}\frac{K(T_S - T_R)\dot{A}}{\sqrt{\pi\alpha}}(t - t_1)^{3/2} \qquad (\text{heat beyond } \dot{A}t_1) \tag{2.31}$$

[5]This example is also discussed in Chapter 3, where it is shown that the temperature T_S does not prevail right to the end of the heated zone.

The cumulative amount of heat supplied to the area At_1 alone, for times greater than t_1, can be calculated from the difference between the heat for the total area calculated from (2.30) and the heat for the excess area from (2.31). The result is

$$Q_c = \frac{4}{3}\frac{K(T_S - T_R)\dot{A}}{\sqrt{\pi\alpha}}[t^{3/2} - (t - t_1)^{3/2}] \quad \text{for } t > t_1 \tag{2.32}$$

The rate at which heat flows to the overburden, for $t > t_1$, is found by differentiating (2.32), giving (2.33).

$$Q = \frac{2K(T_S - T_R)\dot{A}}{\sqrt{\pi\alpha}}[t^{1/2} - (t - t_1)^{1/2}] \tag{2.33}$$

Confinement reduces the heat lost to the overburden; for this reason confined wells, particularly in gravity drainage schemes, can be much more thermally efficient than isolated ones.

Figure 2.4 shows the effect on the heat-loss rate found when a spreading zone encounters its neighbor. One reason for this is simply that when the increase in area stops, it is no longer necessary to continue heating cold overburden. Another factor is that as the overburden becomes hot, the heat flux per unit area decreases. The overall effect is dramatic and significant.

Numerical Problem

In a planned steam flooding project, it is estimated that the steam heated zone will, for each injection well, cover an area of 4 ha after 10 y of steaming (the spacing per injection well is 4 ha).

Calculate the quantity of steam in tonnes required to supply the heat losses to the overburden for each year for the following three cases:

i. The steam-heated zone increases at a steady rate of 0.4 ha/y.

ii. The steam-heated zone grows at a rate of 1 ha/y for 4 y and then stops as it meets its neighboring zones.

iii. The steam-heated zone spreads immediately over the pattern.

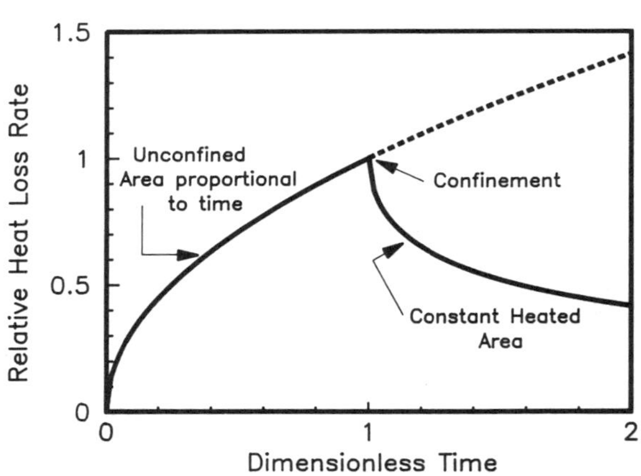

Figure 2.4 Effect of Confinement on Heat Loss to Overburden

Assume the following

Steam pressure	= 5 MPa; Quality = 70%
Overburden conductivity	= 1.7 W/m °C
Overburden heat capacity	= 850 J/kg °C
Overburden density	= 2400 kg/m^3
Initial overburden temperature	= 15°C

Steam condensate is produced at an average temperature of $T_R + 0.75(T_S - T_R)$

For case (iii) plot the temperature, in degrees Celsius, as a function of distance above the top of the reservoir in meters at 2, 5 and 10 years.

Solution The problem will be solved using SI units. From steam tables:

$T_S = 264°\text{C}$ for 5 MPa

$H_S = 2794.3$ kJ/kg — enthalpy of saturated vapor

$H_w = 1154.2$ kJ/kg — enthalpy of boiling liquid

$T_R = 15°\text{C}$

$T_P = 15 + 0.75(264 - 15) = 202°\text{C}$ — production temperature

$H_P = 861.5$ kJ/kg — enthalpy of water at 202°C

$$\text{Net heat per tonne of steam} = (0.7 \times 2794.3 + 0.3 \times 1154.2 - 861.5) \times 10^6$$
$$= 1.441 \times 10^9 \text{ J/t}$$

i. Steam required assuming the steam zone advances at a steady rate of 0.4 ha/y. Use equation 2.27:

$$Q_c = \frac{4}{3}K(T_S - T_R)A\sqrt{\frac{t}{\pi\alpha}} \tag{2.27}$$

$$K = 1.7 \text{ W/m °C}$$

$$A = 4000, 8000, \ldots \text{ m}^2$$

$$t = 1, 2, \ldots \text{ y} = 1 \times (3.1536 \times 10^7)\ldots \text{ s}$$

$$\alpha = \frac{1.7}{850 \times 2400} = 8.333 \times 10^{-7} \text{ m}^2\text{/s}$$

The cumulative heat to the overburden after 1 year is

$$Q_c = \frac{4}{3} \times 1.7(264 - 15)4000\left(\frac{3.1536 \times 10^7}{8.333 \times 10^{-7} \times \pi}\right)^{1/2} = 7.836 \times 10^{12} \text{ J}$$

$$\text{Tonnes of steam required for first year} = \frac{7.836 \times 10^{12}}{1.441 \times 10^9} = 5438$$

The cumulative quantities of steam for successive years are calculated in the same manner, and the yearly quantities are found by subtraction. The results are given in the second column of the following table. The annual heat requirement rises continuously as the heated region expands.

YEAR	ANNUAL STEAM REQUIREMENT IN TONNES PER YEAR CASE (i)	CASE (ii)	CASE (iii)
1	5,438	13,596	81,576
2	9,944	24,858	33,790
3	12,876	32,191	25,928
4	15,248	38,120	21,858
5	17,296	29,643	19,257
6	19,124	22,952	17,411
7	20,792	19,790	16,010
8	22,336	17,719	14,901
9	23,779	16,211	13,997
10	25,140	15,038	13,238
TOTAL	171,973	230,118	257,966
YEARLY AVERAGE kt/y	17.2	23.0	25.8

ii. Steam required assuming that the zone grows at 1 ha/y and then stops.

For the first 4 years, the steam requirement is calculated in the same manner as in (i), using:

$$A = 10{,}000,\ 20{,}000,\ 30{,}000 \quad \text{and} \quad 40{,}000\ \text{m}^2$$

The results are shown in the first four rows of the third column of the preceding table.

For the remaining years, the cumulative quantities of steam are calculated using equation 2.32:

$$Q_c = \frac{4}{3}\frac{K(T_S - T_R)\dot{A}}{\sqrt{\pi\alpha}}[t^{3/2} - (t - t_1)^{3/2}] \tag{2.32}$$

For year 5:

$$t = 5 \times 3.1536 \times 10^7\ \text{s}$$

$$t_1 = 4 \times 3.1536 \times 10^7\ \text{s}$$

$$\dot{A} = \frac{10{,}000}{(3.1536 \times 10^7)} \cdot \frac{\text{m}^2}{\text{s}}$$

$$Q_c = \frac{4}{3} \cdot \frac{1.7(264 - 15)10{,}000}{(\pi 8.333 \times 10^{-7})^{1/2} 3.1536 \times 10^{-7}} \cdot [(5 \times 3.1536 \times 10^{-7})^{3/2} - (3.1536 \times 10^{-7})^{3/2}]$$

$$= 1.994 \times 10^{14}\ \text{J}$$

$$\text{Cumulative steam requirement} = \frac{1.994 \times 10^{14}}{1.441 \times 10^9} = 138{,}408\ \text{t}$$

$$\text{Net steam for 5 y} = 138{,}408 - 108{,}765 = 29{,}643\ \text{t}$$

Annual steam requirements calculated in this manner are given in rows 5 through 10 of the third column of the preceding table. The steam requirement rises to a maximum and then declines after the heated area stops growing.

iii. Steam required assuming heated zone spreads immediately.

The steam required is calculated using equation 2.25 with $A = 40{,}000 \text{ m}^2$ and the remaining parameters as in (i). The results are given in the fourth column of the table. The steam requirement starts out very high (infinite at $t = 0$) and declines with time.

The results of the three calculations are shown in Figure 2.5. The average rates are compared in the following table. These results may be put into a practical perspective by considering the relative costs, with steam costing somewhere in the range of \$5 to \$20 per tonne.

Average Steam Requirement

SPREADING TIME (y)	AVERAGE STEAM RATE (kt/y)
0	25.8
4	23.0
10	17.2

The temperature above the top of the reservoir is calculated using equation 2.22.

$$T = T_R + (T_S - T_R)\,\text{erfc}\left(\frac{x}{2\sqrt{\alpha t}}\right) \qquad (2.22)$$

In SI units x is the distance in meters above the reservoir, t is the time from the start of heating in seconds, and α is the thermal diffusivity in square meters per second.[6]

The calculated temperatures are shown as a function of depth for each of the three times in Figure 2.6.

CONDUCTION AHEAD OF AN ADVANCING FRONT

In previous sections, the unsteady state conduction of heat into a solid from a stationary hot surface was considered. This was extended to the case where the hot surface was expanding in area.

Another one-dimensional heat conduction situation that is of interest in thermal recovery is that where a hot surface or front is advancing through the solid reservoir in a direction approximately normal to its surface. One situation where

[6] For calculations of this type, it will often be more convenient to use the day as the unit of time rather than the second of the SI system. To do this it is necessary to also use the day as the unit of time in each of the other variables. In the present example it would be necessary to convert α to square meters per day; this is also a much more convenient unit.

$$\alpha = 8.333 \times 10^{-7} \text{ m}^2/\text{s} = 0.072 \text{ m}^2/\text{d}$$

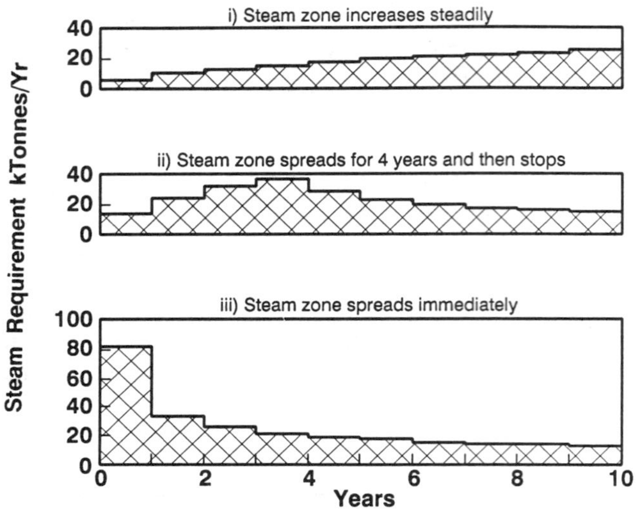

Figure 2.5 Steam Requirements kt/y

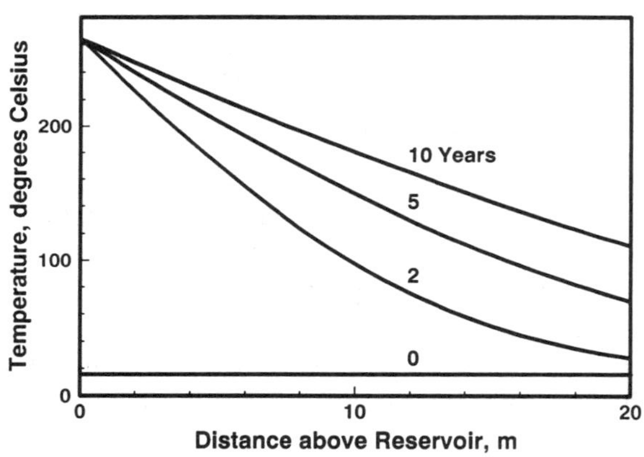

Figure 2.6 Temperature in Overburden above Reservoir

this can occur is at a more-or-less vertical steam front which is advancing as flooding steam is injected behind it. The steam condenses at the front and the condensate runs with the displaced oil through the front and ahead of it where they rapidly attain the temperature of the reservoir matrix. Heat is conducted ahead of the condensation surface and the heat for this preheating must also be supplied by the steam. If the process continues with a steady state of advance the cumulative heat buildup ahead of the front reaches a steady value.

Another example of conduction ahead of an advancing front occurs when oil drains downward across a steam front which is advancing horizontally through a reservoir. This mechanism is important in thermal recovery using gravity drainage. Here the heated oil and condensate are removed continuously by drainage, and the steam-saturated chamber advances into the reservoir. Heat is conducted ahead of the advancing front, and this mobilizes the oil. The situation is depicted in Figure 2.7.

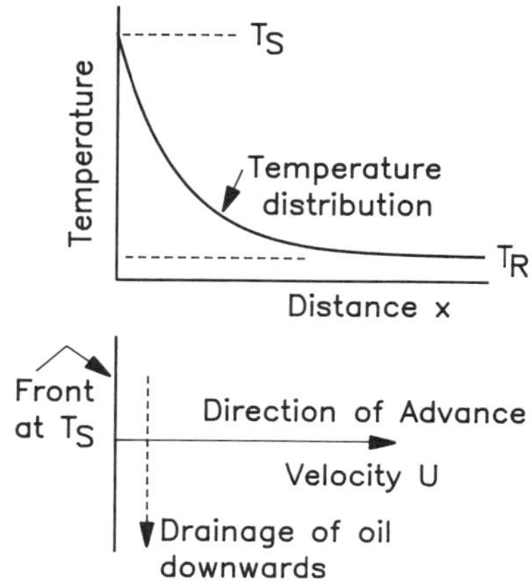

Figure 2.7 Lateral Advance of Steam Front

The one-dimensional Fourier equation 2.34 is applicable.

$$\left(\frac{\partial^2 T}{\partial x^2}\right) = \frac{1}{\alpha}\left(\frac{\partial T}{\partial t}\right) \tag{2.34}$$

The distance x is measured normal to the advancing front. In the analysis of this process, it is convenient to transform the distance variable so that it is measured from the front rather than from a fixed origin. This is done by defining the new variable ξ as in equation 2.35.

$$\xi = x - \int_0^t U dt$$

or, if U is constant,

$$\xi = x - Ut \tag{2.35}$$

In this equation, U is the velocity of the advancing front. The elimination of x from the left-hand side of (2.34) may be accomplished simply by replacing x by ξ.

$$\left(\frac{\partial^2 T}{\partial x^2}\right) = \left(\frac{\partial^2 T}{\partial \xi^2}\right) \tag{2.36}$$

Since T is a function of x and t, a general differential may be written as follows:

$$dT = \left(\frac{\partial T}{\partial x}\right)_t dx + \left(\frac{\partial T}{\partial t}\right)_x dt \tag{2.37}$$

If ξ is maintained constant during the differential change represented by (2.37), then $dx = U\,dt$[7], and (2.37) may be rearranged to give equation 2.38.

$$\left(\frac{\partial T}{\partial t}\right)_\xi = U\left(\frac{\partial T}{\partial t}\right)_t + \left(\frac{\partial T}{\partial t}\right)_x \tag{2.38}$$

[7]This is true even if U is not constant.

Combining (2.36), (2.38), and (2.34) results in equation 2.39. This represents the temperature T as a function of the distance ξ ahead of a front advancing at a velocity U in a reservoir of thermal diffusivity α. It may be made dimensionless by the use of the variables defined by equation 2.40.

$$\left(\frac{\partial^2 T}{\partial \xi^2}\right) + \frac{U}{\alpha}\left(\frac{\partial T}{\partial \xi}\right)_t = \frac{1}{\alpha}\left(\frac{\partial T}{\partial t}\right)_\xi \tag{2.39}$$

$$T^* = \frac{T - T_R}{T_S - T_R}$$

$$\xi^* = \frac{U\xi}{\alpha} \tag{2.40}$$

$$t^* = \frac{U^2 t}{\alpha}$$

When these substitutions are made, the dimensionless equation 2.41 results.

$$\left(\frac{\partial^2 T^*}{\partial \xi^{*2}}\right) + \left(\frac{\partial T^*}{\partial \xi^*}\right) = \left(\frac{\partial T^*}{\partial t^*}\right) \tag{2.41}$$

Imagine a constant temperature heat front of temperature T_S being formed at time $t = 0$ and advancing at a constant velocity U. The reservoir is initially at temperature T_R. Heat conducted ahead of the front warms the reservoir material. Initially there is no heat ahead of the front, but as the front advances and time progresses, heat accumulates ahead of the front.

The process can be imagined in two parts. The first is the transient stage, in which the heat ahead of the front builds up asymptotically to reach an equilibrium level. After this there is a steady state, in which heat ahead of the front is "run over" by the advance of the front; a steady-state, dynamic equilibrium is achieved. In this steady state the temperature at any distance ξ from the front is constant with time.

For this situation the right-hand side of equation 2.41 can be set equal to zero and the equation becomes (2.42):

$$\left(\frac{\partial^2 T^*}{\partial \xi^{*2}}\right) + \left(\frac{\partial T^*}{\partial \xi^*}\right) = 0 \tag{2.42}$$

The boundary conditions are

$$T^* = 0 \qquad \text{for } \xi = \infty$$

and

$$T^* = 1 \qquad \text{for } \xi = 0$$

The solution that satisfies these boundary conditions is given by equation 2.43 or, after substituting the original dimensional variables, by equation 2.44.

$$T^* = e^{-\xi^*} \tag{2.43}$$

$$\frac{T - T_R}{T_S - T_R} = e^{-U\xi/\alpha} = e^{-(U/\alpha)(x - Ut)} \tag{2.44}$$

The total heat stored ahead of the front per unit area can be found by integrating equation 2.44, giving equation 2.45:

$$\frac{Q_c}{A} = \int_0^\infty \rho C(T - T_R)\,d\xi$$

$$= \rho C\alpha\frac{(T_S - T_R)}{U} \tag{2.45}$$

$$= K\frac{(T_S - T_R)}{U}$$

Equation 2.45, since it is derived for the steady state, represents the asymptote to the quantity of heat which can be stored ahead of the front. In many cases, the actual conditions are not far from this steady state.

The amount of heat stored ahead of the front varies inversely with the velocity. Large quantities of heat can be stored ahead of slow-moving fronts and little heat ahead of fast-moving ones. In gravity drainage processes there is equilibrium. Heat stored ahead of the front mobilizes the oil and allows it to drain away, and the oil draining away allows the front to move. The drainage of oil controls the velocity of advance.

TRANSIENT HEAT TRANSFER AHEAD OF AN ADVANCING FRONT

The transient conditions that occur before the steady state is reached can be calculated by solving equation 2.41 without neglecting the time-dependent term. The solution that corresponds to the boundary conditions

$$T^* = 0 \qquad \text{when } t = 0 \quad \text{for all } \xi$$

and

$$T^* = 1 \qquad \text{when } \xi^* = 0 \quad \text{and} \quad t > 0$$

is given by equation 2.46 (Carslaw and Jaeger 1959).

$$T^* = \frac{1}{2}\left[\operatorname{erfc}\left(\frac{\xi^* + t^*}{2\sqrt{t^*}}\right) + e^{-\xi^*}\operatorname{erfc}\left(\frac{\xi^* - t^*}{2\sqrt{t^*}}\right)\right] \tag{2.46}$$

An alternate form of (2.46) is

$$T^* = e^{-\xi^*} + \frac{1}{2}\left[\operatorname{erfc}\left(\frac{\xi^* + t^*}{2\sqrt{t^*}}\right) - e^{-\xi^*}\operatorname{erfc}\left(\frac{\xi^* - t^*}{2\sqrt{t^*}}\right)\right]$$

(*Note:* $T^* \rightarrow e^{-\xi^*}$ as $t^* \rightarrow \infty$ for finite ξ^*). The complementary error function employed in (2.46) has been described previously. Temperature distribution curves calculated from this equation are given in Figure 2.8 and in Table 2.3.

Heat Ahead of Front in Transient Period

The integral of dimensionless temperature with respect to dimensionless distance is given by equation 2.47. Tabulated values of this integral are also given in Table 2.3, where it is referred to as the heat integral.

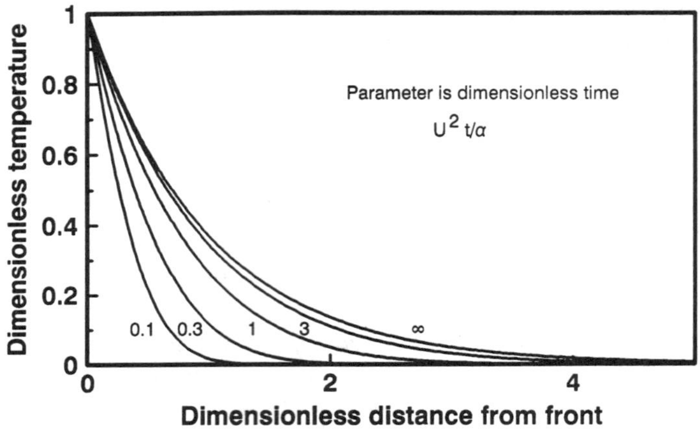

Figure 2.8 Transient Temperatures before an Advancing Front

$$\int_0^\infty T^* d\xi^* = \sqrt{\frac{t^*}{\pi}} e^{-t^*/4} + \left(1 + \frac{t^*}{2}\right) \text{erf}\left(\frac{\sqrt{t^*}}{2}\right) - \frac{t^*}{2} \tag{2.47}$$

The value of this integral ranges from 0 at time zero to 1 at time infinity. Multiplying the value of this integral by the heat accumulation for the steady state given by equation 2.45 gives the net heat accumulation for the transient condition.

Although the heat integral can be evaluated accurately from equation 2.47, it is a cumbersome relation and, for approximate design calculations, a simpler expression is convenient. Equation 2.48 gives estimates that are within 0.03 of the quantities calculated from equation 2.47.

$$\int_0^\infty T^* d\xi^* \simeq 1 - e^{-1.234\sqrt{t^*}} \tag{2.48}$$

Continuation of the Previous Numerical Example

Upon carrying out the steam flood described in the earlier numerical example, it is found that the steam spreads rapidly over the reservoir, as in case (iii) of that example.

It is thought that the production process that occurs involves the downward growth of the steam chamber. Measurements indicate that this occurs during the 10-y period at an average rate of 1.5 m/y.

iv. Calculate how much steam, in tonnes, is required for each year to supply the heat that is stored ahead of the advancing front, assuming that this is the same as if the front were stationary.[8]

v. Calculate the cumulative tonnes of steam required to build up the heat ahead of the front in the steady state.

vi. Repeat the calculation in case (iv), but this time allow for the forward movement of the front using equation 2.47.

[8]This assumption has been proposed by Vogel (1982) as a basis for the conservative design and analysis of steamflooding projects.

TABLE 2.3 Dimensionless Temperatures Beyond an Advancing Front

DIM. TIME	DIMENSIONLESS DISTANCE ξ^* 0.1	0.2	0.3	0.4	0.5	1	2	3	4	5	6	HEAT INTEGRAL
0.01	.456	.142	.029	.004	.000	.000	.000	.000	.000	.000	.000	0.108
0.02	.586	.286	.115	.037	.010	.000	.000	.000	.000	.000	.000	0.150
0.03	.648	.373	.189	.083	.032	.000	.000	.000	.000	.000	.000	0.181
0.04	.687	.432	.247	.128	.060	.000	.000	.000	.000	.000	.000	0.206
0.05	.713	.475	.293	.167	.088	.001	.000	.000	.000	.000	.000	0.228
0.10	.780	.588	.428	.300	.202	.015	.000	.000	.000	.000	.000	0.310
0.20	.827	.673	.538	.423	.327	.067	.001	.000	.000	.000	.000	0.413
0.30	.847	.711	.590	.484	.392	.114	.003	.000	.000	.000	.000	0.483
0.40	.859	.733	.620	.521	.433	.151	.009	.000	.000	.000	.000	0.537
0.50	.867	.748	.641	.546	.462	.180	.015	.001	.000	.000	.000	0.581
1.00	.886	.783	.690	.606	.531	.263	.049	.006	.001	.000	.000	0.720
1.50	.893	.796	.709	.631	.560	.300	.074	.014	.002	.000	.000	0.799
2.00	.897	.804	.719	.643	.575	.321	.090	.022	.004	.001	.000	0.849
2.50	.899	.808	.726	.651	.584	.334	.102	.027	.006	.001	.000	0.884
3.00	.901	.811	.730	.656	.590	.343	.110	.032	.008	.002	.000	0.910
3.50	.902	.813	.732	.660	.594	.349	.116	.036	.010	.003	.001	0.929
4.00	.902	.814	.734	.662	.597	.353	.120	.038	.012	.003	.001	0.943
4.50	.903	.815	.736	.664	.599	.357	.123	.041	.013	.004	.001	0.954
5.00	.903	.816	.737	.665	.601	.359	.126	.042	.014	.004	.001	0.963
5.50	.904	.817	.738	.666	.602	.361	.127	.044	.014	.005	.001	0.970
6.00	.904	.817	.738	.667	.603	.362	.129	.045	.015	.005	.001	0.975
6.50	.904	.817	.739	.668	.603	.363	.130	.046	.016	.005	.002	0.980
7.00	.904	.818	.739	.668	.604	.364	.131	.046	.016	.005	.002	0.983
7.50	.904	.818	.739	.669	.605	.365	.132	.047	.016	.006	.002	0.986
8.00	.904	.818	.740	.669	.605	.365	.132	.048	.017	.006	.002	0.988
8.50	.905	.818	.740	.669	.605	.366	.133	.048	.017	.006	.002	0.991
9.00	.905	.818	.740	.669	.605	.366	.133	.048	.017	.006	.002	0.992
9.50	.905	.818	.740	.670	.606	.366	.134	.048	.017	.006	.002	0.993
10.00	.905	.818	.740	.670	.606	.367	.134	.049	.018	.006	.002	0.995
inf.	.905	.819	.741	.670	.607	.368	.135	.050	.018	.007	.002	1.000

For the purposes of this example assume that the physical properties of the oil sand are the same as those of the overburden.

Solution Steam requirement for the reservoir below the steam chamber:

iv. The quantities of steam required for this case are the same as those calculated in (iii).

v. The heat ahead of a steadily advancing front is calculating using equation 2.45. In the present case the frontal advance rate is

$$U = 1.5 \text{ m/y} = 4.76 \times 10^{-8} \text{ m/s}$$

The stored heat is thus

$$Q_c = \frac{KA(T_S - T_R)}{U} = \frac{1.7 \times 40{,}000(264 - 15)}{4.76 \times 10^{-8}} = 3.56 \times 10^{14} \text{ J}$$

and the corresponding quantity of steam is

$$\frac{3.56 \times 10^{14}}{1.441 \times 10^{9}} = 247 \text{ kt}$$

vi. The buildup of heat ahead of the advancing front can be calculated by multiplying the equilibrium value obtained in (v) by the heat integral factor calculated from equation 2.47 or read directly from Table 2.3. The yearly increments in these heat requirements are shown in the following table; the heat requirement drops off toward zero as the heat accumulates.

Steam Required to Provide Heat Stored Beyond the Front

YEAR	CASE (iv) Fixed Front	CASE (vi) Moving Front
1	81,576	71,583
2	33,790	24,268
3	25,928	16,714
4	21,858	12,891
5	19,257	10,502
6	17,411	8,840
7	16,010	7,608
8	14,901	6,654
9	13,997	5,890
10	13,238	5,265
Total	257,966	170,215
Average Rate, kt/y	25.8	17.0

EFFECT OF CHANGING FRONT VELOCITY

In the preceding section, the temperature distribution ahead of a front moving at constant velocity U was determined. It was shown that a dynamic steady state is approached. In practice, the front velocity is not constant; frequently, it tends to decrease with time.

To study this effect, differential equation 2.41 was integrated numerically using the explicit finite difference procedure. The left-hand side of the equation was evaluated for a given time at equally spaced distances and then the corresponding temperature change at each point was calculated for a small time step, assuming that the right-hand side of the equation did not change significantly.

The results of such a calculation, in which—at a dimensionless time of 2—the front velocity was reduced to half of the initial value (and the dimensionless time using the new value of U became instantaneously 0.5), are shown graphically in Figure 2.9.

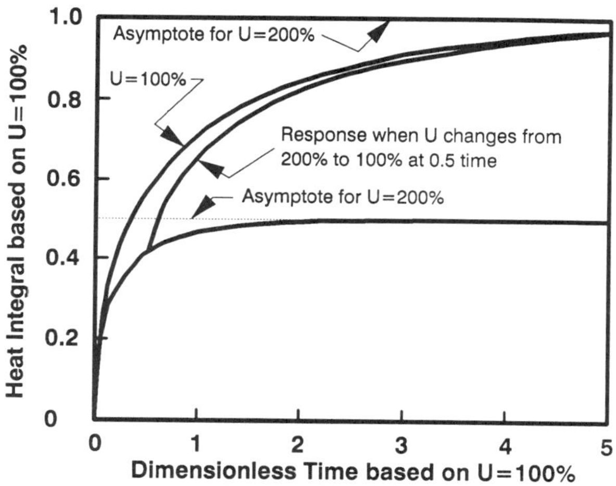

Figure 2.9 Effect of Reduced Front Velocity

At the point at which the velocity was changed, the heat integral had reached 85% of the initial steady-state value. With the decrease in velocity, the potential steady-state value doubled and the heat integral was then only 42.6% of the new steady state. With the reduction in velocity, the "run over" of heat by the interface decreased, and accumulation of heat beyond the interface increased. As a result, the response curve approached the new asymptote.

Also shown in Figure 2.9 is the heat integral curve for the case where the interface started out at the lower velocity. This curve is always above the other, but it may be seen that the difference becomes quite small soon after the change in velocity.

The Situation Where the Front Advance Velocity Is Inversely Proportional to the Square Root of Time

It happens that a simple solution to equation 2.38 can be obtained when the front advance velocity is inversely proportional to the square root of time. It is of interest because it is a simple example of the case where the front velocity falls with time.[9]

The variation of the front velocity U with time is assumed to be given by equation 2.49, where b is a dimensionless constant that determines the rate at which the velocity falls. The velocity starts at infinity unless $b = 0$.

$$U = b\sqrt{\frac{\alpha}{t}} \tag{2.49}$$

The temperature distribution for this situation is given by the remarkably simple equation 2.50; this reduces to equation 2.18 if b is set equal to zero—i.e., if the front is stationary.

[9]A practical situation which corresponds approximately to this condition is that where a cylindrical steam zone is growing radially about a vertical injection well. If the rate of volumetric growth of the chamber is constant, then the radius of the chamber increases at a rate proportional to the reciprocal of the square root of time, as is assumed above.

For this case $b = 0.5R(\alpha t)^{-0.5}$ where R is the radius at time t. However, the situation is not identical to that described above, since the area is also increasing.

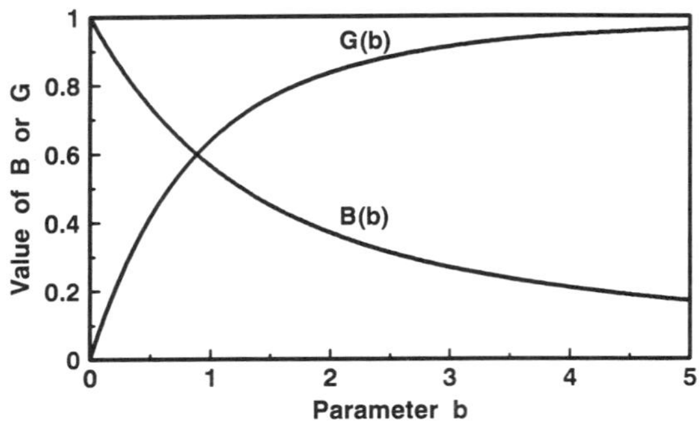

Figure 2.10 Heat Integral Factors

$$T^* = \frac{\text{erfc}(z + b)}{\text{erfc}(b)} \qquad \text{where } z = \frac{\xi}{2\sqrt{\alpha t}} \tag{2.50}$$

The heat integral may be calculated from either equation 2.51, which is similar in form to equation 2.25 for the stationary front, or from equation 2.53, which relates the heat integral to equation 2.45 for the steady state for a constant velocity front.

$$\frac{Q_c}{A} = 2K(T_S - T_R)\sqrt{\frac{t}{\pi\alpha}}B(b) \tag{2.51}$$

where

$$B(b) = \frac{e^{-b^2}}{\text{erfc}(b)} - b\sqrt{\pi} \tag{2.52}$$

$$\frac{Q_c}{A} = \rho C(T_S - T_R)\frac{\alpha}{U}G(b) \qquad \text{where } G(b) = \frac{2b}{\sqrt{\pi}}B(b) \tag{2.53}$$

The factors B and G are shown as functions of the parameter b in Figure 2.10. It is interesting to note that the situation approaches the steady-state case for larger b; i.e., G tends to 1 as b increases.

In many practical cases, it will be found that the heat ahead of an advancing front is not very different from the steady state level. For example, suppose that a front were advancing at a rate that is inversely proportional to the square root of time and that it had progressed a distance of 100 ft in 1 year. Then, using the formula in footnote 9, and assuming a typical value of 0.7 ft^2/d for α gives a value of $0.5 \times 100 \times (0.7 \times 365)^{-0.5} = 3.13$ for b. From Figure 2.10 it can be seen that, for this value of b, the factor $G(b)$ would be nearly 1.

RADIAL HEAT FLOW FROM A WELL

When a fluid such as steam or hot water is injected through a well, there is considerable heat loss from the well bore to the overburden. Fourier's equation for radial flow may be derived from equation 2.6 by substitution. However, it is simpler to start from first principles. The development is similar to that for linear flow given on page 31.

Consider an elemental cylinder about a vertical well such as that shown in Figure 2.11; the height of the cylinder measured normal to the paper is L.

Assume that heat is flowing at a rate Q radially from the center. Then

$$Q = -KA\frac{\partial T}{\partial R} = -2\pi RLK\frac{\partial T}{\partial R} \tag{2.54}$$

and

$$-\frac{\partial Q}{\partial R} = 2\pi RL\rho C\frac{\partial T}{\partial t} = 2\pi LK\frac{\partial T}{\partial R} + 2\pi RLK\frac{\partial^2 T}{\partial R^2} \tag{2.55}$$

or

$$\frac{\partial^2 T}{\partial R^2} + \frac{1}{R}\frac{\partial T}{\partial R} = \frac{1}{\alpha}\frac{\partial T}{\partial t} \tag{2.56}$$

We are interested in the solution of this radial form of Fourier's equation for the case where the temperature at the well bore (i.e., where $R = R_w$) is suddenly raised to the steam temperature T_S and the surrounding ground is initially at T_R.

The temperature distribution will be a function of time and radius. It will depend upon the radius of the well bore, R_w, and the thermal diffusivity, α. It is reasonable to expect that the smaller the radius of the well bore, the less will be the penetration of heat into the surroundings. Also, the heat loss from a unit area of the well surface will be greater than for a plane surface because of the divergent radial heat flow.

The solution for these boundary conditions is developed by Carslaw and Jaeger using the Laplace transformation method. The result is

$$\frac{T - T_R}{T_S - T_R} = 1 + \frac{2}{\pi}\int_0^\infty e^{-(\alpha t/R_w^2)W^2}\frac{\left[J_0\left(W\frac{R}{R_w}\right)Y_0(W) - Y_0\left(W\frac{R}{R_w}\right)J_0(W)\right]}{[J_0^2(W) + Y_0^2(W)]}\,d\ln(W) \tag{2.57}$$

where J_0 and Y_0 are Bessel functions of zero order and of the first and second kinds, respectively, and W is a dummy variable that drops out when the integral is evaluated.

Equation 2.57 is of the form

$$\frac{T - T_R}{T_S - T_R} = F\left(\frac{\alpha t}{R_w^2}, \frac{R}{R_w}\right) \tag{2.58}$$

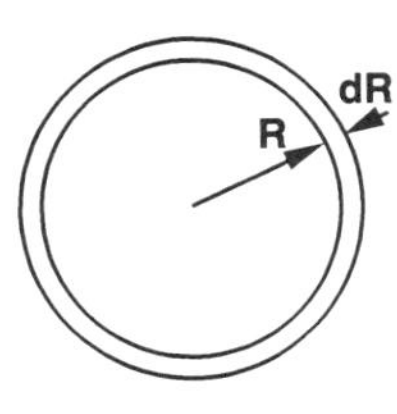

Figure 2.11

The first parameter may be considered to be the dimensionless time and the second, the dimensionless radius.

Although the expression for F (the right-hand side of equation 2.57) is complex, it may be evaluated readily using a computer. A plot of F versus $\log_{10}(R/R_w)$ is given in Figure 2.12.

For large values of dimensionless time, the initial linear part of the curves in Figure 2.12 is given by a straight line drawn from (0, 1) to the point on the horizontal axis where

$$\log_{10}\left(\frac{R}{R_w}\right) = \frac{0.5\left[\log_{10}\left(\frac{\alpha t}{R_w^2}\right) + 0.101\right]^2}{\log_{10}\left(\frac{\alpha t}{R_w^2}\right) - 0.150} \tag{2.58A}$$

This equation is based upon an asymptotic expansion given by Carslaw and Jaeger (1959).

The rate of heat loss from the well at time t is given by

$$Q = -2\pi R_w L K \left(\frac{\partial T}{\partial R}\right)_{R=R_w} \tag{2.59}$$

The temperature gradient at the well may be obtained by differentiating equation 2.57 with respect to R, with the result

$$Q = (2\pi R_w L)\left(\frac{4(T_S - T_R)K}{R_w \pi^2}\right)\int_0^\infty e^{-(\alpha t/R_w^2)W^2}\frac{dW}{W[J_0^2(W) + Y_0^2(W)]} \tag{2.60}$$

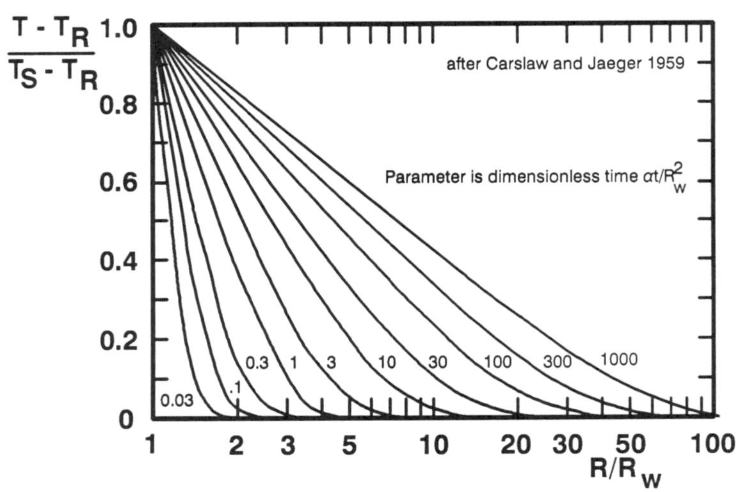

Figure 2.12 Temperature Distribution around a Circular Well Bore as a Function of Time

The heat flow may be written in a dimensionless form as

$$Q^* = \frac{Q}{(2\pi R_w L)(T_S - T_R)} \frac{R_w}{K} = \frac{4}{\pi^2} \int_{-\infty}^{\infty} e^{-(\alpha t/R_w^2)W^2} \frac{dW}{W[J_0^2(W) + Y_0^2(W)]}$$

or

$$Q^* = \frac{Q}{2\pi LK(T_S - T_R)} = F_1\left(\frac{\alpha t}{R_w^2}\right) \tag{2.61}$$

Values of the integral in equation 2.61 are tabulated by Jaeger and Clarke (Jaeger and Clarke 1942), and a plot of Q^* versus the dimensionless time is given in Figure 2.13. The curve in Figure 2.13 may be represented by the following empirical cubic equation.

$$\ln(Q^*) = -0.000629x^3 + 0.0203x^2 - 0.308x - 0.0150$$

where

$$x = \ln\left(\frac{\alpha t}{R_w^2}\right)$$
$$0.1 < \left(\frac{\alpha t}{R_w^2}\right) < 10^5 \tag{2.62}$$

Also shown in Figure 2.13 is the asymptote that corresponds to the heat loss from a portion of a large flat surface having the same heated area as the cylinder.

Cumulative Heat Flow from Well Bore

The total heat lost from the well bore after time t may be found by integrating equation 2.61.

$$Q_c = \int_0^t Q\,dt = 2\pi KL(T_S - T_R)\int_0^t F_1\,dt = 2\pi KL(T_S - T_R)\frac{R_w^2}{\alpha}\int_0^{t^*} F_1\,dt^* \tag{2.63}$$

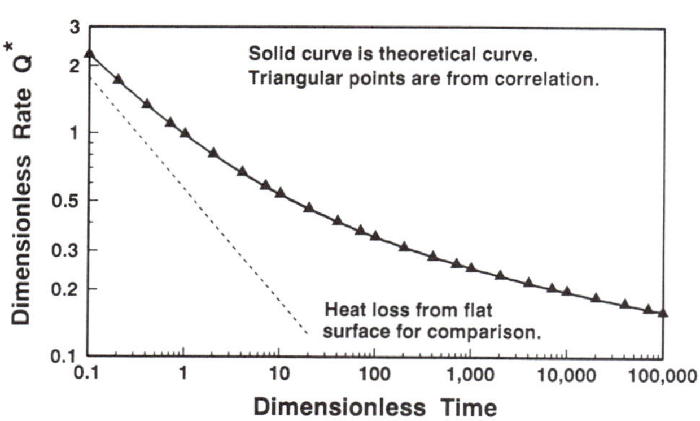

Figure 2.13 Heat Loss from Well Bore. The Value of R_w used in calculating the dimensionless time is the equivalent radius of the well bore after allowing for insulation effects.

where

$$t^* = \frac{\alpha t}{R_w^2}$$

This may be rewritten to give a dimensionless cumulative quantity of heat Q_c^*:

$$Q_c^* = \frac{Q_c \alpha}{2\pi KL(T_S - T_R)R_w^2} = \int_0^{t^*} F_1 \, dt^* \tag{2.64}$$

A plot of Q_c^* versus t^* is given in Figure 2.14. The curve may be represented by the empirical cubic equation

$$\ln(Q_c^*) = -0.000629x^3 + 0.0235x^2 + 0.621x + 0.412 \tag{2.65}$$

where

$$x = \ln\left(\frac{\alpha t}{R_w^2}\right)$$

Factors Affecting Well Bore Heat Loss

The rate of heat loss from a well bore is proportional to the length of the well, to the thermal conductivity of the ground, and to the temperature difference between the well bore and the surroundings. The rate decreases with time, as shown in Figure 2.13. The decrease with time is caused by the decrease of the temperature gradient as the ground around the well becomes heated.

Although the heat loss falls initially at a rate inversely proportional to the square root of time, as it did for linear flow (see equation 2.24), the rate decreases more slowly after the initial period because of the diverging radial flow. A circular well is better cooled by the surroundings than is a portion of a large flat plane having the same area.

It is important to notice the effect of the well radius on the rate of heat loss. For a given value of time in Figure 2.13, the rate of heat loss is less for a smaller radius than for a larger one; this is because its surface area is smaller. A well of zero radius loses no heat!

Insulation of Wells to Reduce Heat Loss

Consider the case of a well having an external radius R_2 and suppose that the well is partly filled with an annulus of material having an inside radius of R_1 and an outside radius of R_2, as in the cross section shown in Figure 2.15.

If the insulation has the same thermal properties[10] as the surroundings, then the effect of the insulation is the same as that of reducing the radius of the well from R_2 to R_1. The heat flow can be determined by using Figure 2.13 if R_1 is substituted for the equivalent well bore radius R_w.

[10]Only the thermal conductivity is of importance, since the amount of heat stored in the insulation will be negligible compared to the total quantities of heat involved. This is equivalent to saying that the transfer of heat through the annulus is equivalent to a steady-state transfer. This is the reason that the plots of temperature versus $\log_{10}(R/R_w)$ become straight in the vicinity of the well in Figure 2.12.

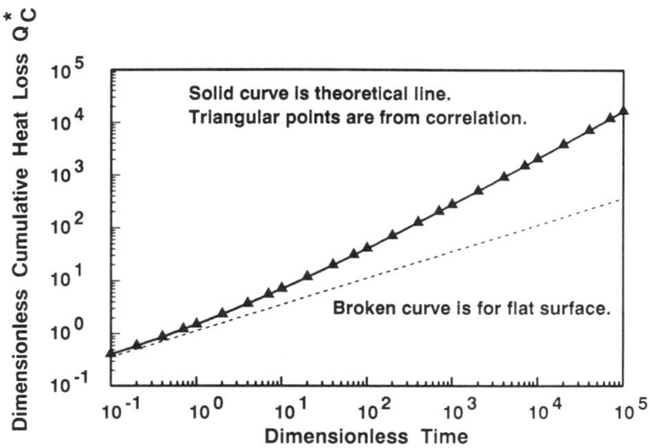

Figure 2.14 Cumulative Heat Loss from Well Bore

If the insulating material has a thermal conductivity different from the surroundings, then an equivalent well bore radius can be determined by making the thermal resistance of the insulating annulus equal to that of a hypothetical annulus having an internal radius R_w and the same conductivity as the ground.

The steady-state thermal resistance of an annular insulator is given by

$$\frac{T_1 - T_2}{Q} = \frac{\ln(R_2/R_1)}{2\pi L K_I} \tag{2.66}$$

where K_I is the thermal conductivity of the insulator. Let this be equivalent to the thermal resistance of an annulus having an internal radius R_w (the *equivalent well radius*) and external radius of R_2 and composed of material like the surroundings. Then

$$\frac{\ln(R_2/R_1)}{2\pi L K_I} = \frac{\ln(R_2/R_w)}{2\pi L K} \tag{2.67}$$

which may be reduced to

$$R_w = R_2\left(\frac{R_1}{R_2}\right)^{K/K_I} \tag{2.68}$$

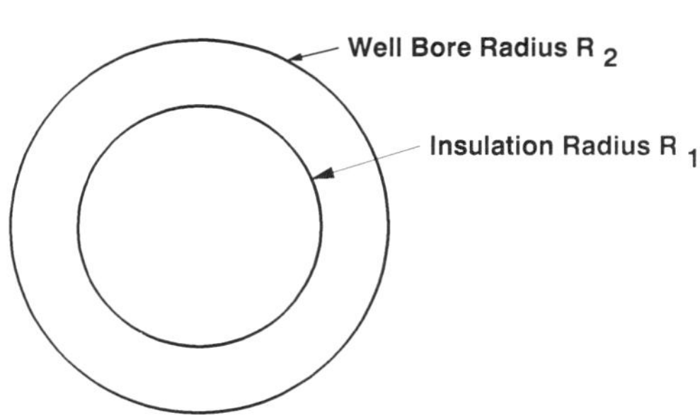

Figure 2.15

Substitution of this value of R_w in the abcissa of Figure 2.13 will make a proper allowance for the effect of the annular insulation.

In most cases the thermal resistance of the annulus, including any insulation, will be defined by an overall heat-transfer coefficient rather than by the thermal conductivity of an insulating material. The estimation of this heat-transfer coefficient is discussed later.

If the transfer from the tubing of radius R_1 to the outer well perimeter is determined by a heat-transfer coefficient U_1 (based on the surface area of the tubing), then the heat-transfer resistance is given by

$$\frac{T_1 - T_2}{Q} = \frac{1}{2\pi R_1 L U_1} \tag{2.69}$$

Equating this to the thermal resistance of the annulus of internal radius R_w filled with material of conductivity K leads to

$$\frac{\ln(R_2/R_w)}{2\pi L K} = \frac{1}{2\pi R_1 L U_1} \tag{2.70}$$

This reduces to the following expression for the effective well radius, R_w:

$$R_w = R_2 e^{-K/U_1 R_1} \tag{2.71}$$

Substitution of this value in the dimensionless time in the abcissa of Figure 2.13 will allow for the barrier to heat-transfer exerted by the annulus. Note that if $U_1 = \infty$, then $R_w = R_2$; also, if U_1 were zero, then R_w would also be zero.

The Equivalent Well Radius with Multiple Resistances

In most cases the heat-transfer coefficient U_1 represents the combined effect of a number of individual resistances to heat flow. A typical case is represented by the following diagram.

Steam $T = T_S$	Steam Film	Tubing Wall	Insulation if Present	Annulus	Casing Wall	Cement	Formation

The resistances of the steam film and of the metal walls are relatively small and may be neglected. The heat-transfer across the annulus occurs by two parallel mechanisms—radiation and convection. The combination of resistances for the overall situation may be represented as follows:

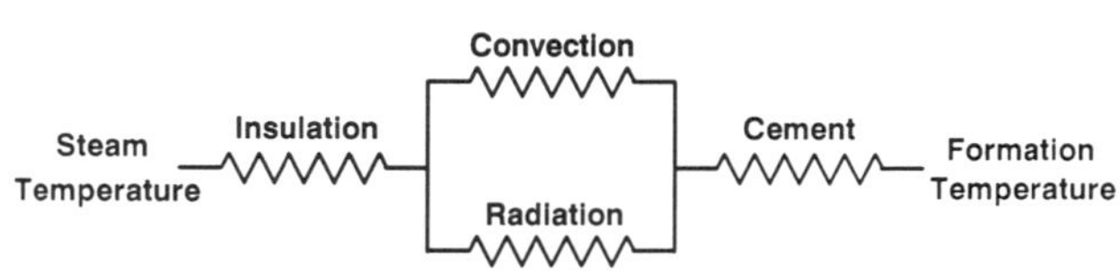

In this diagram the resistances to heat transfer are measured in degrees Celsius per watt, °C W^{-1} (or in English units, °F Btu^{-1} h).

$$r_{\text{cement}} = \frac{\ln(R_2/R_{\text{casing}})}{2\pi L K_{\text{cement}}}$$
$$r_{\text{conv}} = \frac{1}{h_c} \cdot \frac{1}{2\pi R_i} \tag{2.72}$$

where h_c is the convective heat-transfer coefficient for the annulus. Methods for predicting this are discussed later; R_i is the inner radius of the annulus.

$$r_{\text{rad}} = \frac{1}{h_r} \cdot \frac{1}{2\pi R_i} \tag{2.73}$$

$$r_{\text{insul}} = \frac{\ln(R_{\text{casing}}/R_i)}{2\pi L K_{\text{insul}}} \tag{2.74}$$

The overall resistance is

$$r_{\text{total}} = r_{\text{cement}} + \frac{1}{\dfrac{1}{r_{\text{conv}}} + \dfrac{1}{r_{\text{rad}}}} + r_{\text{insul}} \tag{2.75}$$

and the overall heat-transfer coefficient (based on the tubing radius R_1) is given by

$$2\pi R_1 U_1 = \frac{1}{r_{\text{total}}} \tag{2.76}$$

or

$$U_1 = \frac{1}{2\pi R_1 r_{\text{total}}} \tag{2.77}$$

The equivalent radius of the well is given by

$$R_w = R_2 e^{-2\pi K r_{\text{total}}} \tag{2.78}$$

Direct Injection of Steam Down the Well Casing

Direct injection of steam down the well casing is the simplest way to operate an injection well and is frequently practiced. For example, Esso, in its extensive cyclic steam stimulation project at Cold Lake in Alberta, injects steam directly down the annulus. This avoids the need for a high-temperature packer to isolate the annulus from the tubing. The use of such a packer presents difficulties in cyclic wells because of the problem in arranging for venting gases up the anulus during the production pumping period. A major disadvantage of operating with direct steam injection in the annulus is that the casing is raised to the steam temperature, and mechanical stresses due to expansion are very high. High-strength casing (e.g., N80) is normally employed, and even then the compressive yield stress is usually exceeded, and the casing develops a residual tensile stress when it is cooled.

Heat losses are the highest possible because the equivalent well radius is that of the casing (i.e., $U_1 = \infty$). Nevertheless, in relatively shallow projects the heat loss is acceptable. In the numerical example that is developed later, the well bore heat loss for a 17.8 cm-diameter well (7 in.) 460 m long delivering 160 m^3/d of steam is about 5 to 10% of the heat injection rate.

Injection of Steam in the Tubing with the Annulus Full of Gas

In this case there are two mechanisms for heat transfer across the annulus which operate in parallel: radiation and convection.

Radiant heat transfer between two concentric cylinders

The transfer of radiant energy may be calculated from the following equation:

$$Q_R = \frac{\sigma A_i(T_i^4 - T_o^4)}{\dfrac{1}{\epsilon_i} + \dfrac{R_i}{R_o}\left(\dfrac{1}{\epsilon_o} - 1\right)} \tag{2.79}$$

where i refers to the inner cylinder
o refers to the outer cylinder
T is the absolute temperature
ϵ is the surface emissivity
A_i is the area of the inner cylinder
σ is Stefan-Boltzmann constant: $\sigma = 5.669 \times 10^{-8}$ W/m^2 K^4 or 1.714×10^{-9} Btu/h ft^2 (°R)^4

It is assumed that the surfaces behave like "grey" bodies.

The radiation heat-transfer coefficient h_R, based on the area of the inner tube, is given by

$$h_R = \frac{Q_R}{A_i(T_i - T_o)} = \frac{\sigma(T_i^2 + T_o^2)(T_i + T_o)}{\dfrac{1}{\epsilon_i} + \dfrac{R_i}{R_o}\left(\dfrac{1}{\epsilon_o} - 1\right)} \tag{2.80}$$

The emissivity of typical oxidized steel is about 0.8.

A plot of h_R against T_i from equation 2.80 is given in Figure 2.16 for an emissivity of 0.8 and $R_i/R_o = 0.5$. Curves are shown for a range of temperature differences $(T_i - T_o)$ and steam temperature T_i.

Convective Heat Transfer between Two Concentric Vertical Cylinders

The convective heat transfer across the annulus may be written as

$$Q_{\text{conv}} = \frac{2\pi L K_{\text{eff}}\,\Delta T}{\ln(R_o/R_i)} \tag{2.81}$$

or

$$h_C = \frac{Q_{\text{conv}}}{2\pi R_i L\,\Delta T} = \frac{1}{R_i}\frac{K_{\text{eff}}}{\ln(R_o/R_i)}$$

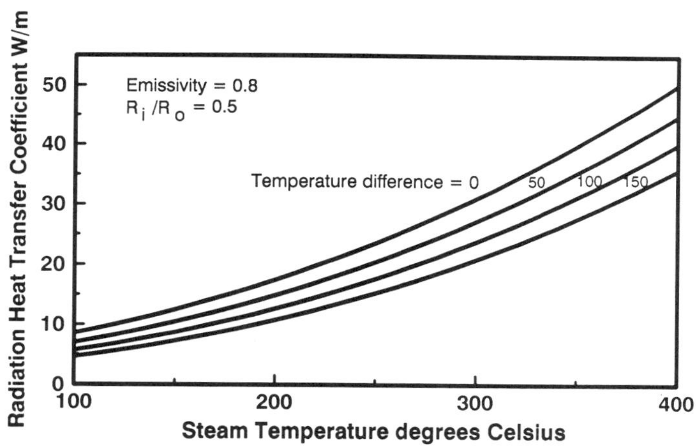

Figure 2.16 Tubing-Casing Radiant Heat Transfer

where K_{eff}, the *effective* thermal conductivity of the gas contents, is higher than the true thermal conductivity, K_g, of the gas because of the fluid motion caused by thermal convection. The ratio K_{eff}/K_g has been studied for heat transfer across the gap between vertical flat plates, but no studies appear to have been made for annuli between vertical cylinders. It is reasonable to assume that the effects will be similar. A mitigating factor is that for most steam injection systems, the transfer by convection is significantly smaller than that for radiation.

For the space between parallel vertical plates, the ratio K_{eff}/K_g is found to be a function of the Rayleigh number,

$$N_{Ra} = \left(\frac{g\beta\rho^2 C_p}{\mu K}\right)\delta^3(T_i - T_o) \tag{2.82}$$

where g is the acceleration due to gravity
β is the thermal coefficient of thermal expansion of the fluid equal to $1/V(\partial V/\partial t)$ (This is $1/T_{abs}$ for an ideal gas.)
ρ is the fluid density
C_p is the fluid specific heat
μ is the fluid viscosity
K is the fluid thermal conductivity
δ is the gap between the vertical surfaces
$T_i - T_o$ is the temperature difference

In this expression the group within the brackets is dependent upon the fluid and its temperature and pressure. It is customary to evaluate these properties at the average temperature between the two surfaces. Numerical values for the group for air at pressures of 1, 10, and 100 bar are shown in Figure 2.17.

The value of the factor declines with increasing temperature and increases with pressure. The pressure effect is largely due to the change in the density ρ; this

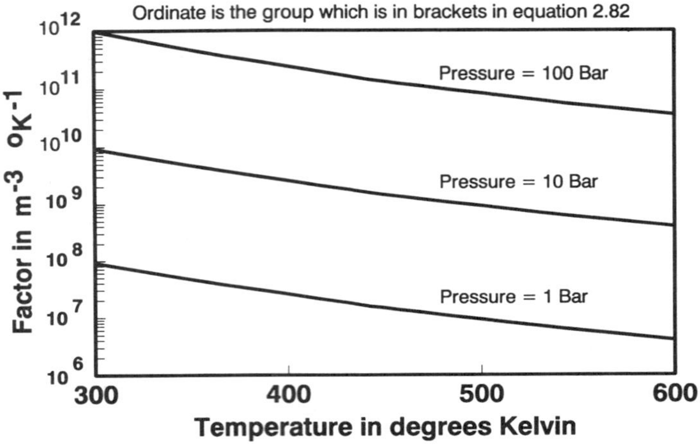

Figure 2.17 Natural Convection Factor for Air

is nearly proportional to pressure. Increasing the pressure by a factor of 10 increases the value of N_{Ra} by a factor of approximately 100. However, since the heat transfer is proportional to the $\frac{1}{3}$ or $\frac{1}{4}$ power of N_{Ra} (see the following), the heat transfer only increases by a factor of approximately $100^{1/4}$ to $100^{1/3}$, or 3.2 to 4.6.

At low values of N_{Ra}, convection has little effect on the transfer of heat. As N_{Ra} is increased, fluid motion increases; eventually, at values between 10^6 and 10^7, the flow becomes turbulent. Holman recommends the following equations (Holman 1981):

$$N_{\text{Ra}} < 6000 \qquad \frac{K_{\text{eff}}}{K_g} = 1$$

$$6000 < N_{\text{Ra}} < 2 \times 10^5 \qquad \frac{K_{\text{eff}}}{K_g} = 0.197 N_{\text{Ra}}^{1/4}\left(\frac{\delta}{L}\right)^{1/9} \tag{2.83}$$

$$2 \times 10^5 < N_{\text{Ra}} < 1.1 \times 10^7 \qquad \frac{K_{\text{eff}}}{K_g} = 0.073 N_{\text{Ra}}^{1/3}\left(\frac{\delta}{L}\right)^{1/9}$$

These equations also include the effect of the ratio of the height of the chamber to the width (L/δ). The range of this variable covered by the correlation is 11 to 42. It does not seem reasonable that the effect of δ/L would increase indefinitely up to the ratios found in injection wells—e.g., values of L/δ of several thousand. It is suggested that an arbitrary value of $L/\delta = 42$ be substituted in the preceding equations for the prediction of convective heat loss in well annuli. That is,

$$6000 < N_{\text{Ra}} < 2 \times 10^5 \qquad \frac{K_{\text{eff}}}{K_g} = 0.13 N_{\text{Ra}}^{1/4}$$

$$2 \times 10^5 < N_{\text{Ra}} < 1.1 \times 10^7 \qquad \frac{K_{\text{eff}}}{K_g} = 0.048 N_{\text{Ra}}^{1/3} \tag{2.84}$$

Values of N_{Ra} may be estimated by multiplying the values of the ordinate from Figure 2.17 by $\delta^3(T_i - T_o)$, as in equation 2.82.

Background Material on Well Bore Heat Loss

The classic papers in this field are those of Ramey (1962, 1965). As well as developing equations for the heat loss from a well at constant temperature, which are similar to those given in the previous section, Ramey also developed an approximate solution for the heat loss from a hot-water injection well. This solution allows for the decrease in temperature as the injected water flows down the well.

Satter (1965) extended these ideas to allow the calculation of heat losses when superheated steam is injected into the well. Willwhite (1966) described methods for predicting the heat-transfer coefficient between the tubing and the well casing. These are basically similar to the methods described here, although they do not use the concept of the equivalent well bore radius.

Fontanilla and Aziz (1982) allowed for the effect of two-phase flow in the tubing of the injection well. Farouq Ali (1981) described a calculation method that combines correlations for the two-phase flow within the well bore and a rigorous treatment of heat loss to the surroundings. The resulting equations were solved by a numerical method. A recent paper by Durrant and Thambynayagam (1986) describes a similar approach that employs a somewhat different computation method.

Numerical Example of Well Bore Heat Loss Calculation

Steam is injected into a reservoir having a depth of 460 m using a well with a 17.8 cm (7 in.) diameter casing. The conditions are as follows:

Reservoir temperature:	10°C
Steam pressure:	10 MPa
Steam quality:	70%
Injection rate:	160 $m^3\ d^{-1}$

Assume that the thermal conductivity of the overburden is 1.7 W m^{-1} $°C^{-1}$ and that its volumetric heat capacity is 2410 kJ m^{-3} $°C^{-1}$. Neglect the effect of the cement around the casing.

(a) Calculate the heat injection rate in megawatts.

(b) Assuming that the steam is injected directly down the casing, calculate the heat loss rate in megawatts after 1, 10, 100, and 1000 days of injection. Express this loss as a percentage of the heat input.

(c) Plot the temperature as a function of distance from the casing surface for each of the times in (b).

(d) Calculate the heat loss rate in watts per square meter for each of the times in (b) and compare to the heat loss that would be expected for a flat surface at the same steam temperature.

(e) Assume that the steam is injected into a 7.3 cm ($2\frac{7}{8}$ in.)-outside-diameter tubing and that this is isolated from the casing by a thermal packer at the bottom so that the annulus is filled with air at atmospheric pressure. Assume that the emissivity of the facing tubing and casing surfaces is 0.8. The internal diameter of the casing may be taken as 16.5 cm (6.5 in.). For these conditions repeat the calculations in parts (b) and (d) and compare the answers using a bar chart. Calculate the heat-transfer coefficients for a casing temperature of 250°C and assume that they do not vary with time. The thermal conductivity of air in the annulus may be taken as 0.0148 W m^{-1} $°C^{-1}$.

(f) Using the heat flows determined in (e), calculate the casing temperature for each time and make separate improved estimates of the heat-transfer coefficient *U*. Repeat the heat-flow calculations of (e) and repeat until consistent values are obtained. Revise the bar chart produced in (e) and also plot a graph showing the casing temperature as a function of the steaming time (use a log scale for time).

Solution

$L = 460$ m; $R_2 = \frac{7}{2}$ in. $= 0.0889$ m; $T_R = 10°C$; $T_S = 311°C$
steam quality = 70%; injection rate = 160 m^3d^{-1} = 160,000 kg d^{-1}
$K = 1.7$ W m^{-1} °C; vol heat cap = 2410 kJ m^{-3} $°C^{-1}$.

(a) Heat-injection rate

From steam tables: vapor enthalpy = 2724.7 kJ kg^{-1}
Liquid enthalpy = 1407.6 kJ kg^{-1}

$$\text{Heat in 70\% quality steam} = 2724.7 \times 0.7 + 1407.6 \times 0.3$$
$$= 2329.57 \text{ kJ kg}^{-1} \text{ above } 0°C$$
$$\text{Heat above } T_R = 2329.57 - 4.2 \times 10 = 2287.6 \text{ kJ kg}^{-1}$$
$$\text{Injection rate} = 160{,}000 \times 2287.6 = 366 \times 10^6 \text{ kJ d}^{-1}$$
$$= 366 \times 10^6 \times 1000/(24 \times 3600 \times 10^6)$$
$$= 4.24 \text{ MW}$$

(b) Heat loss with injection down the casing

Days of injection	1	10	100	1000
Dimensionless time $(\alpha t/R_w^2)^{(1)}$	7.71	77.1	771	7710
ln(dimenionless time)	2.04	4.35	6.65	8.95
Heat loss in megawatts from (2.62)	0.75	0.44	0.29	0.21
Heat loss as percent of input	17.5	10.4	6.8	5.0

(1) $\alpha = 1.7/(2410 \times 1000) = 7.05 \times 10^{-7}$ m^2s^{-1} = 0.060946 m^2d^{-1}

(c) Dimensionless temperatures from Figure 2.12

		Dimensionless Temperatures		
$\log(R/R_w)$	R/R_w	$t^* = 7.71$	$t^* = 77.1$	$t^* = 771$

Days of injection	1	10	100	1000
Heat loss W m^{-2}				
From well	2919	1726	1144	810
Flat surface (2.22)	1169	370	117	37

Corresponding radii and temperatures:

Radius in meters	Temperatures in degrees Celsius 1 d	10 d	100 d
0.09	311	311	311
0.16	209	251	266
0.28	109	188	221
0.50	37	127	176
0.89	10	64	127
1.58	10	25	85
2.81	10	10	52
5.00	10	10	22
8.89	10	10	10

These results are plotted in Figure 2.18.

(d) Calculation of heat-loss rate

$$\text{External area of well casing} = 2\pi R_w L = 256.9 \text{ m}^2$$

Days of injection	1	10	100	1000
Heat loss W m^{-2}				
From well	2919	1726	1144	810
Flat surface (2.22)	1169	370	117	37

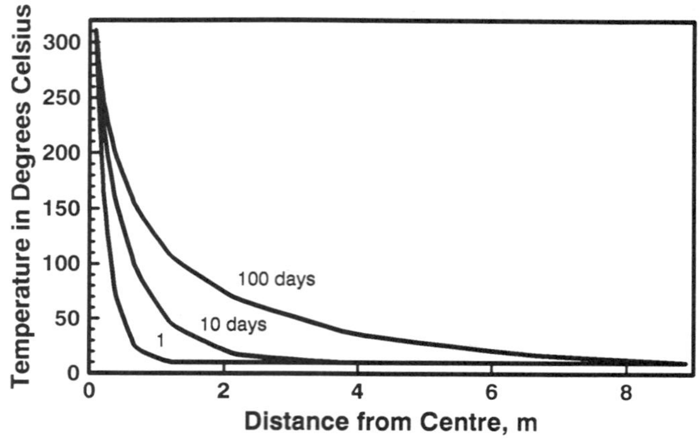

Figure 2.18 Temperatures Around Injection Well

(e) Use equation 2.80 to calculate h_R.

$$\sigma = 5.669 \times 10^{-8} \text{ W m}^{-1} \text{ K}^{-4}$$

$$T_i = 273 + 311 = 584 \text{ K}$$

$$T_o = 273 + 250 = 523 \text{ K}$$

Emissivity for both surfaces equals 0.8.

$$\frac{R_i}{R_o} = \frac{2.875}{6.5} = 0.442; \qquad R_i = 0.442 \times 3.25 \times 0.0254 = 0.0365 \text{ m}$$

$$h_R = 28.4 \text{ W m}^{-2} \text{ °C}^{-1}$$

Use equations 2.84 to calculate h_C.

$$\delta = \frac{0.5(6.5 - 2.875)2.54}{100} = 0.0460 \text{ m}$$

$$\text{Average gas temperature} = 553.5 \text{ K}$$

$$N_{\text{Ra}} \text{ factor from Figure 2.17} = 7.5 \times 10^6$$

$$N_{\text{Ra}} = 44{,}600; \qquad K_{\text{eff}}/K_g = 0.13 N_{\text{Ra}}^{0.25} = 1.89$$

$$h_C = \frac{1.89\, K}{R_i \ln(R_o/R_i)} = 0.9$$

$$U = h_R + h_C = 29.3 \text{ W m}^{-2} \text{ °C}^{-1}$$

$$R_w = 0.01812 \text{ m}$$

Calculation of Loss Rate

Time in days	1	10	100	1,000
Dimensionless time	186	1,860	18,600	186,000
ln(dimensionless time)	5.22	7.53	9.83	12.13
Heat loss in megawatts	0.38	0.26	0.18	0.13

(f) The preceding solution assumes a constant value for the casing temperature. In practice the casing temperature must start initially at T_R and then increase asymptotically toward T_S. As it increases, the value of U also increases and the value of R_w increases. The problem can be approached more exactly by estimating the casing temperature for each time from the heat flows calculated earlier together with the value of U. These temperatures are then used to estimate new values of U, and the calculation is repeated.

Time in days	1	10	100	1,000
First iteration				
Heat flow (MW)	0.375	0.256	0.184	0.132
Casing temperature (°C)	190	228	252	268
Average annual temperature (°C)	250	270	281	290
ΔT (°C)	61	41	30	21
h_R (W m^{-2} °C^{-1})	24.2	26.8	28.5	29.7
N_{Ra} factor $\times 10^{-7}$	1.4	1	0.65	0.62
N_{Ra}	82,925	40,471	18,864	12,922
h_C	2.2	1.8	1.5	1.4
New U (W m^{-2} °C^{-1})	26.4	28.6	30.0	31.1
R_w (m)	0.0152	0.0175	0.0188	0.0199
dimensionless time	263	2,001	17,245	154,030
ln(dimensionless time)	5.570	7.601	9.755	11.945
Second iteration				
New Heat flow (MW)	0.352	0.253	0.186	0.136
Casing temperature (°C)	184	227	252	270
Average annual temperature (°C)	248	269	282	290
ΔT (°C)	63	42	29	21
h_R (W m^{-2} °C^{-1})	23.9	26.7	28.5	29.8
N_{Ra} factor $\times 10^{-7}$	1.4	1	0.65	0.62
N_{Ra}	86,406	40,957	18,619	12,507
h_C	2.2	1.8	1.5	1.4
U (W m^{-2} °C^{-1})	26.1	28.6	30.0	31.2
R_w (m)	0.0149	0.0174	0.0189	0.0200
dimensionless time	273	2,016	17,151	152,638
ln(dimensionless time)	5.611	7.609	9.750	11.936
New heat flow (MW)	0.350	0.253	0.186	0.136

The heat flows just calculated are essentially the same as those at the start of the second iteration.

Summary of Calculations

Heat Loss in MW				
No tubing	0.75	0.44	0.29	0.21
With tubing last iteration	0.350	0.253	0.186	0.136
With tubing first iteration	0.375	0.256	0.184	0.132
Flat surface	0.300	0.095	0.030	0.010
Casing temperature (with tube) (°C)	184	227	252	270
Casing temperature (no tube) (°C)	311	311	311	311

The heat losses summarized in the upper part of the preceding table are compared in Figure 2.19.

The change of casing temperature with time is shown in Figure 2.20. Note that the scale in this figure is logarithmic and that the linear rate of change of temperature becomes very small at the right-hand limit of the figure.

RADIAL CONDUCTIVE HEAT LOSS FROM A BURIED HEATED CYLINDER

Another thermal conduction problem of practical significance is the radial heat loss from a vertical cylinder that is initially at T_S and that is buried within a medium at a lower temperature T_R. For example, when steam is injected into a vertical steam stimulation well, a vertical cylindrical region of the reservoir (radius R_h) can become heated. The pioneering paper by Boberg and Lantz (1966) on cyclic steam stimulation assumes that at the end of steam injection, this situation prevails and that during the steam soaking period that follows, heat is conducted radially into the surrounding colder reservoir. The temperature of the surrounding reservoir is assumed to be uniform at T_R at the end of injection.

The conditions at the start of the soaking period ($t = 0$) are thus

$$T = T_S \qquad \text{for } 0 < R \leq R_h$$

and

$$T = T_R \qquad \text{for } R > R_h$$

also, for

$$t > 0, \qquad T = T_R \quad \text{at } R = \infty .$$

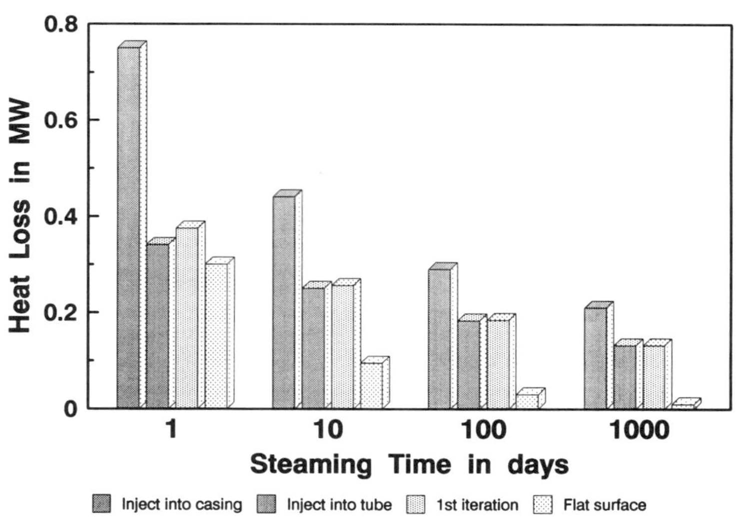

Figure 2.19 Heat Loss from Steam-Injection Well

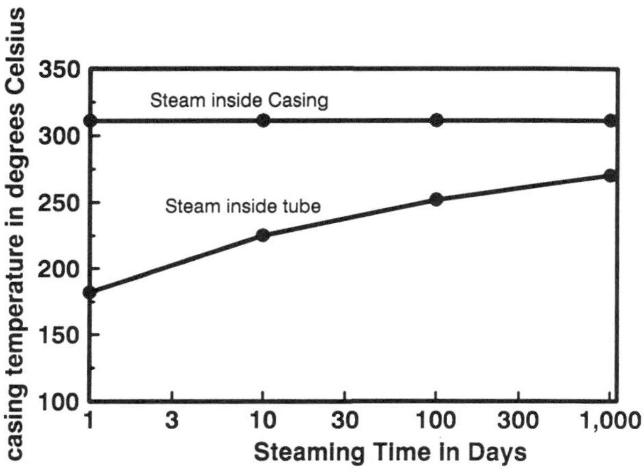

Figure 2.20 Temperature of Well Casing

Two solutions of the radial Fourier equation 2.56 are available which satisfy these conditions.

1. Volumetric heat capacities and thermal conductivities of the cylinder and surroundings are equal.

$$\frac{\overline{T} - T_R}{T_S - T_R} = 2\int_0^{\infty} e^{-(\alpha t/R_h^2)U^2} \frac{J_1(U)\, dU}{U} \tag{2.85}$$

where $\overline{T}$ is the volume average temperature within the cylinder, J_1 is the first order Bessel function of the first kind, and U is a dummy variable (Boberg and Lantz 1966).

The preceding solution is based upon an analysis given by Carslaw and Jaeger (1959, 346). The solution given by these authors is more general and also allows situations where the thermal properties of the cylinder and surroundings differ.

2. The thermal conductivity of the cylinder is infinite.

This case may provide a more realistic representation of a cooling steam chamber. The reason for this is that the interior of a steam chamber tends to remain isothermal as it cools because of the transport of heat from hotter regions to cooler ones by the evaporation and condensation of water.

The solution for this case given by Carslaw and Jaeger (1959, 342) is

$$\frac{\overline{T} - T_R}{T_S - T_R} = \frac{4b}{\pi^2}\int_0^{\infty} e^{-(\alpha t/R_h^2)U^2} \frac{dU}{U\Delta U} \tag{2.86}$$

where

$$b = 2\frac{(\rho C)_{R>R_h}}{(\rho C)_{R\leq R_h}}$$

and $$\Delta U = [UJ_o(U) - bJ_1(U)]^2 + [UY_0(U) - bY_1(U)]^2$$

where J_0, J_1, Y_0, and Y_1 are Bessel functions.

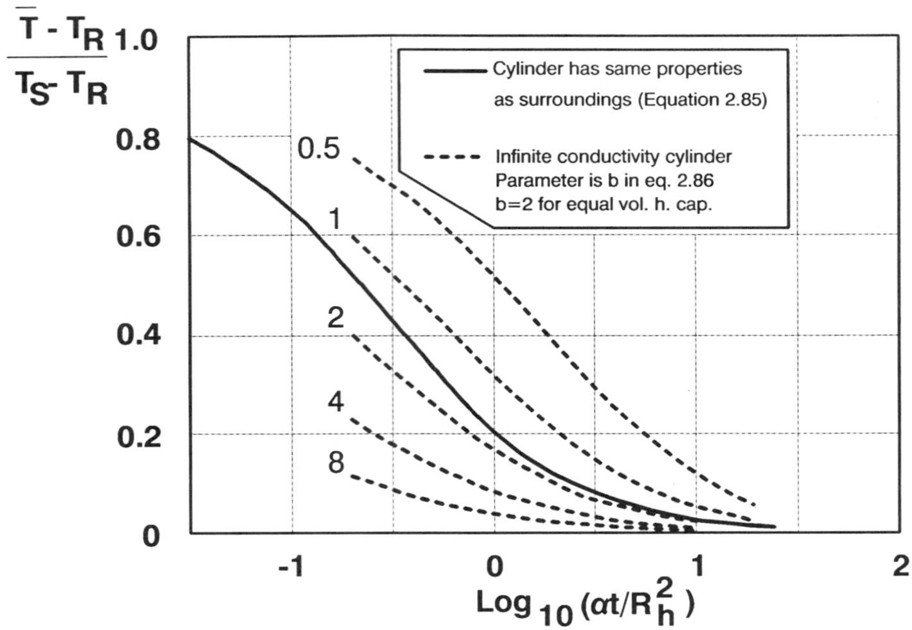

Figure 2.21 Radial Heat Loss from Hot Cylinder

Values of the dimensionless temperature calculated from equations 2.85 and 2.86 are given in Figure 2.21. The broken line is from equation 2.85 for the case

Radial Heat Loss from Hot Cylinder

Equal thermal conductivities and heat capacities (equation 2.85)

$\frac{\alpha t}{R_h^2}$	$\log_{10}\left(\frac{\alpha t}{R_h^2}\right)$	$\frac{\overline{T} - T_R}{T_S - T_R}$
0.0316	−1.50	0.7979
0.0562	−1.25	0.7363
0.1000	−1.00	0.6525
0.1778	−0.75	0.5473
0.3162	−0.50	0.4251
0.5623	−0.25	0.3023
1.0000	0.00	0.1985
1.7783	0.25	0.1229
3.1623	0.50	0.0732
5.6234	0.75	0.0426
10.0000	1.00	0.0244
17.7828	1.25	0.0139
31.6228	1.50	0.0078
56.2341	1.75	0.0044
100.0000	2.00	0.0025

where the thermal properties of the buried cylinder are the same as those of the surroundings. As would be expected, the heat loss for this case is less than that for the cylinder having the same heat capacity as the surroundings but an infinite thermal conductivity (solid line with $b = 2$).

The table on page 70 gives values of the dimensionless temperature calculations for a wider range of times than shown in Figure 2.21 for equation 2.85.

BIBLIOGRAPHY

BOBERG, T. C. and LANTZ, R. B., "Calculation of the Production of a Thermally Stimulated Well," *JPT,* 1613–1623 (December 1966).

CARSLAW, H. S. and JAEGER, J. C.: *Conduction of Heat in Solids,* Oxford: Clarendon Press (1959).

CLOSMAN, P. J. and SMITH, R. A., "Temperature Observations and Steam-Zone Rise in the Vicinity of a Steam-Heated Fracture," *SPEJ,* 575–586 (August 1983).

DURRANT, A. J. and THAMBYNAYAGAM, R. K. M., "Wellbore Heat Transmission and Pressure Drop for Steam/Water Injection and Geothermal Production: A Simple Solution Technique," *SPE Reservoir Engineering,* 148–162 (March 1986).

FAROUQ ALI, S. M., "A Comprehensive Wellbore Steam/Water Flow Model for Steam Injection and Geothermal Applications," *SPEJ,* 527–534 (October 1981).

FONTANILLA, J. P. and AZIZ, K.: "Prediction of Bottom-Hole Conditions for Wet Steam Injection Wells," *JCPT,* 81–88 (March–April 1982).

HASTINGS, C., JR., quoted by ABRAMOWITZ, M. and STEGUN, I. A., *Handbook of Mathematical Functions,* National Bureau of Standards rep. Dover (1965).

HOLMAN, J. P., *Heat Transfer,* 5th Ed., New York: McGraw-Hill, (1981), 286–292.

JAEGER, J. C. and CLARKE, M., *A Short Table . . .,* Proc. Roy. Soc. Edinburgh, A61, (1942) 229–230.

JENSON, V. G. and JEFFREYS, G.V., *Mathematical Methods in Chemical Engineering,* New York: Academic Press (1963), 149–151.

MATTHEWS, C. S. and RUSSELL, D. G., "Pressure Buildup and Flow Tests in Wells," *SPE Monograph* 1 (1967).

RAMEY, H. J.: "Wellbore Heat Transmission," *JPT,* 427–435 (April 1962).

RAMEY, H. J., "How to Calculate Heat Transmission in Hot Fluid Injection" in *Fundamentals of Thermal Oil Recovery,* Dallas, Tex.: Petroleum Engineering Publishing Co. (1965).

SATTER, A., "Heat Losses During Flow of Steam Down a Wellbore," *JPT,* 845–851 (July 1965).

VOGEL, J.V., "Simplified Heat Calculations for Steamfloods," *SPE* 11219, (1982); *JPT,* 1127–1136 (July 1984).

WILLWHITE, G. P., "Overall Heat Transfer Coefficients in Steam and Hot Water Injection Wells," *SPE* 1449 (1966).

3

Convective Heating Within Reservoirs

INTRODUCTION

Chapter 2 discussed the transfer of heat by conduction; this is a slow process, and many years are required for heat to move only a few tens of feet. Approaches such as the heating of reservoirs using wells containing electric or other heaters are ineffective for heating substantial volumes of reservoir because they, too, depend upon conduction. The transfer of significant quantities of heat to the reservoir by conduction through relatively small heating surfaces is simply too slow.

In this chapter the much more practical alternative of heating by the injection of hot fluids (i.e., by forced convective heating) is discussed. For practical reasons the choice of fluid is limited to water: either hot water or steam. Of these, steam is more popular because more heat can be transported per pound and also because steam can produce a much more stable and complete displacement of the oil. Reasons for this second advantage are discussed in a later chapter.

In the processes considered in this chapter, heat is moved into the reservoir as sensible and latent heat in the injected fluid, and, as the fluid passes through the reservoir, heat is given to the cooler surroundings. This transfer of heat produces temperature gradients in the surroundings, and these gradients cause the conduction of heat from the boundaries of the flowing fluid. Thus, while the heat is transported to the reservoir by fluid convection, thermal conduction also plays a significant role. One important mechanism is the conduction of heat away from the heated reservoir into the overburden and underburden. Although conduction tends to be slow, the mechanism is of great importance because of the vast heated areas that develop as a result of hot-fluid injection.

Most of this chapter is concerned with the process of forced convective transport of heat to the reservoir coupled to the dissipation of this heat by conduction through the large heated areas that develop.

SIMPLE CONVECTIVE HEAT TRANSFER WITHOUT CONDUCTIVE HEAT LOSS

Consider the simple, one-dimensional flow of hot water within the reservoir shown in Figure 3.1. It is assumed, in this section, that there is no heat loss from the upper and lower horizontal bounds of the reservoir and that temperature is a function only of the distance, x, and time, t.

It is assumed that the fluid saturations do not change during the process; e.g., the oil saturation is at the residual oil level and the water saturation is constant. The fluid flowing out of the differential element, δx, shown in the figure will, in general, be at a different temperature from that entering. The change in the heat flux will come from or go to the inventory of heat within the element. A balance of the heat rates about the differential element yields

$$-hV_w\rho_w C_w\left(\frac{\partial T}{\partial x}\right)_t \delta x = h\rho_1 C_1\left(\frac{\partial T}{\partial t}\right)_x \delta x \tag{3.1}$$

Heat in − Heat out = accumulation

This may be rearranged as

$$\left(\frac{\partial T}{\partial t}\right)_x + \frac{V_w\rho_w C_w}{\rho_1 C_1}\left(\frac{\partial T}{\partial x}\right)_t = 0 \tag{3.2}$$

where

$$\rho_1 C_1 = (1-\phi)\rho_R C_R + \phi\rho_w C_w S_w + \phi\rho_o C_o S_o$$

The term $\rho_1 C_1$ has the dimensions of heat per unit volume per degree of temperature; it is the volumetric heat capacity of the reservoir and is *not* equal to the average density multiplied by the average heat capacity.

Equation 3.2 shows that for constant positive values of the flow rate, the rate of rise of the temperature at any particular point is of the opposite sign to the temperature gradient if the volumetric heat capacity is constant, and is proportional to the temperature gradient.

Equation 3.3, which is the general solution of 3.2, describes a heat front that moves along the bed at a velocity given by V_f. The solution is such that any existing temperature profile is moved unchanged along the bed.

$$T = F(x - V_f t) \tag{3.3}$$

where F is any function and

$$V_f = \frac{V_w\rho_w C_w}{\rho_1 C_1}$$

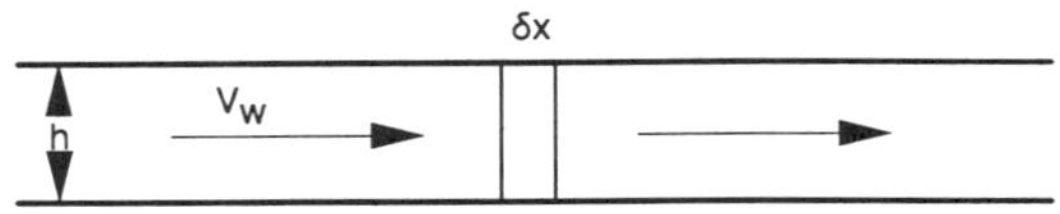

Figure 3.1 In this Figure V_w is equal to the total volumetric flow divided by the cross-sectional area

Figure 3.2 depicts the movement of a heat front along the bed such as would be caused by suddenly raising the temperature of the incoming fluid to a higher temperature.

There are many simplifications in the preceding derivation which will not be realistic in actual operations. The assumption of neglecting the vertical heat losses will be considered later. Another assumption is that the solid and fluid are at the same temperature at a particular location. This assumption is sometimes referred to as thermostatic equilibrium. It is realistic in reservoir scale operations, but it is often not realistic in small process-type packed beds. If heat-transfer resistances between the fluid and the bed and/or within the solids of the bed are included in the theory, sharp fronts tend to become spread out with time. Another phenomenon that causes longitudinal dispersion of heat is that of thermal conduction in the direction of the fluid flow. This becomes important for slow moving fronts (see page 47 for example). Despite these various assumptions, the situation shown in Figures 3.1 and 3.2 has within it the basic concept of a heat front moving through a reservoir at a rate which is less than the fluid velocity.

Overall Heat Balance Approach

Another insight into the preceding problem can be gained by considering an overall heat balance for the process depicted by Figure 3.2. In this process a quantity of hot fluid at T_{inj} has passed into the heat front from the left and has left to the right at temperature T_R. The heat given up by this hot fluid has heated the reservoir in the distance interval $x_1 - x_0$. These two quantities of heat may be equated as follows:

$$hV_w\rho_w C_w(T_{\text{inj}} - T_R)(t_1 - t_0) = h\rho_1 C_1(T_{\text{inj}} - T_R)(x_1 - x_0) \tag{3.4}$$

Equation 3.4 may be rearranged to give

$$V_f = \frac{(x_1 - x_0)}{(t_1 - t_0)} = \frac{V_w\rho_w C_w}{\rho_1 C_1} \tag{3.5}$$

This is the same expression for the velocity that was derived previously.

Another way of writing equation 3.5 is

$$V_f = \frac{(x_1 - x_0)}{(t_1 - t_0)} = \frac{V_w\rho_w C_w(T_{\text{inj}} - T_R)}{\rho_1 C_1(T_{\text{inj}} - T_R)} = \frac{H_0}{\rho_1 C_1(T_{\text{inj}} - T_R)} \tag{3.6}$$

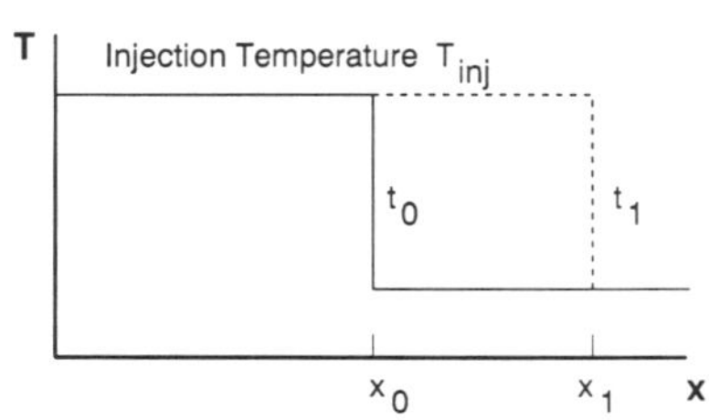

Figure 3.2

In equation 3.6 the numerator, H_0, is the rate of heat injection per unit cross-sectional area and the denominator is the heat content of a unit volume of the heated reservoir above the initial temperature T_R.

Steam Injection

Assume that in Figure 3.1 wet steam having a quality f_s and a temperature T_S is introduced into the reservoir rather than water. The steam flow rate is W_S kilograms per square meter per second and the latent heat is λ kilojoules per kilogram. The water saturation in the steam-saturated zone may be and probably will be lower than in the original reservoir. We will determine the velocity at which the condensation front will advance.

As in the previous section a heat balance may be written that equates the injected heat to that stored in the reservoir. As before, this heat balance uses T_R as the basis temperature.

$$W_s h(f_s\lambda + (T_S - T_R)C_w)(t_1 - t_0) = h(x_1 - x_0)\{(\rho_1 C_1)_s(T_S - T_R) + \lambda\phi\rho_s S_s\}$$

$$V_f = \frac{(x_1 - x_0)}{(t_1 - t_0)} = \frac{W_s(f_s\lambda + (T_S - T_R)C_w)}{(\rho_1 C_1)_s(T_S - T_R) + \lambda\phi\rho_s S_s} = \frac{H_0}{(\rho_1 C_1)_s(T_S - T_R) + \lambda\phi\rho_s S_s} \tag{3.7}$$

In the right-hand side of equation 3.7, the numerator is the total rate of heat injection per unit area and the denominator is the heat per unit volume of hot, steam-saturated reservoir. This is similar to equation 3.6. The term $\lambda\phi\rho_s S_s$ is the latent heat of the steam remaining within a unit volume of steam-swept reservoir; it is normally very small compared to the sensible heat term and can be neglected with little error.

For a given mass rate of injection, H_0 will be larger for steam injection than for hot-water injection. Consequently the rate of advance of a steam condensation front will be greater than that of a thermal front produced by the injection of hot water at the same temperature. The effect may be even larger if the volumetric heat capacity of the steamflooded reservoir $(\rho_1 C_1)_s$ is less than that of the waterflooded reservoir because of the lower water saturation.

An important conclusion which may be drawn from this analysis is that because the steam front moves faster than a water front produced by the same mass flow of water, heat is not carried beyond the condensation front by the condensate from the steam; the steam condensation front catches and keeps pace with the condensate hot-water front, and the two advance together. This phenomenon is discussed further later on.

LAUWERIER'S EQUATION

In many cases in thermal recovery we are interested in the lateral conduction of heat away from a flowing stream of heating fluid. One of the earliest relations that was derived to describe a process of this sort is due to H. A. Lauwerier. He considered the situation shown in Figure 3.3.

Lauwerier considered a situation in which hot water is flowing in a layer of thickness h within an oil sand reservoir. As the water flows through the water-saturated zone, heat is lost to the oil sand above and below. The derivation is also applicable to a permeable sand layer which is being heated by flowing water or other fluid and which is bounded above and below by impermeable rock. Figure 3.3 shows the situation modeled by Lauwerier. The flowing water layer has a height h. Vertical distances are measured from the centerline. It is assumed that there is no vertical temperature gradient within the water layer. Water is injected at a constant rate and temperature.

A heat balance about the region of thickness δx gives equation 3.8. This is the same as equation 3.1 with the addition of the last term, which represents the heat loss to the oil sand above and below.

$$-hV_w\rho_w C_w\left(\frac{\partial T_1^*}{\partial x}\right)\delta x = h\rho_1 C_1\left(\frac{\partial T_1^*}{\partial t}\right)\delta x - 2K_2\left(\frac{\partial T_2^*}{\partial y}\right)_{y=h/2}\delta x \tag{3.8}$$

The term $\rho_1 C_1$ represents the volumetric heat capacity of the water layer, as given by

$$\rho_1 C_1 = (1-\phi)\rho_R C_R + \rho_w C_w S_w + \phi\rho_o C_o S_o \tag{3.9}$$

The conduction of heat within the oil sand is determined by Fourier's equation 3.10, where $\rho_2 C_2$ is the volumetric heat capacity of the oil sand determined by an equation similar to (3.9).

$$K_2\left(\frac{\partial^2 T_2^*}{\partial y^2}\right) = \rho_2 C_2\left(\frac{\partial T_2^*}{\partial t}\right) \tag{3.10}$$

Substituting the dimensionless variables of (3.11),

$$\begin{aligned} x_D &= \frac{4K_2 x}{h^2\rho_w C_w V_w} \\ y_D &= 2y/h \\ t_D &= \frac{4K_2 t}{h^2\rho_1 C_1} \\ \theta &= \frac{\rho_1 C_1}{\rho_2 C_2} \end{aligned} \tag{3.11}$$

results in the following system of equations

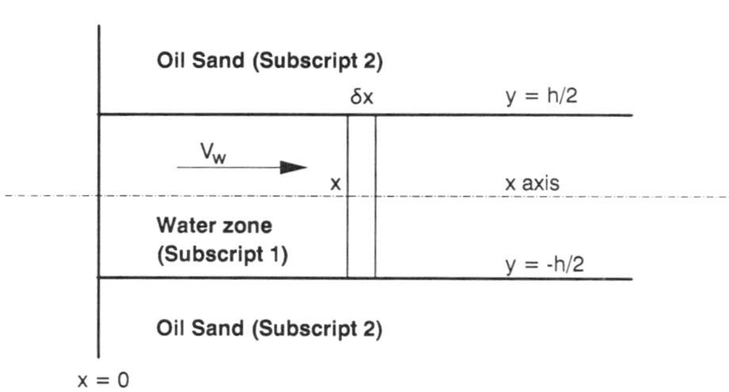

Figure 3.3 Lauwerier's Problem
Initial Boundary Conditions
$T_1^* = T_2^* = 0$ at $t = 0$
$T_1^* = 1$ for $x = 0$ at $t > 0$

For $|y_D| > 1$:

$$\theta\left(\frac{\partial^2 T_2^*}{\partial y_D^2}\right) = \left(\frac{\partial T_2^*}{\partial t_D}\right) \tag{3.12}$$

For $|y_D| = 1$:

$$\left(\frac{\partial T_2^*}{\partial t_D}\right) + \left(\frac{\partial T_2^*}{\partial x_D}\right) - \left(\frac{\partial T_2^*}{\partial y_D}\right) = 0 \tag{3.13}$$

$$\text{and } T_1^* = T_2^*$$

$$\text{For } t_D = 0: \qquad T_1^* = T_2^* = \begin{matrix} 1 & \text{if } x_D < 0 \\ 0 & \text{if } x_D > 0 \end{matrix} \tag{3.14}$$

These equations were solved by Lauwerier to give equation 3.15, which expresses the temperature within the oil sand layer as a function of time and location.

$$\begin{aligned} &\text{If } x_D < t_D \quad \text{then } T_2^* = \operatorname{erfc}\left(\frac{x_D + |y_D| - 1}{2\sqrt{\theta(t_D - x_D)}}\right) \\ &\text{and if } x_D \geq t_D \quad \text{then } T_2^* = 0 \end{aligned} \tag{3.15}$$

the temperature within the water layer is found by substituting $y_D = 1$ to give equation 3.16.

$$\begin{aligned} &\text{If } x_D < t_D \quad \text{then } T_1^* = \operatorname{erfc}\left(\frac{x_D}{2\sqrt{\theta(t_D - x_D)}}\right) \\ &\text{and if } x_D \geq t_D \quad \text{then } T_1^* = 0 \end{aligned} \tag{3.16}$$

Figure 3.4 shows the temperature as a function of distance within the water layer for various values of time plotted using the dimensionless variables just shown with $\theta = 1$.

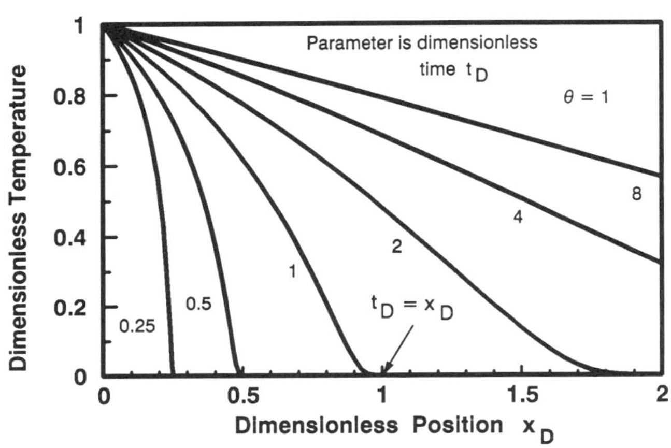

Figure 3.4 Reservoir Temperatures Calculated from Lauwerier's Equation

Numerical Example

Hot water at 200°C is injected into a water layer 4 m thick contained within an oil sand reservoir at a rate of 10 m^3/h. The flow from the well is radial. The following properties may be used:

Porosity 0.30; $T_R = 10$°C
Water layer; $S_w = 1.0$
Oil sand: $S_w = 0.3$; $S_o = 0.7$; $K = 1.2$ Btu/h ft °F
Rock: heat capacity 0.2 Btu/lb °F: SG = 2.2
Oil: heat capacity 0.5; SG = 0.95

Lauwerier's equation, as derived before, can be employed for a radial system if the dimensional distance is redefined. Plot the temperature within the water layer as a function of the radial distance R from the injection well for times of 10, 100, and 1000 days. Also plot the temperature distribution that would have been obtained if there had been no heat loss to the overburden and underburden. Plot the temperature distribution that would be expected in a temperature-observation well located 10 m from the injection well. Include temperatures in the oil sand above and below the water layer.

Solution

1. Lauwerier's equation in radial coordinates

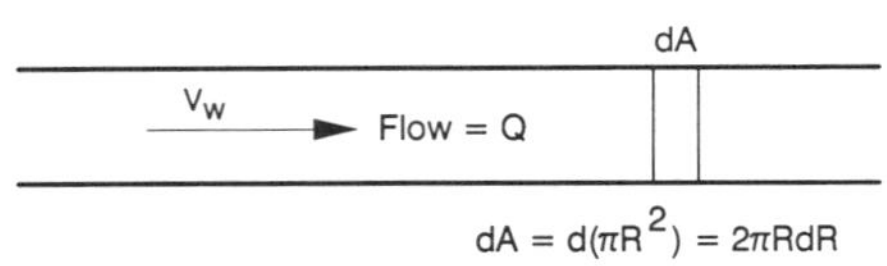

$$-Q_w \rho_w C_w \left(\frac{\partial T}{\partial A}\right) dA = h\rho_1 C_1 \left(\frac{\partial T}{\partial t}\right) dA - 2K_2 \left(\frac{\partial T}{\partial y}\right)_{y=h/2} dA \qquad (3.17)$$

Heat from water — Heat stored — Heat loss

This is similar to equation 3.8. A has been substituted for x and Q_w replaced hV_w. The solution can thus be written immediately by making these substitutions; i.e.,

$$x_D = \frac{4K_2 x}{h^2 \rho_w C_w V_w}$$

becomes

$$x_D = \frac{4K_2 \pi R^2}{h \rho_w C_w Q_w} \qquad (3.18)$$

and y_D, t_D, and θ are unchanged.

Calculation:

The results are given in the following tables.

a. Temperature in the water layer

Column 3

$$x_D = \frac{4K_2 \pi R^2}{(h\rho_w C_w Q_w)}$$

$$K_2 = 1.2 \text{ Btu/h ft °F} = 28.8 \text{ Btu/d ft °F}$$

$$h = 4 \text{ m} = 13.123 \text{ ft}$$

$$\rho_w C_w = 62.4 \text{ Btu/ft}^3 \text{ °F}$$

$$Q_w = 10 \text{ m}^3\text{/h} = 847.4 \text{ ft}^3\text{/d}$$

Column 4

$$t_D = \frac{4K_2 t}{h^2 \rho_1 C_1}$$

$$\rho_r C_r = 62.4 \times 2.2 \times 0.2 = 27.46 \text{ Btu/ft}^3 \text{ °F}$$

$$\rho_1 C_1 = \phi \rho_w C_w + (1 - \phi)\rho_r C_r = 37.94 \text{ Btu/ft}^3 \text{ °F}$$

$$S_w = 0.3; \qquad S_o = 0.7$$

$$\rho_o C_o = 62.4 \times 0.95 \times 0.5 = 29.64 \text{ Btu/ft}^3 \text{ °F}$$

$$\rho_2 C_2 = \phi S_w \rho_w C_w + \phi S_o \rho_o C_o + (1 - \phi)\rho_r C_r = 31.10 \text{ Btu/ft}^3 \text{ °F}$$

$$\theta = \frac{\rho_1 C_1}{\rho_2 C_2} = 1.221$$

$$X = \frac{x_D}{2}[\theta(t_D - x_D)]^{-1/2}$$

$$\text{Temp (in col. 2)} = 10 + 190 \text{ erfc}(X)$$

Radius (ft)	Temperature (°C)	x_D	X
Time = 10 d $t_D = 0.176$			
0	200.0	0.000	0.000
10	198.8	0.005	0.006
20	194.9	0.021	0.024
30	187.4	0.047	0.059
40	173.6	0.083	0.124
50	142.4	0.130	0.275
60	10.0	0.188	—
Time = 100 d $t_D = 1.763$			
0	200.00	0.000	0.000
10	199.6	0.005	0.002
20	198.5	0.021	0.007
30	196.5	0.047	0.016

continued

Radius (ft)	Temperature (°C)	x_D	X
40	193.8	0.083	0.029
50	190.1	0.130	0.046
60	185.5	0.188	0.068
80	173.1	0.334	0.126
100	155.3	0.522	0.212
120	130.3	0.751	0.338
140	95.0	1.022	0.537
160	46.4	1.335	0.923
180	10.0	1.690	2.822
200	10.0	2.086	—
Time = 1000 d $t_D = 17.631$			
0	200.0	0.000	0.000
10	199.9	0.005	0.001
20	199.5	0.021	0.002
30	198.9	0.047	0.005
40	198.1	0.083	0.009
50	197.0	0.130	0.014
60	195.6	0.188	0.020
80	192.2	0.334	0.036
100	187.8	0.522	0.057
120	182.3	0.751	0.083
160	168.2	1.335	0.150
200	149.6	2.086	0.239
250	120.6	3.259	0.389
300	86.7	4.694	0.590
350	52.3	6.389	0.862
400	25.2	8.344	1.239
500	10.0	13.038	2.752
600	10.0	18.774	—

The predicted temperatures in the reservoir are plotted against the distance from the injection well in Figure 3.5 for 10, 100, and 1000 days.

b. Temperature in observation well at $R = 32.8$ ft (10 m)

Column 4

Height above water sand in feet

Column 3

$$y_D - 1 = \frac{\text{height}}{h/2}$$

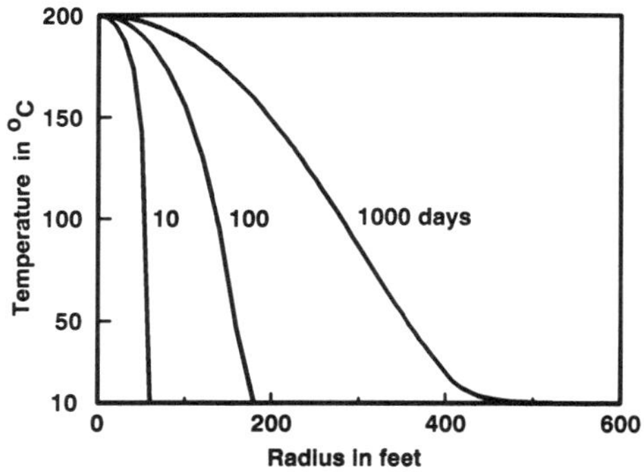

Figure 3.5 Predicted Temperatures in Reservoir as a Function of Distance from Injection Well

Column 2

$$X = \frac{x_D + y_D - 1}{2[\theta(t_D - x_D)]^{1/2}}$$

The other columns are as before.

TEMP °C	X	$y_D - 1$	Height (ft)
	Time = 10 d; $t_D = 0.176$; $x_D = 0.056$		
184.3	0.073	0.000	0
106.0	0.471	0.305	2
51.6	0.869	0.610	4
23.9	1.267	0.914	6
13.5	1.664	1.219	8
10.7	2.062	1.524	10
	Time = 100 d; $t_D = 1.763$; $x_D = 0.056$		
195.8	0.019	0.000	0
151.4	0.231	0.610	4
111.1	0.442	1.219	8
77.6	0.653	1.829	12
52.2	0.864	2.438	16
34.4	1.075	3.048	20
23.1	1.286	3.658	24
16.5	1.497	4.267	28
	Time = 1000 d; $t_D = 17.631$; $x_D = 0.056$		
198.7	0.006	0.000	0
184.6	0.072	0.610	4
170.7	0.138	1.219	8
157.0	0.203	1.829	12
143.6	0.269	2.438	16
130.7	0.335	3.048	20
118.5	0.401	3.658	24
106.8	0.467	4.267	28

The calculated temperatures in the observation well are plotted in Figure 3.6.

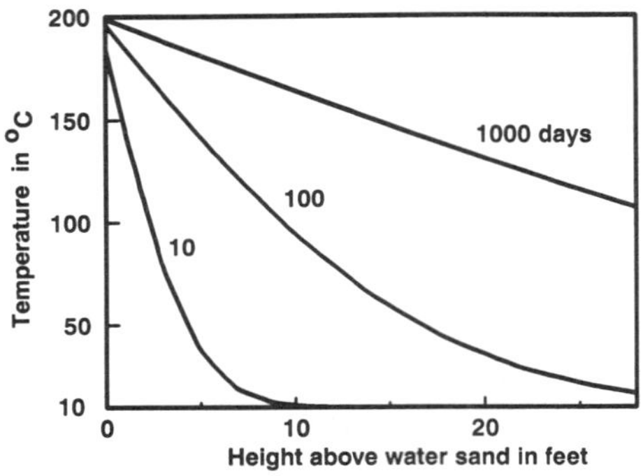

Figure 3.6 Predicted Temperatures in Observation Well

c. Temperature distribution for no loss

Heat Balance

$$Q_w \rho_w C_w (T_S - T_R) t = \pi R^2 h \rho_1 C_1 (T_S - T_R)$$

$$R = \left(\frac{Q_w \rho_w C_w t}{\pi h \rho_1 C_1} \right)^{1/2}$$

This equation yields values of $R = 58.1$, 183.7, and 581.4 ft for 10, 100 and 1000 days. These values have been plotted as dotted lines in Figure 3.5. Note that the no-loss radius occurs when $x_D = t_D$.

THERMAL EFFICIENCY FOR CONSTANT-DISPLACEMENT RATE STEAM-DRIVE

Consider the steam-drive situation shown in Figure 3.7. Steam is injected from the left side at a rate sufficient to cause the heat front to advance at a constant rate $\dot{A}$. The shape of the heated area A is not specified.

The specification of the problem requires that the steam-injection rate be raised continuously to compensate for the increasing heat losses. It is assumed that the temperature T_S in the steam chamber is constant up to the front, where it falls abruptly to T_R. This assumption is reasonable until the time when all the latent heat in the steam is consumed by the heat losses and only sensible heat is available to advance the heated region. The time at which this situation occurs is calculated later on.

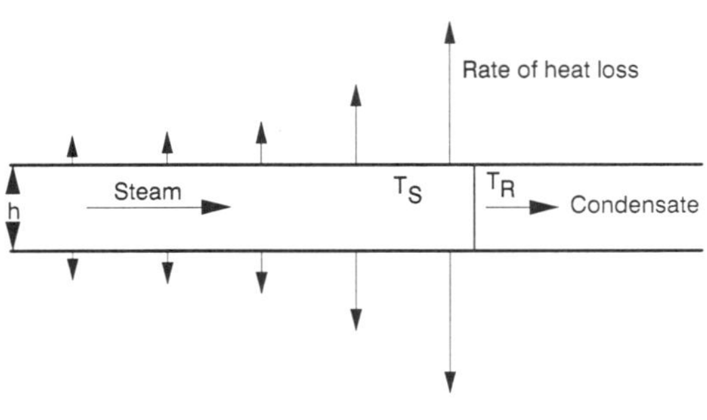

Figure 3.7

The cumulative heat loss from the heated reservoir surface of area $2A$ (A at the top of the reservoir and A below) may be calculated from equation 2.27.

$$Q_{LC} = 2\left(\frac{4}{3} \cdot K_2 A(T_S - T_R)\sqrt{\frac{t}{\pi\alpha_2}}\right) \tag{3.19}$$

where the subscript 2 refers to the over- and underburden.

The cumulative heat required to raise the reservoir and its residual contents from the initial temperature T_R to T_S is

$$Q_{SC} = \rho_1 C_1 h A(T_S - T_R) \tag{3.20}$$

where $\rho_1 C_1$ is the volumetric heat capacity of the steamed reservoir.[1]

The total cumulative heat injected is thus

$$Q_{TC} = Q_{LC} + Q_{SC} = 2\left(\frac{4}{3} \cdot K_2 A(T_S - T_R)\sqrt{\frac{t}{\pi\alpha_2}}\right) + \rho_1 C_1 h A(T_S - T_R) \tag{3.21}$$

The instantaneous rate of heat injection may be found by differentiating equation 3.21 with respect to t.

$$H_0 = \dot{A}(T_S - T_R)\left(4K_2\sqrt{\frac{t}{\pi\alpha_2}} + \rho_1 C_1 h\right) \tag{3.22}$$

The heat-injection rate is equal to the sum of the losses (which increase with the square root of time) plus the constant heat rate to expand the steam chamber.

At the critical time t_c, when latent heat is no longer available at the heat front, the losses are all supplied by the latent heat. When this occurs, the ratio of heat-loss rate to stored-heat rate will be equal to the ratio of the latent heat injection rate to the sensible heat injection rate; i.e.,

$$\frac{\left[4K_2\sqrt{\frac{t_c}{\pi\alpha_2}}\right]}{[\rho_1 C_1 h]} = \frac{H_\lambda}{H_0 - H_\lambda}$$

$$t_c = \pi\alpha_2\left(\frac{H_\lambda}{H_0 - H_\lambda}\frac{\rho_1 C_1 h}{4K_2}\right)^2 \tag{3.23}$$

For times less than t_c, latent heat is available at the heat front, and a sharp temperature gradient is maintained. The critical time is proportional to the square of the reservoir thickness, and it is independent of the rate.

The fraction of the injected heat that remains in the reservoir may be looked upon as a thermal efficiency. It is given by

$$E_h = \frac{Q_{SC}}{Q_{TC}} = \frac{\rho_1 C_1 h}{\rho_1 C_1 h + \frac{8}{3} \cdot K_2\sqrt{\frac{t}{\pi\alpha_2}}} = \frac{1}{1 + \frac{4}{3\sqrt{\pi}} \cdot X} \tag{3.24}$$

[1]The value of $\rho_1 C_1$ used in this equation is for the steamed reservoir—i.e., with fluid saturations corresponding to the depleted reservoir. The displaced fluids are cooled to the reservoir temperature as they pass through the heat front.

where

$$X = \frac{2K_2}{\rho_1 C_1 h}\sqrt{\frac{t}{\alpha_2}}$$

The dimensionless term X in equation 3.24 is the same variable used by Marx and Langenheim; it also arises in the description of their theory in the next section.

Other writers often employ the dimensionless time t_D, which is the square of X.

$$t_D = X^2 = \frac{4K_2^2 t}{(\rho_1 C_1 h)^2 \alpha_2} = \frac{4K_2 \rho_2 C_2 t}{h^2 (\rho_1 C_1)^2} \qquad (3.25)$$

If the volumetric heat capacity of the overburden and underburden are the same as that of the steam zone, then the expression for t_D can be simplified to equation 3.26. This approximation is often sufficiently accurate.

$$\text{If } \rho_1 C_1 = \rho_2 C_2, \qquad \text{then } t_D = \frac{4\alpha_2 t}{h^2} \qquad (3.26)$$

The simplified dimensionless time, t_D, may also be expressed as

$$t_D = \frac{4\alpha_2 A}{h^2 \dot{A}} \qquad (3.26\text{A})$$

The thermal efficiency, E_h, is plotted against the dimensionless time in Figure 3.8. As time increases (or as the flooded area, A, increases) the thermal efficiency decreases; i.e., a larger and larger fraction of the steam is required to supply the increasing heat losses.

The height of the reservoir is a particularly important variable in the dimensionless time because its value is squared. The thermal efficiency is inherently higher for thick reservoirs than for thin ones. Smaller values of A corresponding to closer well spacings also provide higher efficiencies for a given rate as measured by $\dot{A}$. Higher heat injection rates (i.e., higher $\dot{A}$) also give better thermal efficiencies.

If $\dot{A}$ is not constant, Figure 3.8 can still be employed as an approximation if the total steaming time is substituted into (3.26) in order to calculate t_D. The critical

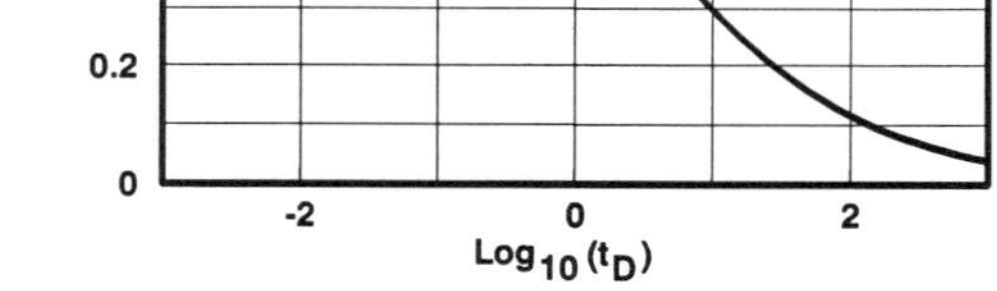

Figure 3.8 Values of Thermal Efficiency Factor E_h for Constant Displacement Rate

time, t_c, from equation 3.23 can also be expressed as an equivalent dimensionless time t_{DC} given by

$$t_{DC} = X_c^2 = \frac{\pi}{4}\left(\frac{H_\lambda}{H_0 - H_\lambda}\right)^2 \tag{3.27}$$

Fraction of Heat in Steam-Saturated Chamber After the Critical Time

$$H_\lambda = \frac{2K_2(T_S - T_R)}{\sqrt{\pi\alpha_2}} \int_0^{A_S} \frac{dA_S}{\sqrt{t - t_0}} = \frac{2K_2(T_S - T_R)}{\sqrt{\pi\alpha_2}} \int_0^{t_S} \frac{\dot{A}\,dt_0}{\sqrt{t - t_0}}$$

where t_0 is the birth time of dA and $\dot{A}$ is constant. Hence,

$$H_\lambda = \frac{4K_2(T_S - T_R)\dot{A}}{\sqrt{\pi\alpha_2}}\left(\sqrt{t} - \sqrt{t - t_S}\right) \tag{3.28}$$

where t_S is the birth time of the limit of A_S at time t.

$$H_0 = \dot{A}(T_S - T_R)\left(4K_2\sqrt{\frac{t}{\pi\alpha_2}} + \rho_1 C_1 h\right) \tag{3.29}$$

$$\frac{H_\lambda}{H_0} = \frac{\sqrt{\dfrac{t}{\pi\alpha_2}} - \sqrt{\dfrac{t - t_S}{\pi\alpha_2}}}{\sqrt{\dfrac{t}{\pi\alpha_2}} + \dfrac{\rho_1 C_1 h}{4K_2}} \tag{3.30}$$

which reduces to

$$\frac{H_\lambda}{H_0} = \frac{1 - \sqrt{1 - \dfrac{A_S}{A}}}{1 + \dfrac{1}{2}\sqrt{\dfrac{\pi}{t_D}}} \tag{3.31}$$

where

$$t_D = \frac{4K_2\rho_2 C_2 t}{h^2(\rho_1 C_1)^2}$$

If $A_S = A$ (3.31) becomes (3.27).

In general, if $t_D \geq t_{DC}$ then (3.31) can be manipulated to become

$$\frac{A_S}{A} = 2n - n^2 \tag{3.32}$$

where

$$n = \frac{1 + \dfrac{1}{2}\sqrt{\dfrac{\pi}{t_D}}}{1 + \dfrac{1}{2}\sqrt{\dfrac{\pi}{t_{DC}}}} \tag{3.33}$$

Asymptote for A_S/A if $t_D = \infty$.

The ratio A_S/A approaches a constant value for large times. This can be calculated as follows.

$$n_\infty = \frac{1}{1 + \frac{1}{2}\sqrt{\frac{\pi}{t_{DC}}}} = \frac{H_\lambda}{H_0} \tag{3.34}$$

and

$$\left(\frac{A_S}{A}\right)_\infty = 2\frac{H_\lambda}{H_0} - \left(\frac{H_\lambda}{H_0}\right)^2 \tag{3.35}$$

THERMAL EFFICIENCY FOR CONSTANT STEAM-INJECTION RATE: MARX AND LANGENHEIM'S THEORY

Marx and Langenheim (1959) developed theoretical relations describing the important case of a growing steam zone that is limited in its growth rate by the loss of heat to the overburden and underburden and by the rate at which steam is introduced. Their theory is similar to that described in the previous section except for the assumption that the steam is introduced at a constant rate rather than at a rate that provides a constant frontal advance rate. Marx and Langenheim's equation has formed a basis for many of the subsequent studies in this field, and it is important to gain an understanding of it.

The situation considered by Marx and Langenheim is shown in Figure 3.9. Steam is introduced at a constant rate into a steam zone that is spreading laterally. The rate of heat injection, H_0, into this zone is specified as input to the problem. The heat goes to increase the heat stored in the growing steam zone and the losses to the overburden and underburden. It is assumed that no heat is transferred ahead of the front. As discussed previously, this assumption is realistic only when the latent heat supplied in the injected steam is sufficient to supply all the losses—i.e., when there is still latent heat arriving at the heat front. This condition will usually be satisfied if a reasonably high oil to steam ratio is being obtained.

The areal shape of the steam zone is not specified. In the original theory it was assumed that the condensation front remains vertical, but this assumption can

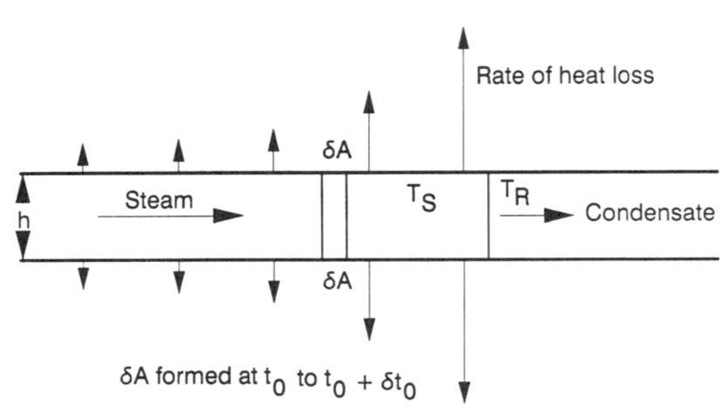

Figure 3.9

be relaxed; this is discussed later in this chapter. Within the steam zone, the temperature is assumed constant at T_S. Outside it is T_R, as shown in Figure 3.10.

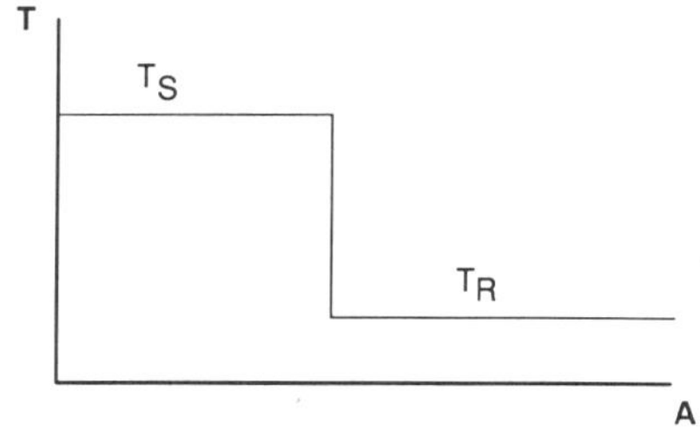

Figure 3.10

At any intermediate time t the vertical heat loss rate per unit area will be larger near the front (see equation 2.24). Equation 3.36 gives the rate of heat loss from the area $2\delta A$ (δA above and δA below), which was initially heated at time t_0. The time t_0 can be looked upon as the "birth time" of any particular heated area.

$$\delta Q_L = 2\delta A \frac{K_2(T_S - T_R)}{\sqrt{\pi\alpha_2(t - t_0)}} \tag{3.36}$$

The total rate of heat loss, Q_L, is found by integrating (3.36) over the whole area, as in (3.37). The time at which the element of area was formed, t_0, is a function of A; alternatively, A may be considered a function of t_0, and the variable of integration may be changed, as shown:

$$dA = \left(\frac{dA}{dt_0}\right) dt_0$$

$$Q_L = 2\int_0^A \frac{K_2(T_S - T_R)}{\sqrt{\pi\alpha_2(t - t_0)}}\, dA = 2\int_0^t \frac{K_2(T_S - T_R)}{\sqrt{\pi\alpha_2(t - t_0)}} \left(\frac{dA}{dt_0}\right) dt_0 \tag{3.37}$$

Equation 3.38 defines Q_s, the rate at which heat is being stored as sensible heat in the reservoir.

$$Q_S = \rho_1 C_1 h(T_S - T_R)\frac{dA}{dt} \tag{3.38}$$

The overall heat balance for the process is given by equation 3.39. It may be solved by Laplace transformation, solution of the resulting algebraic equation for A, and inversion of the transform to give $A(t)$. The inversion of the transform may be carried out by comparison with a table of standard forms.[2]

$$H_0 = 2\int_0^{t_0} \frac{K_2(T_S - T_R)}{\sqrt{\pi\alpha_2(t - t_0)}} \left(\frac{dA}{dt_0}\right) dt_0 + \rho_1 C_1 h(T_S - T_R)\frac{dA}{dt} \tag{3.39}$$

Injection = loss + storage

The solution of (3.39) leads to the result shown in equation 3.40. The theory for this problem is similar to that developed by Carter for the growth of a fracture with side

[2]See CRC *Standard Mathematical Tables* (22nd Ed.), Transform 45, page 510.

leakage and a constant injection rate (Carter 1957), and the mathematical form of the problem is identical.

$$A(t) = \frac{H_0 \rho_1 C_1 h}{4K_2 \rho_2 C_2 (T_S - T_R)} \left(e^{X^2} \operatorname{erfc}(X) + \frac{2X}{\sqrt{\pi}} - 1 \right) \tag{3.40}$$

where

$$X = \frac{2K_2}{\rho_1 C_1 h \sqrt{\alpha_2}} \sqrt{t}$$

Equation 3.40 gives the heated area as a function of time and differentiation gives the rate of area growth:

$$\frac{dA}{dt} = \frac{H_0 e^{X^2} \operatorname{erfc}(X)}{\rho_1 C_1 h (T_S - T_R)} \tag{3.41}$$

The rate at which oil is displaced from the steam zone may be calculated by multiplying the rate of increase of the volume of the steam chamber by its porosity and by the change in oil saturation:

$$q_o = h\phi(S_o - S_{or}) \frac{dA}{dt} \tag{3.42}$$

The residual oil saturation S_{oR} in the steam zone is usually quite small; typical values are 0.05–0.2. A value of 0.15 is a reasonable estimate if no other data are available. In equations 3.40 and 3.41 the dimensionless variable X is employed. This is the variable that was used by Marx and Langenheim. It is the same dimensionless group that was described earlier. Frequently X^2 is referred to as the dimensionless time t_D.

$$t_D = X^2 = \frac{4K_2^2 t}{(\rho_1 C_1)^2 \alpha_2} = \frac{4K_2 \rho_2 C_2 t}{h^2 (\rho_1 C_1)^2} \tag{3.43}$$

If the volumetric heat capacity of the overburden and underburden are the same as that of the steam zone, then the expression for t_D can be simplified to equation 3.44.

$$t_D = \frac{4\alpha_2 t}{h^2} \tag{3.44}$$

The fraction of the injected heat that remains in the reservoir can be determined as a function of the dimensionless time by means of equation 3.45 or from Figure 3.11.

$$E_h = \frac{1}{t_D} \left[e^{t_D} \operatorname{erfc}(\sqrt{t_D}) + 2\sqrt{\frac{t_D}{\pi}} - 1 \right] \tag{3.45}$$

The curve in Figure 3.11 shows that the fraction of the heat lost from the reservoir varies over a large range with the variable t_D; this is proportional to time. As time continues, the fraction of the total heat injected that is lost grows. It should also be noted that there is only a slight difference between the curve drawn for a constant

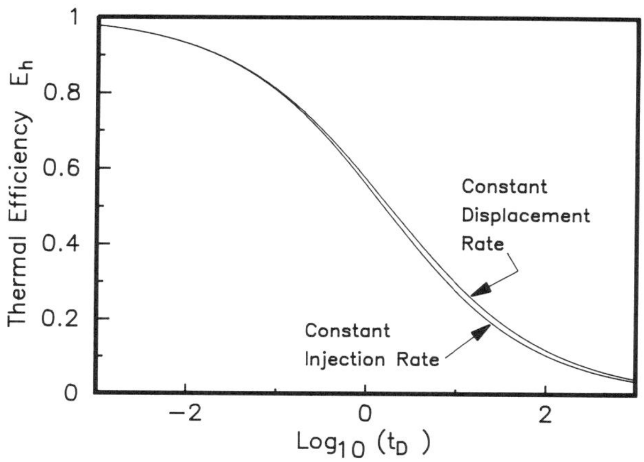

Figure 3.11 Comparison of Thermal Efficiency E_h for Constant Heat-Injection Rate and for Constant Displacement Rate

injection rate (i.e., from equation 3.45) and that for a constant displacement rate from the much simpler equation 3.24.

Numerical values of the term $e^{t_D}\,\text{erfc}(\sqrt{t_D})$ or $e^{X^2}\,\text{erfc}(X)$ in (3.45) can be obtained from Table 3.1. In computer programs the rational approximations developed by Hastings given on page 35 may be employed.

A major factor is that as time goes on and more and more steam is introduced (recall that it is assumed that the injection rate is constant), the steam zone contin-

TABLE 3.1 Values of the Function $e^{X^2}\,\text{erfc}(X)$

X	$e^{X^2}\,\text{erfc}(X)$	X	$e^{X^2}\,\text{erfc}(X)$	X	$e^{X^2}\,\text{erfc}(X)$
0.00	1.000000	1.00	0.427584	2.00	0.255403
0.05	0.945990	1.05	0.414299	2.05	0.250167
0.10	0.896457	1.10	0.401731	2.10	0.245130
0.15	0.850936	1.15	0.389826	2.15	0.240281
0.20	0.809020	1.20	0.378537	2.20	0.235610
0.25	0.770347	1.25	0.367822	2.25	0.231108
0.30	0.734599	1.30	0.357642	2.30	0.226766
0.35	0.701496	1.35	0.347960	2.35	0.222576
0.40	0.670788	1.40	0.338743	2.40	0.218532
0.45	0.642252	1.45	0.329960	2.45	0.214625
0.50	0.615690	1.50	0.321584	2.50	0.210850
0.55	0.590927	1.55	0.313590	2.55	0.207199
0.60	0.567805	1.60	0.305952	2.60	0.203668
0.65	0.546181	1.65	0.298650	2.65	0.200251
0.70	0.525930	1.70	0.291663	2.70	0.196943
0.75	0.506938	1.75	0.284973	2.75	0.193738
0.80	0.489101	1.80	0.278561	2.80	0.190632
0.85	0.472327	1.85	0.272413	2.85	0.187622
0.90	0.456532	1.90	0.266513	2.90	0.184702
0.95	0.441641	1.95	0.260847	2.95	0.181869

ues to grow, and a larger and larger area of overburden is heated. Eventually, nearly all the injected heat is being lost. The thickness of the reservoir, h, is the most significant factor involved in the expression for t_D, since its value is squared.

As an example of the use of Figure 3.11, consider the case of a reservoir 100 ft thick. Assume that the thermal diffusivity of the reservoir and the overburden are both 0.9 ft^2/d. The horizontal scale in the figure corresponds to values of t_D varying from 0.001 to 1000. The corresponding values of real time in days, for this range are given by $h^2 t_D / 4\alpha$, or $2778 t_D = 2.8$ to 2,778,000 d. For a reservoir 10 ft rather than 100 ft thick, the corresponding times are smaller by a factor of 100.

t_D	0.001	0.1	10	1000
t days for $h = 100$	2.8	278	27,780	2,778,000
t days for $h = 10$	0.028	2.8	278	27,780

Numerical Problem Using Marx-Langenheim's Equations

For the two different cases ($h = 10$ and 100 ft and $\rho_1 C_1 = \rho_2 C_2 = 33$ Btu/ft^3 °F), assume:

$$\phi = 0.35$$
$$T_R = 75°\text{F}$$
$$S_o = 0.7$$
$$S_{or} = 0.15 \quad \text{measured at } 75°\text{F}$$

Seventy percent quality steam is injected at a rate of 800 B/d at a pressure of 500 psia. Calculate the following for each reservoir thickness:

1. The area of the steam zone is acres as a function of time
2. The radius of the steam zone, assuming the steam zone is cylindrical
3. The volume of displaced oil
4. The ratio of displaced oil rate to steam injection rate
5. The ratio of cumulative displaced oil to cumulative injected steam

Plot these variables against time for each of the reservoir thicknesses.

For $X > 3$, use

$$e^{X^2} \operatorname{erfc}(X) \simeq \frac{2}{\sqrt{\pi}} \cdot \frac{1}{X + \sqrt{X^2 + 2}} \tag{3.46}$$

Solution The solution to this problem is given in Table 3.2.

TABLE 3.2

YEARS	t_D	ACRES	RADIUS (ft)	CUMULATIVE VOLUME (B)	BPD	OSR	CUMULATIVE OSR
Height = 100 ft							
0.00	0	0	0	0	694.20	0.868	0.868
0.50	0.07	0.71	99	105,887	531.54	0.664	0.725
1.00	0.13	1.33	136	197,965	481.52	0.602	0.678
1.50	0.20	1.89	162	282,630	448.18	0.560	0.645
2.00	0.26	2.42	183	362,011	422.96	0.529	0.620
2.50	0.33	2.93	201	437,267	402.66	0.503	0.599
3.00	0.39	3.41	217	509,139	385.68	0.482	0.581
4.00	0.53	4.32	245	644,675	358.39	0.448	0.552
5.00	0.66	5.17	268	771,442	336.99	0.421	0.528
6.00	0.79	5.97	288	891,128	319.48	0.399	0.509
7.00	0.92	6.73	305	1,004,989	304.74	0.381	0.492
8.00	1.05	7.46	322	1,113,876	292.07	0.365	0.477
9.00	1.18	8.16	336	1,218,450	281.00	0.351	0.464
10.00	1.31	8.83	350	1,319,229	271.20	0.339	0.452
Height = 10 ft							
0.00	0	0	0	0	694.20	0.868	0.868
0.50	6.57	2.71	194	40,495	144.21	0.180	0.277
1.00	13.14	4.19	241	62,547	106.43	0.133	0.214
1.50	19.71	5.34	272	79,786	88.86	0.111	0.182
2.00	26.28	6.32	296	94,433	78.17	0.098	0.162
2.50	32.85	7.19	316	107,393	70.77	0.088	0.147
3.00	39.42	7.98	333	119,143	65.25	0.082	0.136
4.00	52.56	9.38	361	140,059	57.40	0.072	0.120
5.00	65.70	10.62	384	158,528	51.97	0.065	0.109
6.00	78.84	11.74	403	175,249	47.92	0.060	0.100
7.00	91.98	12.77	421	190,640	44.74	0.056	0.093
8.00	105.12	13.73	436	204,977	42.15	0.053	0.088
9.00	118.26	14.63	450	218,449	39.99	0.050	0.083
10.00	131.40	15.48	463	231,198	38.15	0.048	0.079

Data:

Thermal diffusivity: $\alpha = 0.9\ \text{ft}^2/\text{d}$

Heat capacity: $\rho C = 33\ \text{Btu/ft}^3\ {}^\circ\text{F}$

Porosity: $\phi = 0.35$

$T_R = 75^\circ\text{F}$

$T_S = 467^\circ\text{F}$ from steam table

$S_o = 0.7$

$S_{or} = 0.15$

$$\text{Steam rate} = 800\ \text{BPD} = 800 \times 350 = 280{,}000\ \text{lb/d}$$

Height h = 100 ft and 10 ft two cases

Heat-Injection Rate		
Enthalpy of vapor	1205 Btu/lb	from steam tables
Enthalpy of liquid at T_S	450 Btu/lb	
Enthalpy of liquid at T_R	43 Btu/lb	
Quality = 70%		

$$\text{Heat-injection rate} = 280{,}000(0.7 \times 1205 + 0.3 \times 450 - 43)$$
$$= 2.6194 \times 10^8 \text{ Btu/d}$$

Column 1 in table: Time in years

Column 2: Dimensionless time (equation 3.44)

$$t_D = 4 \times 0.9 \times \text{years} \times 365/h^2$$

Column 3: Area in acres (equation 3.40)

$$\text{Area}[(2.6194 \times 10^8 \text{ h})/(4 \times 29.7 \times 392 \times 43{,}560)]f(t_D) \qquad (43{,}560 \text{ ft}^2 = 1 \text{ acre})$$

$$f(t_D) = \left(e^{X^2} \operatorname{erfc}(X) + \frac{2X}{\sqrt{\pi}} - 1\right) \quad \text{and} \quad X = \sqrt{t_D}$$

Column 4: Radius in feet

$$\text{Radius} = [(\text{Col } 3 \times 43{,}560)/\pi]^{0.5}$$

Column 5: Cumulative barrels displaced

$$A \times 43{,}560 \times h \times 0.35 \times 0.55/5.615 \qquad (5.615 \text{ ft}^3 = 1 \text{ B})$$

Column 6: Barrels displaced per day (equation 3.42 using equation 3.41)

Column 7: Oil-steam ratio is column 6 divided by 800.

Column 8: Cumulative OSR is column 5 divided by 800 × years × 365.

The heated areas are shown as functions of time in Figure 3.12 and the instantaneous and cumulative oil-steam ratios are shown in Figure 3.13. The heated area grows only about 70% faster for the thinner reservoir, even though the thickness is reduced by a factor of 10. For the thinner reservoir the vertical heat losses are a much larger fraction of the heat input. This is shown clearly by the curves in

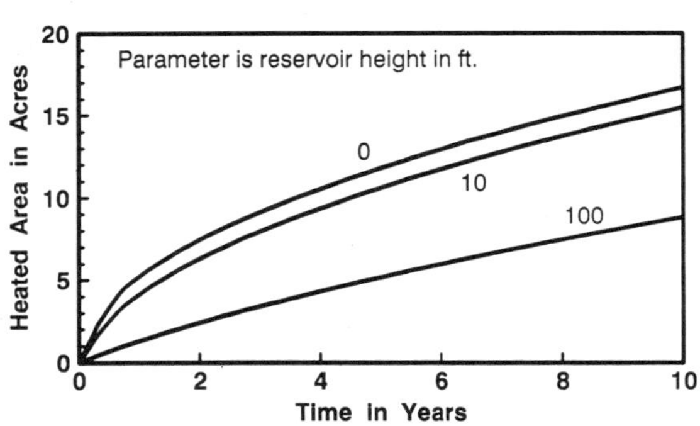

Figure 3.12 Heated Area as a Function of Time

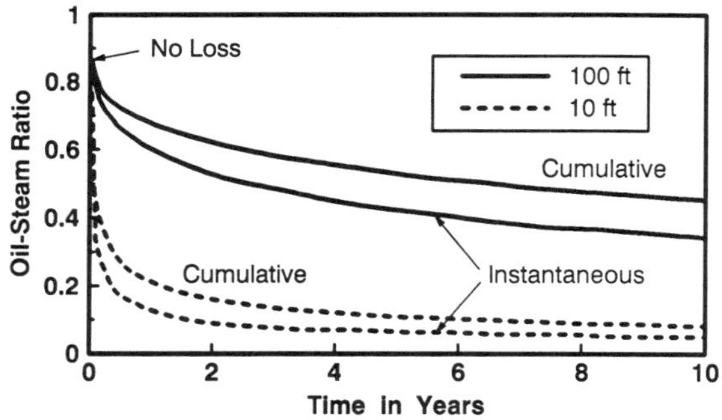

Figure 3.13 Displaced Oil-Steam Ratios

Figure 3.13. The point on the vertical axis marked "No Loss" corresponds to the maximum possible oil-to-steam ratio required by a heat balance with no heat losses. It can be calculated by equation 3.47. While both sets of curves start at the same no-loss point, the curves for the thinner reservoir drop very much more rapidly.

$$\text{OSR}_{\max} = \frac{H_S \phi (S_o - S_{or})}{\rho_1 C_1 (T_S - T_R)} \tag{3.47}$$

In the equation, H_S is the net heat per unit volume of steam measured as water.

SIMPLE FORMULAS FOR ESTIMATION OF THE OIL-STEAM RATIO

Often the physical properties of reservoirs are known only approximately, and simpler formulas may suffice to provide initial estimates. One approach is to base the estimate upon equation 3.21. The cumulative injected heat given by (3.21), corresponds to a cumulative quantity of displaced oil of $\phi \Delta S_o h A$. The oil-steam ratio is thus

$$\begin{aligned} \text{OSR} &= \frac{H_S \phi (S_o - S_{or})}{\rho_1 C_1 (T_S - T_R)} \left(\frac{1}{1 + \dfrac{8}{3} \dfrac{K_2}{\rho_1 C_1 h} \sqrt{\dfrac{t}{\pi \alpha_2}}} \right) \\ &= \text{OSR}_{\max} \left(\frac{1}{1 + \dfrac{8}{3} \dfrac{K_2}{\rho_1 C_1 h} \sqrt{\dfrac{t}{\pi \alpha_2}}} \right) \end{aligned} \tag{3.48}$$

If it is assumed that the thermal properties of the reservoir and of the overburden are equal, then (3.48) becomes

$$\text{OSR} = \text{OSR}_{\max} \left(\frac{1}{1 + \dfrac{8}{3} \sqrt{\dfrac{\alpha t}{\pi h^2}}} \right) \tag{3.49}$$

If numerical values for H_S, ρC, and α equal to the typical ones used in the previous example are employed, namely,

$$\rho C = 33 \text{ Btu/ft}^3\ °\text{F}$$

$$H_S = 58{,}375 \text{ Btu/ft}^3$$

$$\alpha = 0.9 \text{ ft}^2/\text{d}$$

then equation 3.49 may be written

$$\text{OSR} = \frac{1769\phi\,\Delta S_o}{(T_S - T_R)(1 + 1.43\sqrt{t/h^2})} \tag{3.50}$$

where ϕ and ΔS_o are fractions
T_S and T_R are in degrees Farenheit
t is in days
h is in feet

The following values have been calculated using equation 3.50 with $h = 100$ ft for the previous numerical example; they are compared to values taken from Table 3.2.

Years	OSR from (3.42)	OSR from Table 3.2
0	0.868	0.868
3	0.590	0.581
7	0.504	0.492
10	0.466	0.452

As would be expected from the comparison of the efficiency curves shown in Figure 3.11, the results from the simple formulas are quite close to those from the Marx-Langenheim equation.

In Chapter 4 a paper by Vogel is discussed in which he suggests that heat losses should be calculated by assuming that the steam chamber spreads immediately across the top of the reservoir. This results in an equation similar to (3.49) in which the factor 8/3 is replaced by 4.

$$\text{OSR} = \text{OSR}_{\text{max}}\left(\frac{1}{1 + 4\sqrt{\dfrac{\alpha t}{\pi h^2}}}\right) \quad \text{for immediate steam spreading} \tag{3.51}$$

If the same numerical values are used for the physical properties, this becomes

$$\text{OSR} = \left[\frac{1769 \cdot \phi\,\Delta S_o}{(T_S - T_R)\left(1 + 2.14\sqrt{\dfrac{t}{h^2}}\right)}\right] \tag{3.52}$$

where the units for the variables must be the same as in equation 3.50.

CONVECTIVE TRANSFER OF HEAT BEYOND THE CONDENSATION FRONT

In the Marx-Langenheim theory and in the previous numerical example, it is assumed that no heat is transferred beyond the heat front—i.e., that at the front the steam gives up all its heat, both latent and sensible. This idea is consistent with the ideas developed earlier, where it was shown that the velocity of a condensation front is greater than that of a thermal front carried forward by the sensible heat of the condensate alone. Thus any heat carried before the condensation front tends to be "overrun," and the two fronts remain combined. The assumption of a single front made by Marx and Langenheim is reasonable if there is still vapor left to condense at the front. However, the situation changes as the front becomes more remote, since eventually all the steam is condensed by heat losses before the heat front.

Since the steam zone is at the steam-saturation temperature throughout,[3] the only source of heat to supply the vertical losses above and below the zone is the latent heat of the steam. At the point where the latent heat has been completely consumed to supply losses, the only remaining heat to be carried forward is the sensible heat in the liquid water, and the process becomes rather like that of Lauwerier, which was discussed earlier. It is not identical, however, because as the heat losses are transferred to the overburden and to the underburden, the rate at which steam must be condensed to supply these losses decreases and excess steam becomes available to advance the condensation front further. Thus even though essentially all the latent heat of the steam is being used to supply the vertical losses, the condensation front still advances; at the same time, substantial quantities of heat are carried beyond the condensation front by the sensible heat of the condensate and the heat front passes beyond the condensation front. Figure 3.14 is a qualitative representation of the situation.

Mandl and Volek (1969) and Hearn (1969) were the first to recognize this phenomenon; they each developed an equation that predicts the time, t_c, at which it occurs.

Let H_λ be the rate at which latent heat is injected into the reservoir. The total heat-injection rate is still taken as H_0. At the critical time the latent heat-injection rate is just equal to the vertical heat-loss rate. If this is so, then the rate at which heat is being stored must be equal to the rate at which sensible heat is being injected—i.e., to $H_0 - H_\lambda$. This is shown by equation 3.53, which reduces to equation 3.54.

$$H_0 - H_\lambda = Q_S = \rho_1 C_1 h(T_S - T_R)\frac{dA}{dt} = \rho_1 C_1 h(T_S - T_R)\frac{H_0 e^{X^2}\,\mathrm{erfc}(X)}{\rho_1 C_1 h(T_S - T_R)} \qquad (3.53)$$

$$1 - \frac{H_\lambda}{H_0} = e^{X_c^2}\,\mathrm{erfc}(X_c) \qquad (3.54)$$

[3]The steam temperature is assumed constant in this chapter. In actual practice there tends to be a small reduction in steam saturation temperature, which is caused by the pressure drop as the steam flows away from the injection well.

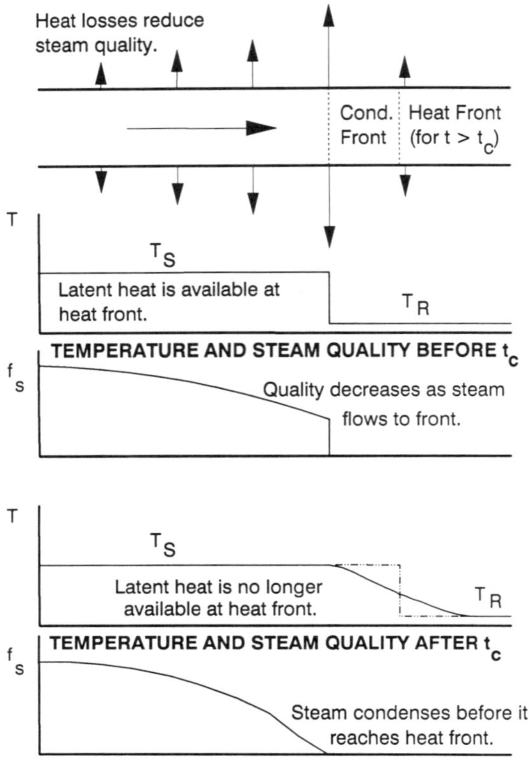

Figure 3.14

The critical time may be found by solving equation 3.54 for the value of X_c that corresponds to the particular value of H_λ/H_0 and then obtaining the time by the use of equation (3.43) or (3.44).

Equation 3.54 was derived by Mandl and Volek (1969) and almost simultaneously by Hearn (1969). For given values of H_λ and H_0, the critical dimensionless time can be found by interpolation of Table 3.1.

SIZE OF STEAM ZONE FOR TIME GREATER THAN THE MANDL AND VOLEK'S CRITICAL TIME

Beyond the critical time, the vertical loss from the steam zone is given by equation 3.55, where A_S refers to the area of the steam zone.

$$H_\lambda = \int_0^{A_s} \frac{2K_2(T_S - T_R)}{\sqrt{\pi\alpha_2(t - t_A)}}\, dA \tag{3.55}$$

The time t_A at which the area dA of the steam zone was formed is found by calculating time as a functon of area from equation 3.56.

$$A(t_A) = \frac{H_0\rho_1 C_1 h}{4K_2\rho_2 C_2(T_S - T_R)} f(X) \tag{3.56}$$

where

$$X = \frac{2K_2}{\rho_1 C_1 h \sqrt{\alpha_2}} \sqrt{t_A}$$

$$f(X) = \left(e^{X^2} \operatorname{erfc}(X) + \frac{2X}{\sqrt{\pi}} - 1\right)$$

This time may then be substituted into (3.55) and H_λ calculated as a functon of A_S by evaluating the integral. From this the value of A_S may be determined for specific values of H_λ. This procedure was followed by Hearn, who expressed his results by equation 3.57.

$$A_S = \frac{H_0 \rho_1 C_1 h}{4K_2^2(T_S - T_R)} F\left(X, \frac{H_\lambda}{H_0}\right) \tag{3.57}$$

or

$$A_S = \frac{H_0 \rho_1 C_1 h}{4K_2 \rho_2 C_2 (T_S - T_R)} F\left(X, \frac{H_\lambda}{H_0}\right)$$

Hearn's function F is given by Table 3.3. For times less than the critical, F is identical to f.

TABLE 3.3 Values of $f(X)$ and $F\left(X, \frac{H_\lambda}{H_0}\right)$

		Values of F for various H_λ/H_0						
X	$f(X)$	0.2	0.3	0.4	0.5	0.6	0.7	0.8
0.2	0.035	0.035	0.035	0.035	0.035	0.035	0.035	0.035
0.4	0.122	0.103	0.121	0.122	0.122	0.122	0.122	0.122
0.6	0.245	0.171	0.221	0.243	0.245	0.245	0.245	0.245
0.8	0.392	0.240	0.321	0.372	0.392	0.392	0.392	0.392
1.0	0.556	0.309	0.422	0.502	0.546	0.556	0.556	0.556
1.2	0.733	0.378	0.523	0.632	0.702	0.732	0.733	0.733
1.4	0.918	0.448	0.624	0.762	0.858	0.910	0.918	0.918
1.6	1.111	0.517	0.726	0.894	1.016	1.089	1.111	1.111
1.8	1.310	0.586	0.827	1.025	1.173	1.268	1.308	1.310
2.0	1.512	0.656	0.929	1.156	1.330	1.448	1.506	1.512
2.5	2.032	0.829	1.184	1.485	1.726	1.899	2.002	2.032
3.0	2.566	1.003	1.439	1.814	2.122	2.352	2.501	2.563
3.5	3.105	1.177	1.694	2.144	2.518	2.806	2.999	3.094
4.0	3.650	1.351	1.949	2.475	2.915	3.260	3.499	3.628
4.5	4.200	1.525	2.204	2.805	3.313	3.714	3.999	4.162
5.0	4.753	1.699	2.460	3.136	3.710	4.169	4.500	4.696
6.0	5.863	2.047	2.971	3.798	4.506	5.079	5.503	5.766
7.0	6.978	2.396	3.483	4.460	5.303	5.990	6.506	6.837
8.0	8.097	2.744	3.995	5.122	6.099	6.902	7.510	7.908
9.0	9.218	3.092	4.506	5.785	6.896	7.814	8.514	8.980
10.0	10.340	3.441	5.018	6.447	7.693	8.726	9.518	10.052

(from Hearn 1969)

The fraction of the total heat injected that remains within the steam zone may be calculated by equation 3.58.

$$E_h = \frac{A_S h \rho_1 C_1 (T_S - T_R)}{H_0 t} = \left(\frac{1}{X^2}\right) \cdot F\left(X, \frac{H_\lambda}{H_0}\right) \tag{3.58}$$

The results are shown in Table 3.4 and plotted in Figure 3.15.

A similar figure has been derived by Myhill and Stegemeier (1978) based on the theory of Mandl and Volek (1969) and modified by unpublished work of Prats and Vogiatzis. It is reproduced in Chapter 4 as Figure 4.10.

EFFECT OF A NONVERTICAL FRONT

Myhill and Stegemeier (1978) point out that for the heat-loss equation used by Marx and Langenheim (and the extensions of it) to be applicable, it is not necessary for the heat front to be vertical. It is necessary only that the total volume of the steam zone can be represented by the expression $(Ah/2)$, where A represents the sum of the upper and lower surface areas.

Examples where this is true include

1. A sloped, but straight, front that is advancing linearly,
2. An inclined front that is straight but advancing only at the top, and
3. Cylindrical fronts.

TABLE 3.4 Reservoir Heating Efficiency Calculated from Hearn (1969)

		EFFICIENCY FOR VARIOUS VALUES OF H_λ/H_0							
X	X^2	0.2	0.3	0.4	0.5	0.6	0.7	0.8	1.0
0.2	0.04	0.875	0.875	0.875	0.875	0.875	0.875	0.875	0.875
0.4	0.16	0.644	0.756	0.763	0.763	0.763	0.763	0.763	0.763
0.6	0.36	0.475	0.614	0.675	0.681	0.681	0.681	0.681	0.681
0.8	0.64	0.375	0.502	0.581	0.613	0.613	0.613	0.613	0.613
1.0	1.00	0.309	0.422	0.502	0.546	0.556	0.556	0.556	0.556
1.2	1.44	0.262	0.363	0.439	0.487	0.508	0.509	0.509	0.509
1.4	1.96	0.229	0.318	0.389	0.438	0.464	0.468	0.468	0.468
1.6	2.56	0.202	0.284	0.349	0.397	0.425	0.434	0.434	0.434
1.8	3.24	0.181	0.255	0.316	0.362	0.391	0.404	0.404	0.404
2.0	4.00	0.164	0.232	0.289	0.333	0.362	0.377	0.378	0.378
2.5	6.25	0.133	0.189	0.238	0.276	0.304	0.320	0.325	0.325
3.0	9.00	0.111	0.160	0.202	0.236	0.261	0.278	0.285	0.285
3.5	12.25	0.096	0.138	0.175	0.206	0.229	0.245	0.253	0.253
4.0	16.00	0.084	0.122	0.155	0.182	0.204	0.219	0.227	0.228
4.5	20.25	0.075	0.109	0.139	0.164	0.183	0.197	0.206	0.207
5.0	25.00	0.068	0.098	0.125	0.148	0.167	0.180	0.188	0.190
6.0	36.00	0.057	0.083	0.105	0.125	0.141	0.153	0.160	0.163
7.0	49.00	0.049	0.071	0.091	0.108	0.122	0.133	0.140	0.142
8.0	64.00	0.043	0.062	0.080	0.095	0.108	0.117	0.124	0.127
9.0	81.00	0.038	0.056	0.071	0.085	0.096	0.105	0.111	0.114
10.0	100.00	0.034	0.050	0.064	0.077	0.087	0.095	0.101	0.103

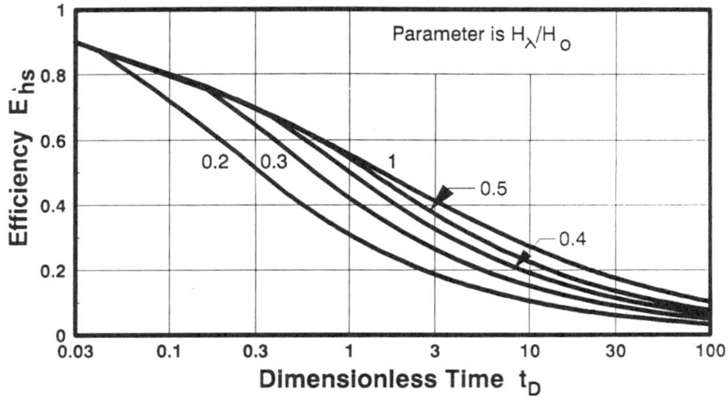

Figure 3.15 Reservoir Heating Efficiency (based on Hearn 1969)

For conical fronts the volume varies with the degree of truncation from an extreme of $Ah/3$ to $Ah/2$ as the shape approaches a cylinder. Even this variation changes the dimensionless time by a factor of only 4/9. As may be seen from the horizontal scale of Figure 3.15, changing t_D by a factor of this magnitude does not have a large effect on the predicted thermal efficiency.

STEAM INJECTION INTO A THIN CHANNEL OR FRACTURE

In his paper, Hearn (1969) discusses the limiting case where h is assumed to approach zero. His result may be obtained by allowing h, in the right-hand side of equation 3.40, to approach zero. If the multiplier h is combined with each of the terms inside the brackets of equation 3.40, only the central term remains as h approaches zero and the equation becomes

$$A(t) = \frac{H_0\sqrt{\alpha_2 t}}{K_2(T_S - T_R)\sqrt{\pi}} \tag{3.59}$$

Substituting t_A for t in equation 3.59, solving the resulting equation for t_A, substituting the result into equation 3.55, and rearranging leads to

$$H_\lambda = \frac{2H_0}{\pi}\int_0^{A_S} \frac{dA}{\sqrt{\left(\dfrac{H_0}{K_2(T_S - T_R)}\right)^2 \dfrac{\alpha_2 t}{\pi} - A^2}} \tag{3.60}$$

This may be integrated to give

$$H_\lambda = \frac{2H_0}{\pi}\left|\sin^{-1}\left(A\sqrt{\frac{\pi}{\alpha_2 t}}\cdot\frac{K_2(T_S - T_R)}{H_0}\right)\right|_0^{A_S} \tag{3.61}$$

which leads to

$$A_S = \frac{H_0\sqrt{\alpha_2 t}}{K_2(T_S - T_R)\sqrt{\pi}}\sin\left(\frac{\pi}{2}\cdot\frac{H_\lambda}{H_0}\right) \tag{3.62}$$

This remarkably simple expression indicates that the steam zone remains a constant fraction of the total with the fraction being a sine function involving the ratio

H_λ/H_0. Equation 3.59 was also derived in Chapter 2 as equation 2.29. Its use is discussed in a later numerical example.

Comparison of Fracture Filled with Steam for Constant Injection Rate and for Constant Areal Growth Rate

For a constant steam injection rate into a fraction it can be seen from equations 3.59 and 3.62 that the ratio of the steam saturated area to the total heated area is given by,

$$\frac{A_S}{A} = \sin\left(\frac{\pi}{2} \cdot \frac{H_\lambda}{H_0}\right) \tag{3.63}$$

For a constant injection rate it was shown that, for long injection times the same ratio is given by,

$$\left(\frac{A_S}{A}\right)_\infty = 2\frac{H_\lambda}{H_0} - \left(\frac{H_\lambda}{H_0}\right)^2 \tag{3.64}$$

These two values are quite similar.

$\frac{H_\lambda}{H_0}$	Values of A_S/A	
	Constant Injection	Constant Displacement $t = \infty$
1	1	1
0.8	0.95	0.96
0.6	0.81	0.84
0.4	0.59	0.64
0.2	0.31	0.36
0	0	0

Calculation of the Mandl-Volek Critical Time for a Numerical Example

The solution to the numerical example of the Marx-Langenheim theory ignored the possibility of the steam being completely condensed by the vertical heat losses. As has been explained, this occurs after the Mandl-Volek critical time.

X_c and the dimensionless time, t_{Dc}, are calculated later.

$$H_\lambda = 800 \times 350 \times 755 \times 0.7 = 148 \times 10^6 \text{ Btu/d}$$

$$H_0 = H_\lambda + 800 \times 350 \times 407 = 262 \times 10^6 \text{ Btu/d}$$

Find X_c from Table 3.1.

$$1 - \frac{H_\lambda}{H_0} = 0.435 = \exp(X_c^2)\ \text{erfc}(X_c)$$

The root is $X_c = 0.973$; this is obtained by interpolation of Table 3.1.

$$t_{DC} = X_c^2 = 0.947$$

The corresponding actual times are calculated as follows:

$$t_c = \frac{h^2 t_{DC}}{4\alpha} = 26 \text{ d} \quad \text{for } h = 10 \text{ ft}$$

or

$$t_c = 2630 \text{ d} \qquad \text{for } h = 100 \text{ ft}$$

For the thinner reservoir the critical time corresponds to only 26 days (0.07 years); for the thicker reservoir the time is 2630 days, or 7.2 years. At times later than these the condensation front lags behind the heat front. This will tend to reduce the quantity of oil displaced below that calculated previously because the residual oil in the waterflooded region will be larger than in the steamflooded zone. The displacement of oil by steam is compared to that by water in Chapter 5.

The size of the steam zone (as distinct from the larger heated zone) can be obtained by calculating the volume (and hence the area) of the steam zone that would be obtained if there were no heat losses at all and then multiplying this by the efficiency read from Figure 3.15. This is done for the previous numerical example in Table 3.5.

Extension of Numerical Example to Injection Into a Very Thin Horizontal Layer or Fracture

In the previous calculation, it was seen that the heated area increased more rapidly for the thinner reservoir. The reason for this is, of course, that less heat is needed to heat a given area of reservoir when it is not as thick. The limiting case corresponds to that of a very thin reservoir or fracture. This case may be viewed in two ways:

1. The limiting situation for the injection of steam into progressively thinner reservoirs, or
2. The injection of steam into a narrow fracture within a thick reservoir.

In the first case the injected heat is essentially lost, whereas in the second it heats the adjacent tar sands. The total heated area and the steam zone area can be calculated for the data of the previous example using Hearn's equations 3.59 and 3.62, respectively.

$$H_0 = 262 \times 10^6 \text{ Btu/d}$$

$$t = 365 t_{\text{year}}$$

$$\alpha_2 = 0.9 \text{ ft}^2/\text{d}$$

$$K_2 = 29.7 \text{ Btu/ft d °F}$$

$$T_S = 467\text{°F};\ T_R = 75\text{°F}$$

TABLE 3.5 Calculation of Steam Zone Area Allowing for Mandl-Volek Effect

Time		Steam Zone	Heated Area in Acres		
Years	Dimension	Efficiency[2]	No Loss[1]	With Loss	Steam Zone
Reservoir thickness = 10 ft:					
1	13	0.20	17	4.2	3.4
2	26	0.14	34	6.3	4.8
3	39	0.12	51	8.0	6.1
4	53	0.10	68	9.4	6.8
5	66	0.09	85	10.6	7.6
6	79	0.08	102	11.7	8.2
7	92	0.078	119	12.8	9.3
8	105	—	136	13.7	—
9	118	—	153	14.6	—
10	131	—	170	15.5	—
Reservoir thickness = 100 ft:					
1	0.13	0.76	1.7	1.3	1.3
2	0.26	0.71	3.4	2.4	2.4
3	0.39	0.67	5.1	3.4	3.4
4	0.53	0.63	6.8	4.3	4.3
5	0.66	0.61	8.5	5.2	5.2
6	0.79	0.59	10.2	6.0	6.0
7	0.92	0.56	11.9	6.7	6.7
		Critical time = 7.2 y			
8	1.05	0.535	13.6	7.5	7.3
9	1.18	0.515	15.3	8.2	7.9
10	1.31	0.495	17.0	8.8	8.4

[1] $H_0^* t/(\rho_1 C_1 h(T_S - T_R))$ converted to acres.
[2] From Figure 3.15.

$$\text{Area in acres} = \frac{H_0\sqrt{\alpha_2 t}}{43{,}560\sqrt{\pi}K_2(T_S - T_R)} = 5.283\sqrt{t_{\text{year}}}$$

$$H_\lambda = 148 \times 10^6 \text{ Btu/d}$$

$$A_S = A \sin\left[\left(\frac{\pi}{2}\right)\left(\frac{H_\lambda}{H_0}\right)\right] = 0.775 \text{ of the total heated area}$$

The results are given in Table 3.6. Also shown in Table 3.6 are calculated average values for the total thickness of the heated zone above and below the heated zone. These were obtained by dividing the total volume of the heated zone corresponding to the quantity of steam injected by the total heated area. It will be noted that the heat penetrates about 50 ft (102/2) on either side of the heated zone in 10 years. With heat-penetration distances of this order, it is easy to see why steaming thin reservoirs for long periods is inefficient.

TABLE 3.6 Calculated Heated Area and Steam Zone Area for Injection into a Very Thin Layer

Years	1	2	3	4	5	6	7	8	9	10
Hot zone in acres										
	5.3	7.5	9.1	10.6	11.8	12.9	14.0	14.9	15.8	16.7
Steam zone in acres										
	4.1	5.8	7.1	8.2	9.2	10.0	10.8	11.6	12.3	12.9
Average heated zone thickness in feet[(1)]										
	32	45	56	64	72	79	85	91	96	102

[(1)]Assuming that all of the injected heat remains in a zone of uniform thickness having the same area as that calculated for the hot zone.

BIBLIOGRAPHY

CARTER, R. D., Appendix I of *Optimum Fluid Characteristics for Fracture Extension* by G. C. Howard and G. R. Fast, Drill. and Prod. Prac., API (1957), 267–268.

HEARN, C. L., "Effect of Latent Heat Content of Injected Steam in a Steam Drive," *JPT,* 374–375 (April 1969). © 1969 SPE.

LAUWERIER, H. A., "The Transport of Heat in an Oil Layer Caused by the Injection of Hot Fluid," *Appl. Sci. Res.* A, 5: 145–150 (1955).

MANDL, G. and VOLEK, C.W., "Heat and Mass Transport in Steam Drive Processes," *SPEJ,* 59–79 (March 1969).

MARX, J.W., and LANGENHEIM, R. N., "Reservoir Heating by Hot Fluid Injection," *Pet. Trans. AIME,* 216: 312–315 (1959).

MYHILL, N. A., and STEGEMEIER, G. L., "Steam-Drive Correlation and Prediction," *JPT,* 173–182 (February 1978).

RAMEY, H. J., "How to Calculate Heat Transmission in Hot Fluid Injection," in *Fundamentals of Thermal Oil Recovery,* Dallas, Tex.: Petroleum Engineer Publishing Company. (1965).

VOGEL, J. H., "Simplified Heat Calculations for Steamfloods," *SPE* 11219, (1982); *JPT,* 1127–1136 (July 1984).

4

Steamflooding

INTRODUCTION

In this chapter the ideas introduced in Chapter 3 are expanded, and it is shown how they may be used as the basis for the analysis of field projects. The chapter also discusses important factors that were not included in the development of the ideas in Chapter 3. These include the effect of gravity in causing override of the steam, the effect of steaming upon the permeability of the matrix, depletion, and steam distillation.

A QUALITATIVE DISCUSSION OF STEAM-INJECTION PROCESSES

Steam-injection processes for the recovery of heavy oils are divided into two categories:

1. Cyclic stimulation
2. Steamflooding

Cyclic Stimulation

In stimulation, steam is injected into the reservoir at a rate of up to about 1000 B/d (160 t/d or m^3/d) for a period ranging from one to several weeks, and then the well is produced by allowing fluids to flow back. When the pressure at the bottom of the well drops, the well is pumped. During the pumping period, the well temperature continues to fall.

Over a period that can range from several months to a year or more, the oil-production rate falls to the point where it is no longer advantageous to continue, and the well is restimulated by injecting more steam. This cyclic process is continued until the quantity of oil recovered is no longer sufficient to justify further steaming. At this time the recovery is typically of the order of 15%; the recovery depends on the nature of the reservoir, the economic value of the produced oil, the well spacing, and other variables.

Steamflooding

In the flooding process, steam is injected continuously into one or more wells and oil is driven to separate production wells. Usually the wells are placed in regular patterns. The steamflooding process is also referred to as steam drive. In the examples described in this chapter, the objective is to drive the oil sideways toward production wells. If the reservoir dips, it is advantageous to drive the oil downward in order to utilize gravity to keep the steam from bypassing the oil.[1]

Frequently the two methods of steam injection are combined and wells are produced by stimulation before flooding is started. When it is desired to produce very viscous oils such as from oil sands, stimulation before flooding is almost essential in order to achieve flow communication between the injection and production wells.

Communication can be established between pairs of wells even in cold tar sand by creating a fracture between them. This can be done by injecting steam at a sufficiently high pressure. In tar sands deeper than about 1000 ft, such fractures are usually vertical, and they tend to have a definite azimuthal (compass) orientation. In much of Alberta this is approximately SW/NE. At shallower depths, horizontal fractures tend to be formed.

If steam is injected into a vertical fracture in cold tar sand, heating will occur and condensate will flow to the connecting production well. There is a tendency for the steam to override, and the fracture can become heated along the top without much heat penetrating to the lower parts. The pressure gradient along a steamed communicating fracture tends to be small because of the need to prevent excessive steam bypassing. As a result, while heat is transferred to the adjacent reservoir, oil production is slow because there is little driving force available to move the heated oil. Because of these difficulties and because of the attractiveness of early oil production, the preheating of the reservoir before steam flooding appears to be the preferable route to achieving conventional steamflooding in bitumen reservoirs.

Although little practical field experience is available, vertical steam flooding using the steam-assisted gravity drainage approach, which is described in Chapter 7, may be preferable in many circumstances—particularly for projects involving thick reservoirs. In many cases, steam flood projects that were started with the idea of driving oil horizontally have ended up with more and more attention being paid to the importance of gravity in providing drive. It is recognized that downward steam drive in dipping reservoirs is a practical means of achieving high injection and production rates. In commercial operations, steam stimulation is often economically at-

[1]It is being realized more and more that downward steamflooding offers considerable advantages. One way of accomplishing this is to use horizontal production wells located near the base of the reservoir, with the steam introduced above. This approach, which has become known as *steam-assisted gravity drainage,* is discussed in Chapter 7.

tractive because it enables rapid production of oil with acceptable and sometimes very high oil-to-steam ratios.

While the short-term economics of stimulation are frequently satisfactory, only about 15 to 20% of the oil can be produced economically. After this, the oil-steam ratio becomes relatively poor. At this stage it is common, at least in fields containing mobile oil (particularly those in California), to convert the steam stimulation operation to a steamflood.

Steamfloods can produce recoveries of the order of 50% of the original oil in place with oil-steam ratios of the order of 0.2. Volumes of steam are traditionally measured in terms of the volume of the equivalent water; a barrel of steam is thus 350 lb of steam (including any liquid water associated with the steam) and 1 m^3 is 1 t.

The oil-steam ratio is dependent upon the nature of the reservoir. Very deep reservoirs (deeper than 5000 ft at the very most) are uneconomic for conventional steam stimulation and flooding because of the very high steam pressures and correspondingly high temperatures required. Heat losses and the quantity of steam required to raise the reservoir-to-steam temperature become excessive. Another factor is the excessive heat loss that can occur from the well bore to the overburden. There is an increase in the well bore heat losses as the depth of the reservoir is increased. As was discussed in Chapter 2, this increase is caused by the extra length of the well and also by the higher steam temperature associated with the higher pressures. Thermal insulation can be used to extend the practical depth for steam injection, but this tends to be expensive.

The next most important criterion for a successful steam recovery project is that the reservoir should be thick—certainly at least 10 ft thick and preferably much thicker. The reason for this is that the heat losses to the overburden and underburden represent an excessive proportion of the total heat requirement for thin reservoirs. This idea was discussed in the last chapter.

Typical successful steam-drive projects are in relatively shallow, fairly thick reservoirs—e.g., 1000 to 2000 ft in depth and 100 ft thick. Usually these reservoirs consist of unconsolidated or loosely consolidated sand having reasonably high permeability and porosity (e.g., 1 D and 30% porosity) and high oil saturation. It is usual to produce oil by stimulation from both the injection and the production wells before the drive commences. Stimulation is often continued, even during the drive, if the temperature of the produced fluids tends to fall. It is also becoming common, as steamflooded fields become depleted, to recover some of the remaining oil by waterflooding. In this situation it is still desirable to stimulate the producers periodically if the production tends to fall in temperature.

Very shallow reservoirs are usually not suitable for steamflooding (or for stimulation, either). The reason for this is that the steam pressure that can be utilized would have to be kept low to avoid fracturing to the surface of the ground above the reservoir. With the lower temperatures that correspond to the lower-pressure steam, the oil (particularly if it is bitumen) may not become sufficiently fluid to make recovery practicable. Fracture pressure is, to a first approximation, equal to 1 psi/ft of depth from the surface.

The use of horizontal wells in place of conventional ones makes the use of steamflooding processes in shallow reservoirs more practical. Their greater contact

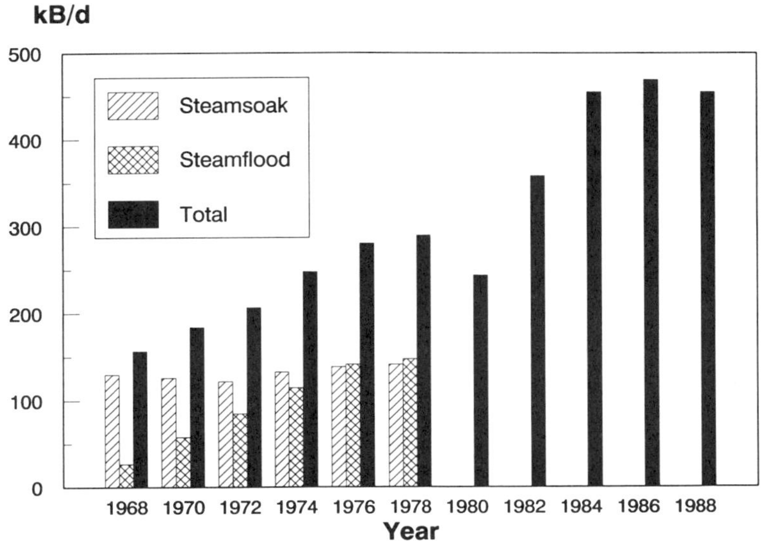

Figure 4.1 Heavy Oil Recovery with Steam in the United States (Source of Data *Oil and Gas Journal*)

with the reservoir allows more viscous oils to be produced at a useful rate. This is discussed in Chapter 7.

During recent years the trend in the heavy oil fields in California has been to switch from steam stimulation to flooding, and most heavy oil from there is now produced by steamflooding. The main reason for this is the economic incentive to improve the recovery. Figure 4.1 compares the historical recovery of oil by steam stimulation with that by steamflooding in California.

In addition to providing a higher recovery, steamflooding—with its continuous injection of heat—can produce oil significantly faster than can the cyclic stimulation process. This, too, can have a significant economic impact. The main disadvantages of steamflooding compared to stimulation are the following:

- There is a lower oil-steam ratio. In steamflooding it is necessary to heat a larger part of the reservoir, whereas in stimulation, at least in the early cycles, the heating is confined to a smaller region around the well.
- There is a longer period of time before significant production starts.
- Often flooding is not possible initially because of the lack of flow communication.

SUITABILITY OF SPECIFIC RESERVOIRS FOR STEAMFLOODING

The choice of steamflooding as a means for the recovery of petroleum has been discussed by a number of authors (including Farouq Ali 1974, Farouq Ali and Meldau 1979, Geffen 1973, Matthews 1983, and Chu 1985). Whereas the suitability of a reservoir for production by steam stimulation can be determined relatively simply by means of single well tests, field experimentation to determine its suitability for steamflooding is much more costly. Even the simplest test must involve multiple wells and long periods of operation.

Several quantitative guidelines have been developed to indicate whether a reservoir property might respond favorably to steamfloods. Table 4.1 is a summary of such screening guides; it is taken from Chu (1985).

Matthews lists the following factors that are unfavorable for steamflooding.

1. Oil saturation less than 40%
2. Porosity less than 20%
3. Oil-zone thickness less than 30 ft
4. Permeability less than 100 mD
5. Ratio of net to gross pay less than 50%
6. Layers of very low oil saturation and high permeability in the oil zone that act as thief zones
7. Extremely high viscosity
8. Fractures[2]
9. Large permeability variations in the oil zone
10. Poor reservoir continuity between injectors and producers
11. Deep high-pressure reservoirs and shallow reservoirs with insufficient overburden to permit steam injection without fracturing

He points out that steamfloods may be successful even if one or two of the above conditions are not met, provided that the remaining factors are highly favorable.

Chu (1985) describes an empirical correlation that predicts the oil-steam ratio (OSR) or its reciprocal, the steam-oil ratio (SOR). His correlation equations are given next; note that the units employed are, in some cases, not the customary ones.

If SOR ≤ 5.0 (OSR ≥ 0.20):

English units (as defined shortly):

$$\text{SOR} = 18.744 + 0.001453D - 0.05088h - 0.0008864k - 0.0005915\mu - 14.79S_o - 0.0002938\frac{kh}{\mu}$$

Metric units (as defined shortly):

$$\text{SOR} = 18.744 + 0.004767D - 0.16693h - 0.8981k - 0.5915\mu - 14.79S_o - 0.0009767\frac{kh}{\mu}$$

[2]Fractures may be undesirable because they promote bypassing of the steam. In the steam-assisted gravity drainage process, however, which is operated below the critical steam-coning rate, fractures enhance the process if they are vertical and have little effect if they are not.

TABLE 4.1 Screening Guides for Steamflood Projects (Chu 1985)

	ϕS_o	ϕ	S_o	Density (°API)	Thickness h (ft)	Depth D (ft)	Permeability k (mD)	Viscosity μ (cp)	kh/μ (mD-ft/cp)
Farouq Ali (1979)	0.15 to 0.22	0.30		12 to 15	30	< 3000	~ 1000	< 1000	
Geffen (1973)	> 0.10			> 10	> 20	< 4000			> 20
Lewin (1976)	> 0.065		> 0.50	> 10	> 20	< 5000			> 100
Iyoho (1978)	> 0.065	> 0.30	> 0.50	10 to 20	30 to 400	2500 to 5000	> 1000	200 to 1000	> 50
Chu (1985)	> 0.08	> 0.20	> 0.40	< 36	> 10	> 400			

If SOR > 5.0 (OSR < 0.20):

English units (as defined shortly):

$$\text{OSR} = -0.011253 + 0.00002779D + 0.0001579h - 0.001357\theta + 0.000007232\mu + 0.00001043\frac{kh}{\mu} + 0.5120\phi S_o$$

Metric units (as defined shortly):

$$\text{OSR} = -0.011253 + 0.00009117D + 0.0005180h - 0.077756\theta + 0.007232\mu + 0.00003467\frac{kh}{\mu} + 0.5120\phi S_o$$

	English	Metric
D = depth	ft	m
h = thickness	ft	m
k = permeability	mD	mD
S_o = oil saturation at start	fraction of pore volume	
S_{oi} = initial oil saturation	fraction of pore volume	
S_{or} = residual oil saturation	fraction of pore volume	
T = temperature	°F	°C
θ = dip angle	degree	rad
μ = viscosity	cp	Pa s
ϕ = porosity	fraction of bulk volume	

These equations may be used quite simply because all they require are fairly basic measurements or estimates of the reservoir properties. Two forms of the equations are given, one for English units and one for metric. Chu's paper contains summary data for 28 different steamflood field projects, including references. The equations just given were found to correlate well with the data. Chu recommends that the equation for SOR ≤ 5 be tried first and that the second equation be used only if the answer from the first indicates the SOR to be greater than 5.

Tables 4.2, 4.3, 4.4, and 4.5 reproduce the field project data summaries collected by Chu. References to the sources of the data are listed in Chu's paper. It is interesting to note that of the 28 projects studied by Chu, only 7 gave oil-steam ratios higher than 0.2.

THE PROPERTIES OF STEAM

The most important properties of steam for thermal recovery processes are those involving enthalpy. Enthalpy is defined as

$$\mathbf{H} = \mathbf{U} + \mathbf{PV}$$

where $\mathbf{H}$ is the enthalpy in units of energy per unit mass, e.g., kJ/kg (or Btu/lb)
$\mathbf{U}$ is the internal energy, kJ/kg (or Btu/lb)
P is the pressure, kPa (or Btu/ft^3)
$\mathbf{V}$ is the specific volume, m^3/kg (lb/ft^3)

The term $\mathbf{PV}$ has the dimensions of energy per unit mass; it is the work that must be done per unit mass of material to introduce it at pressure P into a continuous flow system operated in a steady state. Similarly, material leaving the system can do an equivalent amount of work. The total heat effect in a continuous-flow, isolated system operating in a steady state is thus

$$\text{Heat added} = \sum \mathbf{H}_P W_P - \sum \mathbf{H}_F W_F$$

where the terms $\mathbf{H}_P$ and W_P refer to the enthalpies and masses of the products leaving the system and $\mathbf{H}_F$ and W_F are the corresponding terms for the feed.

For example, the heat added in a boiler to convert the feedwater to steam is $W_F(\mathbf{H}_P - \mathbf{H}_F)$ if the mass of steam product is equal to the mass of feed. The preceding formulation of the law of conservation of energy neglects terms such as potential energy, kinetic energy, electrical energy, and work other than that included in the $\mathbf{PV}$ terms; this is justifiable in the calculations described in this book. In other cases where these other energy terms are significant, they must be included as additional terms in the energy balance.

At any particular pressure, the temperature and the enthalpy of liquid water increase steadily as heat is added until the boiling point of the water is reached. If further heat is added, the water boils at a constant temperature until an additional amount of heat equal to the latent heat of evaporation has been added. As this heat is added, the liquid is continuously transformed into vapor until eventually only water vapor is present; a very large increase in volume accompanies this transformation. If further heat is added, the steam becomes superheated.

The steam employed for process heating and power generation in process plants such as refineries and power stations is usually dry and superheated. In the oil fields, wet steam (i.e., a mixture of water and vapor) is employed. The parameter that defines the condition of such a mixture is the steam quality, f_S; it is defined as the weight fraction of the steam mixture that is vapor. A weight fraction $(1 - f_S)$ is thus liquid. The steam quality is often expressed as a percentage rather than as a numerical fraction.

The enthalpy of steam of quality f_S can be calculated from the data of Table 4.6 as

$$\mathbf{H}_S = (1 - f_S)\mathbf{H}_L + f_S\mathbf{H}_V \tag{4.1}$$

or, since

$$\mathbf{H}_V = \mathbf{H}_L + \lambda$$

TABLE 4.2 Steamflood Field Projects—Reservoir Characteristics I

Project Number	Field, Location (Operator)	Year Started	Pilot (P) or Fieldwide (F)	Formation	D (ft)	Gross Pay (ft)	Net Pay, h (ft)	Θ (degrees)	ϕ (fraction)
1	Smackover, AR (Phillips)	1964	P	Nacatoch	1,940	—	15	—	0.36
2	Smackover, AR (Phillips)	1971	F	Nacatoch	1,920	130	25	0 to 5 (2.5 average)	0.35
3	Brea, CA (Shell)	1964	P	Lower "B" sands east	4,600 to 5,000	300 to 800 (550 average) normal to bedding	189 to 504 (346.5 average) normal to bedding	66	0.22
4	Coalinga, CA (Shell)	1977	P	Temblor Zone 1	800 to 1,700 (1,250 average)	200	150	21	0.30
5	E. Coalinga, CA (Shell)	1963	P, F	Temblor Zone 1	900 to 2,200 (1,500 average)	300	50	14	0.31 to 0.36 (0.31 average)
6	Huntington Beach, CA (Socal)	1965	P	Upper Tar Lower Tar	2,000 2,400	— —	100 50	0	0.38 0.38
7	Inglewood, CA (Chevron-Socal)	1965	P	Upper Investment	1,000	—	43	0	0.39
8	Kern River, CA (Chevron)	1968	P, F	Kern River series	800	85	70	4	0.35
9	Kern River, CA (Getty)	1971	F	Kern River series	930	80	70	3 to 6 (4.5 average)	0.35
10	Midway Sunset, CA (Chevron)	1975	P	Monarch sand	976	0 to 600 (350 average)	0 to 500 (260 average)	—	0.27
11	Midway Sunset, CA (Santa Fe-DOE)	1975	P	"200" sand, Spellacy	400	200	150	5 to 15 (10 average)	0.30
12	Midway Sunset, CA (Tenneco)	1970	F	Metson	1,050	—	120 to 1,050 (354 average)	70	0.34
13	Mount Poso, CA (Shell)	1970	F	Upper Vedder	1,800	70	55	5 to 15 (10 average)	0.33
14	Shiells Canyon, CA (Texaco)	1973	P	203 zone	850	—	160	35	0.205

15	S. Belridge, CA (Mobil)	1969	F	Tulare, D and E zones	1,009 to 1,197 (1,103 average)	210	91	7	0.35
16	Yorba Linda, CA (Shell)	1971	F	Upper Conglomeration	650	—	325	12	0.30
17	El Dorado, KA (Cities)	1965	P	El Dorado	650	—	10 to 20 17	—	0.25 to 0.30 0.26
18	Deerfield, MO (Esso-Humble)	1956	P	Warner	160	—	26	—	0.235
19	Tatums, OK (Shell)	1964	F	Des Moines Zone VIII	850 to 2,000 (1,425 average)	66	37	45	0.28
20	Franklin Heavy Pool, PA (Pennzoil)	1965	P	First Venango	450 to 600 (525 average)	70	50	—	0.12
21	Charco Redondo, TX (Texaco)	1965	P		200	—	10		0.30 to 0.35 (0.325 average)
22	Slocum, TX (Shell)	1967	F	Carrizo	520	34	32	0 to 5 (2.5 average)	0.34
23	Winkleman Dome, WY (Pan American)	1964	F	Nugget	1,225	180	57	> 25	0.236
24	Schoonebeek, The Netherlands (Nederlandse)	1961	P	Valanginian	2,450 to 2,650 (2,550 average)	—	74 to 90 (82 average)	6.5	0.30
25	Schoonebeek, The Netherlands (Nederlandse)	1972	F	Middle and Lower Bentheim sandstone	2,900	—	92 to 154 (123 average)	6	0.32
26	Tia Juana, Venezuela C-3/C-4 (Shell)	1961	P	Lower Langunillas	U 1,450 L 1,700	U 150 L 100	U 120 L 80	3	0.33
27	Tia Juana, Venezuela M-6 (Maraven)	1975	F	Lower Langunillas	1,624	250	125	0 to 5 (2.5 average)	0.381
28	Georgsdorf, W. Germany (BEB)	1975	P	Valanginian	2,130 to 2,790 (2,460 average)	140	125	14	0.25

(from C. Chu 1985)

TABLE 4.3 Steamflood Field Projects—Reservoir Characteristics II

Project Number	k (mD)	T (°F)	p at Start (psi)	Density (°API)	μ at T (cp)	S_o at Start	kh/μ (mD-ft/cp)	ϕS_o	OIP at Start (STB/acre-ft)
1	5,000	110	5	19	75	0.80	1,000	0.288	
2	2,000	110	7	20	75	0.50	667	0.175	2,234
3	70	175	110	24	6	0.49	4,042	0.108	1,400
4	1,000	95	147 to 400	12	3,000	0.60	50	0.180	895
5	900 to 10,000 (3,000 average)	100	—	16 to 30 (23 average)	10 to 300 (54.8 average)	0.57	2,737	0.177	1,397 1,370
6	2,300	110	600	13	700	0.61	329	0.232	
	2,300	115	1,200	12	2,000	0.61	58	0.232	1,800
7	5,900	100	120	14.5	1,200	0.64	211	0.250	1,800
8	1,900	85	60	14	2,710	0.50	49	0.175	1,930
9	2,400	90	80	14	4,500	0.47	37	0.165	1,344
10	520	105	75	14	1,500	0.59	90	0.159	1,167
11	1,050 to 3,440 (2,245 average)	90	100	11	6,500	0.59	52	0.177	1,236 1,373
12	2,000	100	0 to 130	12	3,000	0.75	236	0.255	1,978
13	24,000	110	100	15	280	0.58	4,714	0.191	1,490
14	140	105	85	34	6	0.45	3,733	0.092	716
15	3,000	95	—	13.0	1,600	0.69 (estimated)	170	0.241	1,874

16	600	85	—	12 to 15 (13.5 average)	6,400	0.46	30	0.138	1,071
17	20 to 1,000 (141 average)	70	—	37	4	0.48	599	0.125	968
18	450	60	50	18	1,000	0.29	12	0.068	535
19	500	75	—	13 to 15	1,600	0.78	12	0.218	1,694
20	800	58	—	43.9 (estimated)	47	0.40	851	0.048	372
21	2,500	72	0 to 20	18	95	0.15 to 0.45 (0.30 average)	263	0.098	756
22	3,500	80	120	18 to 19 (18.5 average)	1,000 to 3,000 (2,000 average)	0.68	56	0.231	1,787
23	481	81	—	14	800 to 1,000 (900 average)	0.796	30	0.188	1,457
24	1,000 to 10,000 (3,160 average)	100	120	25	200	0.85	1,296	0.255	1,978
25	1,000 to 10,000 (3,160 average)	104	—	25	160	0.60 to 0.95 (0.775 average)	2,429	0.248	1,924
26	1,000 to 3,000 (2,000 average)	113	U 175 L 350	12	U 1,300 L 5,000	0.71	U 185 L 32	0.234	1,818
27	2,800	113	100	12	600 to 6,000 (1,897 average)	0.67	185	0.257	1,992
28	1,150	104	465	27	160	0.80	898	0.200	1,551

(from C. Chu 1985)

TABLE 4.4 Steamflood Field Projects—Design Operation and Performance I

Project Number	Field, Location (Operator)	Pattern Type[1]	Pattern Size (acre)	Total Area (acre)	Number of Wells: Injection	Number of Wells: Production	Steam Injection Rate (B/d)	Average Steam Rate (B/d/Injector)
1	Smackover, AR (Phillips)	I-5	10 22	10 22	1	4 8	3,086	3,086
2	Smackover, AR (Phillips)	I-6, I-9 etc.	40	985	9	120	7,179	798
3	Brea, CA (Shell)	UD	—	5 estimated along bedding	1 to 2	13	1,050 (1969–70)	1,050
4	Coalinga, CA (Shell)	I-5	2.5	10.0	4	9	3,000	750
5	E. Coalinga, CA (Shell)	5S	—	—	1.3 DD	30	4,000 (1964–69)	
6	Huntington Beach, CA (Socal)	I-11, I-14	17	34	2	23	2,220 (4/65–4/67)	1,300 to 1,350
7	Inglewood, CA (Chevron-Socal)	I-6	2.6	2.6	1	5	1,080	1,080
8	Kern River, CA (Chevron)	I-5, I-7	—	100	25	50	11,000	440
9	Kern River, CA (Getty)	5S	2.7	310	114	85 corner interior	19,000 2,900	167 483
10	Midway Sunset, CA (Chevron)	I-5	3.8	23	6	15	1,500	375
11	Midway Sunset, CA (Santa Fe-DOE)	I-7	2.35	9.4	4	10 —	(6/78–12/79) 6,000 (estimated)	—
12	Midway Sunset, CA (Tenneco)	UD	—	—	18		(6/73–12/75)	
13	Mount Poso, CA (Shell)	UD + DD	—	290	41	159	74,000 (7/78–12/79)	1,919
14	Shiells Canyon, CA (Texaco)	UD	—	—	2	10	760	560 + 200

15	S. Belridge, CA (Mobil)	DD	—	204	15	41	6,300 (estimated) (1/70–9/73)	—
16	Yorba Linda, CA (Shell)	I-9	—	—	16	74	—	850
17	El Dorado, KA (Cities)	N5	1.6	6.4	9	4	1,170	130
18	Deerfield, MO (Esso-Humble)	I-5	0.625	5.625	9	16	450 (estimated)	50 (estimated)
19	Tatums, OK (Shell)	UD	—	60	4	20	2,570	643
20	Franklin Heavy Pool, PA (Pennzoil)	N5	1	1	4	1	1,200	300
21	Charco Redondo, TX (Texaco)	I-5	2.5	2.5	1	4	577	577
22	Slocum, TX (Shell)	I-13	5.3	258	49	258	17,000 (5/75–4/76)	347
23	Winkleman Dome, WY (Pan American)	I-5	10	110	15	21	4,100 (1972–74)	273
24	Schoonebeek, The Netherlands (Nederlandse)	3 I-5 + 1 I-4	16.4	65.5	4	8	5,300	1,325
25	Schoonebeek, The Netherlands (Nederlandse)	DD	—	—	2	6(?)	3,010	—
26	Tia Juana, Venezuela C-3/C-4 (Shell)	I-7	11.42	79.94	7	24	9,405 (1963–66)	1,344 (1963–66)
27	Tia Juana, Venezuela M-6 (Maraven)	I-7	25.2	1,831	19	131[2]	45,000[2] (1979)	—
28	Georgsdorf, W. Germany (BEB)	IRR	—	160	2	12	4,600	2,300

[1]I-5 = inverted five-spot; N5 = normal five-spot; 5S = five-spot; UD = updip, DD = downdip, IRR = irregular.
[2]Includes steam-soak wells.
(from C. Chu 1985)

TABLE 4.5 Steamflood Field Projects—Design Operation and Performance II

Project Number	Steam Pressure (psia)	Steam Quality (%)	Thermal Oil Production (B/d)	Average Oil Rate (B/d/Producer)	Oil/Steam Ratio (B/B)	Steam/Oil Ratio (B/B)	Remarks	References
1	405	100	100	12.5	0.24	4.14	I5	18
2	—	—	1,167	9.7	0.16	6.2	1975–77 data cumulative 1979, SOR = 5.7	19
3	2,000	53 to 55 (bottom)	240 (1969–70)	—	0.16	6.3	UD injection	20
4	600 to 900	—	600	66.7	0.19	5.1		21
5	—	—	667 (1964–69)	—	0.17	6.0	5S DD area only	22
6	800 to 1,000	—	440 (4/65–4/67)	19.1	0.20	5.0		23
7	350 to 425	—	129	25.7	0.12 0.27*	8.4 3.7*		24
8	—	—	1,800	36	0.16	6.1		25, 26
9	200	70	5,000	59	0.21	4.8		8, 15, 27
10	250 to 550	40 to 80	520	—	0.16 (cumulative)	6.1 (cumulative)		28, 29
11	—	—	160 (6/79–12/79)	16.0	0.11	9.4		30
12	—	—	1,050 total 650 thermal (6/73–12/75)	—	0.13 (cumulative)	8.0 (cumulative)		31, 32

13	350	—	18,000	113	0.19 (cumulative 1979)	5.3	33
14	480 to 680	57 to 76	230	23	0.25	4.0	34
15	—	—	1,100** (estimated) (1/70–9/73)	—	0.17	5.7	35
16	200	70	7,000	—	0.17	5.8	36
17	465	75	20	5	0.04	26.0	37
18	125	—	6.7	0.4	0.02	50.4	38
19	1,300	60	650	32.5	0.10	10.0	39
20	400	—	14	14	0.05	21.7	40
21	156	—	30.0	7.5	0.05	20.0	41
22	500	80	1,400	5.4	0.14 (maximum 0.17)	7.1 5.9	42, 43
23	1,150	—	850	40.5	0.21	4.8	35, 36
24	600	—	1,900	237.5	0.38	2.7	46
25	1,233	—	2,100	—	0.70	1.43	47
26	U 220 L 350	—	6,338 (1963–66)	264 (1963–66)	0.62 (cumulative)	1.62 (cumulative)	48
27	—	—	17,000† (1979)	—	0.16 (end 1979)	6.3 (end 1979)	13, 49, 50
28	—	—	1,250	—	0.19	5.3	51

*Includes postinjection production (1972–74).
**Steamflood only.
†Includes steam-soak wells.
(from C. Chu 1985)

TABLE 4.6 Enthalpy of Water and Steam at Saturation Conditions

P (MPa)	T (°C)	Enthalpy kJ/kg Water	Evap.	Steam	P (MPa)	T (°C)	Enthalpy kJ/kg Water	Evap.	Steam
0.006	0.0	0.0	2501.6	2501.6	1.50	198.3	844.6	1945.3	2789.9
0.10	99.6	417.5	2257.9	2675.4	1.60	201.4	858.5	1933.2	2791.7
0.11	102.3	428.8	2250.8	2679.6	1.70	204.3	871.8	1921.6	2793.4
0.12	104.8	439.4	2244.1	2683.4	1.80	207.1	884.5	1910.3	2794.8
0.13	107.1	449.2	2237.8	2687.0	1.90	209.8	896.8	1899.3	2796.1
0.14	109.3	458.4	2231.9	2690.3	2.00	212.4	908.6	1888.7	2797.2
0.15	111.4	467.1	2226.3	2693.4	2.50	223.9	961.9	1839.0	2800.9
0.16	113.3	475.4	2220.8	2696.2	3.00	233.8	1008.3	1794.0	2802.3
0.17	115.2	483.2	2215.8	2699.0	3.50	242.5	1049.7	1752.2	2801.9
0.18	116.9	490.7	2210.8	2701.5	4.00	250.3	1087.4	1712.9	2800.3
0.19	118.6	497.9	2206.1	2704.0	4.50	257.4	1122.1	1675.6	2797.7
0.20	120.2	504.7	2201.6	2706.3	5.00	263.9	1154.5	1639.7	2794.2
0.21	121.8	511.3	2197.2	2708.5	5.50	269.9	1184.9	1605.0	2789.9
0.22	123.3	517.6	2193.0	2710.6	6.00	275.6	1213.7	1571.3	2785.0
0.23	124.7	523.7	2188.9	2712.6	6.50	280.8	1241.2	1538.3	2779.5
0.24	126.1	529.6	2184.9	2714.5	7.00	285.8	1267.5	1506.0	2773.5
0.25	127.4	535.4	2181.0	2716.4	7.50	290.5	1292.7	1474.1	2766.8
0.26	128.7	540.9	2177.3	2718.2	8.00	295.0	1317.2	1442.7	2759.9
0.27	130.0	546.2	2173.7	2719.9	8.50	299.2	1340.8	1411.6	2752.4
0.28	131.2	551.5	2170.0	2721.5	9.00	303.3	1363.8	1380.8	2744.6
0.29	132.4	556.5	2166.6	2723.1	9.50	307.2	1386.2	1350.2	2736.4
0.30	133.5	561.4	2163.2	2724.7	10.00	311.0	1408.1	1319.7	2727.8
0.35	138.9	584.3	2147.3	2731.6	11.00	318.0	1450.6	1258.8	2709.4
0.40	143.6	604.7	2132.9	2737.6	12.00	324.6	1491.7	1197.5	2689.2
0.45	147.9	623.2	2119.7	2742.9	13.00	330.8	1531.9	1135.1	2667.0
0.50	151.8	640.1	2107.4	2747.5	14.00	336.6	1571.5	1070.9	2642.4
0.55	155.5	655.8	2095.9	2751.7	15.00	342.1	1610.9	1004.2	2615.1
0.60	158.8	670.4	2085.0	2755.5	16.00	347.3	1650.4	934.5	2584.9
0.70	165.0	697.1	2064.9	2762.0	17.00	352.3	1691.6	860.0	2551.6
0.80	170.4	720.9	2046.5	2767.5	18.00	357.0	1734.8	779.0	2513.8
0.90	175.4	742.6	2029.5	2772.1	19.00	361.4	1778.7	691.8	2470.5
1.00	179.9	762.6	2013.6	2776.2	20.00	365.7	1826.6	591.6	2418.2
1.10	184.1	781.1	1998.6	2779.7	21.00	369.8	1886.3	461.2	2347.5
1.20	188.0	798.4	1984.3	2782.7	22.00	373.7	2010.3	186.3	2196.6
1.30	191.6	814.7	1970.7	2785.4	22.12	374.2	2107.4	0.0	2107.4
1.40	195.0	830.1	1957.7	2787.8					

Abstracted from "U.K. Steam Tables in SI Units 1970", United Kingdom Committee on Properties of Steam, Edward Arnold, London 1970.

$\mathbf{H}_S$ can also be expressed as

$$\mathbf{H}_S = \mathbf{H}_L + f_S\lambda \tag{4.2}$$

where λ is the latent heat of evaporation and the subscripts, S, L, and V refer to the steam mixture, boiling liquid, and saturated vapor, respectively.

Table 4.6 and Figure 4.2 give the enthalpy of boiling water, the latent heat of evaporation, and the enthalpy of saturated steam as a function of pressure and temperature.

The values in Table 4.6 are in S.I. units. They may be converted to English units by the following conversions:

$$(\text{Pressure in psia}) = 145(\text{pressure in MPa})$$

$$(\text{Temperature in °F}) = 1.8(\text{temperature in °C}) + 32$$

$$(\text{Enthalpy in Btu/lb}) = (\text{enthalpy in kJ/kg})/2.326$$

Simple relations for calculating the enthalpies of saturated liquid and vapor as functions of temperature and pressure are given in Appendix 9.

Specific points that should be noted with respect to Table 4.6 are as follows: The latent heat of evaporation decreases as the pressure is raised and disappears at the critical point. Less latent heat is available per unit mass of steam when the operating pressure is higher. Above the critical point (22.1 MPa, 374°C or 3208 psia, 706°F), none is available. The enthalpy of saturated steam reaches a maximum of 2802 kJ/kg at 236°C and 3 MPa (1205 Btu/lb at 457°F and 435 psia).

In the field, steam is generated at pressures up to about 15 MPa (2200 psi). It is usual to generate wet steam—i.e., a mixture of saturated steam vapor and water. Typically a quality of 70 to 80% is employed. The main reason for using wet rather than dry or superheated steam is to reduce the purity requirements for the feedwater for the steam generators.

Oil field steam generators usually contain a single boiler tube coiled around the combustion zone. Water is pumped at high pressure into one end and a mixture of about 70 to 80% vapor and 20 to 30% liquid leaves the other. The water usually contains high concentrations of dissolved solids. These remain dissolved in the residual liquid water and are removed continuously with the steam product. More

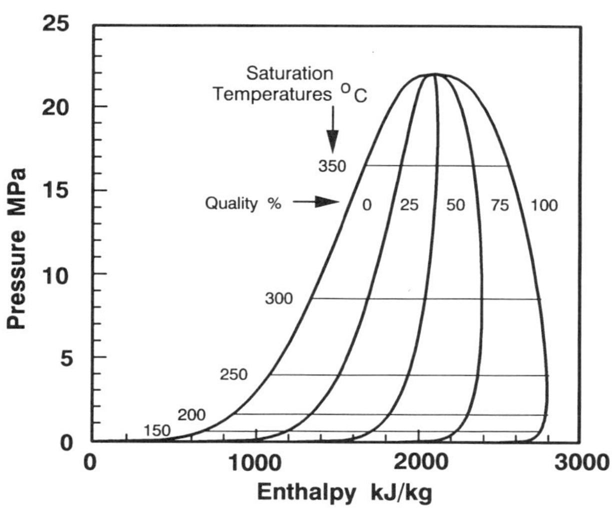

Figure 4.2 Pressure-Enthalpy Diagram for Steam-Water

information on water treating, steam generation, and steam distribution is given in Chapter 8.

TEMPERATURE DISTRIBUTION IN STEAMFLOODING

Figure 4.3 shows, in an idealized way, one concept of the conditions around a steam-injection well. The temperature in the vicinity of the well is nearly constant and is equal to the saturation temperature of the steam. This temperature prevails to the point where the last of the steam condenses.

Beyond the condensation front there is a hot-water zone in which the temperature falls. The temperature gradient just beyond the front may be relatively abrupt or more gentle, depending upon the conditions. This was discussed in the last chapter in connection with the works of Mandl and Volek and of Hearn.

Much of the heat introduced with the steam is lost to the overburden and to the underburden by thermal conduction in the manner discussed earlier.

In the situation shown in the figure it is assumed that the hot zone has reached the overburden and underburden. In practice, it is possible that conditions may exist in which the steam zone has not yet extended to the upper and lower limits of the reservoir.

A particularly common and important situation is that where the steam zone has risen, because of gravity effects, to the top of the reservoir but has extended only part of the way to the bottom. Under these conditions the oil below the steam zone is being heated but is produced slowly, and the potential thermal advantage of having a thick reservoir to heat may not be realized. It is a challenge of thermal recovery engineering to devise systems by which the maximum available thickness of reservoir material is produced in order to minimize the area of over- and underburden to which heat is being lost. Steam stimulation tends to do this initially.

In any case, as time goes on the steam zone expands, and the area that is being heated above and below increases. As a result the heat losses also increase, and

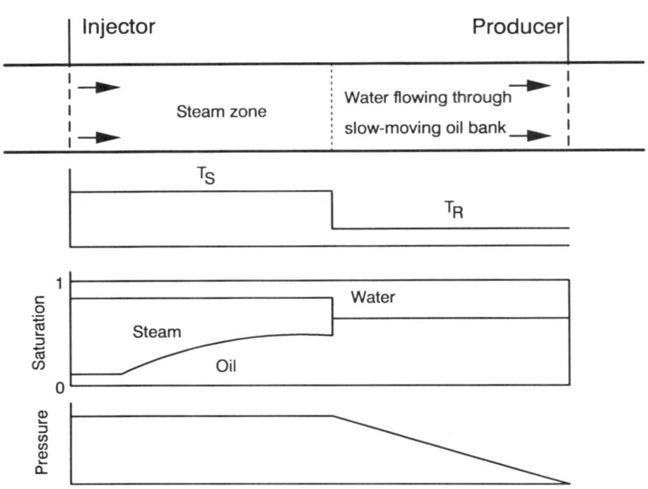

Figure 4.3 Diagram showing the Distribution of Temperature, Pressure, and Saturations in a Hypothetical One-Dimensional Steamflood

a smaller portion of the heat in the injected steam is employed in useful reservoir heating. The heat losses increase up to the point where areal growth becomes limited by interference with the neighboring patterns (see Figure 2.4).

The spacing between injectors and producers is an important factor in determining the utilization of heat. Large spacings result in large areas of overburden and underburden having to be kept hot for longer periods of time. For a given flow it takes longer to drain the oil between the injector and the producer if the spacing is greater.

The design of a steamflood involves an economic balance between the thermal efficiency of close spacing and the lower well investment required for the fewer wells involved with larger spacings. Another factor, particularly with tar sands, is the difficulty in establishing communication. Sometimes there is also the difficulty in maintaining communication, since interconnecting flow paths may tend to block when cold viscous oil drains into them by gravity drainage. (See the discussion on fingering in Chapter 5).

Typical commercial steamflood projects have production wells with spacings of 2 to 6 acres with either one injection well per production well (inverted five-spot pattern in Figure 4.4) or one injection well for every two production wells (inverted seven-spot pattern in Figure 4.5). Line-drive configurations are also common (see Figure 4.5). Another popular configuration is the inverted nine-spot; this results in three producers per injector. In this arrangement, the producing wells are not all equal—the "corner" wells (2, 4, 6, and 8 in Figure 4.4) have different surroundings from the "side" wells (3, 5, 7, and 9). (*Note:* 1 acre = 43,560 ft^2 = 0.405 ha and 1 ha = 10,000 m^2.)

A feature that is common in many steamfloods is the addition of infill wells as the project matures. These are frequently added when steam breaks through to the producers as a result of gravity override. Infill wells allow the recovery of some of the bypassed oil which lies below the steam zone. This is discussed further later on in this chapter.

When there is a dip in the reservoir it is usually advantageous to drive the oil downdip in order to make use of the gravitational force to stabilize the displacement front. This is discussed in the next chapter.

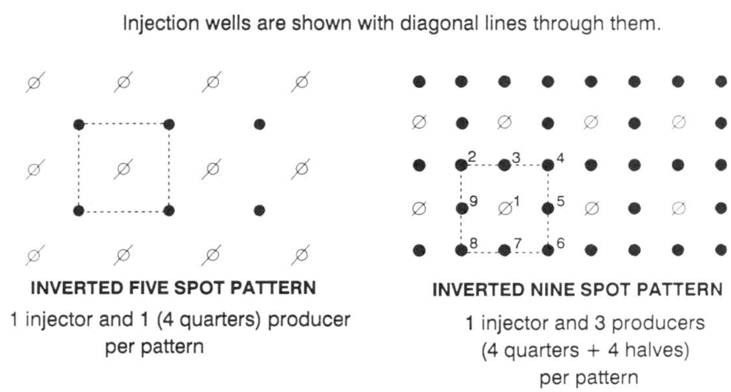

Figure 4.4 Inverted Five- and Nine-Spot Well Patterns

Injection wells are shown with diagonal lines through them.

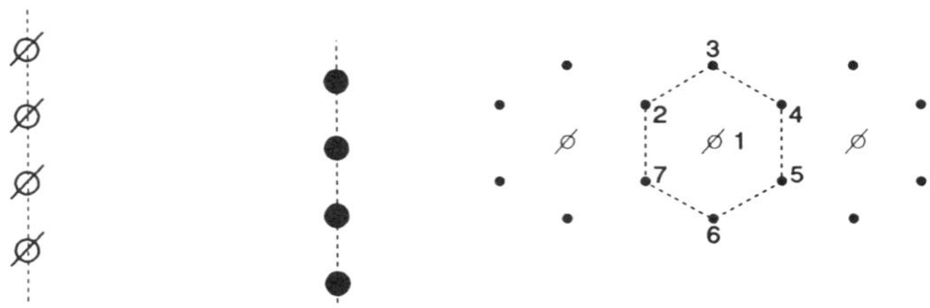

Figure 4.5 Inverted Seven-Spot and Staggered-Line Drive Patterns

FINGERING

In the displacement process shown in Figure 4.3, the condensed water runs more rapidly than the oil to the production well because it is much less viscous than the oil that it is displacing. Frequently the water runs as separate rivulets, or fingers, through the oil; the flow pattern can be visualized as oil and water running together along separate flow paths, with the water velocity being much higher than that of the oil. Thus, rather than dry oil, a mixture containing very substantial quantities of water is produced. The fingering of water through the oil may also be promoted by heterogeneities within the reservoir, including those created by the fracturing that results from steam injection at pressures above the minimum in situ stress. Passage of the water must occur if steam is to continue to supply heat to the reservoir. If the removal of condensate is not possible with the available pressure drops, then the process will be slowed greatly.

Even if there were no fingering due to the formation of unstable water/oil displacement fronts, the water would still run through the oil layer, with an early breakthrough because of the adverse viscosity ratio. It is shown in the next chapter that when an attempt is made to displace a viscous oil with water, breakthrough of the water occurs rapidly, because of the relative permeability and viscosity characteristics—even if the flow is diffuse rather than segregated (i.e., even if the water does not run as fingers).

GRAVITY OVERRIDE

A major difference between the practical situation and the flow depicted in Figure 4.3 is that the difference in density between the steam and the liquids in the reservoir causes the steam to override—i.e., to flow above the oil; the situation is as depicted in Figure 4.6. Eventually steam breaks through at the production well.

The upper steam-swept region has a much lower residual oil saturation than the lower water-flooded region. For example, Blevins and Billingsley (1975) report a

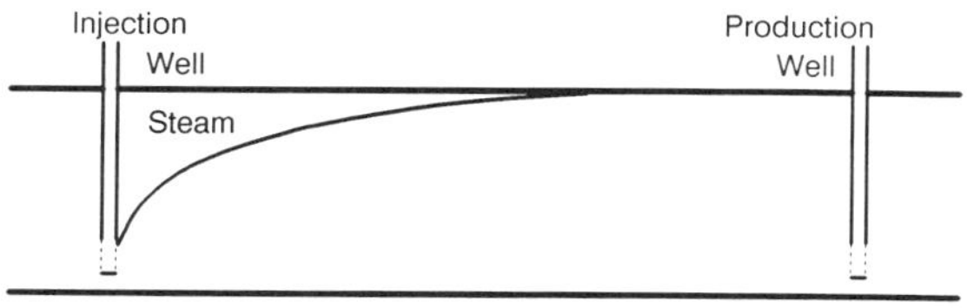

Figure 4.6 Gravity Override of Steam

6% oil saturation in the steam-swept zone versus 23% for the water-swept zone for a steamflood in the Kern River field in California. The upper steam-swept zone in this project represented about one-third of the swept volume and the waterflooded zone represented two-thirds. Once steam has broken through, there is little pressure gradient to remove the oil, particularly since it is necessary to reduce the injection rate in order to control steam bypassing (i.e. steam "coning").

A significant effort is being made currently to develop steam additives that will restrict the flow of steam within the steam zone, thereby increasing the pressure drop and causing more rapid encroachment of the steam into the waterflooded zone. A popular approach is the addition of surfactant materials to the steam, which cause the formation of foam within the steam zone (e.g., Dilgren et al. 1982, Al-Kahaafji et al. 1982, and Eson and O'Nesky 1982). For foaming materials to be effective, it is necessary that they be chemically stable at the steam temperature and that their cost be low enough for them to be economic in the quantities required. Foam additives can also reduce the bypassing of steam through high permeability strata in heterogeneous reservoirs.

Promising results were obtained by Ploeg and Duerksen (1985) in field tests in the Midway-Sunset field in California in which sulphonate solutions were added to the injected steam. These authors concluded that incremental oil was produced and that the process was economical.

Friedmann and Jensen (1986) have reported an experimental study of the flooding of Berea cores with foams prepared using Chevron Chaser SD1000 surfactant. They found that the surfactant reduced the relative permeability to gas. High oil saturations reduced the degree of foam formation and propagation. Foams, preformed at high velocities in sandpacks, could be propagated through reservoir material at low velocities. However, it was not possible to generate foams in situ at the same low velocities.

Mohammadi, van Slyke, and Ganong (1989) reported that in a steamflooding test involving four five-spot patterns in the Potter sand in the Midway-Sunset field in California, the addition of NaCl, alpha olefin sodium sulphonate, and nitrogen resulted in the incremental production of 207 kB of oil in 2 years. Four million pounds of AOS were injected.

Results from a surfactant/steam-injection field test in the Guadalupe heavy oil field in California have been reported by Mohammadi and McCallum in California. The test involved the addition of alkyl toluene sulphonate and nitrogen to the steam. Positive results were obtained, although the test was stopped abruptly because of the unavailability of steam. An incremental production of 29,400 B of oil was reported as the result of the injection of 257,000 lb of active ATS.

The addition of *thin film spreading agents* (TFSA) to the steam is another related approach in which there is interest. These materials are made by treating phenol with formaldehyde and then reacting the resulting polyols with ethylene oxide or propylene oxide.

Products of this type are frequently used as demulsifiers to treat heavy crudes. In this application they are thought to work by being adsorbed at the water-oil interface and displacing the bulky asphaltene-type materials that stabilize the water in the oil emulsion. With the thinner demulsifier molecules at the interface, water droplets are thought to approach each other more closely and then to coalesce. It is thought that the effect of the TFSA in steam recovery is to promote the water wetting of the rock—i.e., to detach oil from oil-wet portions of the surface.

Blair, Scribner, and Stout (1982) describe tests in California in which indications of significantly improved performance were obtained for such a chemical in cyclic steam stimulation operations. Further results (Stout, Blair, and Scribner 1983) have shown that the effects of the TFSA appear to persist into subsequent cycles even though addition is stopped.

STEAMFLOODING MECHANISMS

Reduction of Oil Viscosity

The main physical effect of steam that promotes the recovery of heavy oil is the reversible reduction in viscosity that results from increasing the temperature. This reduction in viscosity is very dramatic; with oil sand bitumen, it is almost of the nature of the melting of a solid to form a fluid liquid.

Figure 1.9 shows typical viscosity-temperature relationships for a variety of bitumens and heavy oils and also for lighter oils. Figure 4.7 shows the effect of temperature on the ratio of the viscosity of various oils to that of water. The reduction in the viscosity of the oil makes it easier to push the oil at appreciable rates with the pressure gradients available. There are also other effects that promote the mobility of the oil.

The first of these effects is due to the improvement in the ratio of the viscosity of the oil to that of the water. This makes water percolation able to drag oil at a

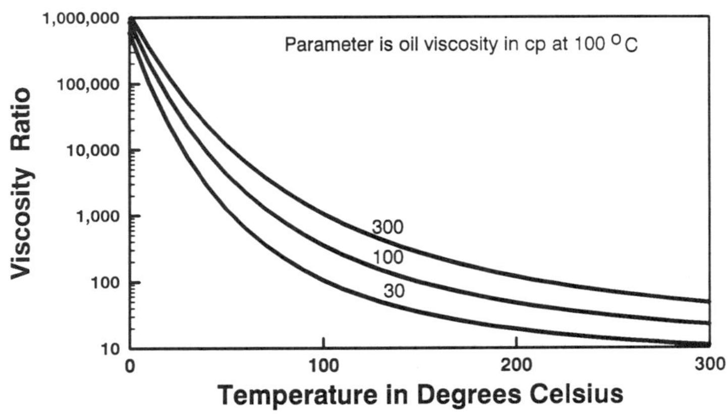

Figure 4.7 The Effect of Temperature on the Ratio of Oil Viscosity to Water Viscosity

faster rate to the production well, which results in more effective depletion in the waterflooded region for a given volume of water (condensate).

Even after heating, water still fingers through the oil because of the adverse mobility ratio. However, as will be discussed in the next chapter, there is much less tendency for steam to do so. It seems likely that in most steamflood circumstances, the steam condensation front advances in a stable manner (i.e., without fingers) except when it is moving upward. While the condensation front advances in a stable manner, the condensate drains through the oil to the production well, often in fingers.

This mechanism removes the relatively large volumes of condensate and thus allows further condensation of the steam. Often the condensate that must be removed is larger in volume than the volume of the oil produced. In order to heat a volume of reservoir to steam temperature, more than one pore volume of steam (measured as water) is required.

The calculated quantity of steam required to raise a high-quality reservoir from a reservoir temperature of 10°C to the steam temperature is given in Table 4.7. The quantity of steam is expressed as the volume of steam required to heat that volume of reservoir which contains a unit volume of oil. The calculated quantities in the table represent the heat required solely to raise the reservoir and its contents to steam temperature. It is necessary, in addition, to provide steam to supply the heat losses. In relatively efficient situations, this will have the effect of approximately doubling the steam requirements shown.

Changes in Relative Permeability

Another phenomenon that plays a role in increasing the effectiveness of steamfloods is that the relative permeability curves change with temperature. Although there is not a consensus on this, experimenters have generally found that raising the temperature changes relative permeability curves. The main effect is that the relative permeability for oil flow tends to be increased, and the residual oil saturation is reduced. Another factor is that the relative permeability for liquid water flow seems to be lower under steamflooding conditions than it is with ordinary oils having

TABLE 4.7 Quantity of Steam Required to Raise a High-Quality Reservoir to Steam Temperature

Basis: Porosity	32%
Oil Saturation	80%
Reservoir Temperature	10°C
Steam Quality	70%

Steam Temperature (°C)	Ratio of Steam to Oil (m^3/m^3)[1]
100	0.52
150	0.86
200	1.27
250	1.81

[1]Or in B/B.

viscosities at room temperature similar to that of the heavy oil at steamflood conditions.

A possible partial explanation for these effects is that water has a tendency to form water-oil emulsions, within the reservoir, with bituminous oils under steaming conditions. This can explain the lower residual oil, since the residual oil droplets are "diluted" with micron-size droplets of water. In a way, a steamflood can be visualized as being partially miscible. Another reason for a lower residual oil saturation which is applicable when there is steam saturation is the steam distillation effect; this is considered later. Emulsification also has the effect of reducing the apparent water-relative permeability because some of the water is tied up with the slow moving oil phase.

If in situ emulsification does play a role in the displacement of heavy oils, then it seems likely that the conditions of the experiment—such as those involved in the preconditioning of the core or sand pack—as well as the measured saturations, will play an important role. For example, changes that affect the wettability of the core, the preflooding conditions, and whether steam has contacted the oil may be expected to have important influences. Experiments (Chung and Butler 1988, Jamaluddin and Butler 1988) have shown that water in oil emulsification is promoted by the direct condensation of steam on colder bitumen and also by an oil-wetted reservoir matrix. There is less emulsion production, if any, when oil and water flow together as condensed phases. The effect of emulsification upon the relative permeability of the oil and water is thus intertwined with the conditions in the steam-saturated regions of the reservoir, particularly at the condensation interface. Although oil and water flowing together probably do not emulsify, water in oil emulsion formed at the condensation interface can be pushed ahead of the steam chamber and then flow in the absence of steam.

Results from some published studies of the effect of temperature on relative permeability are given in Figures 4.8 and 4.9.

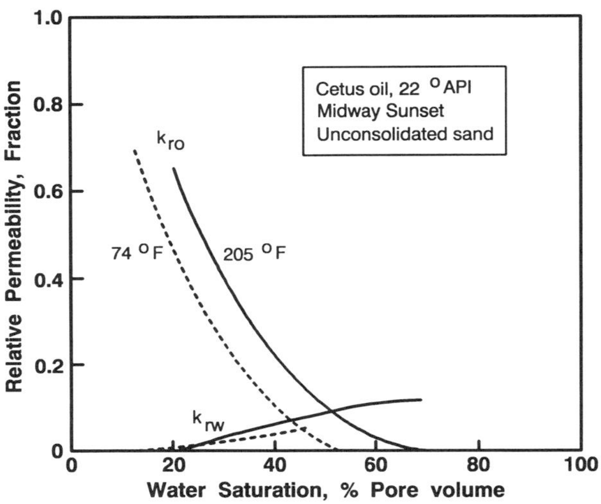

Figure 4.8 The Effect of Temperature on Relative Permeability (Data of Montgomery reported in Wu 1977)

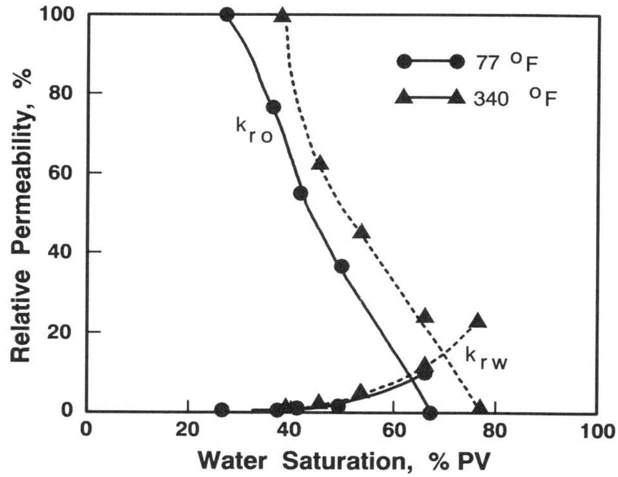

Figure 4.9 Relative Permeability Curves for Berea Sandstone Core (from Lo and Mungan 1973)

A paper by Bennion, Moore, and Thomas (1983) indicates that vastly different relative permeability curves are obtained if heavy oil cores are extracted with a solvent and then restored than if they are preserved with the original reservoir fluid in them. It appears that the relative permeabilities of cores containing heavy oils are very dependent upon the state of wetting of the porous solid. This lead should be followed further.[3]

The effect of steam treatment and temperature on relative permeabilities is complicated and not understood. Overall, however, it appears that steaming does effect an improvement.

MYHILL AND STEGEMEIER'S APPROACH TO STEAMFLOODING

The paper by Myhill and Stegemeier (1978) should be read by all workers in this field. It uses the heat conduction and heat convection ideas of the previous chapter to provide an estimate of the efficiency of a steamflood.

The basic idea used by Myhill and Stegemeier involves the calculation of the size of the steam zone from a simple energy balance using the Marx-Langenheim (1959) approach modified by the ideas of Mandl and Volek (1969) in order to include the possibility of all the steam being condensed before it reaches the heat front. These methods were discussed in the previous chapter.

The objective is to calculate the volume of the steam zone from the amount of heat injected into the reservoir, the heat needed to raise a unit volume of steam zone to the saturation temperature, and the heat lost to the overburden and underburden.

The method is simple to use, is rapid, and gives a useful idea of the practicality of using a steam flood in a particular situation.

[3]See also the discussion of the work by M. Kwan (1988) in Chapter 1, page 18.

Summary of Myhill and Stegemeier's Assumptions

The basic assumptions for the calculation are as follows:

1. The reservoir contains a uniform amount of oil per unit bulk volume as defined by the product of porosity, net to gross thickness, and oil saturation in the net pay. Gross thickness and area per injector are also constant throughout the reservoir.
2. Thermal properties, including initial formation temperature, heat capacity of reservoir rock, and heat capacity and conductivity of cap and base rock are assumed constant throughout the zone.
3. Steam is injected at a constant pressure, quality, and rate per injector.
4. Vertical temperature gradients in the reservoir are zero.
5. Heat losses from the steam zone are by conduction only and occur normal to the reservoir into the cap and base rock. Heat is transferred in the reservoir by convection only, and heat passes through the condensation front only after Mandl and Volek's critical time.
6. The quantity of residual oil remaining in the steamed chamber can be represented by an average, assumed residual oil saturation.

Outline of Method

The heart of the method is Figure 4.10. It allows the thermal efficiency of the heating to be obtained from a knowledge of the variables in the dimensionless time number and the steam-condition parameter, which is called f_{hv}.

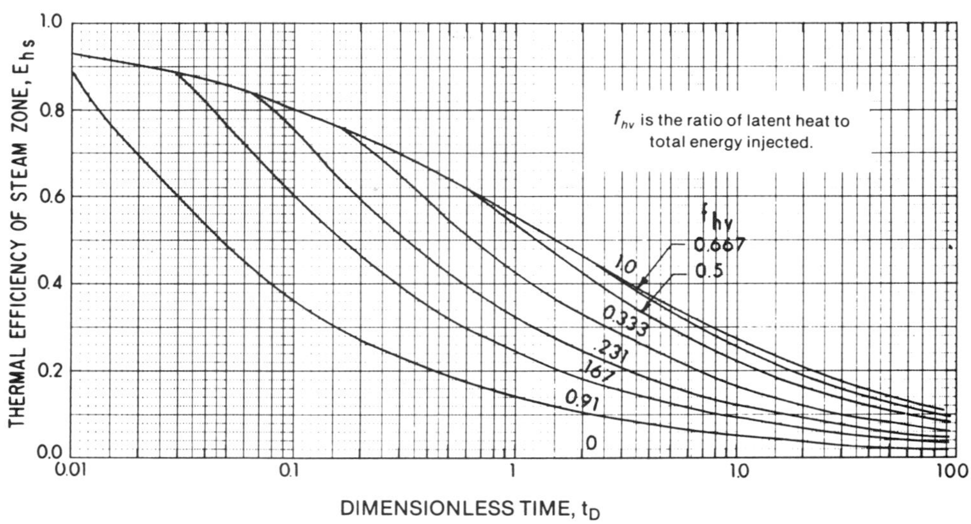

Figure 4.10 Fraction of Heat Injected in Steamflood that Remains in Steam Zone (from Prats 1982)

This parameter is the ratio of injected latent heat to injected total heat:

$$f_{hv} = \frac{f_s\lambda}{\mathbf{H}_s - \mathbf{H}_{wr}} = \frac{H_\lambda}{H_0}$$

where f_s is the injected steam quality measured at the bottom of the injection well
λ is the latent heat of evaporation of water
$\mathbf{H}_s$ is the enthalpy of the injected steam
$\mathbf{H}_{wr}$ is the enthalpy of liquid water at reservoir temperature

Once the thermal efficiency is known, the volume of the steam chamber can be calculated for the injection of a given amount of steam and a knowledge of the heat capacity of a unit volume of the chamber.

$$\text{Heat in steam chamber} = H_0 t E_{hs} = V_C(\rho C)_C(T_S - T_R)$$

or

$$V_C = \frac{H_0 t E_{hs}}{(\rho C)_C(T_S - T_R)} \tag{4.3}$$

where V_C is the volume of the steam chamber
H_0 is the average heat injection rate
$(\rho C)_C$ is the volumetric heat capacity of the steam chamber after the oil has been displaced

Myhill and Stegemeier relate the volume of the steam chamber to the volume of oil displaced from the steam zone. To do this, they assume a value for the residual oil saturation within the steam zone:

$$q_c = V_C\phi(S_o - S_{or})$$
$$= \frac{H_0\phi(S_o - S_{or})E_{hs}t}{(\rho C)_C(T_S - T_R)} \tag{4.4}$$

where q_c is the cumulative volume of oil displaced
ϕ is the porosity
S_o is the initial oil saturation
S_{or} is the residual oil saturation

The rate of oil displacement at time t is obtained from equation 3.42, which may be rewritten for times before t_c as

$$q = \frac{H_0\phi(S_o - S_{or})}{(\rho C)_C(T_S - T_R)} e^{t_D} \operatorname{erfc}(\sqrt{t_D}) \tag{4.5}$$

The value of the function of t_D may be obtained from Table 3.1.

Limitations

This approach neglects the oil removed ahead of the steam zone by the flowing aqueous condensate. This amount is often quite small, but it can become significant, particularly where significant heat is carried past the condensation front, for injection times greater than the Mandl-Volek critical time.

Unless an allowance is made for it in choosing the value of S_{or}, the approach also neglects the small oil bank (see Figure 4.3 and Chapter 5) that builds up behind the condensate front. The oil saturation in the steam zone tends to be reduced further by the action of the flowing steam behind the front. The effect is due both to the sweeping action of the steam in moving the oil and also to steam distillation. The latter mechanism removes the lighter fraction of the oil selectively, leaving behind a reduced saturation of oil which is heavier than the original crude.

Figure 4.11 (Farouq Ali 1982) shows experimental values for the residual oil saturation taken from a number of experiments and literature data. The mean value appears to lie in the range 10 to 15%.[4] There is a trend for lower residual oil saturation to be obtained with lower initial oil viscosities and with higher steam temperatures (pressures). The data are scattered, probably because of the variation of other factors such as the properties of the reservoir matrix.

Myhill and Stegemeier assume that the volume of the oil displaced is also equal to that produced. This is a weak part of their method, particularly if an attempt is made to predict the oil production during the early part of the flood. Also, oil may be displaced elsewhere than to the production well, particularly in unconfined or only partially confined pilots. It may also be left behind in the chamber as bypassed oil.

The strict application of the Myhill-Stegemeier approach would predict the highest rate of production (for a constant steam injection rate) at the start of the

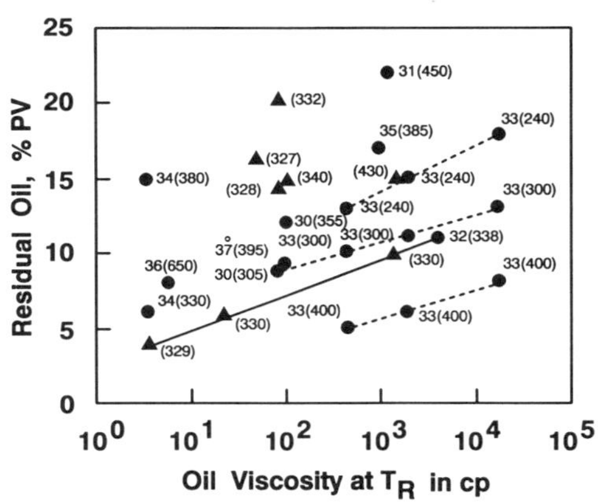

Figure 4.11 Steamflood Residual Oil as a Function of Temperature and Oil Viscosity (from Farouq Ali 1982). Some of the data (the solid circles) in this figure are from literature references and some (the open circles) from work reported for the first time in Farouq Ali (1982). The numbers in brackets are the steam temperatures in degrees Fahrenheit. The numbers without brackets are literature references as follows:

31 Blevins et al. (1969)
32 Bursell G. G. (1979)
33 Bursell, G. G. and Pitmann, G. M. (1975)
34 Ozen, A. S. and Farouq Ali, S. M. (1969)
35 Valleroy, V.V. et al. (1967)

[4]The tendency of the steam to override introduces a difficulty in applying the Myhill and Stegemeier theory, particularly in thick reservoirs. At the point of steam breakthrough, the average steam zone thickness is less than the height of the reservoir. After breakthrough, there is a tendency for heated oil to be bypassed because of insufficient pressure gradient to move it to the production well. In this circumstance, the average residual oil saturation within the heated region is higher than that found in one-dimensional steam displacement.

flood since the predicted thermal efficiency is highest then. The rate would then be predicted to fall with time due to the increasing proportion of the heat needed to supply the losses above and below the growing chamber.

Another weakness in the Myhill-Stegemeier approach is that the theory does not predict what the experimental conditions will be. For example, one has to specify in the calculation both the steam pressure and the injection rate, whereas, in practice, the injection pressure is dependent upon the rate. In many cases, it may be impossible to inject steam at the desired rate without fracturing the reservoir. Often the economics will depend upon the rate at which the process can be conducted.

The Myhill-Stegemeier method leads to the conclusion that high injection rates are most efficient because they allow production with less heat loss to the overburden and underburden. However there are practical limitations to the use of high rates. Nevertheless, the method does rationalize the results found toward the end of steamfloods when most of the displaced oil has been recovered; for such a simple approach, the agreement between the predictions and the results make it useful. It is also useful for prediction if some experimental data are available for the rate of injection that may be achieved.

Comparisons of Theoretical Predictions with Data

Figure 4.12 shows the oil-to-steam ratios predicted by Myhill and Stegemeier for a number of scaled laboratory steamfloods compared with the experimental values. Each of these points represents conditions well on into the flood, and the agreement is generally good.

Figure 4.13 shows a similar comparison for field steam drives. For many of the fields, two experimental points are shown. The lower circles correspond to the additional production ascribed to the use of steam, whereas the upper triangles represent the total production; i.e., they include the production that would have been obtained without steam.

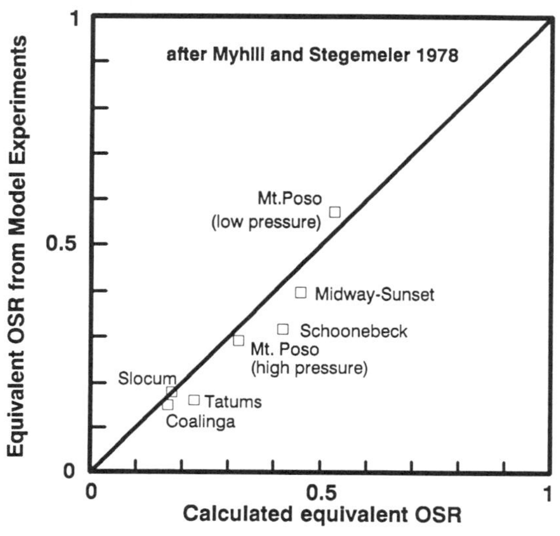

Figure 4.12 Comparison of Experimental Model Results with Calculated Values

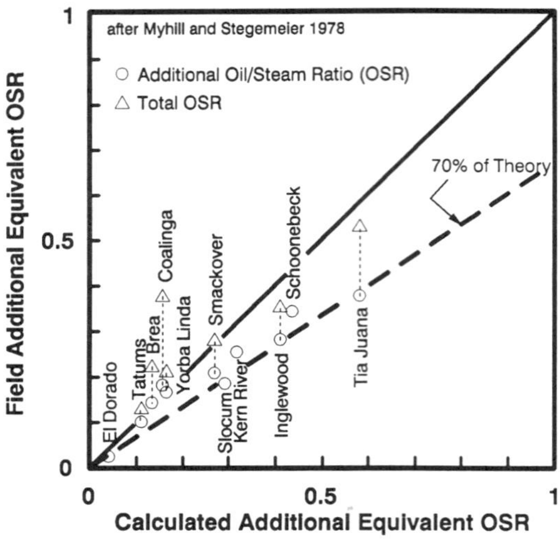

Figure 4.13 Comparison of Field Steam-Drive Results with Calculations

In general, the experimental field project data in Figure 4.13 tend to fall below the solid theoretical line and lie mostly in the range of 70 to 100% of the theoretical. The broken line represents 70% of the theoretical prediction. Myhill and Stegemeier point out that there are several reasons why field data might be expected to be below the theoretical, including the fact that much of the field data come from patterns that are unconfined. In such patterns, some of the mobilized oil may be driven outside of the pattern. Another reason is that steam override may result in the average thickness of the steam zone being less than the reservoir thickness.

As was shown in Chapter 3, the following equation predicts almost the same OSR as does the more complicated Marx-Langenheim expression.

$$\text{OSR} = \text{OSR}_{\max}\left\{\frac{1}{1 + \frac{8}{3}\sqrt{\frac{\alpha t}{\pi h^2}}}\right\} \tag{3.49}$$

$$\cong \frac{1769\phi\,\Delta S_o}{(T_S - T_R)(1 + 1.43\sqrt{t/h^2})} \tag{3.50}$$

T_S and T_R in °F, t in d, and h in ft.

It predicts higher oil-steam ratios for the following conditions:

- Higher values of ΔS_o—i.e., higher S_o or lower S_{or}
- Higher porosity
- More rapid recovery, lower t
- Thicker reservoirs, high h

The lower oil-steam ratios found in practice as compared to those that may be expected from equation 3.49 result from the mechanism being different from that postulated.

Override of the steam results in undisplaced oil remaining in the reservoir. Another way of looking at this is to say that h (in practice) is less than the reservoir thickness. The heat that has penetrated below the steam chamber is equivalent to a loss, even though it results in heating the oil below. Even if much of this lower oil is later heated to the steam temperature, it tends to stay within the steam chamber because it has been bypassed by the advancing front and little pressure gradient is available to move it.

Another way of looking at the problem of bypassed oil is to say that the average residual oil saturation in the steam-heated region is greater than that corresponding to the value for a one-dimensional steamflood. From this point of view, one can consider the reservoir height to be the appropriate value for h, but a much lower value of ΔS_o is required to allow for the bypassed oil.

Ten-Pattern Steamflood

As an example of this idea, we will consider the Chevron "Ten-Pattern Steamflood" of the Kern River Field in California, which has been discussed by Blevin and Billingsley (1975) and by Oglesby et al. (1982). The project consisted of a steamflood in a pattern consisting of ten contiguous inverted seven-spots. The area per seven-spot averaged 6.1 acres to give an average spacing of 320 ft between injectors and producers and also between adjacent producers. Characteristics of the reservoir are given in the following table.

Reservoir Characteristics: Ten-Pattern Flood, Kern River

Depth	700–797 ft
Oil gravity	14°API
Net sand thickness	97 ft
T_R	90°F
T_S	Approx. 310°F
S_o	0.52 (after primary production)
ϕ	0.34
k	4000 mD

(Oglesby et al. 1982)

The reservoir was steamed for 7 y; 18.58×10^6 B of steam were injected to produce 3.02×10^6 B of oil (i.e., OSR = 0.16; SOR = 6.15). Performance data are shown in Figure 4.14.

The oil in place in the reservoir initially is given by

$$\phi S_o A h = 0.34 \times 0.52 \times 61 \times 43560 \times 97$$

$$= 45.6 \times 10^6 \text{ ft}^3 \quad \text{or} \quad 8.1 \times 10^6 \text{ B}$$

The recovery was thus 37% of the oil at the end of the steamflood, and the average oil saturation remaining was $0.52 \times 0.63 = 0.328$. This oil was made up of oil remaining within the steam-swept zone and of bypassed oil, such as that beneath the steam-swept zone.

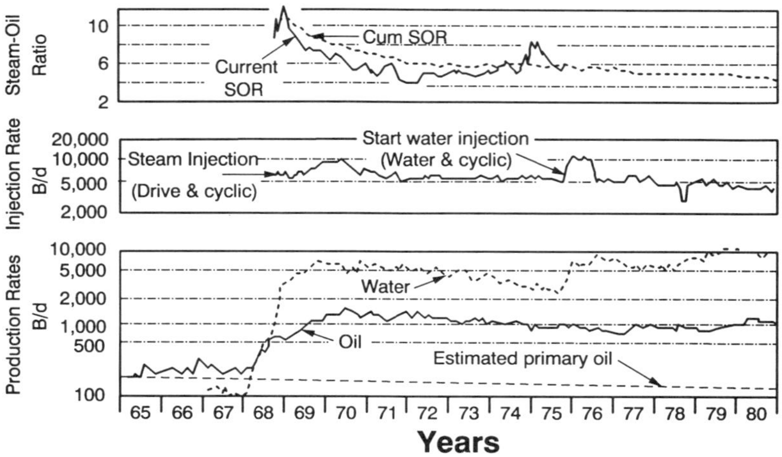

Figure 4.14 Performance of Ten-Pattern Steamflood (from Oglesby et al. 1982)

Calculating the expected OSR using equation 3.50 and a residual oil saturation of 0.328 leads to

$$\text{OSR} = \frac{1769 \times 0.34 \times (0.52 - 0.328)}{(310 - 90)(1 + 1.43\sqrt{7 \times 365/97^2}}$$

$$= 0.30$$

This value is much higher than the value of 0.16 found in the field and, of course, very much higher than would be found if a lower residual oil fraction had been substituted in the equation. Part of the reason for the high prediction is that some of the injected heat bypassed directly to the production wells. It was estimated by Blevins and Billingsley that this would reach 18% of the injected heat.[5] If allowance is made for this bypassed heat, then the expected OSR would be $0.82 \times 0.30 = 0.246$. This is still significantly higher than the value of 0.16 observed in the field.

It is possible that the injected steam may have had a lower quality than was assumed in deriving equation 3.50; steam quality data are not available in the published information. Another similar factor is that no allowance is made for heat losses in the well bore in the preceding calculation. However, it is unlikely that these factors will account for the whole discrepancy.

Another possibility is that the spreading of the heated zone across the patterns may have been much more rapid and that the heat losses are underestimated. Equation 3.52 is similar to equation 3.50 but is based on the assumption that the steam zone spreads immediately across the flooded area. Using equation 3.52 instead of 3.50 for the preceding example leads to a calculated oil-steam ratio of 0.225 or 0.184 if allowance is made for the bypassed heat. This is much closer to the observed rates.

[5]It will be noted from Figure 4.14 that the steam-injection rate was lowered from about 10,000 to 6000 B/d during the period 1970 to 1975 in order to conserve steam after breakthrough.

It is clear from the preceding calculations that considerable hot oil was bypassed in this steamflood. This has been recognized in the Chevron Kern River project, and much of the remaining heated oil has been recovered by the waterflood that has been operated since 1975. During this waterflood, cyclic steam stimulation of the production wells has been used. As will be seen from Figure 4.14, this operation has been very successful, and the cumulative SOR has fallen from 6 to almost 4 by the end of 1980. During this period, the oil production rate remained at about 1000 B/d (about 50 B/d per production well). It is estimated that 78% of the OOIP will have been recovered by the end of the flood.

Ten-Pattern Steamflood—Oil Recovery

	% Recovery OOIP
Primary Production	10
Steamflood	34
Waterflood	34 (20 by end of 1980)
	78

San Ardo Steamflood and Infill Drilling

Another large, successful California steamflood in which a high recovery is being achieved is the Texaco project in the San Ardo field (Traverse, Deibert, and Sustek 1983). Most of the steamflood recovery has been from the Aurignac zone. This has characteristics similar to the Kern River field. Although it is much deeper, steam can be injected with a bottom hole pressure of only 125 psig at 1300 B/d per well.

Properties of Aurignac Zone—San Ardo

Area	1755 acres
h	97 ft
Depth	2300 ft
ϕ	0.349
k	1000–3000 mD
T_R	100°F
Oil gravity	13°API

(Traverse et al. 1982)

The field has been developed using repeated, inverted nine-spot patterns with a pattern area of 20 acres. It has been concluded that a 50% recovery is achievable with this spacing. In order to achieve a higher recovery and to maintain the production rate, an infill drilling has been initiated. Four infill wells are added to each pattern, as shown in Figure 4.15.

These infill wells have the objective of removing the oil from below the steam zone, as shown in the cross-sectional drawing at the right of Figure 4.15.

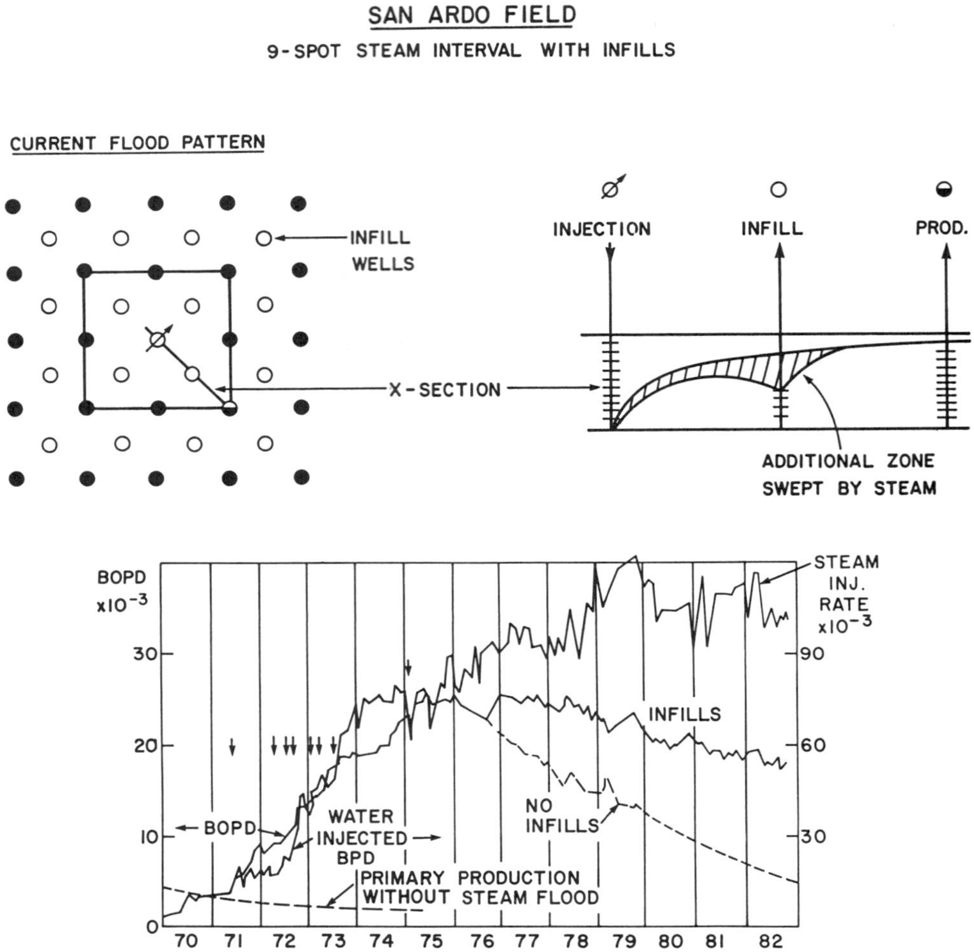

Figure 4.15 Addition of Infill Wells to San Ardo 9-Spot Pattern (from Traverse et al. 1983)

In a nine-spot pattern there are three producers per injection well. In the infilled pattern shown in Figure 4.15, there are seven producers per injection well. Texaco plans to reduce this ratio and to promote recovery of additional oil by the conversion of the corner wells of the original nine-spot pattern from producers to injectors. This idea is shown in Figure 4.16; it has been called *cross-flooding* by Texaco. Also shown in this figure is the concept of how this conversion will recover additional oil from the bank which has accumulated around these producers. The conversion of the corner wells will result in two injectors per original 20-acre, nine-spot pattern and six producers, or a ratio of three producers per injector. This converts the pattern to a repeated 10-acre inverted nine-spot pattern. Texaco estimates that the recovery from their project will increase from 50% for the original pattern to 60% for the pattern with infill drilling. An important economic consideration is that the production rate is maintained.

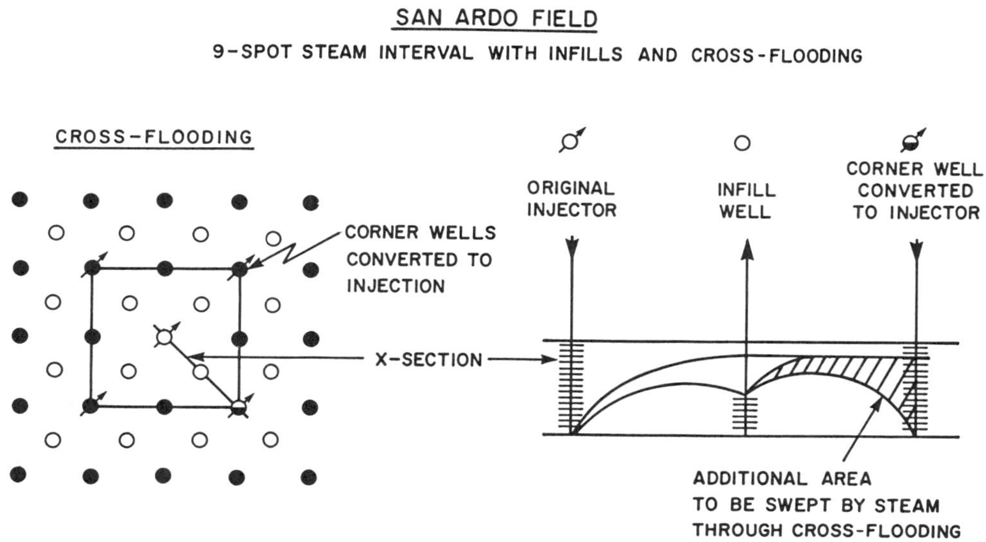

Figure 4.16 Cross-Flooding at San Ardo Pattern (from Traverse et al. 1983)

With further operational changes such as waterflooding after steamflooding, the use of foam additives, and the selective recompletion of wells, Texaco anticipates that a recovery of 79% will be achievable at San Ardo.

COMPARISON OF STEAMFLOOD AND STEAM SOAK

An interesting part of Myhill and Stegemeier's paper is concerned with the effect of switching from steam stimulation to a steamflood.

Figure 4.17 shows the injection rates, from a scaled model, for a steam soak on 2.5-acre spacing that was converted to a steam drive after 4.5 years compared to

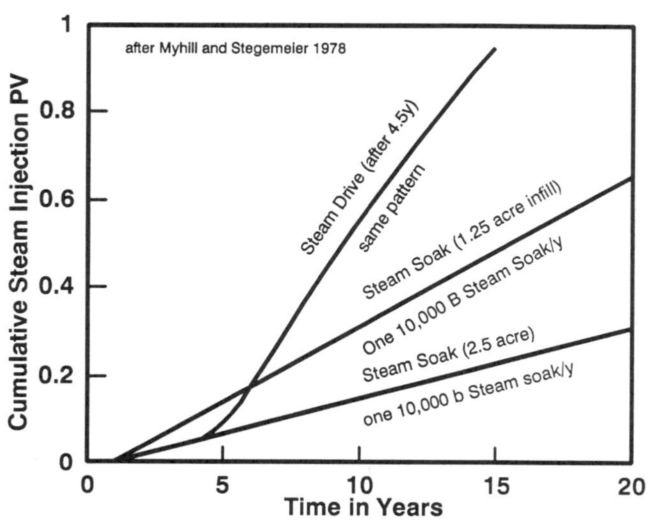

Figure 4.17 Cumulative Steam Injection-Midway-Sunset Model Experiments

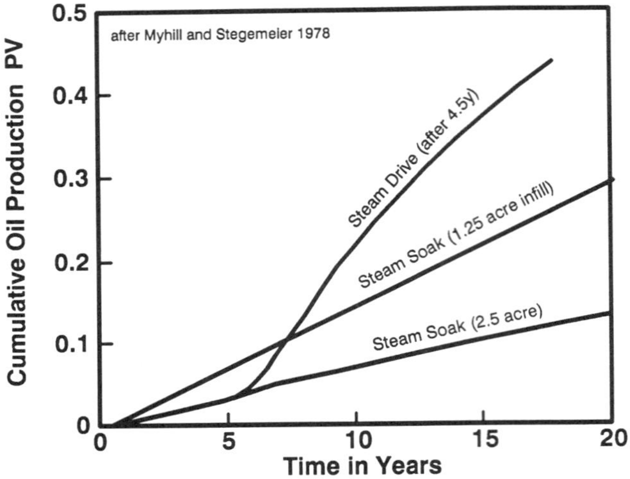

Figure 4.18 Cumulative Oil Production in Midway-Sunset Model Experiments

those for steam soaking and to those for a soak project with closer spacing. It was possible to inject more steam with the flood than with the steam soak even with infilling.

Figure 4.18 shows a comparison of the cumulative oil production for the same experiments; the parallel with the injection curves is very striking. Higher injection rates give higher production rates. The convergence of the cumulative oil-to-steam ratio curves shown in Figure 4.19 is also very interesting. One can see from this study why there has been a general trend to switch from steam stimulation to flooding in California as a field matures. One may presume that the same tendency will develop in the Canadian bitumen fields as the projects mature, as virgin high-quality tar sand reservoirs become more scarce, and as practical experience is obtained in the recovery of tar sand oil by flooding. It is probably desirable to extend the cyclic steam stimulation phase in Alberta because of the generally higher initial oil viscosity. Also, the use of vertical steamflooding with horizontal wells (steam-assisted gravity drainage; see Chapter 7) will probably prove to be a superior alternative to conventional horizontal steamflooding in many projects.

STEAMFLOODING MULTILAYER RESERVOIRS

In some cases, multiple reservoirs separated by impermeable barriers may be steamflooded sequentially. In these cases, some of the heat lost during the flooding of one layer may be present in the layer above (or below) when it is flooded. A study by Restine (1983) for two such operations in Getty's Kern River Field shows, as might be expected, considerable improvements in the oil-steam ratio and higher production rates for the preheated oil sand. This effect affords greater economy for the production of oil from stacked reservoirs than from single ones.

One wonders, for example, whether the extensive steaming of the Clearwater sands in the Cold Lake field will result in more economic production from the higher Grand Rapids formation, perhaps using the same wells that were drilled to exploit the Clearwater formation.

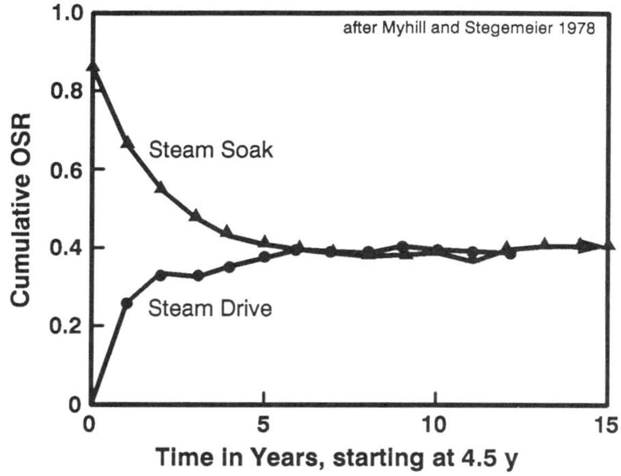

Figure 4.19 Cumulative Oil-Steam Ratio as a Function of Time from Start of Steam-Drive for Midway-Sunset Experiments

JONES' STEAM DRIVE MODEL

There has been progress in modifying the Myhill-Stegemeier approach so as to include factors which were ignored in the original treatment.

Jones (1981) describes an empirical approach that is simple and realistic. Figure 4.20 from his paper compares the reported recovery from the Kern River field with that predicted using the Myhill-Stegemeier theory. A curve for Jones' model is also shown.

The Myhill-Stegemeier method gives unrealistically high production rates early in the flood, reasonable ones in the middle, and as the flood reaches its end, rates which are several times too high. Jones considers that there are three major periods in a steamflood and that the dominant factors change as the process moves from one stage to the next.

Stage 1

During this first stage the dominant factor is the very high viscosity of the cold oil and, in some cases, the need to build an oil bank—i.e., to fill gas saturation with

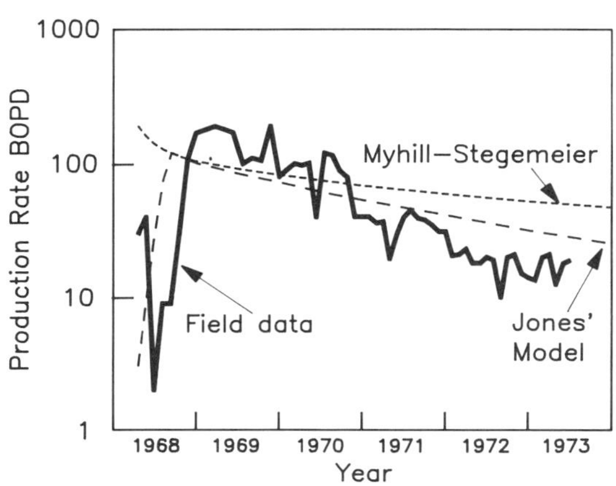

Figure 4.20 Comparison for Kern River, California, Steamflood Field Data with Theoretical Predictions (after Jones 1981)

oil. During this period, water channels through the oil, and there is little production until warm oil can approach the production well. It is during this period that steam stimulation of the producer(s) can be particularly valuable and with heavy bitumens, almost essential. (See later discussion of tar sand flooding.)

Stage 2

In the second stage hot oil is moved to the production well relatively easily, and the production rate is about equal to the rate of growth of the steam chamber. The Myhill-Stegemeier assumptions are reasonably valid. The peak production occurs early in this stage.

Stage 3

The Myhill-Stegemeier theory would allow the second stage to continue indefinitely, with the production rate dropping asymptotically to zero as the area for vertical heat losses gradually increased. In practice the drainage area is finite, and the production rate becomes limited because of the depletion of the reservoir. No allowance is made for depletion in the Myhill-Stegemeier theory.

Jones' Empirical Adjustment Factors

Jones allows for the effects just described by multiplying the production rates predicted from the Myhill-Stegemeier theory by three empirical factors: V_{PD}, A_{CD}, and V_{OD}.

$$q = V_{PD} A_{CD} V_{OD} \frac{H_0 \phi (S_o - S_{or})}{(\rho C)_C (T_S - T_R)} e^{t_D} \operatorname{erfc}(\sqrt{t_D}) \tag{4.6}$$

V_{PD} allows for the effect of the initial gas saturation. It is given by equation 4.7.

$$V_{PD} = \left(\frac{V_{s,\mathrm{inj}} 5.62}{43{,}560 A h_n \phi S_g} \right)^2$$

$$0 \le V_{PD} \le 1; \quad \text{or else} \quad V_{PD} = 1$$

where A is effective pattern area in acres
h_n is net zone thickness in feet
ϕ is porosity (4.7)
S_g is initial gas saturation
$V_{s,\mathrm{inj}}$ is injected steam in barrels

V_{PD} is equal to the square of the injected steam volume, measured as liquid water, divided by the initial volume of free gas in the reservoir. When this ratio becomes unity, then V_{PD} is forced to be 1. A_{CD} allows for the effect of the initial oil viscosity. It is calculated from equation 4.8. It will be noted that the higher the value of μ_{oi}, the lower is A_{CD}. As the steam zone increases in area, the value of A_{CD} increases up to the forced limit of 1. For an initial oil viscosity of 10^6 cp, the square-root term is just equal to unity.

$$A_{CD} = \left(\frac{A_s}{A\sqrt{0.11 \ln \frac{\mu_{oi}}{100}}} \right)^2 \tag{4.8}$$

$$0 \le A_{CD} \le 1; \quad \text{or else} \quad A_{CD} = 1$$

where A_s is steam zone area in acres
A is effective pattern area in acres
μ_{oi} is initial reservoir oil viscosity in cp

Equation 4.8 predicts that for an initial viscosity of 10^6 cp, the viscosity effect does not vanish until the steam chamber occupies the entire pattern area. For lower viscosities, the influence is predicted to disappear part of the way through the flood. A viscosity of 10^6 cp is about equal to the value found in the Athabasca tar sands; this is a very high value.

A conceptual problem that arises with Jones' A_{CD} factor is that of the overall volumetric balance in the reservoir. It is not clear where the displaced oil from the steam chamber has gone, if it has not been pushed to the production well(s). One must imagine the oil as still remaining within the steam chamber—i.e., as being bypassed—until the oil beyond becomes sufficiently fluid to allow it to flow out.[6]

V_{OD} is the factor that allows for the depletion effect. It is given by equation 4.9.

$$V_{OD} = \sqrt{1 - \frac{N_P}{N}\frac{S_{oi}}{\Delta S_o}} \tag{4.9}$$

$$0 \le V_{OD} \le 1; \quad \text{or else} \quad V_{OD} = 1$$

where N_P is cumulative oil production in barrels
N is equal to OOIP $= \dfrac{43{,}560 A h_n \phi S_{oi}}{5.62}$ B
S_{oi} is initial oil saturation
ΔS_o is equal to $S_{oi} - S_{or}$

Equation 4.9 appeared with N_D in place of N_P in the original JPT publication, but this was changed in the version of the paper which was bound in the SPE transactions.

The effect of these empirical factors on the predicted oil production rate for a particular example is shown by the curves of Figure 4.21, which is taken from Jones' paper. This figure also shows the value of the three individual correction factors, which were calculated as just described.

Jones' method is a very practical one; in his paper, Jones shows how it may be applied to a wide range of field examples with considerable success, although judg-

[6]An extreme example of this is where the reservoir fractures as a result of the high steam injection pressure. In this case, condensate followed by steam passes down the fracture. The reservoir is heated either side of the fracture but the heated oil remains because of the low driving force which is available.

Under less extreme conditions, steam fingers may grow into the cold reservoir which is saturated with low-mobility, viscous oil. In either case heated oil is bypassed rather than produced.

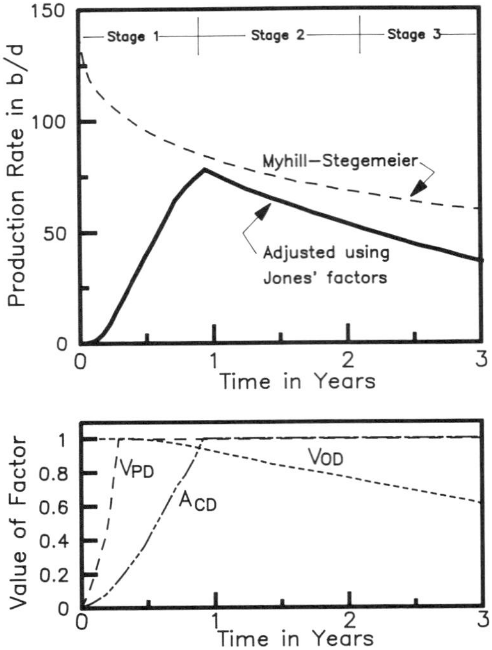

Figure 4.21 Prediction of Recovery rate using Jones' Method (after Jones 1981)

ment is required in the interpretation of the field conditions. The method may be used easily with programmable calculators or personal computers.

INJECTIVITY

The rate at which steam can be injected into the reservoir is not predicted by Myhill and Stegemeier's theory nor by its modifications. Nevertheless, it is of prime importance. The economics of a steamflooding project are largely dependent upon the rates at which steam is injected and oil is recovered. Low injection rates imply slow production rates and low cash flows. Also, as was discussed in Chapter 3 and in the earlier parts of this chapter, the terminal efficiency of steamflooding is determined largely by the rate at which it can be conducted. Slow injection rates result in a large proportion of the injected heat being lost because of the longer times required.

Much of the value of precommercial field-pilot experimentation lies in the determination of the practical steam-injection rates that are achievable. Once these are known, the Myhill-Stegemeier approach and its modifications can yield reasonable estimates of the performance which is to be expected.

In this section, the rate at which injection can be achieved in reservoirs is approached by considering the steady-state flow between injection and production wells, for various geometries, assuming that the injected fluid has the same properties as the displaced oil. At the start of the process this is a reasonable representation, provided that the oil is the only mobile phase. Assuming that the injected fluid is more mobile than the displaced one, one would expect that the injectivity would improve as the displacement proceeded. Advantage can be taken of this improved injectivity by increasing the injection rate for a given injection pressure or by allowing the pressure to fall for a given injection rate.

When steam is injected, the volume displaced is related to the volume of injected steam and to the thermal properties of the fluids and reservoir material. The quantity of steam required to displace a unit volume of oil is determined by a heat balance; this heat balance has been considered in Chapter 3 and is discussed in much more detail in Chapter 5. For the present, it is sufficient to note that the volume of steam (measured as the equivalent volume of liquid water) injected is usually equal to several times the volume of oil displaced. Since there is usually little resistance to the flow of steam in the steam chamber or to the flow of water through the displaced oil, the injectivity for steam may be expected to be considerably larger than calculated for simple oil flow. A reasonable approximation is to multiply the calculated injectivity by the estimated SOR and by a factor of about 1.5 to allow for the effective increase of the injector well bore radius due to heating.

Steady-state Displacement Between an Isolated Pair of Vertical Wells

The radial flow of oil to an isolated vertical producer or from an isolated injector was discussed in Chapter 1, where it was shown that the pressure could be calculated using equations such as

$$P = P_o - \frac{q\mu}{2\pi kh} \ln R \tag{4.10}$$

where P is the pressure at a radius R from a well in which the injection rate is q. For a production well, q is negative.

Consider a well pair consisting of an injector and a producer whose centers are separated by a distance L. Assume that the system is operating in a steady state and that the flow of injection fluid, q, is equal to the flow of produced fluid. Two equations similar to 4.10 can be written, one for the injector and the second for the producer.

$$\text{Injector} \qquad P_i = P_{oi} - \frac{q\mu}{2\pi kh} \ln R_i \tag{4.11}$$

$$\text{Producer} \qquad P_p = P_{op} + \frac{q\mu}{2\pi kh} \ln R_p \tag{4.12}$$

The plus sign appears in equation 4.12 because the flow is toward the well. These two equations represent the reservoir pressures for the cases where the wells are operating in a steady state and individually (i.e., there is injection without production and vice versa).

Each of these equations is a solution of Laplace's equation.

$$\frac{\partial^2 P}{\partial x^2} + \frac{\partial^2 P}{\partial y^2} = 0 \tag{4.13}$$

A combined solution that represents the combined pressures for the cases where the two wells are operating together is obtained by adding the two solutions (4.11) and (4.12) to give

$$P = P_i + P_p = P_o - \frac{q\mu}{2\pi kh} \ln \frac{R_i}{R_p} \tag{4.14}$$

At the midpoint between the wells, $R_i = R_p$ and the pressure becomes P_o. At the injector well bore,[7]

$$R_i = R_{wi}, \qquad R_p \simeq L$$

and

$$P_{wi} = P_o + \frac{q\mu}{2\pi kh} \ln \frac{L}{R_{wi}} \tag{4.15}$$

At the producer well bore,

$$R_p = R_{wp}, \qquad R_i \simeq L$$

and

$$P_{wp} = P_o - \frac{q\mu}{2\pi kh} \ln \frac{L}{R_{wp}} \tag{4.16}$$

The difference in pressure between the injection and production wells is given by

$$\Delta P = P_{wi} - P_{wp} = \frac{q\mu}{2\pi kh} \ln\left(\frac{L^2}{R_{wi} R_{wp}}\right)$$

If $R_{wi} = R_{wp}$, this becomes:

$$\Delta P = P_{wi} - P_{wp} = \frac{q\mu}{\pi kh} \ln \frac{L}{R_w} \tag{4.17}$$

If $R_{wi} \neq R_{wp}$, then the geometric mean, $R_w = \sqrt{R_{wi} R_{wp}}$, can be used in (4.17).

This may be rearranged to give the normalized injectivity (in consistent units).

$$\text{Normalized injectivity} = \frac{q\mu}{kh\,\Delta P} = \frac{\pi}{\ln\left(\dfrac{L}{R_w}\right)} \tag{4.18}$$

Equations for the normalized injectivity for a variety of geometric arrangements are given by Morel-Seytoux (1966).[8] In his paper, Morel-Seytoux describes the normalized injectivity just given as the *conductivity* and uses the term *normalized injectiv-*

[7]Equation 4.14 is an exact solution of the Laplace equation for a line source and sink. It is, however, only an approximation for the flow between cylindrical wells, since the constant pressure lines are only approximately circles near the line sources and sinks. However, for practical situations where $L \gg R_w$, it is an accurate approximation.

[8]Analytical equations for the flows between wells have been discussed by a number of authors. The general scheme of superimposing the pressure distribution of a number of wells was developed by Muskat (1937) and subsequent work is based to a large extent on this pioneering effort.

ity to refer to a dimensional value in which q is measured in barrels per day, μ in centipoise, k in millidarcies, h in feet, and ΔP in pounds per square inch (these units are often termed *field* units). Morel-Seytoux normalized injectivity is thus given by

$$\frac{(q \text{ in B/d})(\mu \text{ in cp})}{(k \text{ in mD})(h \text{ in ft})(\Delta P \text{ in psi})} = 1.127 \times 10^{-3} \times \left(\frac{q\mu}{kh\,\Delta P}\right)_{\text{consistent}} \tag{4.19}$$

Equation 4.19 may be used to calculate the injectivity in barrels per day from the normalized values given here.

It can be seen from equation 4.18 that the injectivity decreases as L increases; however, the effect is not very great because of the nature of the logarithmic function. Calculated values are shown in the following table.

$\frac{L}{R_w}$	Normalized Injectivity in Consistent Units
100	0.682
200	0.593
500	0.506
1000	0.455
2000	0.413
5000	0.369

Most of the resistance to flow occurs in the immediate vicinities of the two well bores, and the resistance added by increasing L is not very great. In steamflooding, this means that once the resistance around the injection well decreases as a result of the invasion of the low-viscosity steam, then the injectivity may be expected to increase considerably. For example, suppose that steam flow has increased the effective radius of the injection well from 0.1 to 10 m. For a case where L is 100 m, then L/R_w will change from 1000 to $100/\sqrt{10 \times 0.1} = 100$ and $\ln(L/R_w)$ will change from 6.91 to 4.61; the injectivity will increase by a factor of $6.91/4.61 = 1.5$. Stimulation of the production well can also result in an effective larger well bore radius, and this too will increase the injectivity.

Time for Breakthrough

Consider the straight streamline that joins the two wells just discussed. The pressure gradient along this streamline can be obtained by setting $R_p = L - R_i$ in equation 4.14 and differentiating P with respect to R_i.

$$\left(\frac{\partial P}{\partial R_i}\right)_{x=0} = -\frac{q\mu}{2\pi kh}\left(\frac{1}{R_i} + \frac{1}{L - R_i}\right) \tag{4.20}$$

The average fluid velocity along the central streamline is given by $q/A\phi\,\Delta S_o$ and, using Darcy's equation, is

$$\frac{dS}{dt} = V_{x=0} = \frac{q}{A\phi\,\Delta S_o} = -\frac{k}{\mu\phi\,\Delta S_o}\left(\frac{\partial P}{\partial R_i}\right)_{x=0} = \frac{q}{2\pi h\phi\,\Delta S_o}\cdot\frac{L}{S(L - S)} \tag{4.21}$$

In this equation, S is the distance that a particle of the fluid moves from the injection well in time t. The time for breakthrough can be obtained by integrating equation 4.21, with the result

$$t_{BT} = \frac{\pi h L^2 \phi\,\Delta S_o}{3q} \tag{4.22}$$

Substituting for the value of q from equation 4.18 leads to the expression

$$t_{BT} = \frac{\phi\,\Delta S_o \mu L^2 \ln(L/R_w)}{3k\,\Delta P} \tag{4.23}$$

The volume of oil that is displaced during the breakthrough period is given by rearranging equation 4.22:

$$qt_{BT} = \frac{\pi h L^2 \phi\,\Delta S_o}{3} \tag{4.24}$$

This volume of oil is independent of the rate of injection. It is equal to one-third of the volume of mobile oil contained within a cylinder of reservoir of height h and radius L.

Isolated Injection Well Surrounded by a Circle of Equally Spaced Producers

The general procedure just described can be extended to the case of an injector surrounded by a circle containing equally spaced producers, as shown in the left part of Figure 4.22.

Results of this analysis are tabulated next (Morel-Seytoux 1966):

$$\frac{q\mu}{kh\,\Delta P} = \frac{2\pi N}{(N + 1)\ln\left(\frac{L}{R_w}\right) - \ln(N)} \tag{4.25}$$

$$t_{BT} = \frac{\phi\,\Delta S_o \mu L^2\left[(N + 1)\ln\left(\frac{L}{R_w}\right) - \ln(N)\right]}{2(N + 2)k\,\Delta P} \tag{4.26}$$

$$qt_{BT} = \frac{N}{N + 2}\pi h L^2 \phi\,\Delta S_o \tag{4.27}$$

In these equations, N represents the number of wells surrounding the injector. Thus, for example, for an isolated five-spot there is one injector with four producers arranged around it and N is equal to 4. The radius of the surrounding circle, L, in this case is half of the diagonal length of the square pattern. The injectivities for a number of values of N are shown in Figure 4.23.

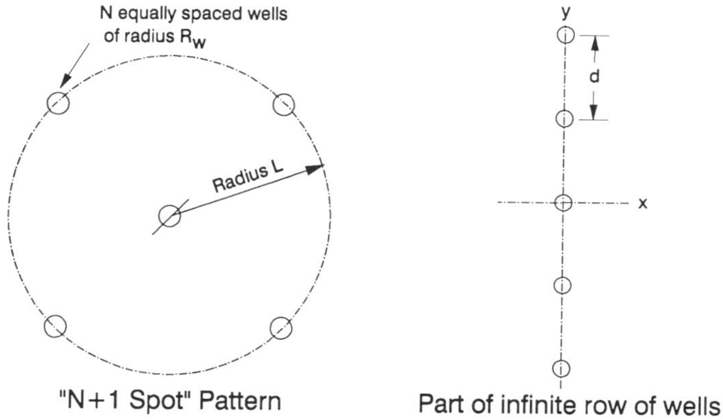

Figure 4.22 Typical Isolated Well Patterns

The dimensionless breakthrough time for $N = 1, \ldots, 6$ and for a continuous circle of production wells is shown in Figure 4.24. In this figure the dimensionless breakthrough time is defined as

$$\text{Dimensionless breakthrough time} = \frac{k\,\Delta P t_{BT}}{\phi\,\Delta S_o \mu L^2} \tag{4.28}$$

It is a function of the system geometry—in this case, of N and L/R_w.

Confined Patterns

In a repeated regular pattern, planes of symmetry between wells become no-flow boundaries. The flow in such patterns can be computed by extending the well pattern indefinitely and summing up the pressure terms for each well as one or more infinite series. In some cases the resulting answers can be surprisingly simple. For example, the pressure produced by the series of equal injection wells uniformly

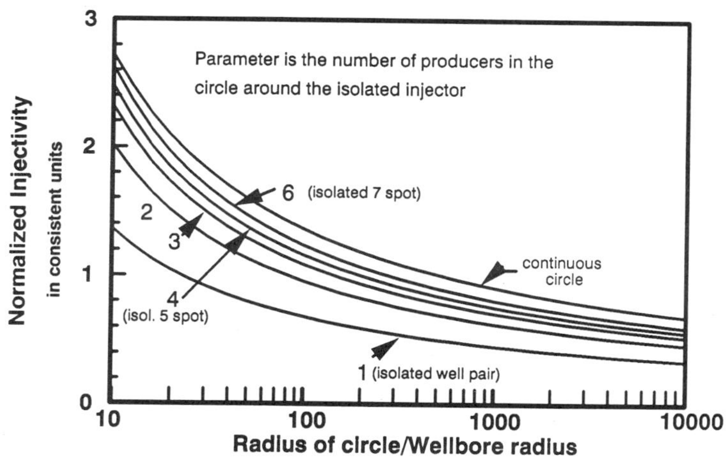

Figure 4.23 Injectivity of Isolated Well Surrounded by a Circle of Equally-Spaced Producers

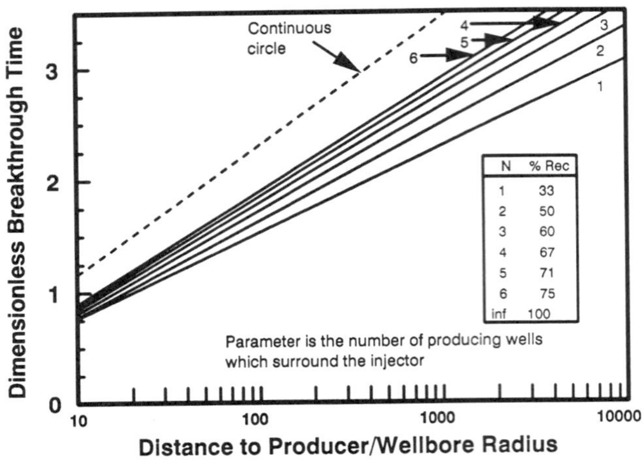

Figure 4.24 Breakthrough Time for Isolated patterns

spaced along a straight line, which is shown in the right hand part of Figure 4.22 and also in Figure 4.25, is given by (Muskat 1937)

$$P = P_o - \frac{q\mu}{4\pi k} \ln\left[\cosh\frac{2\pi x}{d} - \cos\frac{2\pi y}{d}\right] \tag{4.29}$$

In equations 4.29 and 4.30 the flow, q, is per unit length of well. Equation 4.29 can be used to predict the flows between a horizontal injector and a horizontal producer, as shown in the small diagram in Figure 4.26. To do this it is necessary to write equations for the contribution of four series of regularly spaced wells and then to combine these with the result shown in equation 4.30.

Confined Horizontal Well Pair

$$\frac{q\mu}{k\,\Delta P} = \frac{2\pi}{\ln\left\{\dfrac{\left[1-\cos\left(\dfrac{\pi(L-R_w)}{h}\right)\right]\left[1-\cos\left(\dfrac{\pi(L-R_w+2C)}{h}\right)\right]}{\left[1-\cos\left(\dfrac{\pi R_w}{h}\right)\right]\left[1-\cos\left(\dfrac{\pi(R_w+2C)}{h}\right)\right]}\right\}} \tag{4.30}$$

The dimensionless injectivity from equation 4.30 has been plotted against the logarithm of C/L in Figure 4.26 for a constant ratio of $R_w/L = 0.002$. The injectivity rises from a low value of 0.3 to the asymptote corresponding to a pair of isolated wells (i.e., the injectivity given by equation 4.18).

C is equal to R_w for the case where the two wells are immediately below and above the reservoir boundaries. In this case, if it is also assumed that $L \gg R_w$, equation 4.30 can be reduced to

$$\frac{q\mu}{k\,\Delta P} = \frac{0.5\pi}{\ln\left(\dfrac{L}{R_w}\right) - 1.0009} \tag{4.31}$$

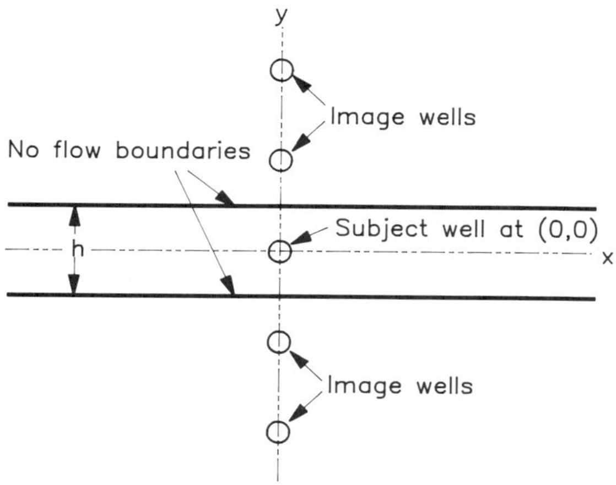

Figure 4.25 Infinite Vertical Column of Horizontal Wells to Represent Well within Horizontal Boundaries

Extending this approach to repeated inverted five-spot and seven-spot patterns of vertical wells results in the injectivities given by the following two equations.

Repeated Five-spot

$$\frac{q\mu}{kh\,\Delta P} = \frac{\pi}{\ln\left(\dfrac{L}{R_w}\right) - 0.6174} \tag{4.32}$$

L = distance from injector to producer

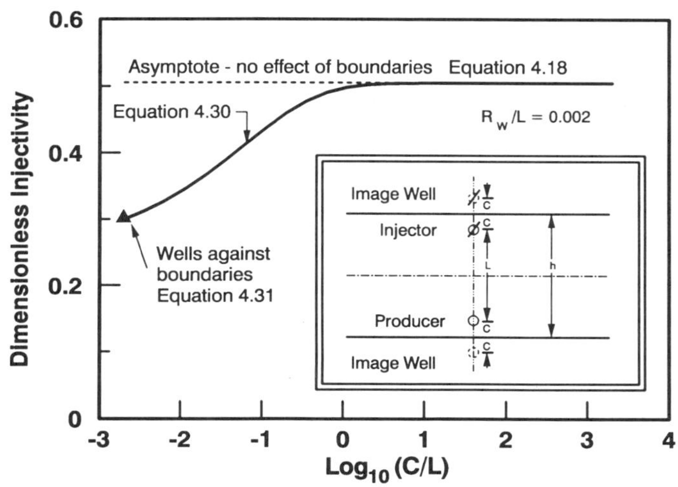

Figure 4.26 Effect of Proximity to Reservoir Boundaries on Injectivity for Horizontal Wells

Repeated Seven-spot

$$\frac{q\mu}{kh\,\Delta P} = \frac{4\pi}{3\left[\ln\left(\frac{L}{R_w}\right) - 0.5691\right]}$$

These are plotted in Figure 4.27.[9] Also shown in this figure is a curve for an injection well located within a continuous row of producers at the same radius. Comparing Figure 4.27 with Figure 4.23 shows that the injector in the repeated pattern has a significantly lower injectivity than that in an isolated pattern. A considerable fraction of the oil in the isolated pattern flows outside of the pattern and then back toward the production wells. The quantity of oil that is produced at breakthrough is also considerably larger for the isolated pattern; much of the produced oil has come from outside the pattern. This is shown in the following table.

Volume of Oil Produced at Breakthrough Divided by Volume of Oil Initially Within Pattern

		Five-Spot	Seven-Spot
Initial mobile oil		$2\phi\,\Delta S_o L^2$	$2.598\phi\,\Delta S_o L^2$
Fraction produced at breakthrough	Isolated	1.0472	0.9069
	Repeated	0.7178	0.7437

The dimensionless breakthrough times (as defined by equation 4.28) are plotted in Figure 4.28 for isolated and repeated 5- and 7-spot patterns.

STEAM ZONE SHAPE: VAN LOOKEREN'S EQUATIONS

Van Lookeren (1977) developed equations that describe the degree of override that may be expected in a steamflood. These equations are based upon fundamental principles such as Darcy's law and make use of the assumption of segregated flow

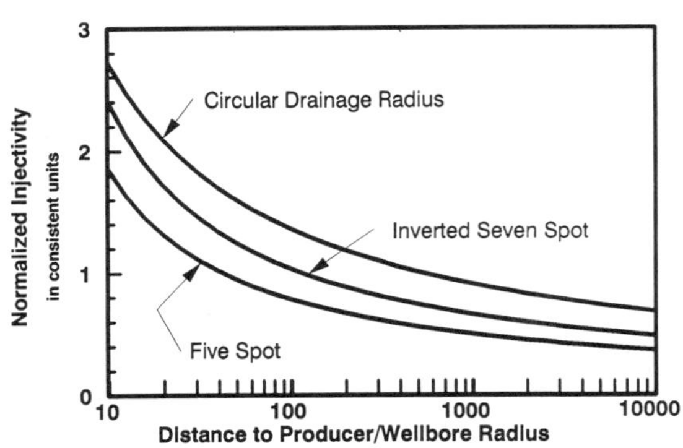

Figure 4.27 Injectivity in Confined Patterns

[9]Similar equations are given by Deppe (1961), who also gives equations for the inverted nine-spot pattern and for patterns at the boundary of a field development.

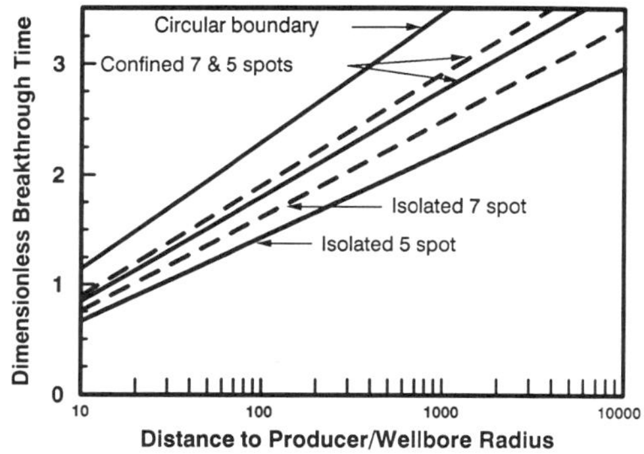

Figure 4.28 Breakthrough Times for Confined and Isolated Patterns

(flow normal to bedding plane of the reservoir is neglected). He developed equations for the case of linear flow and also the case of radial flow. The two are similar; only the radial flow equations are described here.

A basic finding in van Lookeren's paper is that the degree of override may be characterized by a dimensionless number, which he terms A_R; it is given by equation 4.33 and is proportional to the square root of the ratio of the viscous forces within the steam zone to the gravity forces.

$$A_R = \sqrt{\frac{\nu_s W_{si}}{\pi \Delta\rho g h^2 k_s}} \cdot (1 - M^*) \tag{4.33}$$

where the pseudomobility

$$M^* = \frac{\mu_o^* k_s \rho_s}{\mu_s k_o \rho_o} \cdot (\mathrm{OSR})_i$$

and where ν_s is kinematic viscosity of steam, m^2/s
W_{si} is steam injection rate, kg/s
$\Delta\rho$ is ρ_o/ρ_s, kg/m^3
g is gravity, m/s^2
h is thickness, m
k_s is permeability of steam zone to steam, μm^2
μ_o^* is effective viscosity of oil, Pa · s
$(\mathrm{OSR})_i$ is instantaneous oil-steam ratio

The pseudomobility ratio is analogous to the mobility ratio used to characterize the stability of water floods. Values of M^* less than unity lead to stable interfaces, whereas those greater than unity tend to lead to unstable interfaces and the development of steam tongues.

In many cases, where the oil in the reservoir is not extremely viscous or where it has been preheated by stimulation before flooding, M^* may be relatively small compared to unity. On the other hand, where attempts are made to drive cold viscous oil with steam directly, M^* will be high and it will control the situation; steam fingering will then occur.[10]

[10]This is related to the situation where, in Jones' method, the value of A_{CD} is lower because of an initial high oil viscosity. In both cases, oil that is heated does not flow readily to the producer because of bypassing.

If M^* is relatively small, then the value of A_R is controlled by the square-root term in equation 4.33. Of the variables in this term, the only one that is in the direct control of the operator is the injection rate W_{si}. Higher rates give higher A_R—i.e., the viscous forces increase while the gravity forces remain the same.

Lower values of the permeability to steam will also give higher values of A_R; as will be seen, this leads to steeper fronts and thicker steam zones. Accompanying this will be an increase in the oil recovery and in the oil-steam ratio. There is considerable research and development activity that has the objective of reducing the permeability of the steam zone to steam in order to increase the pressure gradient below the steam zone.[11] A promising method involves the addition of surfactants to cause the formation of foams.

The effect of A_R on the predicted shape of the steam interface can be seen from Figure 4.29. With low values of A_R, the steam tends to be confined to the top of the reservoir, and the front is inclined at a low angle. As A_R increases, the front approaches the vertical.

With a situation such as that shown in the top drawing in Figure 4.29, it is apparent that steam will break through early and that, for the amount of oil that will be produced, the heat losses to the overburden and to the unswept reservoir below the steam zone may be excessive. In this case the advantage that would have been expected for a thick reservoir from Myhill and Stegemeier's model will not be obtained. The production rate and oil-to-steam ratio will be almost independent of

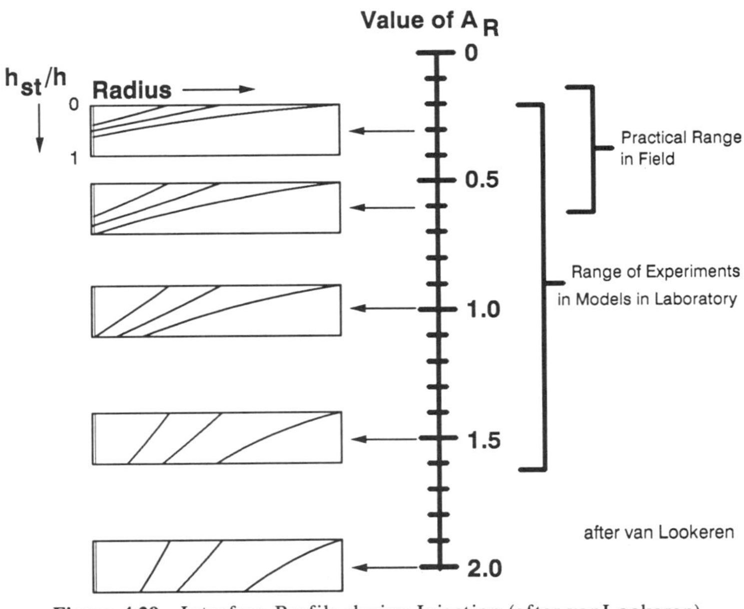

Figure 4.29 Interface Profile during Injection (after vanLookeren)

[11]Not only does this increase the recovery of oil from below the steam zone but it also improves the recovery within the steam zone. The apparent viscosity of the steam is increased and it is better able to displace oil from the steam zone; this aspect is discussed further towards the end of Chapter 5.

reservoir thickness. It is of interest to note (equation 4.33) that A_R is proportional to the square root of the ratio $(W_{si}/h)/h$—i.e., to the rate of steam injection per unit height of reservoir divided by h. In thicker reservoirs the override is thus greater even if the rate of steam injection per unit height is maintained.

Doscher and Ghassemi (1981) and Doscher, Omoregie, and Ghassemi (1982), consider that, in many practical cases, the high oil-to-steam ratios expected for thick reservoirs are never obtained because only the oil from the top of the reservoir is removed. In these cases, according to van Lookeren's theory, the average thickness of the steam zone will be

$$\bar{h}_{st} = 0.5hA_R = 0.5\sqrt{\frac{\nu_s W_{si}}{\pi \Delta\rho g k_s}}(1 - M^*) \tag{4.34}$$

Another problem, which is apparent from the override shown in the upper diagrams of Figure 4.29, is that after breakthrough there is a tendency for the steam to have to push remaining oil up the slope. Gravity is playing a role, but it is not helping the movement of the oil. What would be more desirable is a system whereby the oil is allowed to drain downwards. This is one of the thoughts behind the steam-assisted gravity drainage process to horizontal wells discussed in Chapter 7.

With the situation shown in the upper drawings in Figure 4.29, the only driving force that is moving oil to the production well comes from the difference between the injection pressure of the steam and the pressure at the production well. At the start this is very large, but so is the resisting force of the oil-saturated reservoir. Once breakthrough occurs, the resistance to flow through the steam zone becomes relatively low, and the driving force required to maintain a given flow of steam decreases. Under these conditions much less drive is available to move the oil below.

For low values of A_R, Figure 4.29 shows that the steam-liquid interface does not extend to the base of the injection well. The steam escapes from the top of the well or, if the well is perforated only at the bottom, rises very rapidly to the top of the formation. The reason for this is not, as might be assumed at first, that there is a particularly favorable open streak at the top but simply that the steam, being light, tends to float to the surface. Figure 4.30, which is from van Lookeren's paper, allows one to estimate the liquid level within the well.

The parameter LNTM depends upon the values of the drainage radius, R_e, the well bore radius, R_w, and the skin factor for the well, S. For a given value of A_R, decreasing R_w or increasing S has the effect of causing the steam-water interface to move lower down the well.

In order to estimate the vertical conformance of a steamflood, it is useful to be able to calculate the average, area-weighted steam zone thickness; this is given in Figure 4.31. The curves in this figure are drawn for two extreme theoretical cases, 1 and 2, which are developed by van Lookeren. Actual cases are expected to fall in between these two theoretical curves, and it is suggested by van Lookeren that the average curve should be used.

Low values of A_R correspond to cases where the steam zone either does not reach or only barely reaches the base rock; see Figure 4.29. When A_R is less than

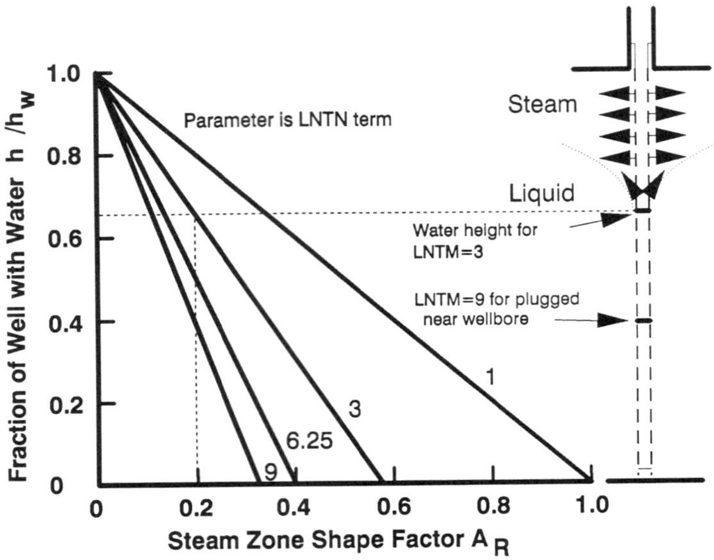

Figure 4.30 Predicted Water Level in Injection Well

$$\text{LNTM} = \ln(R_e/R_w) - 1/2 + R_w^2/2R_e^2 + S$$

S is skin factor; see Chapter 6 (after van Lookeren)

1.0, the mean steam zone thickness, as a fraction of the total thickness, is simply equal to half the A_R. This is also equal to the vertical conformance.

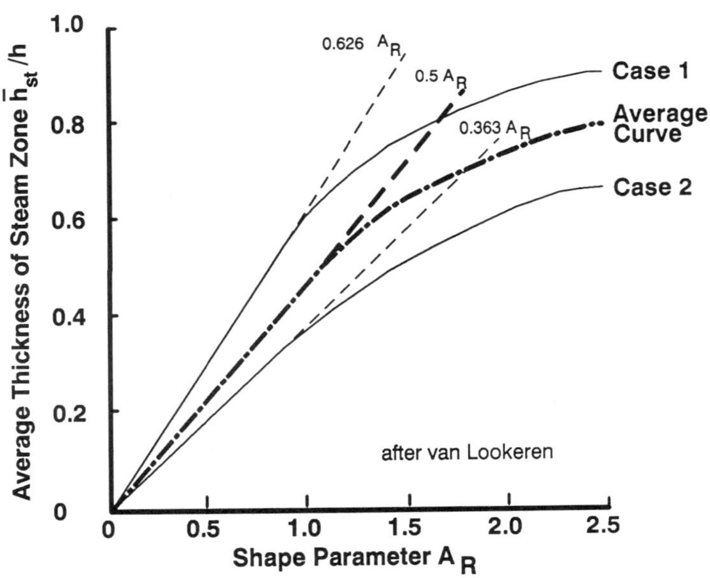

Figure 4.31 Relative Average Steam-Zone Thickness as a Function of A_R (after van Lookeren)

Since the reservoir height occurs in the denominator of A_R, this means that, for low values of A_R, the mean steam zone height is independent of the reservoir thickness and also of time.

Although the steam height within the well depends upon R_w and S, as shown by Figure 4.30, these parameters do not affect the average steam zone thickness as given by Figure 4.31. The effect of R_w and S is confined to the immediate vicinity of the injection well. Injecting steam only at the bottom of the reservoir tends to affect the thickness of the steam zone only in the region close to the injection well.

Numerical Example of the Use of van Lookeren's Theory

As an example of the use of van Lookeren's theory, consider the ten-pattern Kern River steamflood discussed previously. The values of the variables required to calculate A_R are as follows:

Steam injection rate: 18.58×10^6 B of steam were injected over a period of 7 y, or 727 B/d per injection well average

$$\text{Rate per injector} = W_{si} = 727 \times 350 \times 0.4536/86{,}400 = 1.3 \text{ kg/s}$$

$$\nu_s \text{ at } 310°\text{F} = 5.0 \times 10^{-6} \text{ m}^2/\text{s}$$

$$\Delta\rho \cong 960 \text{ kg/m}^3$$

$$g = 9.81 \text{ m/s}^2$$

$$h = 97 \times 0.3048 = 29.6 \text{ m}$$

$$k_s = 0.4 \times 10^{-12} \text{ m}^2 \text{ (assuming } k_R \cong 0.4)$$

$$A_R = \left(\frac{5 \times 10^{-6} \times 1.3}{\pi \times 960 \times 9.81 \times 29.6^2 \times 0.4 \times 10^{-12}}\right)^{1/2} = 0.792$$

The average steam chamber height would thus be

$$\bar{h}_{st} = 0.5 A_R h = 11.7 \text{ m}, \quad \text{or} \quad 38.5 \text{ ft}$$

Using this height and an expected residual oil saturation of about 0.15 would give an expected recovery of

$$\frac{0.52 - 0.15}{0.52} \times \frac{38.5}{97} \times 100\% = 28.2\%$$

This is less than the reported recovery of 37%. One reason for this difference is that the production of oil by waterflooding beneath the steam zone has not been included.

FAROUQ ALI'S UNIFIED APPROACH

Farouq Ali (1982) has presented a description of an approach that encompasses many of the concepts described previously in this chapter and unites them into a simplified approximate model.

His procedure involves the calculation of the steam zone thickness $\bar{h}_{st}$ from van Lookeren's theory and then, using the Mandl-Volek method, the calculation of the steam chamber volume for successive time steps. At each time step, the flow of oil and water from the region below the steam zone is estimated assuming that the temperature is uniform at a value determined by the heat content for the heat loss calculated. Relative permeabilities from Gomaa's correlation are used (Figure 4.31). The procedure is repeated until the steam chamber volume grows to the breakthrough volume calculated at the start. At this point the steam injection rate can be adjusted to control the amount of steam bypass. Farouq Ali provides an encouraging comparison in his paper of the results of his calculation for the Kern River data with the same curves drawn by Jones.

GOMAA'S CORRELATIONS FOR PREDICTING OIL RECOVERY

Gomaa (1980) developed a set of correlation charts for the prediction of steamflood oil recovery and oil-to-steam ratio as a function of reservoir characteristics and operating conditions. The correlations are based upon a series of numerical simulation studies.

Although Gomaa's study is limited to a particular set of fluid and rock properties and is dependent on the assumptions inherent in the numerical simulation method employed, it develops interesting conclusions and ideas. The study considers a reservoir with the relative permeability curves shown in Figure 4.32. These curves were found to give a satisfactory history match for an actual Kern River steamflood. Compared to the values which are commonly found for conventional oils, the relative permeability of water is very low. It has been found necessary to employ relative permeability curves of this type to simulate heavy oil steamfloods in numerical simulators. If conventional curves are employed it is found that water is produced much too quickly. The distorted relative permeabilities employed compensate for other problems which are involved in the simulation such as the extremely large temperature and viscosity gradients which occur in the vicinity of the condensation front.

The gravity of the oil in the study was 14° API and the reservoir temperature was 90°F.

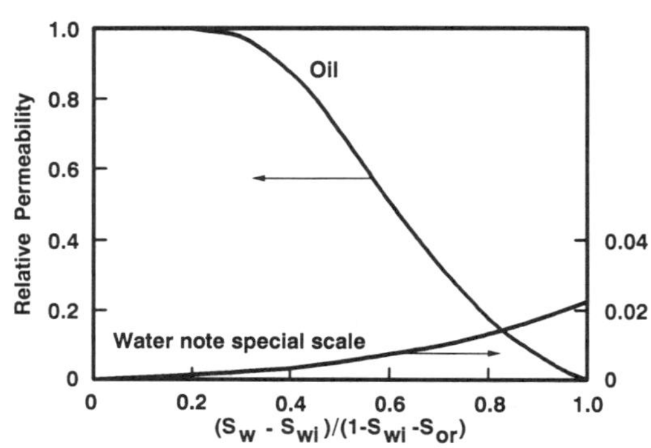

Figure 4.32 Normalized Oil-Water Relative Permeabilities (from Gomaa 1980)

Figure 4.33 shows the oil recovery for reservoirs of various thicknesses as a function of time with a constant steam-injection rate of 1.7 B/d per acre foot of reservoir. For example, the steam-injection rate for the case of the reservoir 300 ft thick was 30 times greater than that for the reservoir 10 ft thick. Higher recoveries were found for thick reservoirs. This might be expected because of the smaller proportion of the total injected heat that would be expected to be lost vertically.

A significant finding from this study was that if the heat injection rates are corrected for the vertical heat losses to give the net heat injected to the reservoir, then the diverse curves of Figure 4.33 all fall on the single curve of Figure 4.34. It was also found that neither the pattern shape nor the pattern size affected the results if the steam-injection rate per unit volume of reservoir was maintained constant. There was a small effect of the rate per unit volume parameter that was neither investigated nor included in the correlation.

An interesting feature of these studies is that an intermediate steam quality (40%) was found to give the highest thermal efficiency. More heat was required for a given production with steam of lower or higher quality.

Decreasing the steam quality from 100% increases the amount of hot waterflooding that occurs beneath the steam zone and delays the time of steam breakthrough (Figure 4.35); with the assumptions made in the simulation calculations, this increases the recovery. Figure 4.36 shows the calculated oil recovery as a function of the net heat injected for various steam qualities.

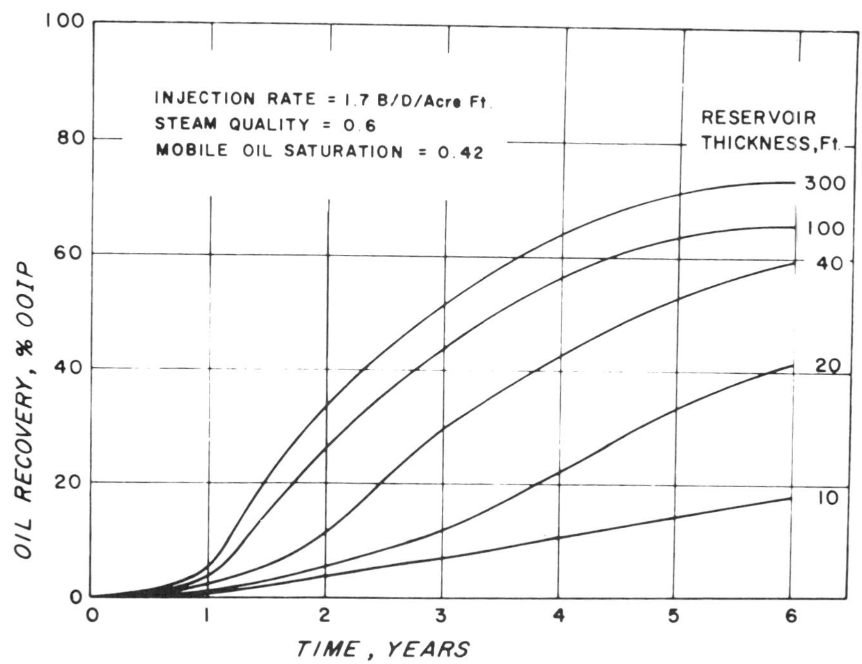

Figure 4.33 Effect of Reservoir Thickness on Steamflood Oil Recovery (from Gomaa 1980)

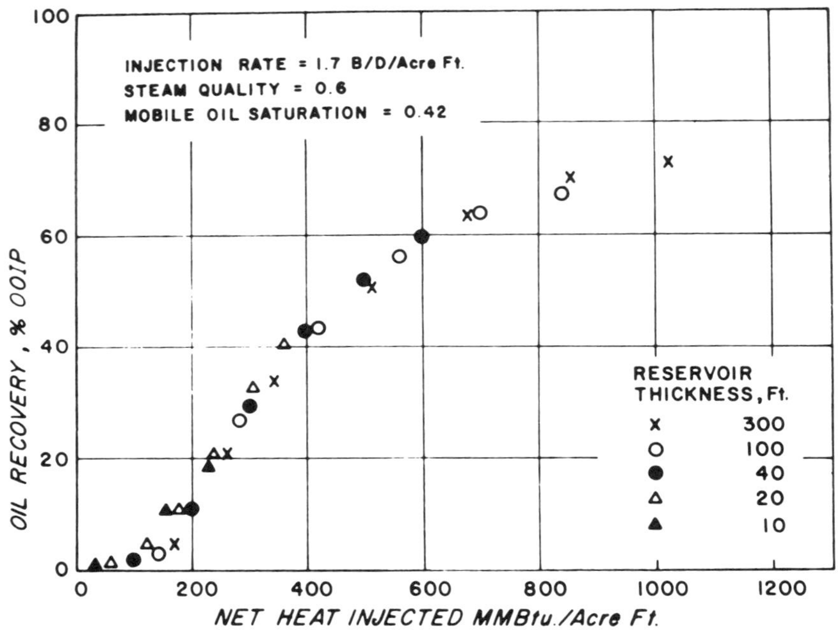

Figure 4.34 Oil Recovery as a Function of Net Heat Injected for Various Reservoir Thickness (from Gomaa 1980)

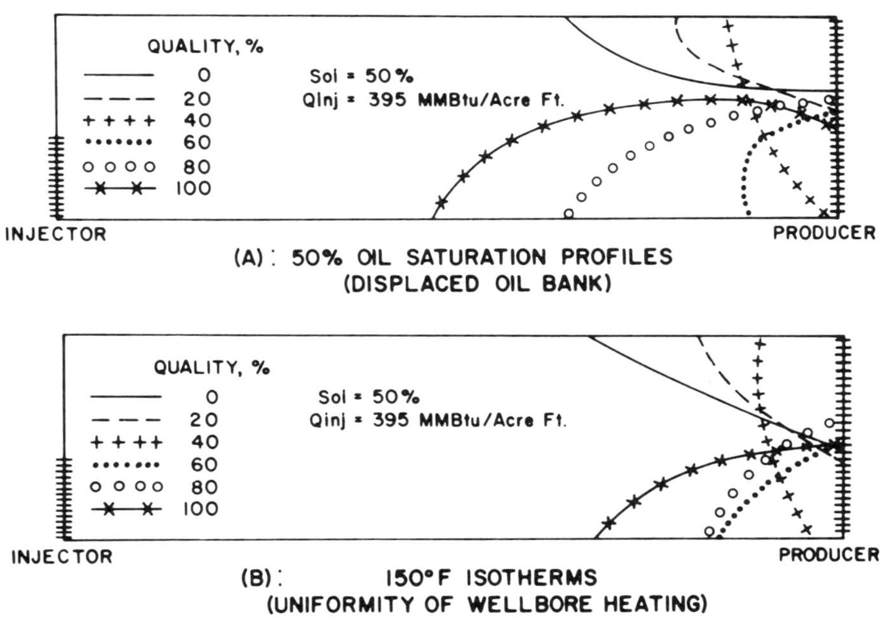

Figure 4.35 Effect of Steam Quality on Displacement Parameters (from Gomaa 1980)

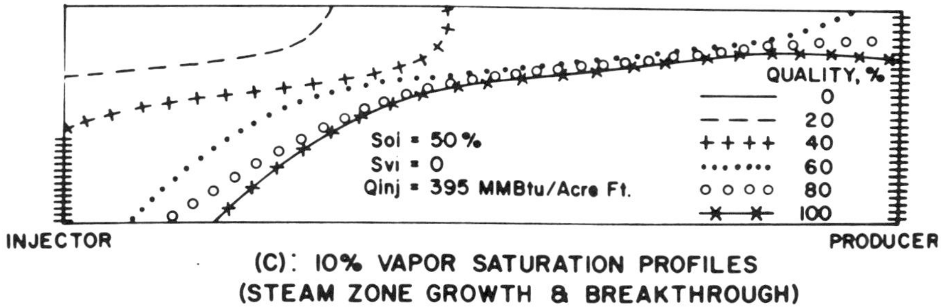

Figure 4.35 *(continued)*

In using Gomaa's correlation, one first calculates the net heat injected into the reservoir up until the end of the current time step. This is corrected for the vertical heat loss read from Figure 4.37, and this net heat injection is adjusted to allow for the effect of steam quality using the factor read from Figure 4.38.

Using the *effective heat injection* just calculated, the oil recovery as a percentage of the original mobile oil (i.e., the oil saturation at the start of the flood minus the estimated residual oil saturation after steaming) is obtained using Figure 4.39. If the reservoir has been produced by steam stimulation prior to the flood, then the initial oil saturation should be adjusted for the oil production during the steam stimulation before using Gomaa's correlation.

Figure 4.40 shows some typical results from the correlation; the importance of reservoir thickness, oil saturation and the net-gross pay ratio are quite evident.

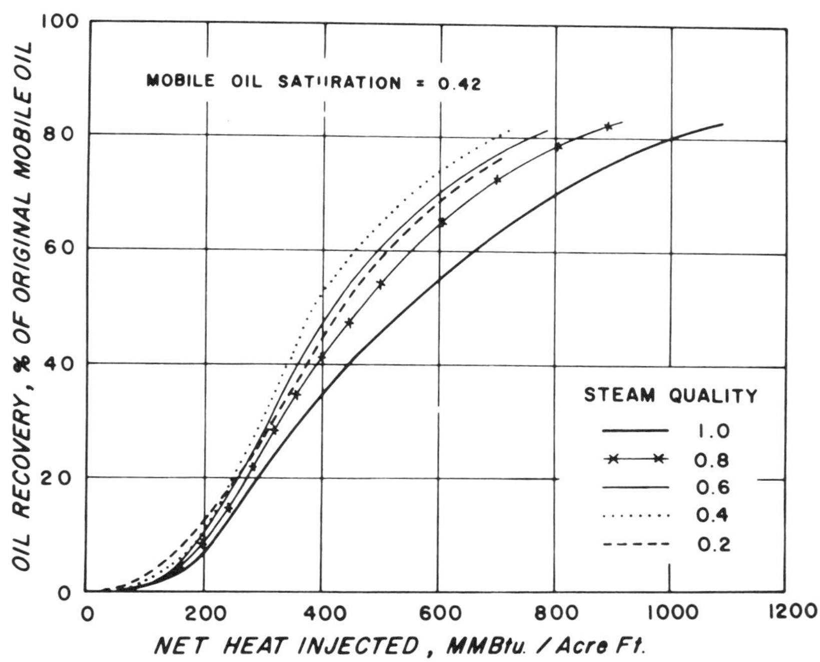

Figure 4.36 Effect of Steam Quality on Oil Recovery (from Gomaa 1980)

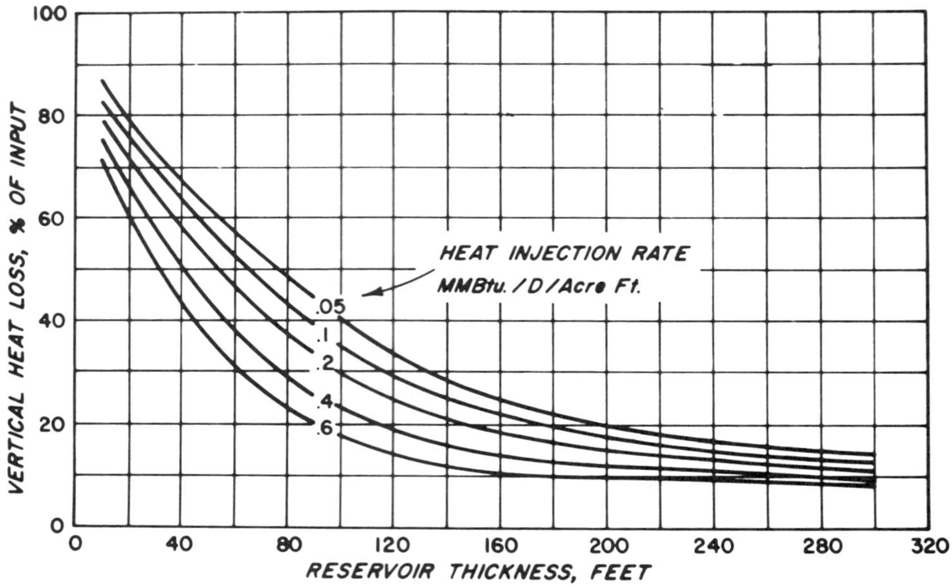

Figure 4.37 Heat Loss to Overlying and Underlying Strata (from Gomaa 1980)

VOGEL'S SIMPLIFIED HEAT CALCULATION FOR STEAMFLOODS

Vogel (1984) has presented an approach to the calculation of the steam requirements for a steamflood that is simple, practical, and conservative. The case that Vogel considers is the one in which override of the steam chamber occurs rapidly and the production of oil is by gravity drainage, assisted by "steam drag." As production proceeds, the steam chamber thickens. The general concept is shown in Figure 4.41.

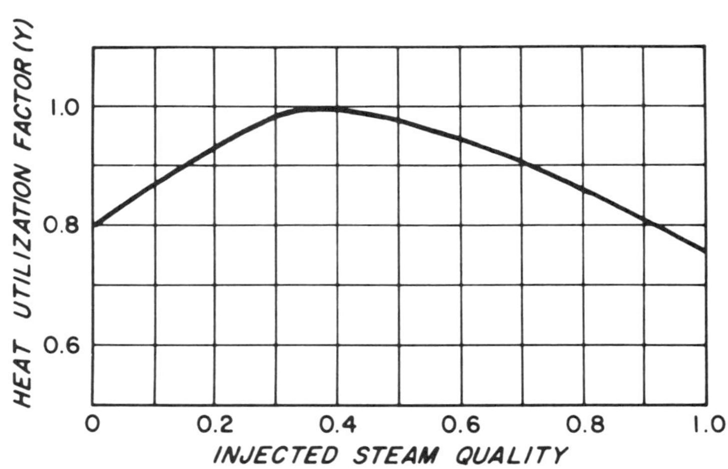

Figure 4.38 Heat-Utilization Factor as a Function of Steam Quality (from Gomaa 1980)

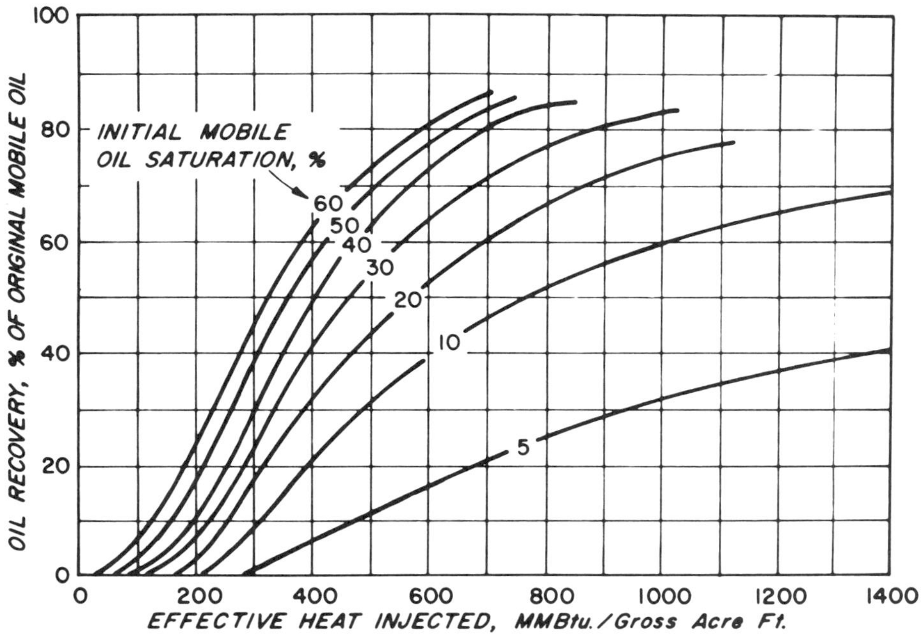

Figure 4.39 Steamflood Recovery as a Function of Effective Heat Injected and Mobile Oil Saturation (from Gomaa 1980)

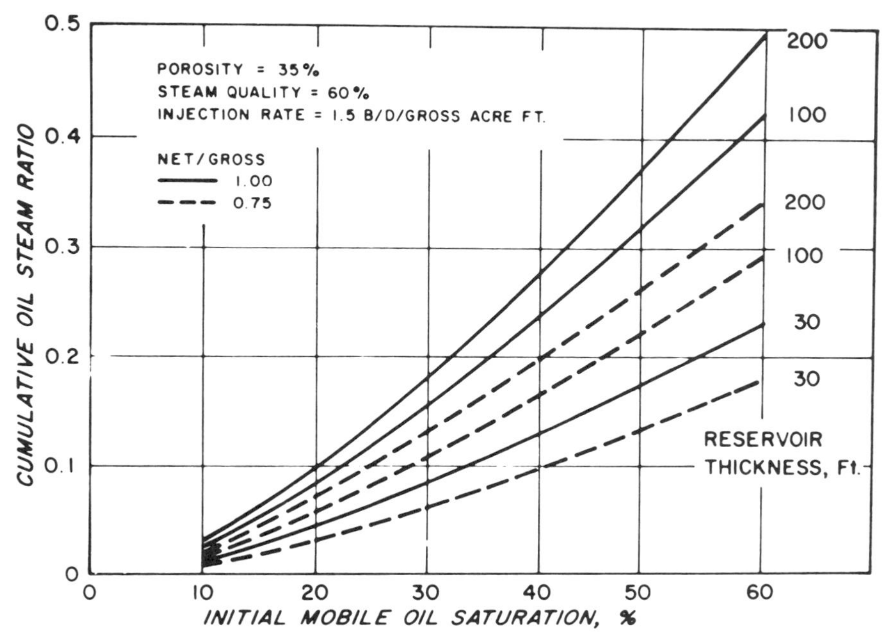

Figure 4.40 Effect of Oil Saturation, Reservoir Thickness, and Net-Gross Ratio on Cumulative Oil-Steam Ratio (from Gomaa 1980)

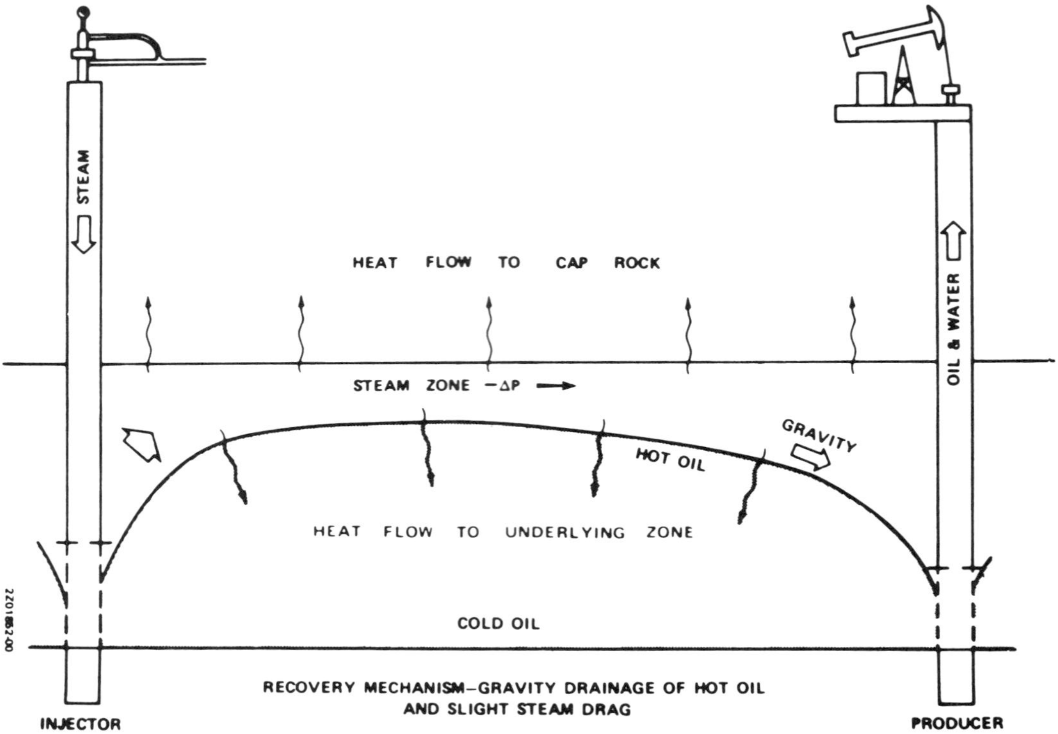

Figure 4.41 Schematic Cross Section of Continuous Steam-Injection Recovery Process (from Vogel 1982)

In Vogel's approach there is no way of predicting how rapidly the drainage from the reservoir will occur, and it is necessary to assume a lifetime for the steamflood. The method can also be used to analyze the status of an ongoing flood. In either case it is necessary to estimate for the time of interest (i.e., for time t) the corresponding average height of the steam chamber, h.

The heat stored within the drainage steam chamber is given by equation 4.35.

$$Q_s = Ah\rho_1 C_1(T_S - T_R) \tag{4.35}$$

Vogel assumes that the steam chamber spreads immediately across the top of the whole reservoir pattern. He then calculates the vertical heat losses to the overburden and also to the material below the steam chamber (this may be underburden and/or undrained reservoir). To calculate these heat losses, he uses equation 4.36 (same as equation 2.25) for the heat loss upward and also for the heat loss downward.

$$Q_L = 2K_2 A(T_S - T_R)\sqrt{\frac{t}{\pi\alpha_2}} \tag{4.36}$$

The heat loss to the overburden, calculated in this way, is conservative because the whole upper surface of the reservoir cannot heat up immediately (see equations 2.25 and 2.27).

The heat loss to the underburden tends to be overestimated even more because the hot interface advances during the drainage process. Both these errors tend to result in a pessimistic estimate of the steam requirements.

Vogel points out that if it is found for a field steamflood that more heat is being consumed during the process than would be calculated by his approach, then it is likely that there are additional heat losses, such as to other regions or to a water layer. This is a very useful feature of his approach.

In comparing the results calculated by adding the heat stored in the steam zone from (4.35) and the vertical heat losses from (4.36) to the heat injected in the steam, one also has to make allowances for heat loss from surface lines, heat loss from the well bore, and the heat in the produced fluids.

Comparison of Vogel's Predictions with Myhill-Stegemeier

By combining Equations 4.35 and 4.36, it is simple to derive an expression for the thermal efficiency, E_h, which is the fraction of the net injected heat that remains in the steam chamber; this is shown by equation 4.37.

$$E_h = \frac{1}{1 + \frac{2}{\sqrt{\pi}} X} \tag{4.37}$$

where

$$X = \sqrt{t_D}$$

(see equation 3.25)

This equation is similar to that derived in Chapter 3 for a steadily advancing steam front (equation 3.24). The difference lies in the factor of $\frac{2}{3}$ in the expression for the vertical heat losses. In equation 3.24 it was assumed that the area of the hot surfaces of the overburden and underburden increased at a constant rate. In equation 4.37 it is assumed that the surface of the overburden and underburden is heated immediately.

Vogel shows that this simple expression predicts thermal efficiencies that are lower than those predicted by the Myhill-Stegemeier frontal displacement approach and which are in close agreement with the field results given in their paper. Figure 4.42 compares the efficiency calculated from equation 4.37 with that from the Marx-Langenheim approach.

Also shown in the figure is a curve for 70% of the Marx-Langenheim efficiency. Recall that the Myhill-Stegemeier analysis showed that the field oil-steam ratios were generally in the range of 70 to 100% of the Marx-Langenheim theoretical efficiency. They are in agreement with the much simpler Vogel equation over the range of most practical interest.

Vogel suggests that in a steamflood it is desirable to inject steam at a higher rate initially and then to reduce the rate to compensate for the reduced vertical heat flux. He also makes the point that in a mature steamflood—i.e., where the steam has reached the breakthrough point—the rate of production is essentially independent of the rate of steam injection; additional steam tends to blow through the reservoir with little incremental production of oil.

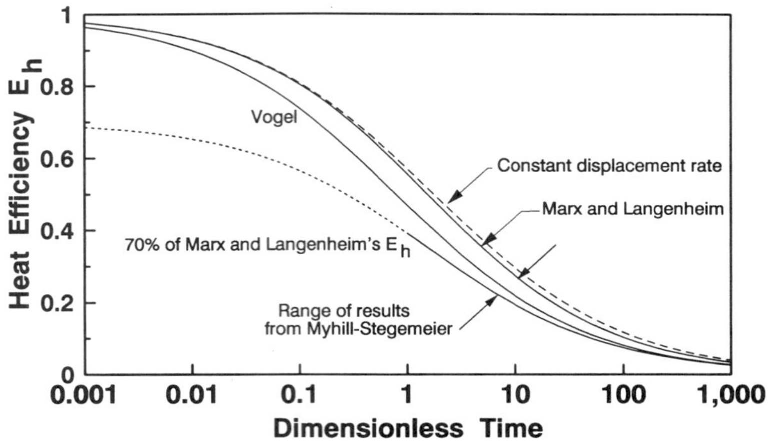

Figure 4.42 Comparison of Calculated Heat Efficiencies

The simple approach suggested by Vogel is practical and useful. Its weakness is that it does not give means for the estimation of the rate at which oil will drain. In Chapter 7 the calculation of the rate at which oil drains from around a steam chamber to a horizontal well is described. Such a system is a logical extension of the process shown in Figure 4.41.

The preceding equations due to Vogel have been previously described in a form that allows the direct calculation of the OSR. (See equations 3.51 and 3.52.)

Numerical Example

Following the discussion of van Lookeren's equations, a numerical example was given on page 157 in which it was predicted that for the conditions of the ten-pattern Kern River steamflood, the average depth of the steam chamber would be 38.5 ft.

At steam breakthrough for the 61-acre, ten-pattern project, the oil displaced from the steam chamber would have a volume of

$$A\phi\,\Delta S_o \bar{h}_{st} = \frac{61 \times 43{,}560 \times 0.34 \times (0.52 - 0.15) \times 38.5}{5.62}$$

$$= 2.29 \times 10^6 \text{ B}$$

if it is assumed that the residual oil saturation within the steam chamber was 0.15. In the field production it was found that the production was 3.02×10^6 B.

The OSR for the project is predicted by equation 3.52 as

$$\text{OSR} = \frac{1769 \times 0.34 \times (0.52 - 0.15)}{(310 - 90)(1 + 2.14\sqrt{7 \times 365/38.5^2})} = 0.26$$

Blevins and Billingsley report that 18% of the injected heat appeared in the production. This would reduce the expected OSR to $0.26 \times 0.82 = 0.21$. This is still higher than the experimental ratio of 0.16. The discrepancy is really larger than it appears because the actual production was larger than that just calculated, and the 0.16 ratio from the field includes this effect. A possible explanation is the indication in the paper that steam was lost to upper sand levels in part of the project area.

THE FAST PROCESS

A somewhat different and interesting approach to steamflooding has been described by Conoco (Britton et al. 1983) for the recovery of very viscous tar from the San Miguel tar sand deposit in the Street Ranch in South Texas. The tar in this deposit is of a very low quality:

Specific gravity (60°F)	1.080–1.093
Viscosity at 175°F cs	520,000
Sulphur content wt. %	9.5–11.0
CCR wt%	24.5

It is thought that this very viscous and dense oil results from the extensive bacterial degradation of a lighter original crude oil—presumably like the process that was described on page 10 for Athabasca but more severe.

The San Miguel tar sand is at a depth of 1500 ft (457 m) and is divided horizontally by a nonpermeable limestone barrier. The test was confined to a layer 26 ft (8.3 m) in thickness. A five-spot, 5-acre (2 ha) pattern was used.

The four producers were fractured using cold water under conditions thought to produce horizontal fractures. This was possible because of the in situ stress in this particular reservoir. In many other reservoirs, vertical fractures would be expected. Next they were steam stimulated, perforated, and resteamed. The center injector was fractured hydraulically with fresh water. This was followed by hot water and steam injection at a very high rate and with an injection pressure that exceeded the horizontal fracture pressure.

Performance data for the pilot are shown in Figures 4.43 and 4.44. During the first 174 d of operation, the average injection rate was about 3000 B/d (477 m^3/d). Comparison of this rate with the data in Table 4.4 shows that this rate is several times higher than conventional practice—particularly when the small thickness of the reservoir is considered. This drastic treatment resulted in the production of oil,

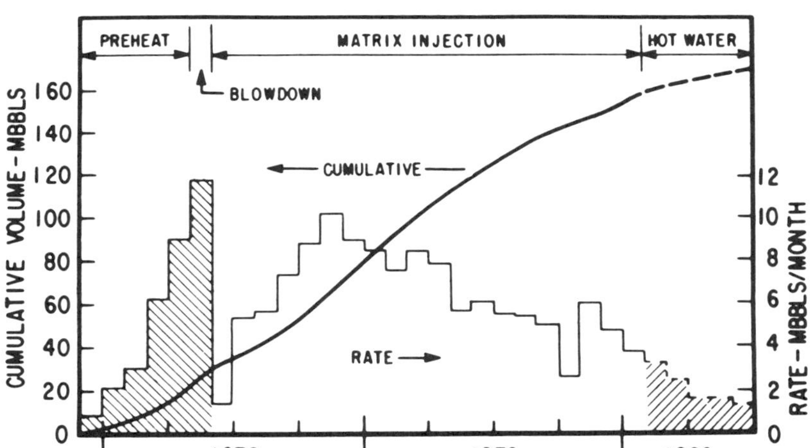

Figure 4.43 Street Ranch Tar Production (from Britton et al. 1983)

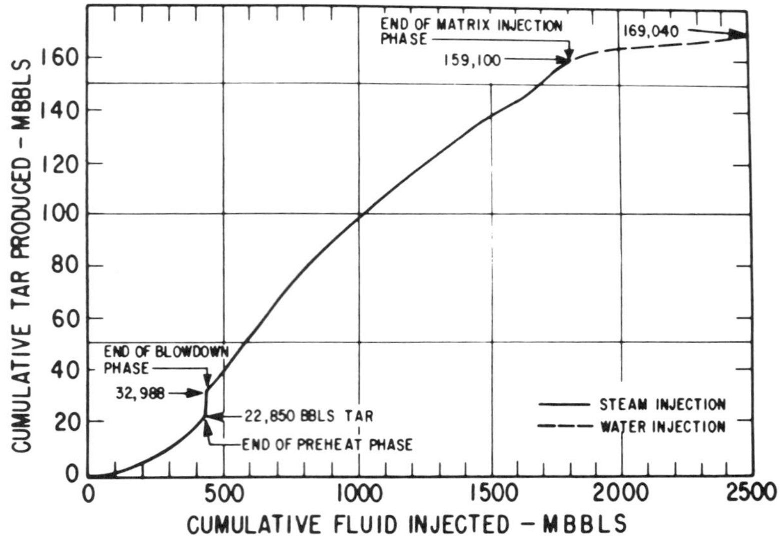

Figure 4.44 Street Ranch Pilot Performance Data (from Britton et al. 1983)

and it was demonstrated that each of the four producers could flow tar at a rate greater than 100 barrels per day.

Steam injection was stopped at 174 d to allow the reservoir pressure to drop so that additional observation wells could be drilled. Following this, a prolonged period of steam injection at a pressure just below the fracture pressure was carried out, and considerable oil was produced. In the final stage of the project, a waterflooding operation was carried out and little additional oil was removed.

Overall, about 170,000 B of tar were produced from the injection of 1.8 million barrels of steam. On a cumulative basis, the steam-to-tar ratio was reported as 10.9. Even though this figure is very high, it is remarkable in view of the very viscous nature of the oil, the relatively thin reservoir, and the low initial tar saturation of the reservoir (about 55%).

OTHER MECHANISMS IN STEAMFLOODING

The material discussed previously concentrates on the effect of heat in lowering the viscosity of the oil and thus making it more mobile. Other mechanisms that play an important role in steamflooding include steam distillation of the lighter components from the residual oil within the steam chamber and also the thermal expansion of the oil.[12]

Wu and Brown (1975; also reported in Wu 1977) have measured the yield of hydrocarbon distillates produced by contacting steam with a series of crude oils. They found that the volume of distillate was a function of the quantity of steam employed (measured as the corresponding volume of liquid water) but was essentially

[12]The thermal expansion of the oil is largely ignored in this book. Its effect is, however, included indirectly because the residual oil saturations are measured at ambient temperature. At steam conditions S_{or} would be larger by as much as 10%.

independent of the steam pressure (or temperature). Figure 4.45 shows the yields of distillates that they obtained.

Distillation occurs to a much larger extent with light crude oils than with heavy ones. However, with even the heaviest oil shown in Figure 4.45, there is significant distillate produced.

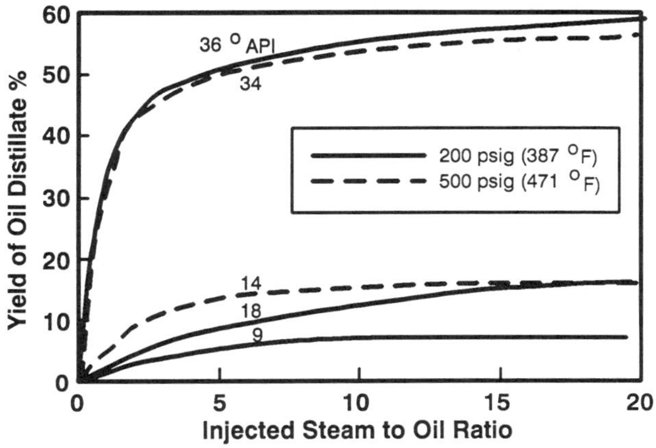

Figure 4.45 Steam Distillation of Heavy Crude Oils (after Wu and Brown 1975 and Wu 1977)

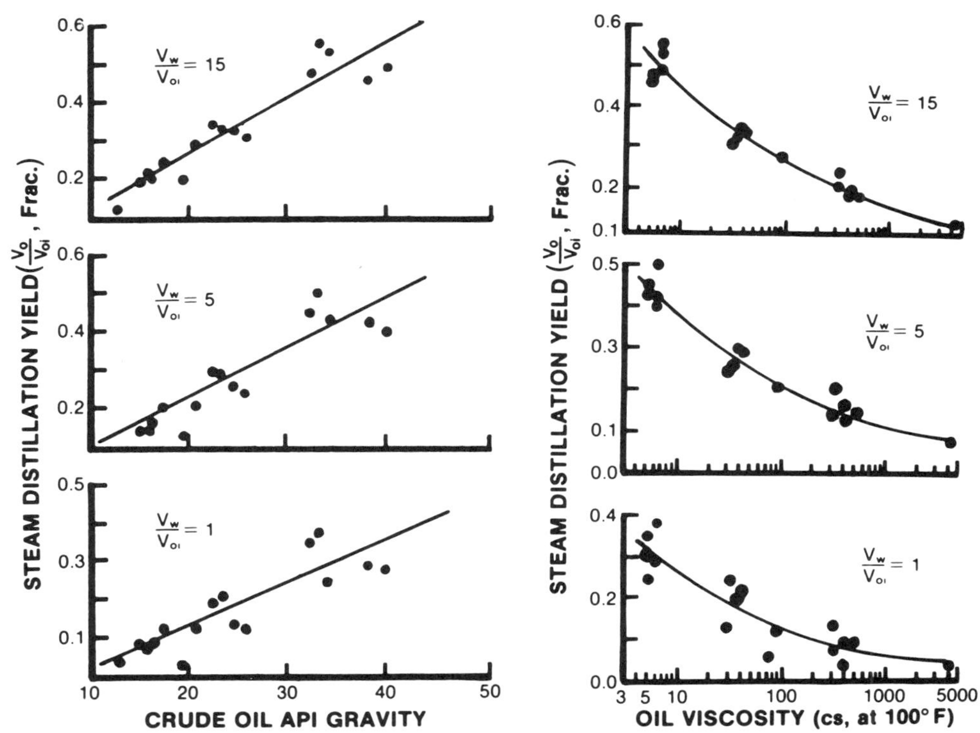

Figure 4.46 Correlation of Steam Distillation Yields with API Gravity and Oil Viscosity (from Wu and Elder 1983)

The effect of steam distillation on the recovery process is greater than that of the simple production of the distillate that is removed from the residual oil, because the distillate is an effective solvent that reduces the viscosity of the oil beyond the steam zone as it condenses and mixes with the reservoir oil.

Wu and Elder reported steam distillation results on a range of 16 crude oils and correlated the hydrocarbon distillate yields against the gravity of the crude oil and the volume of steam used (left-hand side of Figure 4.46) and against the viscosity of the crude oil measured at 100°F (right-hand side of Figure 4.46).

In the left side of Figure 4.47 the distillate yields are correlated against the simulated distillation temperatures of the crude oils for 20 volume percent overhead. The correlation on the right-hand side of this figure shows the steam distillate yield as a function of the yields at various cut points for the simulated distillation of the various crude oils. Each of these correlations allows the prediction of the steam distillate yield of crude oil as a function of the properties of the crude oil and the quantity of steam employed.

Hsueh, Hong, and Duerksen (1984) showed how steam distillation curves such as those in Figure 4.45 can be predicted accurately using a simulated true boiling-point curve obtained by gas chromatographic analysis of the crude and a thermodynamic calculation based on the Peng-Robinson equation of state.

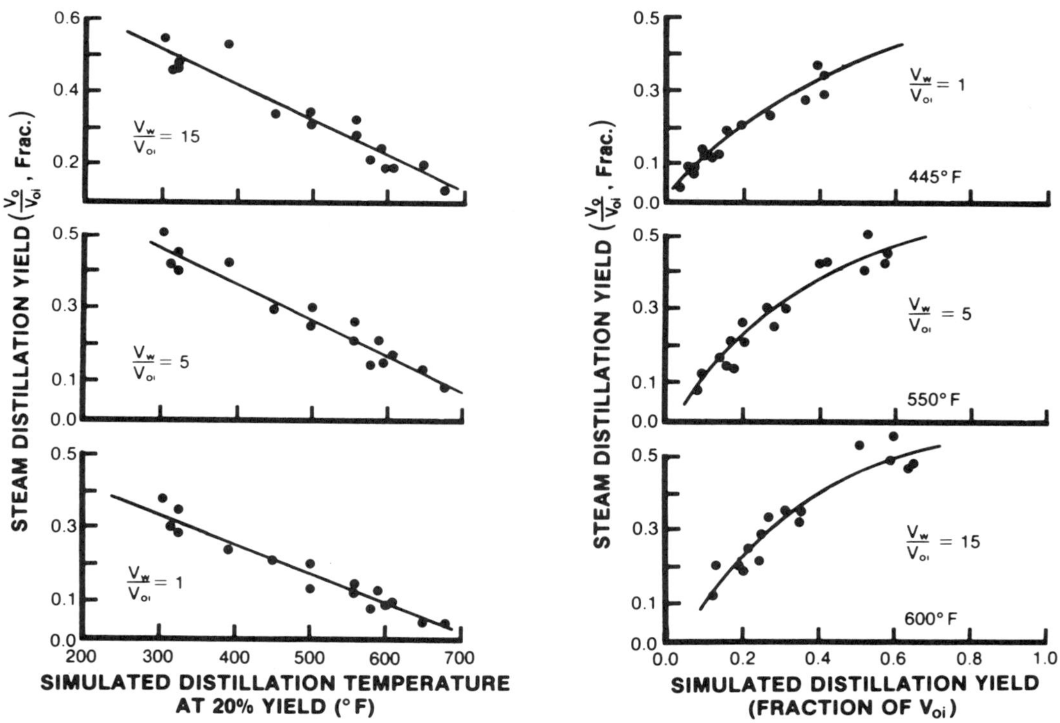

Figure 4.47 Correlation of Steam Distillation Yields with Data from Simulated Distillations (from Wu and Elder 1983)

In a pioneering, experimental laboratory study, Willman and his co-workers (Willman et al. 1961) carried out a series of linear hot-water and steamfloods. They found that steamfloods recovered more oil than hot waterfloods, which, in turn, recovered more oil than cold waterfloods. Figure 4.48 shows the results of one of their series of experiments.

With both waterfloods and steamfloods, water breakthrough occurred relatively quickly in each experiment. However, steam breakthrough did not occur until about 2.3 pore volumes of total fluids were produced.

In another experiment it was found that a still higher recovery could be obtained by using high-pressure steam (84% recovery with steam at 520°F, 800 psig versus 77.6% with steam at 327°F, 84 psig). This additional recovery was thought to be due largely to the lower oil viscosity that resulted from the higher temperature.

The increased recovery found with steam was greater with light oils than with heavy oils. The differences could be explained approximately from distillation data for the oils. From the studies it was concluded that the extra recovery from thermal floods comes from several factors.

For both hot water and steam

- Reduction in oil viscosity because of higher temperatures
- Thermal swelling of the oil

For steam drive, the preceding plus

- Removal of additional material from the residual oil by steam distillation
- Exhaustive gas drive resulting from the steam flush
- A solvent dilution effect caused by the condensation of the light ends in front of the steam zone

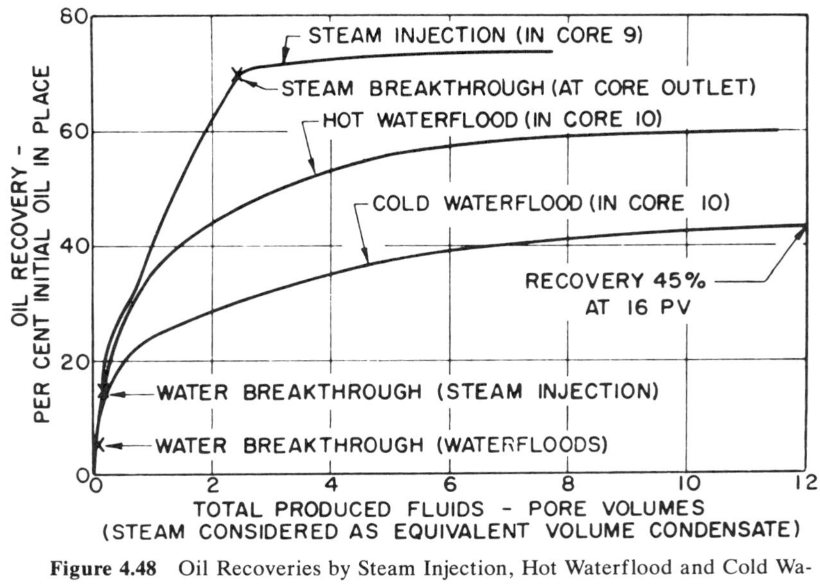

Figure 4.48 Oil Recoveries by Steam Injection, Hot Waterflood and Cold Waterflood, in Cores containing 12.2 °API Bachaquero Crude Oil and Connate Water. Cold water 80°F; Hot water 330°F; Steam 327°F (from Willman et al. 1961)

Table 4.8 shows the magnitude of these effects as evaluated by Willman et al. for experiments with three oils having different amounts of volatile material. These oils were blends of a very low volatility white oil and a heavy naphtha.

The recoveries shown on the first line of the table are for a hot waterflood. The second line shows the additional recovery that was obtained with the "nondistillable" oil using a hot nitrogen flood; this represents the additional recovery that would be obtained by the gas-water-drive effect during the steamflood. Added to this are the steam distillation effects of line 3. Thus, for example, for the 25% distillable oil, the residual oil without distillation would be 42.2% (100 − 54.8 − 3.0) of the original oil in place. It would be expected that 25% of this residual oil, or 10.5%, would be recovered by distillation. The total of the first three lines is somewhat less than the recovery by steamflooding (line 5). The difference (line 6) is the *unexplained recovery.*

It is thought that much of the recovery described as unexplained in the table was the result of the solvent-extraction effect mentioned previously. In the case of the nonvolatile oil, this difference was very small, as might be expected. Although the extra recovery was smaller for the oil with 50% volatiles than that for the oil with 25%, the level of recovery was much larger in this case and the opportunity for recovering additional oil was smaller. Willman's paper was one of the first in this field and must be regarded as a major contribution to the subject.

The production of a 24°API crude from the Brea Field in California by means of a downward steamflood was described by Volek and Pryor (1972). They describe both laboratory recovery experiments and the field project. The reservoir was 4600 ft deep, and very high steam temperatures were involved because of the pressure; reservoir temperatures as high as 623°F were recorded in observation wells. Initially, the high temperatures resulted in excessive well bore heat losses and mechanical problems. These were largely overcome by employing insulated tubing.

The reservoir was steeply dipping (66° from the horizontal). Steam was injected near the top of the formation, and a stable condensation front formed, which advanced downward. As a result of the steam distillation effect, a residual oil saturation of less than 8% was obtained, and it was found that the product oil contained

TABLE 4.8 Actual Steam Recoveries Compared with Predicted Recoveries without Solvent Extraction

	Recovery—Percent of Oil in Place		
	Nondistillable	25 Percent Distillable	50 Percent Distillable
1. Hot waterflood recovery (includes viscosity reduction and swelling)	54.8	54.8	58.0
2. Extra recovery due to gas-drive effects	3.0	3.0	3.0
3. Extra recovery due to distillation[1]	—	10.5	19.5
4. Predicted recovery based on above mechanisms	57.8	68.3	80.5
5. Actual recovery by steam	59.0	76.0	83.9
6. Unexplained recovery (line 5 minus line 4)	1.2	7.7	3.4

[1](Percent of oil that is steam-distillable) × [100 − (line 1 + line 2)]
(Willman et al. 1961)

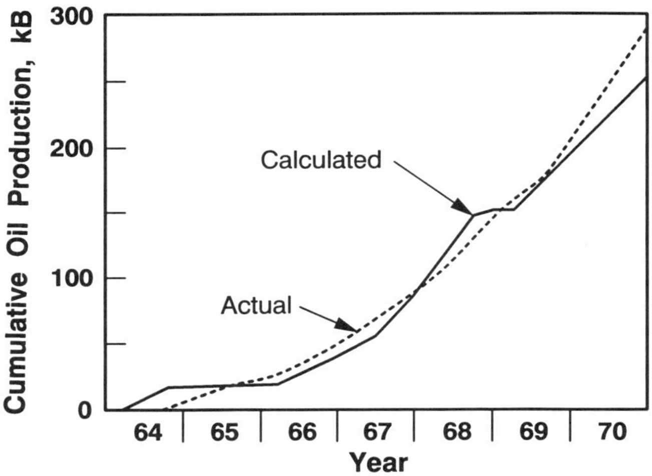

Figure 4.49 Comparison of Actual Oil Production with that Calculated from Estimated Steam Chamber Volume (from Volek and Pryor 1972)

significantly higher proportions of light components than did the original crude. The quantity of oil produced agreed quite well with that calculated to be displaced based on estimates of the steam chamber volume using the Mandl and Volek approach. A comparison is shown in Figure 4.49.

In later years, the actual production tended to exceed that calculated from the steam chamber volume. It was thought that this was due to the effect of hot steam condensate producing oil by waterflooding ahead of the descending condensation front.

CONVERSION OF MATURE STEAMFLOODS TO HOT WATERFLOODING

During a steamflood it is found that the rate of production of oil increases initially to a maximum and then declines. At the same time, the steam oil ratio starts very high, falls to a minimum near to the time when the oil production reaches its maximum, and then gradually rises as the reservoir becomes depleted.

Predicted simulation data for the Kern River ten-pattern project (Hong 1985) are shown in Figures 4.50 and 4.51.

It is found that considerable economy can be achieved in such projects by converting from high-quality steam injection to low-quality steam or hot water in the later stages of production. Ault, Johnson, and Kamilas (1985) show that in the Kern River ten-pattern steamflood this procedure resulted in considerable energy savings with the same production of oil (see Figure 4.14).

In a related paper, Hong lists a number of guidelines to indicate when a steamflood project should be converted to a waterflood. The oil-production rate should have passed its peak and be in a decline. The SOR (steam-oil ratio) and WOR (water-oil ratio) may be expected to be rising steadily. The reservoir pressure will also have peaked and be close to its initial value. Total fluid production will likely be 85 to 95% of the steam injection rate. The cumulative heat injection (injected minus produced) will be about 450 to 500 million Btu per gross ac-ft. Alternatively, the calculated heat stored in the reservoir (as distinct from that lost ver-

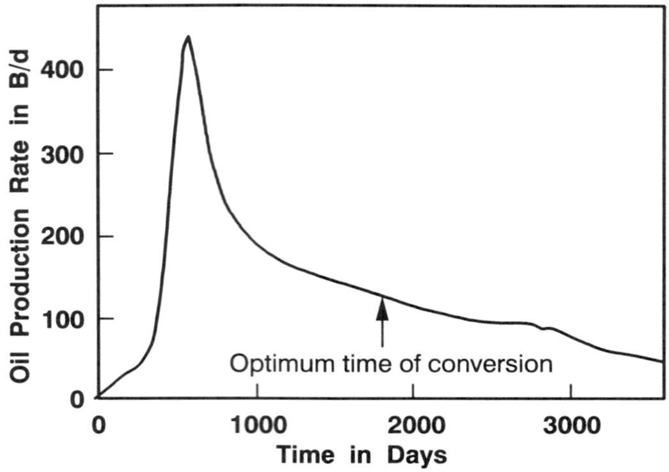

Figure 4.50 Predicted Oil Production Rate, Kern River Ten-Pattern Project (from Hong 1985)

tically and that produced) will have reached 280 to 300 million Btu per gross ac-ft (240 to 260 MJ/m^3).

QUALITATIVE REVIEW OF STEAMFLOODING

The theme of this chapter is the quantitative analysis and prediction of steamflooding performance using the conductive and convective heat transfer equations developed in Chapters 2 and 3. In approaches such as that of Myhill and Stegemeier the volume of steam-saturated reservoir is calculated from the heat content of the injected steam and the volumetric heat capacity of the reservoir with an allowance for the heat that is lost by conduction, vertically to the overburden and underburden.

A major problem is found in applying this method because frequently, and probably usually, the steam-saturated zone does not grow around the injector as a continuous region with a near vertical condensation surface.

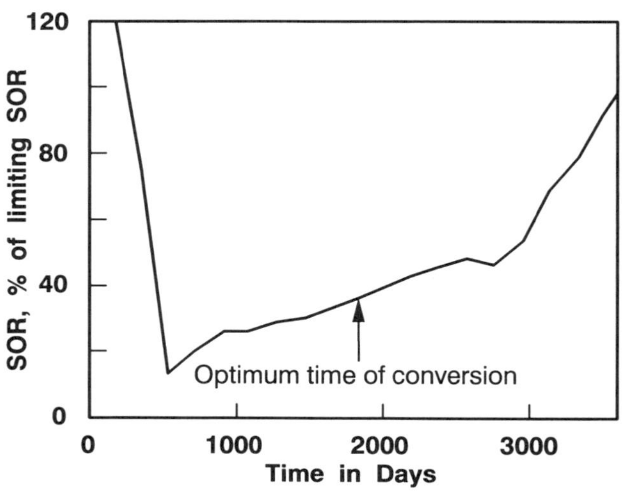

Figure 4.51 Predicted Steam-Oil Ratio and Optimum Time of Conversion (from Hong 1985)

Often, in cold bitumen-containing reservoirs steam can be injected only by employing injection pressures high enough to fracture the reservoir. In these circumstances the steam heats the reservoir adjacent to the fractures but there is little driving force to move the oil. Heated oil is bypassed.

In reservoirs containing more mobile oil the steam chamber does grow in a continuous fashion but, as described by van Lookeren, the steam tends to override and the condensation surface becomes relatively flat. This effect is more pronounced in thicker reservoirs. As a result much of the advantage for thicker reservoirs which would be predicted from the Marx-Langenheim theory is not found; the system behaves as if the reservoir height were no more than that of the overriding steam chamber. The heat losses for a given volume of steam chamber are larger than would be expected from the simple Marx-Langenheim or Myhill-Stegemeier approach because of the greatly extended surface of the steam zone.

Another problem which accompanies steam override is that steam breaks through at the production wells and the pressure differential that was driving the oil is largely lost. Again heated oil is bypassed. This is rather similar to the problem of gas coning which occurs in the production of conventional oil from beneath a gas cap. In conventional oil production the problem can sometimes be controlled by producing at a rate low enough to prevent drawing gas to the well. With heavy oil and steam this rate is often too low to be economic although, as will be seen in Chapter 7, this limitation can be improved by using horizontal wells.

Useful approaches, which have been discussed, to overcome the problem of premature steam breakthrough with conventional steamfloods include

i) The use of infill wells
ii) Waterflooding after breakthrough
iii) Using low quality steam
iv) Foaming additives which will increase the apparent viscosity of the steam

The use of horizontal wells as an alternative to conventional wells is a production method which is becoming increasingly attractive and which is discussed in Chapter 7.

BIBLIOGRAPHY

Al-Khafaji, A. H., Wang, P. F., Castanier, L. M., and Brigham, W. E., "Steam Surfactant Systems at Reservoir Conditions," *SPE* 10777 (1982).

Ault, J.W., Johnson, W. M., and Kamilas, G. N., "Conversion of Mature Steamfloods to Low Quality Steam and/or Hot-Water Injection Projects," SPE 13604, Bakersfield, Calif.: California Regional Meeting, (March 1985), 149–166.

Bennion, D.W., Moore, R. G., and Thomas, F. B., "Effect of Relative Permeability on the Numerical Simulation of the Steam Stimulation Process," Paper 83-34-46, Pet. Soc. of CIM, 34th Annual Technical Meeting, Banff, Alberta (1983).

Blair, C. M., Scribner, R. E., and Stout, C. A. "Chemical Enhancement of Oil Production by Cyclic Steam Injection," SPE 10700 (1982).

BLEVINS, T. R., ASELTINE, R. J., and KIRK, R. S.: "Analysis of a Steam Drive Project, Inglewood Field, California," *JPT,* 1141–1150 (September 1969).

BLEVINS, T. R. and BILLINGSLEY, R. H., "The Ten Pattern Steamflood, Kern River Field, California," *JPT,* 1505–1514 (December 1975).

BRITTON, M.W., MARTIN, W. L., LEIBRECHT, R. J., and HARMON, R. A., "The Street Ranch Pilot Test of Fracture-Assisted Steamflood Technology," *JPT,* 511–522 (March 1983). © 1983 SPE.

BUCKLES, R. S., "Steam Stimulation Heavy Oil Recovery at Cold Lake, Alberta," *SPE* 7994 (1979).

BURGER, J. and CHAMPLON, D., "How to Estimate Production Cost by Steam Drive," *Pet. Eng. Intl.*, 56–70 (June 1983). (This reference is not mentioned in the text but is a very practical approach to the use of the Marx-Langenheim method.)

BURSELL, C. G.: "Steam Displacement—Kern River Field", *JPT* 1225–1231 (October 1970).

BURSELL, C. G. and PITTMAN, G. M.: "Performance of Steam Displacement in the Kern River Field," *JPT* 997–1004 (August 1975).

CHU, C., "State-of-the-Art Review of Steamflood Field Projects," *JPT,* 1887–1902 (October 1985). © 1985 SPE.

CHUNG, K. H. and BUTLER, R. M., "Geometrical Effect of Steam Injection on the Formation of Emulsions in the Steam-Assisted Gravity Drainage Process," *JCPT,* 27, no. 1: 36–42 (January–February 1988).

DEPPE, J. C., "Injection Rates—The Effect of Mobility Ratio, Area Swept and Pattern," *SPEJ,* 81–91 (June 1961).

DILGREN, R. E., DEEMER, A. R., and OWENS, K. B., "The Laboratory Development and Field Testing of Steam/Noncondensible Gas Foams for Mobility Control in Heavy Oil Recovery," *SPE* 10774, 1982.

DOSCHER, T. M. and GHASSEMI, F., "The Effect of Reservoir Thickness and Low Viscosity Fluid on the Steam Drag Process," *SPE* 9897 (1981).

DOSCHER, T. M., OMOREGIE, O. S., and GHASSEMI, F., "Steam Drive Definition and Enhancement," *JPT,* 1543–1545 (July 1982).

ESON, R. L., and O'NESKY, S. K., "Evaluation of a Conventional Steam Drive with Ancillary Materials: North Kern Front Field," *SPE* 10775, 1982.

FAROUQ ALI, S. M., "Current Status of Steam Injection as a Heavy Oil Recovery Method," *JCPT* (January–March 1974).

FAROUQ ALI, S. M., "Steam Injection Theories A Unified Approach," *SPE* 10746 (1982). © 1982 SPE.

FAROUQ ALI, S. M. and MELDAU, R. F., "Current Steamflood Technology," *JPT,* 1332–42, (Oct. 1979).

FRIEDMANN, F. and JENSEN, J. A., "Some Factors Influencing the Formation and Propogation of Foams in Porous Media," *SPE* 15087, 56th California Regional Meeting of the SPE, Oakland, Calif. (April 2–4, 1986).

GEFFEN, T. M., "Oil Production to Expect from Known Technology," *Oil Gas J.,* 66–76, May 7, 1973.

GOMAA, E. E., "Correlations for Predicting Oil Recovery by Steamflood," *JPT,* 325–332 (February 1980). © 1980 SPE.

HONG, K. C., "Guidelines for Converting Steamflood to Waterflood," *SPE* 13605, Bakersfield, Calif.: California Regional Meeting (March 1985), 167–179. © 1985 SPE.

Hong, K. C., "Steamflood Strategies for a Steeply Dipping Reservoir," *SPERE,* 431–439 (May 1988).

Hsueh, L., Hong, K. C., and Duerksen, J. H., "Simulation of High Pressure and High Temperature Steam Distillation of Crude Oils," 2d UNITAR Conference, reported in *The Future of Heavy Crudes and Tar Sands,* Caracas, Venezuela, Feb. 7–17, 1982 New York: McGraw-Hill (1984), 924–935.

Jamaluddin, A. K. M. and Butler, R. M., "Factors Affecting the Formation of Water-in-Oil Emulsions During Thermal Recovery," *AOSTRA J. of Research,* 4, no. 2: 109 (1988).

Jones, J., "Steam Drive Model for Hand-Held Programmable Calculators," *JPT,* 1583–1598 (September 1981). © 1981 SPE.

Lo, H.Y. and Mungan, N., "Effect of Temperature on Water-Oil Relative Permeabilities in Oil-Wet and Water-Wet Systems," *SPE* 4505 (1973). © 1973 SPE.

Mandl, G. and Volek, C.W., "Heat and Mass Transport in Steam Drive Processes," *SPEJ,* 59–79 (March 1969).

Marx, J.W. and Langenheim, R. N., "Reservoir Heating by Hot Fluid Injection," *Pet. Trans. AIME,* 216: 312–315 (1959).

Matthews, C. S., "Steamflooding," *JPT,* 465–471 (March 1983).

Mohammadi, S. S., van Slyke, D. C., and Ganong, B. L., "Steam-Foam Pilot Project in Dome Tumbador, Midway-Sunset Field," *SPE Res. Eng.,* 7–16 (February 1989).

Mohammadi, S. S. and McCallum, T. J., "Steam-Foam Pilot Project in Guadalupe Field, California," *SPE Res. Eng.,* 17–23 (February 1989).

Morel-Seytoux, H. J., "Unit Mobility Displacement Calculations for Pattern Floods in Homogeneous Medium," *SPEJ,* 217–227 (September 1966).

Muskat, M., *The Flow of Homogeneous Fluids Through Porous Media,* New York: McGraw-Hill (1937); Reprinted by IHRDC, Boston (1982).

Myhill, N. A. and Stegemeier, G. L., "Steam-Drive Correlation and Prediction," *JPT,* 173–182 (February 1978). © 1978 SPE.

Oglesby, K. D., Blevins, T. R., Rogers, E. E. and Johnson, W. M., "Status of the 10-Pattern Steamflood, Kern River Field, California," *JPT,* 2251–2257 (October 1982). © 1982 SPE.

Ozen, A. S. and Farouq Ali, S. M.: "An Investigation of the Recovery of the Bradford Crude by Steam Injection," *JPT* 692–698 (June 1969).

Ploeg, J. F. and Duerksen, J. H., "Two Successful Steam/Foam Field Tests, Sections 15A and 26C, Midway-Sunset Field," *SPE* 13609, California Regional Meeting, Bakersfield (March 27–29, 1985).

Prats, M., "Thermal Recovery," *SPE Monograph Volume 7,* SPE, Dallas (1982).

Restine, J. L., "Effect of Preheating on Kern River Field Steam Drive," *JPT,* 523–529 (March 1983).

Stout, C. A., Blair, C. M., Jr., and Scribner, R. E., "Continuing Beneficial Action of Thin Film Spreading Agents Injected during Cyclic Steam Simulation," Paper 83-34-31, 34th Annual Technical Meeting, Petroleum Society of CIM, Banff, Alberta (1983).

Traverse, E. F., Deibert, A. D. and Sustek, A. J., "San Ardo—A Case History of a Successful Steamflood," *SPE* 11737, Ventura, Calif.: California Regional Meeting (1983).

Valleroy, V. V., Willman, B.T., Campbell, J. B., and Powers, L.W.: "Deerfield Pilot Test of Recovery by Steam-Drive," *JPT* 956–964 (July 1967).

van Lookeren, J., "Calculation Methods for Linear and Radial Steam Flow in Oil Reservoirs," *SPEJ,* 427–439 (June 1983). The paper was presented originally in 1977 as SPE 6788. © 1977 SPE.

VOGEL, J.V., "Simplified Heat Calculations for Steamfloods," *JPT,* 1127–1136 (July 1984).

VOLEK, C.W. and PRYOR, J. A., "Steam Distillation Drive—Brea Field California," *JPT,* 899–906 (August 1972). © 1972 SPE.

WILLMAN, B.T., VALLEROY, V. V., RUNBERG, G.W., CORNELIUS, A. J., and POWERS, L.W., "Laboratory Studies of Oil Recovery by Steam Injection," *JPT,* 681–690 (July 1961). © 1961 SPE.

WU, C. H., "A Critical Review of Steamflood Mechanisms," *SPE* 6550, 1977. © 1977 SPE.

WU, C. H. and BROWN, A., "A Laboratory Study on Steam Distillation in Porous Media," *SPE* 5569 (1975). © 1975 SPE.

WU, C. H. and ELDER, R. B., "Correlation of Crude Oil Steam Distillation Yields with Basic Crude Oil Properties," *SPEJ,* 937–945 (December 1983). © 1983 SPE.

5

The Displacement of Heavy Oil

INTRODUCTION

The material presented in the previous chapters is concerned with the mechanism by which heat is transferred in the reservoir and it concentrates upon the size of the steam zone which is formed. It was assumed that the heated oil was displaced from the chamber and that most of it found its way in some manner to production wells.

It was also assumed that it was possible to inject fluids into the reservoir at specified rates and pressures. In actual practice, injection at appreciable rates at below fracture pressure may not even be possible without previous heating of the reservoir by steam stimulation or otherwise. Frequently steam stimulation may involve reservoir expansion by fracturing.

The heat balance and heat transfer considerations of the previous chapters are only part of the problem. In this chapter consideration is given to the problem of the displacement of oil by water, by hot water and by steam and to the mechanisms by which this displacement occurs.

FACTORS AFFECTING DISPLACEMENT

There are many factors which influence the efficiency and rate at which oil may be displaced from the reservoir by water and steam. The most important of these are:

PROPERTIES OF THE RESERVOIR MATRIX

Permeability, wetting, homogeneity, capillarity

Fracturing—either natural or, more importantly, in many thermal recovery situations, fracturing or reservoir expansion created by high pressure fluid injection

DISPLACED FLUID PROPERTIES

Particularly viscosity

DISPLACING FLUID PROPERTIES

Viscosity
Tendency to change phase (both condensation and evaporation)
Potential to transfer heat

CONDITIONS AND GEOMETRY

Flood velocity (injection & production rates)
Geometric parameters such as dip, spacing and pattern

DISPLACEMENT CONCEPTS

Piston Displacement

This is an ideal concept in which the displacing fluid flows through the reservoir in a direction which is normal to the front and moves reservoir fluids to the production well at an equal rate. The concept is that of "pushing" the oil by the injected fluid.

Breakthrough

The arrival of the displacing fluid at the production well is termed *breakthrough*. For piston displacement, this is the end of the process. However, in reality, further oil in admixture with the displacing fluid is often produced long after breakthrough. With very heavy oils nearly all the production from waterflooding operations contains a preponderance of water and may be considered to occur following breakthrough.

Override

When steam is injected, it tends to rise and advance along the top of the reservoir. Thus the interface between the steam and reservoir liquids becomes inclined, and steam will arrive at the production well before the reservoir is completely swept. This effect reduces the vertical conformance and tends to waste steam by bypassing. The "short circuit" of the steam to the production well reduces the pressure gradient available to move the oil.

Frontal Stability

Under some conditions, particularly when a viscous oil is displaced by pushing it with a fluid of lower viscosity (e.g., heavy oil with water), the front between the fluids becomes unstable, and fingers of the low viscosity material penetrate into the fluid being displaced. There are a number of factors that affect frontal stability: the viscosities of the fluids, the direction of displacement relative to gravity, velocity, condensation of steam and imbibition of water. Figure 5.1 shows fingers that developed during the displacement of a viscous oil with water in a laboratory experi-

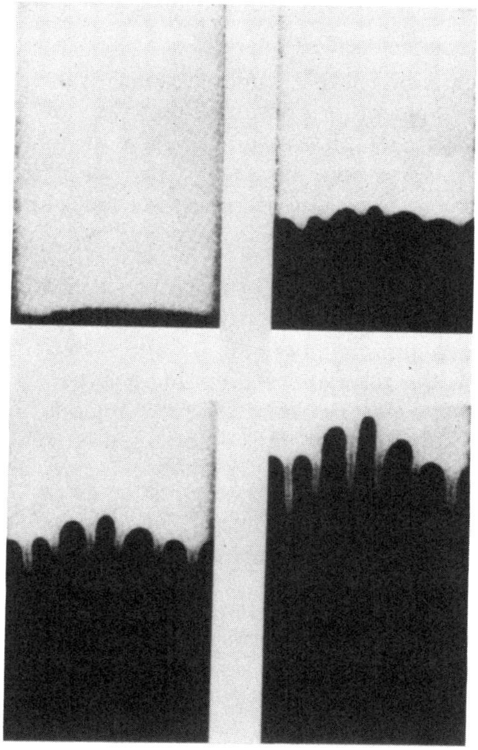

Figure 5.1 Viscous Fingering. Photographs Showing the Development of Viscous Fingers Resulting from the Displacement of Oil by Water in a Hele Shaw Cell. Velocity is 1.8 Times the Critical One (from Chuoke et al. 1959)

ment carried out in a glass Hele Shaw cell. A Hele Shaw cell is constructed from two closely spaced glass plates with the fluids in between. It has been used to provide a two-dimensional model of flow within a porous solid.

In steamflooding, the interface at which the steam condenses is frequently stable and advances regularly—particularly if the advance direction is downwards. However, the condensate that is formed flows as fingers through the oil being displaced.

THE THEORETICAL APPROACHES TO DISPLACEMENT

In the sections that follow, the theory of the mechanism of displacement is developed from three different points of view:

1. Displacement assuming the existence of a sharp interface between the displacing fluid (e.g., water) and the oil. As the water advances, the oil is displaced, and it is assumed that there is no mingling of the oil and water. This is known as *Muskat's model.* Its main use here is to develop concepts on the stability of advancing displacement fronts (interfaces).
2. *The Buckley-Leverett theory.* In this theory, allowance is made for the possibility that there can be an intermingling of oil and displacing fluid, which changes the front of the Muskat model to an advancing interfacial region with

a thickness that increases as the process proceeds. This intermingling can also be accompanied by larger-scale fingering.

3. A heavy oil displacement concept that combines some of the characteristics of both (1) and (2). In this concept, it is assumed that the flow of the displacing fluid occurs as numerous fingers. It is assumed that at any section in the reservoir, the flowing fluid in some pores consists entirely of oil and that in others it is entirely water or steam. This assumption is sometimes known as *stratified flow.* It is also related to the capillary displacement theory of C.W. Nutt.

Finally, the one-dimensional displacement of oil by steam is considered. This is a coupled process. Within the steam chamber there is an isothermal displacement of oil by steam. Beyond the steam chamber, the oil and water that have left the steam chamber flow through the reservoir. In this region, there is also some displacement of oil by the flowing water. Between these two regions lies the advancing condensation interface, the position of which is determined largely by heat balance considerations.

FLOOD INTERFACE STABILITY—MUSKAT'S MODEL

In this section the problem of the stability of the front between the displacing and displaced fluid is analyzed using the assumptions known as the Muskat model (Muskat 1937). It is assumed that one fluid (usually oil) is displaced by the other, and the residual fluids are ignored. On one side of the moving interface, only the displaced fluid is assumed to move and on the other, only the displacing fluid.

Darcy's Law and Interfacial Stability

Consider the inclined reservoir shown in Figure 5.2. Phase 1 is displacing phase 2. The flow is assumed to be one-dimensional and inclined at an angle θ to the horizontal. It is assumed that only phase 1 flows behind the front and only phase 2 flows beyond it.

Darcy's law may be written for each phase as[1]

$$\left(\frac{dP}{dL}\right)_1 = -\frac{V_1\mu_1}{k_1} - \rho_1 g \sin\theta \tag{5.1}$$

$$\left(\frac{dP}{dL}\right)_2 = -\frac{V_2\mu_2}{k_2} - \rho_2 g \sin\theta$$

If a small penetration such as that shown forms by some random perturbation, then it can grow forward if the absolute pressure gradient within phase 1 is less than that within phase 2. With this condition true, the pressure at the end of the protuberance can overwhelm the pressure in phase 2. This condition may be written as equation 5.2.

[1]In equations 5.1 the velocity V_i is equal to the volumetric fluid flow divided by the cross-sectional area for flow, i.e., q_i/A. It is equal to the average velocity within the pores of the matrix multiplied by the porosity, ϕ.

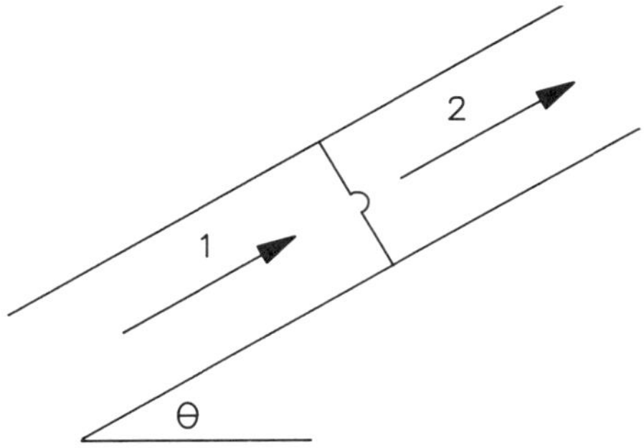

Figure 5.2 One-Dimensional Flooding with an Incipient Finger

Penetration can occur if

$$-\left(\frac{dP}{dL}\right)_1 < -\left(\frac{dP}{dL}\right)_2$$

i.e., if

$$\frac{V_1\mu_1}{k_1} - \frac{V_2\mu_2}{k_2} + (\rho_1 - \rho_2)g \sin\theta < 0 \tag{5.2}$$

Usually conservation will require that the velocity in phase 1 be equal to the velocity within phase 2. This will not be the case if a phase change such as condensation is occurring at the interface. We consider this case later. For the present it is assumed that $V_1 = V_2$ in equation 5.2.

If this is the case, then it is apparent that the flow will be stable if

$$\frac{\mu_1}{k_1} > \frac{\mu_2}{k_2} \quad \text{and} \quad \rho_1 \sin\theta > \rho_2 \sin\theta \tag{5.3}$$

Conversely, the flow is always unstable if both the viscosity and the gravity terms are unfavorable, i.e., if,

$$\frac{\mu_1}{k_1} < \frac{\mu_2}{k_2} \quad \text{and} \quad \rho_1 \sin\theta < \rho_2 \sin\theta \tag{5.4}$$

In between these conditions there are combinations of conditions that can result in stability or instability depending upon the relative magnitude of the counteracting terms.

The term involving gravity in Equation 5.2 will be zero if the flow is horizontal or if the densities of the two phases are equal. In this case the condition for stability is determined by the mobility ratio, as given in equation 5.5. If $\rho_1 \sin\theta = \rho_2 \sin\theta$—i.e., if $\rho_1 = \rho_2$ or $\theta = 0$, then the flow is stable if

$$\frac{\mu_1}{k_1} > \frac{\mu_2}{k_2} \tag{5.5}$$

or

$$\text{Mobility ratio} = M = \frac{k_1/\mu_1}{k_2/\mu_2} < 1$$

If M is less than 1—as, for example, when the oil viscosity is low compared to that of the flooding material—then the front is stable. Equation 5.5 is quoted very commonly in the literature.

Even if the viscosity ratio term is unfavorable, it can be overcome by a favorable gravity term, provided that the flooding velocity is sufficiently low. The gravity term tends to stabilize the flow if the densest stream is below the other—e.g., if light gas is displacing dense oil downward or water is displacing a lighter oil upward.

The gravity term can stabilize the effect of an unfavorable mobility ratio if

$$V < \frac{(\rho_1 - \rho_2)g \sin \theta}{\dfrac{\mu_2}{k_2} - \dfrac{\mu_1}{k_1}} \tag{5.6}$$

Similarly, a favorable mobility ratio can stabilize a flood having a destabilizing gravity term if the flooding velocity is high enough to overcome the gravity effect, i.e., if

$$V > V_c = \frac{-(\rho_1 - \rho_2)g \sin \theta}{\dfrac{\mu_1}{k_1} - \dfrac{\mu_2}{k_2}} \tag{5.7}$$

As will be shown later, the Buckley-Leverett effects tend to improve the stability; it is shown on page 209 that the condition for stability for a horizontal system then becomes

$$M_s = \frac{\dfrac{k_{rof}}{\mu_o} + \dfrac{k_{rwf}}{\mu_w}}{\dfrac{k'_{ro}}{\mu_o}} < 1$$

M_s is less than 1 for stable displacement. This can be visualized by saying that the flooding fluid is not really just water but a mixture of water and oil, which behaves as if it has a much higher viscosity. This advantage is really largely illusionary, however, because, as will be seen, the Buckley-Leverett mechanism can leave vast quantities of oil behind the front when the oil is very viscous.

Effect of Interfacial Tension

There are other stabilizing effects that tend to reduce viscous fingering. One of these is the effect of interfacial tension; this tends to stabilize because it acts to shorten the interface.

This stabilizing effect is the largest for small fingers because the pressure gradient across an interface is inversely proportional to the radius of curvature. As a result, very small perturbations tend to shrink; with the right conditions, larger ones can grow.

Fingering can be initiated by fluctuations caused by nonuniformities in the reservoir matrix; then those fingers that are larger than the critical dimension can grow.

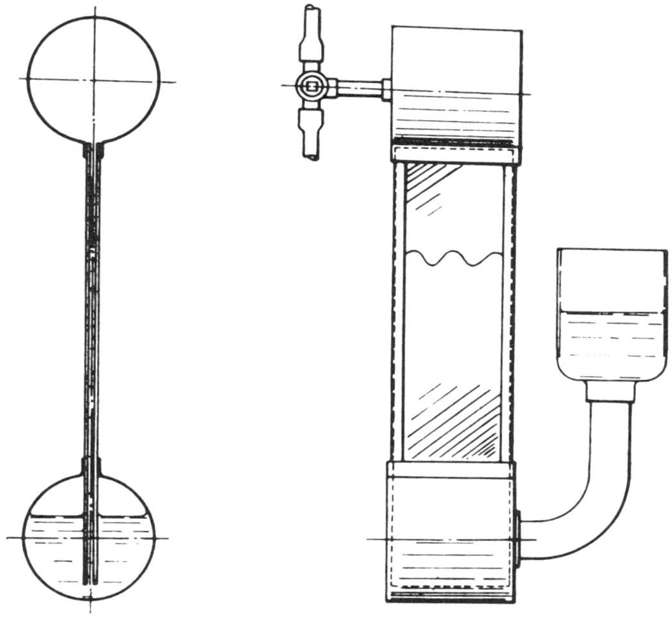

Figure 5.3 Hele-Shaw Model Apparatus. The Cell Could be Adjusted so That it was Either Vertical or Horizontal (from Saffman and Taylor 1958)

The effect of interfacial tension has been studied in Hele-Shaw laboratory models by Chuoke, van Meurs, and van der Poel (1959) and by Saffman and Taylor (1958). Models such as that shown in Figure 5.3 consist of two parallel, closely spaced glass plates. The equations of motion for a single fluid flowing in the two dimensions of such a model are the same as in a homogeneous porous solid with a permeability of $b^2/12$ in two dimensions, where b is the distance of separation between the plates. The permeability is zero in the third dimension.

Figure 5.4 shows the interface in such a model in an experiment that exhibits instability.

A Simple Theory for Stabilization by Interfacial Tension

Figure 5.5 represents an interface within a porous solid at a point where a penetration, which is assumed to be hemispherical and of radius R, has formed. Also shown is a similar negative penetration. The pressure within the protuberance is higher than otherwise because of the tension within the interface. This excess pressure tends to oppose the growth of both positive and negative interfacial protuberances. If the incremental pressure due to interfacial tension is of the order[2] of $2\sigma/R$, then

[2]It can be shown that the excess pressure within a bubble or droplet of radius R is equal to $2\sigma/R$ by balancing the pressure force on a midplane of the bubble $\Delta P(\pi R^2)$ to the tensile stress holding it together along the perimeter of the midplane $\sigma(2\pi R)$.

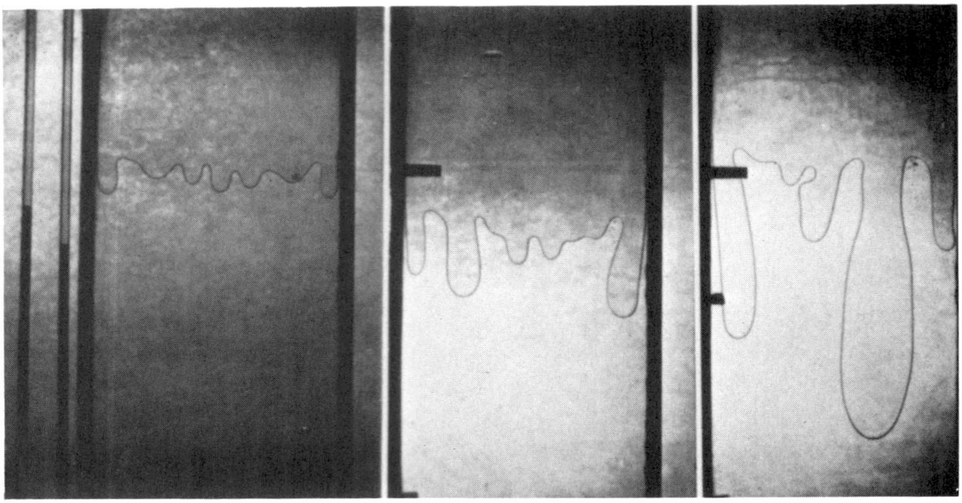

Figure 5.4 Photograph of Progressive Fingering in Hele-Shaw Model. Air is Displacing Glycerine Downwards at a Velocity Greater Than the Critical One (from Saffman and Taylor 1958)

the incremental pressure gradient necessary to make it grow is approximately $2\sigma/R^2$ (equation 5.8).

$$\frac{dP}{dL_1} - \frac{dP}{dL_2} > \frac{2\sigma}{R^2} \tag{5.8}$$

Obtaining the pressure gradient terms from 5.1 and substituting in 5.8 gives, with some rearrangement, the minimum radius R of the protuberance which will be able to grow (equation 5.9). The critical wavelength λ_c (looking on the two adjacent hemispheres as a wave) is about $4R$. The condition that the protuberances should grow is thus

$$\begin{aligned} \lambda_c = 4R &> 4\left[\frac{2\sigma}{V\left(\dfrac{\mu_2}{k_2} - \dfrac{\mu_1}{k_1}\right) + (\rho_2 - \rho_1)g \sin\theta}\right]^{1/2} \\ &= 4\left[\frac{2\sigma}{\left(\dfrac{\mu_2}{k_2} - \dfrac{\mu_1}{k_1}\right)(V - V_c)}\right]^{1/2} \end{aligned} \tag{5.9}$$

A more accurate and sophisticated analysis of this problem was published by Chuoke, van Meurs, and van der Poel (1959) with the result given by equation 5.10. This is the same as 5.9 except for the constant 2π, which replaces $4\sqrt{2}$. Essentially the same equation was published by Saffman and Taylor (1958).

$$\lambda_c = 2\pi\left[\frac{\sigma}{\left(\dfrac{\mu_2}{k_2} - \dfrac{\mu_1}{k_1}\right)(V - V_c)}\right]^{1/2} \tag{5.10}$$

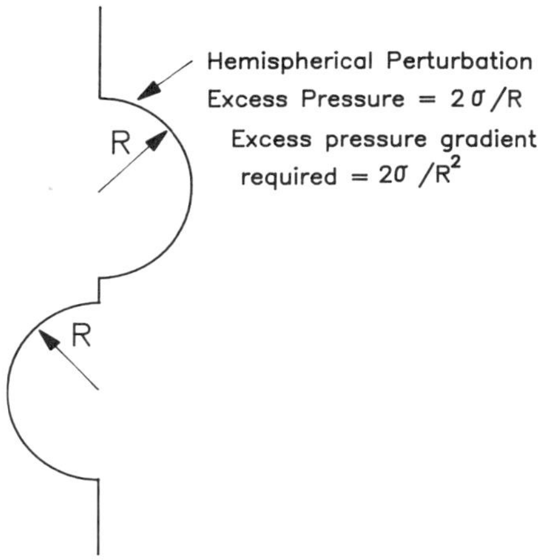

Figure 5.5

and

$$\lambda_m = \lambda_c \sqrt{3}$$

In addition to the critical wavelength for finger growth, Chuoke calculated the wavelength of "maximum instability," λ_m. This is the perturbation wavelength (the pitch of repeated fingers), which will grow at the fastest rate; it should be the prevalent wavelength. It is shown with equation 5.10.

An important concept in this theory is that for fingering of this type to occur, the dimensions of the reservoir must be substantially larger than the wavelength of a critical perturbation. For example, while large fingers can grow in field reservoirs, they might not be possible in laboratory-scale equipment. As a result, laboratory flooding results may lead to optimistic predictions.

Chuoke et al. extended the idea just described to represent *interfaces* within porous solids. To do this, they substituted σ^*, which they defined as the effective surface tension, for σ. The effective surface tension σ^* depends on the capillary properties and wetting of the matrix as well as on the interfacial tension. Its use allows the representation of the idea that when a protuberance forms, much new interfacial area is created, particularly if the matrix is wetted by the displacing fluid.

This idea is related to the concept of imbibition. When a protruding water finger tends to form, the water within it is drawn away by imbibition into the surrounding reservoir if this is water wet and at the irreducible saturation. Removal of water by imbibition tends to reduce the rate of growth of the finger by increasing the absolute pressure differential required to transfer the increased flow of water into it.

If σ^* is assumed proportional to σ and k_1 and k_2 are assumed to be equal to k, then equation 5.10 becomes

$$\lambda_m = C\left[\frac{\sigma^* k}{(\mu_2 - \mu_1)(V - V_c)}\right]^{1/2} \tag{5.11}$$

Chuoke assumed that the effective surface tension is proportional to the interfacial tension and used equation 5.11 to predict the most probable perturbation wavelength. In this equation C is a constant (Chuoke's constant) for a particular reservoir material that includes some of the preceding numerical constants as well as a proportionality constant for the relationship between the effective interfacial tension and the actual surface tension.

This idea was studied further by Peters and Flock at the University of Alberta (1981). Values of C taken from the literature are given in Table 5.1. They show the large effect that the wettability of the matrix has upon stability. The displacement of oil by water is stabilized considerably by the imbibition effect if the reservoir is water-wet.

TABLE 5.1 Values of C, Chuoke's Constant

	Oil-wet	Water-wet
Peters and Flock (1981)	25.4	190.5
Chuoke	30	200

When an automobile is driven in the rain, there is a very noticeable tendency for water to stream down the windshield if it is dirty (non-water-wetted), whereas the water will run in continuous curtains if the windshield is clean and water-wet. This effect is approximately analogous to that just described.

Effect of Condensation upon Interfacial Stability

Equation 5.12 is the condition for stability that was developed earlier.

$$-\left(\frac{dP}{dL}\right)_1 < -\left(\frac{dP}{dL}\right)_2$$

or

$$\frac{V_1\mu_1}{k_1} - \frac{V_2\mu_2}{k_2} + (\rho_1 - \rho_2)g \sin\theta < 0 \tag{5.12}$$

Consider a one-dimensional flood in which steam is introduced into a cold-water-saturated reservoir. This is shown in Figure 5.6. It is assumed that there are no transverse heat losses and that the water saturation in the steam-swept zone is constant and equal to S_{wi}. Although the diagram shows the reservoir as being horizontal, this is not necessarily so.

Steam passes through the steam zone and condenses at the interface. The latent and sensible heat raise more of the reservoir to the steam-saturation temperature, and the condensate joins the water already present and is pushed forward by the steam. We will consider the stability of the interface. As was discussed in Chapter 3, the temperature falls quickly from T_S to T_R at the condensation front if it is assumed that there are no lateral heat losses, that the flow velocity is uniform across the section, that the temperature of the solid is equal to the fluid temperature at each point, and if thermal conduction is ignored.

From what has gone before one might suspect that the interface would tend to be unstable because the steam is much less viscous than the water. However there is another very important factor: the velocity within the steam zone is much higher

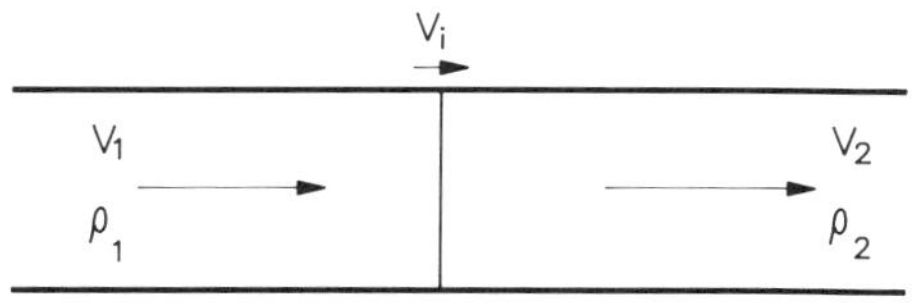

Figure 5.6 Advancing Condensation Front

than that in the water zone because of the shrinkage that occurs during condensation. This tends to stabilize the process by requiring a higher absolute pressure gradient within the steam zone.

The ratio of the velocity of the water to that of the steam may be obtained by combining a material balance (5.13) with a heat balance (5.14).

MATERIAL BALANCE

$$\rho_2 V_2 = \rho_1 V_1 + \phi V_i(1 - S_{wi})(\rho_2 - \rho_1) \tag{5.13}$$

HEAT BALANCE ABOVE T_2

$$\rho_1 V_1(\mathbf{H}_1 - \mathbf{h}_2) = V_i[(1 - \phi)\rho_r C_r(T_1 - T_2) + \phi\rho_w S_{wi}(\mathbf{h}_1 - \mathbf{h}_2) + \phi\rho_1(1 - S_{wi})(\mathbf{H}_1 - \mathbf{h}_2)] \tag{5.14}$$

Equation 5.14 may be rearranged to provide the ratio of the velocity of the interface to that of the steam, namely,

$$\frac{V_i}{V_1} = \frac{\rho_1(\mathbf{H}_1 - \mathbf{h}_2)}{\phi\rho_1(1 - S_{wi})(\mathbf{H}_1 - \mathbf{h}_2) + \phi\rho_w S_{wi}(\mathbf{h}_1 - \mathbf{h}_2) + (1 - \phi)\rho_r C_r(T_1 - T_2)} \tag{5.14a}$$

In the preceding, the fluid velocities are defined as the volumetric flows divided by the total area. The actual average fluid velocities within the pores will be larger by a factor of $1/\phi$.

For practical cases, the terms involving the heat capacities of the rock and residual water in the denominator of equation 5.14a will be much larger than that for the steam. For example, consider the following assumed typical numerical data:

Steam pressure		515 psia	(3.55 MPa)
Steam temperature	T_1	470°F	(243°C)
Steam enthalpy	$\mathbf{H}_1$	1205 Btu/lb	(2803 kJ/kg)
Steam density	ρ_1	1.11 lb/ft^3	(17.8 kg/m^3)
Hot-water enthalpy	$\mathbf{h}_1$	453 Btu/lb	(1054 kJ/kg)
Hot-water density	ρ_w	50 lb/ft^3	(801 kg/m^3)
Reservoir temperature	T_2	50°F	(10°C)
Cold-water enthalpy	$\mathbf{h}_2$	18 Btu/lb	(42 kJ/kg)
Rock heat capacity	C_r	0.23 Btu/lb °F	(0.96 kJ/kg °C)
Rock density	ρ_r	165 lb/ft^3	(2643 kg/m^3)
Porosity	ϕ	0.3	
Irreducible water saturation	S_{wi}	0.25	

Then, by substitution,

$$\frac{V_i}{V_1} = \frac{1318}{296 + 1631 + 11157} = 0.1007$$

The ratio of the interface velocity to the actual steam velocity within the pores is, for this example, given by

$$\frac{V_i}{V_1}\phi(1 - S_{wi}) = 0.1007 \times 0.3 \times 0.75 = 0.0227$$

The ratio V_2/V_1 can be calculated by rearranging equation 5.13 to give

$$\frac{V_2}{V_1} = \frac{\rho_1}{\rho_2} + \frac{V_i}{V_1}\frac{\phi(1 - S_{wi})(\rho_2 - \rho_1)}{\rho_2} \tag{5.15}$$

This ratio can be substituted into equation 5.2 to give the condition for stability. If it is assumed that $k_1 = k_2 = k$, then equation 5.2 can be rearranged to give the following condition for stability.

$$\frac{V_2\mu_2}{V_1\mu_1} < 1 + \frac{k(\rho_1 - \rho_2)g \sin\theta}{V_1\mu_1} \tag{5.16}$$

The gravity term on the right-hand side of equation 5.16 can either help stabilize or destabilize the front depending on whether the steam is flowing from above or from below.

The left-hand side of equation 5.16 has been computed for the conditions given in the previous example for a number of assumed steam temperatures with the results shown in Table 5.2.

TABLE 5.2 Values of Stability Factor for Adiabatic Displacement of Water by Steam

Steam pressure (psia)	100	400	700	1000	1300	1600
T_1 (°F)	328	445	503	545	578	605
μ_1/μ_2	11.5	7.0	5.7	5.0	4.5	3.9
$V_2\mu_2/V_1\mu_1$	0.12	0.23	0.30	0.36	0.42	0.45

The values of the factor are less than unity, although they tend to increase with temperature. This means that unless the gravity term in 5.16 is quite unfavorable, the heating/displacement flood should be stable. The shrinkage that occurs during the condensation of steam is thus able to stabilize the steam-water interface. In evaluating the left-hand side of equation 5.16, the viscosity of water at steam temperature rather than the much higher value at reservoir temperature has been used for μ_2. The logic behind this is that right at the front, the water is in equilibrium with the steam and is at the same temperature. For a protuberance to grow as depicted in Figure 5.2, it is necessary for the pressure gradient in the steam to overcome that in the water right at the interface. If such a protuberance does start, there is an additional stabilizing influence because of the higher steam velocity required to supply the heat loss from the growing protuberance. This additional flow rate in the steam side will require an additional pressure gradient, and this will result in additional stabilization, as is discussed in the next section.

In a number of simple steam waterfloods that were carried out in the laboratory by Baker (1973) in a three-dimensional reservoir model, it was found that the condensation interface was indeed stable.

In steamfloods, the condensation front is usually stable if the flow is sideways or downwards, i.e., if gravity tends to stabilize the front. However, the condensate from the steam runs through the oil which is being displaced and is produced with it. The condensation interface is stable, but the water-oil interface is usually not.

Miller's Temperature Gradient Stabilization

Another mechanism has been described by Miller (1975) and Armento and Miller (1977) that tends to stabilize steam fronts. This is depicted in Figure 5.7. The figure shows a steam condensation front in which a perturbation has formed at *P*.

There is a sharp temperature gradient at the front and, because *P* is indented, the heat transfer from it to the neighboring reservoir will be at a greater rate than elsewhere along the front. This results in increased flow at *P* above the interface, with the result that the finger will tend to fill up with water. Also, the increased flow to *P* will tend to increase the local pressure drop and reduce the amount of "push" available at *P*.

As in the discussion of the stabilizing effect of interfacial tension, it is apparent from Figure 5.7 that the heat-transport effect will tend to stabilize small fingers more than it will large ones. Again, there is the possibility that laboratory experiments will show stable displacements, which may not be found on a larger scale in the field. Also, heterogeneities in the field tend to promote fingering and frontal instability. An extreme case of this is the flow of steam along either a natural or arti-

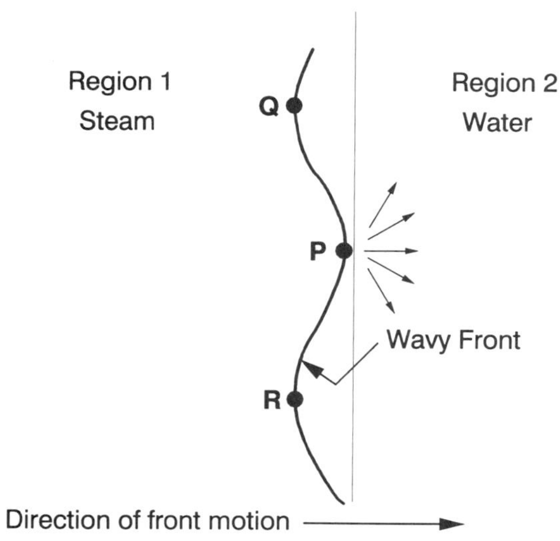

Figure 5.7 Increased Heat Loss from Frontal Indentation (Armento and Miller 1977)

ficial fracture. In this situation the condensation interface can advance, even though vast quantities of heat are transferred through the walls of the fracture.

DARCY'S LAW FOR TWO-PHASE FLOW

If two or more separate phases flow simultaneously through a porous solid, then the flow of each phase is less than that which would be produced by the same pressure gradient if it were the only phase present. The individual fluids compete with each other as they flow through the medium and impede each other's progress. Allowance is made for this by introducing new variables, relative permeabilities, into Darcy's equation. These empirical correction factors are obtained for any particular porous medium by experiment.

For two-phase flow, Darcy's law is modified by the introduction of the relative permeability coefficients k_{ro} and k_{rw}. Equations 5.17 and 5.18 are the Darcy equations for the flow rate of oil, q_o, and water, q_w, for one-dimensional flow in a bed of cross-sectional area A that is inclined at an angle θ to the horizontal (see Figure 5.8). If $\sin\theta$ and the flow are both positive, then the flow is uphill. In general, the pressure in the oil phase, P_o, is not the same as the pressure in the water phase, P_w, because of the effect of capillary pressure P_c. The effective permeability for each phase is equal to the absolute permeability k multiplied by the relative permeability k_{ro} or k_{rw}.

$$q_o = -\frac{kk_{ro}A}{\mu_o}\left(\frac{\partial P_o}{\partial x} + \rho_o g \sin\theta\right) \tag{5.17}$$

$$q_w = -\frac{kk_{rw}A}{\mu_w}\left(\frac{\partial P_w}{\partial x} + \rho_w g \sin\theta\right) \tag{5.18}$$

Relative Permeability Curves

The advantage of the relative permeability concept as a means for allowing for the effect of the competitive flow of immiscible fluids in porous solids is that in many practical situations, it is found that the relative permeability values are, to a first approximation at least, functions of the fluid saturations (i.e., the volume fractions of the individual fluids present in the pore space) alone. In most situations, relative permeabilities are largely independent of flow velocity and of the fluid viscosities.

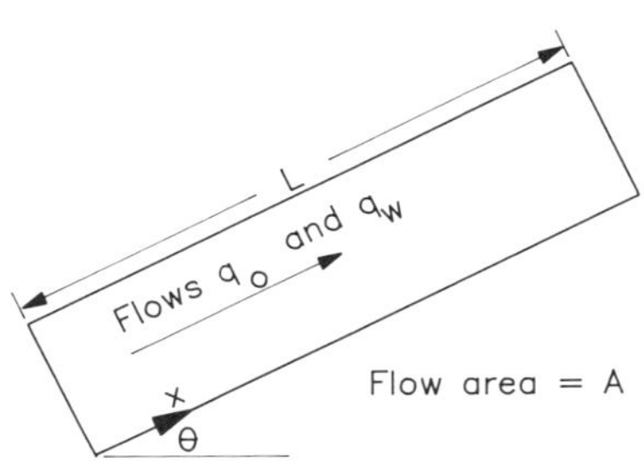

Figure 5.8 Darcy's Law for Two Phase Flow

Measurements of the relative permeabilities for binary oil-water systems covering a wide range of oil viscosity for a particular reservoir rock are shown in Figure 5.9.

The pioneering paper by Leverett, from which this figure is copied, also includes data that show that over a wide range of interfacial tension (but a range that excludes the extremely low interfacial tensions obtainable during surfactant flooding), the relative permeabilities are essentially independent of interfacial tension.

The relative permeabilities are monotonic functions of the saturation of the reservoir material; i.e., the relative permeability of either phase increases with its saturation. Often the curves are qualitatively of the general shape shown in Figure 5.10.

The curves in this figure are used in the illustrative examples that follow. They do not represent any particular reservoir situation, but they are similar in general shape to curves for the flow of conventional oils through water-wet, unconsolidated sands. The particular curves shown correspond to the simple, arbitrary, algebraic equations in the figure.

The relative permeability curve for water starts at an irreducible water saturation S_{wi} and rises to 1 at $S_w = 1$. It is tangent to the x axis. This is common when the rock is water-wet. At S_{wi} the irreducible water is held in place by interfacial tension forces. While the water phase is still continuous, the connections between water pockets are vanishingly thin in places.

The relative permeability curve for oil is generally similar in shape to that for water but falls rather more sharply to the residual oil saturation S_{or}. In a water-wet system, the residual oil saturation is not as well defined as the irreducible water saturation, and its value may be more dependent upon the history of how the core was flooded.

For the purposes of this analysis it is assumed that definite, simple, saturation-dependent relative permeability curves exist and that, for the purpose of the numerical examples developed later, they may be represented as in Figure 5.10.

THE FRACTIONAL FLOW EQUATION

For what follows, it is useful to use the equations defining the relative permeabilities (5.17 and 5.18) to derive an equation that relates the composition of the flowing

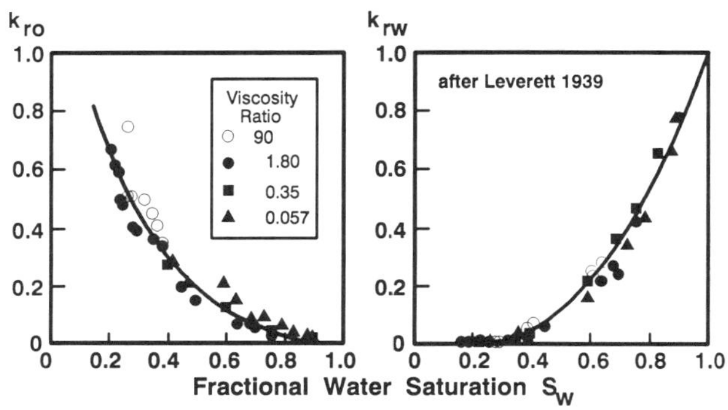

Figure 5.9 Effect of Water Saturation on the Relative Permeability to Oil (left) and to Water (right) for a Particular Core with a Wide Range of Oil to Water Viscosity Ratios

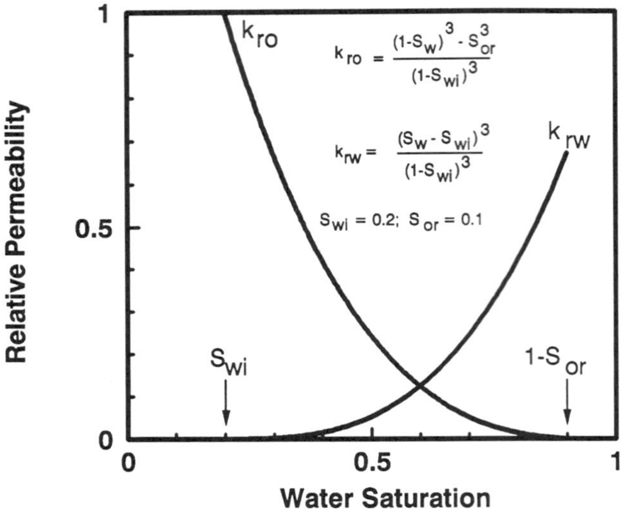

Figure 5.10 Hypothetical Relative Permeability Curves

stream to the saturation in the matrix. We will derive an equation to determine the fraction of water f_w in the flowing stream as a function of the water saturation S_w in the matrix.[3] The choice of water fraction and saturation rather than oil fraction and saturation is arbitrary; the corresponding oil fraction and saturation can be obtained readily from the water fraction and saturation.

We write the oil flow as the difference between the total flow and the water flow as in equation 5.19 and substitute in 5.17 to give 5.20.

$$q_o = q_t - q_w \tag{5.19}$$

$$\frac{q_t \mu_o}{kk_{ro}} - \frac{q_w \mu_o}{kk_{ro}} = -A\left(\frac{\partial P_o}{\partial x} + \rho_o g \sin\theta\right) \tag{5.20}$$

Rearrange equation 5.18 to give 5.21, subtract equation 5.20 from equation 5.21, and, making use of the definition of the capillary pressure 5.22, rearrange to give equation 5.23.

$$\frac{q_w \mu_w}{kk_{rw}} = -A\left(\frac{\partial P_w}{\partial x} + \rho_w g \sin\theta\right) \tag{5.21}$$

$$P_c = P_o - P_w; \qquad \Delta\rho = \rho_w - \rho_o \tag{5.22}$$

$$q_w\left(\frac{\mu_w}{kk_{rw}} + \frac{\mu_o}{kk_{ro}}\right) = \frac{q_t \mu_o}{kk_{ro}} + A\left(\frac{\partial P_c}{\partial x} - \Delta\rho g \sin\theta\right) \tag{5.23}$$

Equation 5.23 may be rearranged to give the general fractional flow equation 5.24.

$$f_w = \frac{q_w}{q_t} = \frac{1 + \dfrac{kk_{ro}A}{q_t \mu_o}\left(\dfrac{\partial P_c}{\partial x} - \Delta\rho g \sin\theta\right)}{1 + \dfrac{\mu_w}{k_{rw}}\dfrac{k_{ro}}{\mu_o}} \tag{5.24}$$

[3]This derivation follows that of Dake (1978).

This equation contains within it the effects of four different factors upon the fractional flow:

1. The ratio of the viscosities of the two fluids.
2. The dependence of the relative permeabilities upon saturation.
3. The effect of gravity. For upward displacement (as in Figure 5.1) of a light oil with water, $\Delta\rho$ is positive; since $\sin\theta$ is also positive, the term containing gravity has the effect of decreasing the water fractional flow. This is usually a desirable effect.
4. The effect of the capillary pressure term. The effect of this term can be seen by writing it as in Equation 5.25. Both of the factors on the right-hand side of equation 5.25 are negative for water displacing oil in a water-wet system. The term $\partial P_c/\partial S_w$ is negative because, for water-wet systems, capillary pressure is monotonic with P_c decreasing with increasing S_w; a typical curve is shown in Figure 5.11. The term $\partial S_w/\partial_x$ is negative when water is displacing oil from a core having a high initial oil saturation. It follows that the effect of the capillary term, in this case, will be to increase the water flow—i.e., water tends to be drawn ahead into the oil-rich zone by capillary attraction.

$$\frac{\partial P_c}{\partial x} = \frac{\partial P_c}{\partial S_w} \cdot \frac{\partial S_w}{\partial x}; \qquad \frac{\partial P_c}{\partial S_w} \text{ and } \frac{\partial S_w}{\partial x} \text{ are both } \quad -ve \tag{5.25}$$

For the present we will assume that the gravity term is zero (either the system is horizontal, $\theta = 0$, or the two phases have the same density, $\Delta\rho = 0$) and assume that the capillary pressure term can be neglected. Except right in the immediate vicinity of the front, this is often a reasonable assumption, since the magnitude of the saturation gradient is small.

With these assumptions[4] the fractional flow equation 5.24 reduces to equation 5.26:

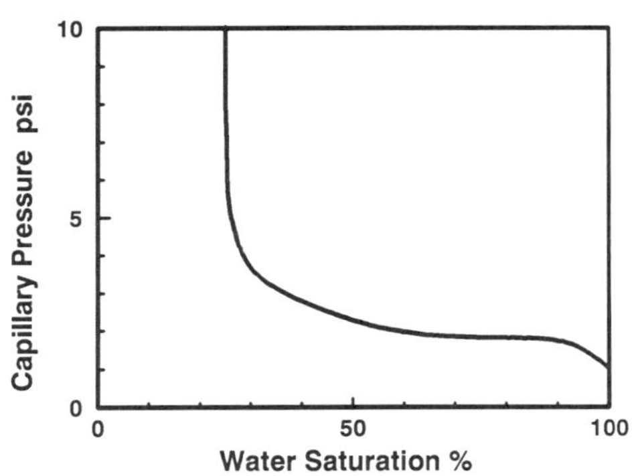

Figure 5.11 Typical Capillary Pressure Curve

[4]Equation 5.26 can be derived very simply if these assumptions are made initially. If $\sin\theta = 0$ and $P_o = P_w = P$, then equations 5.17 and 5.18 become $-kA\frac{\partial P}{\partial x} = \frac{q_o\mu_o}{k_{ro}} = \frac{q_w\mu_w}{k_{rw}}$. Hence $\frac{q_w}{q_o} = \frac{k_{rw}\mu_o}{\mu_w k_{ro}}$, which can be manipulated to give equation 5.26.

$$f_w = \frac{1}{1 + \dfrac{\mu_w}{k_{rw}} \cdot \dfrac{k_{ro}}{\mu_o}} \tag{5.26}$$

This shows that the fraction of water in the flowing stream is a function of the ratio of the relative permeabilities for the flow of the two phases and that it is also equally dependent upon the ratio of the viscosities of the two phases. By inspection, we can see that for a particular saturation (which fixes the relative permeabilities), an increase in oil viscosity increases the flow of water.

Figure 5.12 shows curves of f_w that are calculated from the relative permeability curves of Figure 5.10 for various constant ratios of water viscosity to oil viscosity. The curves demonstrate the increasing tendency for water to flow through the reservoir (even at relatively low water saturations or high oil saturations) when the oil is very viscous.

Effect of the Gravity Term on Fractional Flow

The dimensionless gravity term in equation 5.24 is written in equation 5.27. It is more significant when the difference in densities is higher, when the reservoir is steeply inclined, when the total flow is low, and when the ratio k/μ_o is high.

$$\beta = \frac{kk_{ro}A\,\Delta\rho g \sin\theta}{q_t\mu_o} \tag{5.27}$$

The term β is proportional to the ratio of the gravity potential gradient $\Delta\,\rho g \sin\theta$ to the viscous potential gradient $q_t\mu_o/(kk_{ro}A)$.

High velocities (i.e., high viscosity forces) tend to overcome the effect of gravity. Gravity tends to stabilize the flood if the heavier fluid is below. For a water flood where the water is more dense than the oil, upward displacement reduces the fractional water flow. With gas as the displacing fluid, the flow of gas is reduced by downward flow. For waterfloods of heavy oils where the densities of the two fluids are very similar and when μ_o is very large, gravity has little effect.

Figure 5.13 shows the calculated effect of the gravity term on the fractional flow curves for a matrix with the same relative permeability curves as in the previous example and with a water-to-oil viscosity ratio of 0.1. As the gravity term is

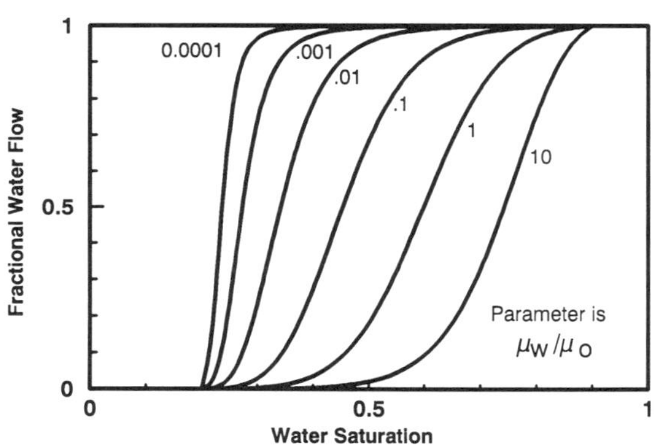

Figure 5.12 Effect of Viscosity Ratio on Fractional Water Flow

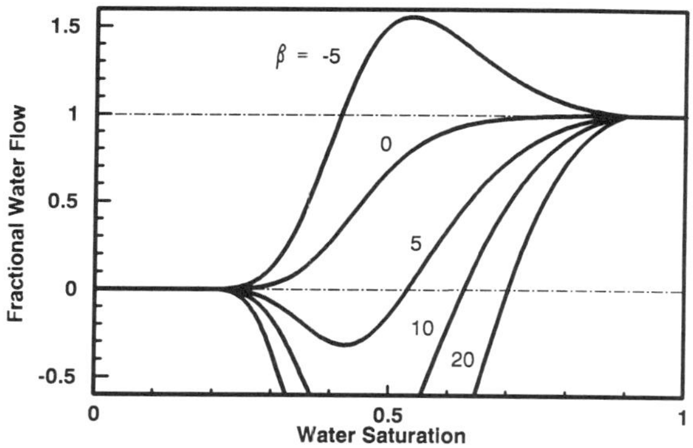

Figure 5.13 Effect of Gravity Term on Fractional Water Flow

increased in value, the fraction of water in the flow for a given rock saturation decreases markedly. The curve for $\beta = -5$ in the diagram corresponds to a case where the water flow is downward; there is countercurrent flow in the range where f_w is greater than 1—i.e., the oil is rising through the falling water. Similarly, with the positive values of β, there is a range where f_w is negative. In this region, water is falling countercurrently through a rising oil stream.

Effect of Segregated Flow on Apparent Relative Permeability and Fractional Flow

The relative permeability concept assumes that at each point in the reservoir, there are both a water and an oil saturation and that the flows of the two phases are related to these saturations by relative permeability curves and flow equations 5.17 and 5.18. This concept cannot be true on a microscopic scale, since, with immiscible fluids, any infinitesimal volume must be filled with either water, oil, or—failing that—rock.

On a larger scale, it is possible to imagine that a volume (of, for example, 1 cc) could have an average composition essentially the same as that of a similar neighboring volume and that the flow within it might also be similar. In this circumstance, the saturations and flowing stream composition must be looked on as statistical averages. This concept is similar to our acceptance of the intensive thermodynamic fluid variables such as temperature and density, which, because of the atomic nature of matter, have meaning only as statistical averages.

It is possible—and usually very likely—that some volumes carry essentially water and others carry oil, i.e., that the flow through the reservoir consists of separate streams or rivulets of water and oil moving beside each other but separately. Such rivulets or fingers may be on a small scale or they may be on a much larger scale. For example, we might imagine individual streams having cross-sectional areas measured in square feet or very small streams, each of only a few pores in size. It can also happen that the rivulets meander and change with time.

One example of large individual streams occurring is when steam segregates above the liquid streams in a steamflood. Here the upper part of the entire reservoir may consist of steam moving separately above oil and water, which are frequently moving in the same general direction below but at different velocities.

An interesting and limiting model can be developed by assuming that within the cross-section of a reservoir in which a one-dimensional waterflood is occurring, the pores contain either oil at the residual oil saturation with water flowing past or water at the irreducible water saturation with oil flowing past. We may also imagine pores that have water flowing in at one end and displaced oil leaving at the other (see the discussion of C.W. Nutt's capillary bundle theory that is described later).

In any one pore in such a system, the saturations are either

$$S_w = S_{wi} \qquad \text{and} \quad S_o = 1 - S_{wi}$$

or

$$S_w = 1 - S_{or} \quad \text{and} \quad S_o = S_{or} \tag{5.28}$$

The corresponding relative permeabilities of water and oil for the two types of pores are

$$\begin{aligned} k_{ro} &= k'_{ro} \quad \text{and} \quad k_{rw} = 0 \\ k_{ro} &= 0 \qquad \text{and} \quad k_{rw} = k'_{rw} \end{aligned} \tag{5.29}$$

where k'_{ro} is defined as the relative permeability to oil at S_{wi} and k'_{rw} is defined as the relative permeability to water at $S_w = 1 - S_{ro}$. k'_{ro} and k'_{rw} are often referred to as the *endpoint* relative permeabilities.

The area fraction, a, of the pores carrying water at any particular cross section ($1 - a$ is carrying oil) is given by

$$a = \frac{S_w^* - S_{wi}}{1 - S_{or} - S_{wi}} \tag{5.30}$$

where S_w^* is the volume average water saturation.

The flows of water and oil are given by equations 5.31 and 5.32 if the effect of gravity is neglected. Comparison of these equations with 5.17 and 5.18 leads to the expressions 5.33 and 5.34 for the apparent relative permeability of the reservoir in segregated flow.

$$q_w = \frac{-kk'_{rw}aA\left(\dfrac{\partial P}{\partial x}\right)}{\mu_w} \tag{5.31}$$

$$q_o = \frac{-kk'_{ro}(1 - a)A\left(\dfrac{\partial P}{\partial x}\right)}{\mu_o} \tag{5.32}$$

$$k_{rw}^* = k'_{rw}a = k'_{rw}\left(\frac{S_w^* - S_{wi}}{1 - S_{or} - S_{wi}}\right) \tag{5.33}$$

$$k_{ro}^* = k'_{ro}(1 - a) = k'_{ro}\left(\frac{1 - S_{or} - S_w^*}{1 - S_{or} - S_{wi}}\right) \tag{5.34}$$

Equations 5.33 and 5.34 show that the apparent relative permeabilities are linear functions of saturation. They are the equations of straight lines that join the practical endpoints of the conventional relative permeability curves with the points where the relative permeabilities become zero. This is shown in Figure 5.14.

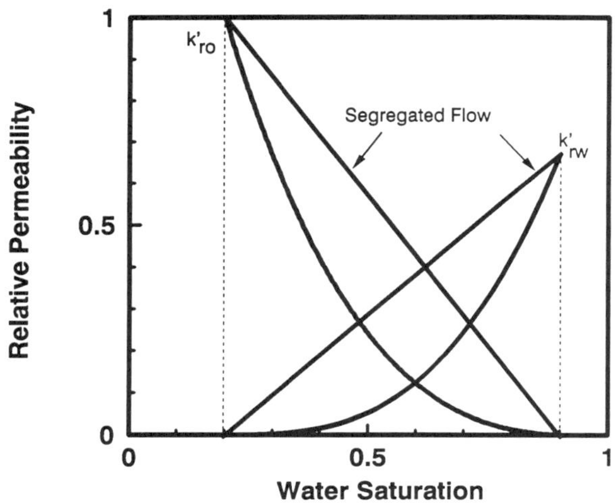

Figure 5.14 Effect of Segregated Flow on Relative Permeabilities

The ratio of the two relative permeabilities for stratified flow is given by equation 5.35, which is derived by dividing 5.34 by 5.33. Fractional flow curves for segregated flow are different in shape from those for diffuse flow, as may be seen by the example in Figure 5.15.

$$\frac{k_{ro}^*}{k_{rw}^*} = \frac{k_{ro}'}{k_{rw}'}\left(\frac{1 - S_{or} - S_w^*}{S_w^* - S_{wi}}\right) \tag{5.35}$$

At first sight it would seem from Figure 5.15 that the displacement of oil by water is much less effective in the system with segregated flow. However, it is not apparent from the figure that the two fractional flow curves cross at high water saturation. As a result of this, the displacement of oil actually becomes more efficient at high water saturations for the segregated flow case than for the diffuse flow one. This difference is demonstrated in later examples.

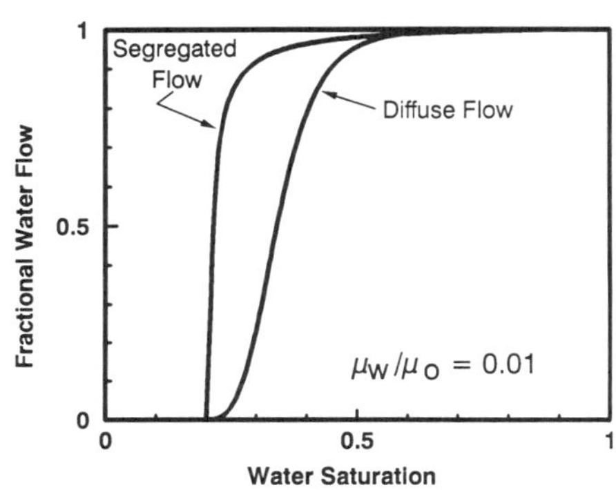

Figure 5.15 Comparison of Fractional Flow Curves for Segregated and Diffuse Flow

THE BUCKLEY-LEVERETT DISPLACEMENT THEORY

Buckley and Leverett (1942) developed a theory that provides a quantitative description of the displacement of one fluid from a porous matrix by an immiscible flood. This theory introduced the idea that water intermingles with the oil as it is being displaced, so that the interfacial surface of the Muskat model becomes a zone with a varying water saturation. This concept applies to the case where the relative permeabilities are nonlinear functions of saturation, as, for example, in the curves of Figure 5.14, and also to the case where the effective permeabilities are linear functions.

The theory makes use of the fractional flow concept. It is assumed that a fractional flow curve is available and that this curve represents the flow process that is taking place in the reservoir. This curve may or may not be similar to that which occurs in laboratory flood tests using small-diameter cores. For example, flow segregation can produce fractional flow curves very different from those for the flow found in laboratory core floods. Dake (1978) discusses a number of factors that can affect the fractional flow curves.

The Velocity of the Shock Front

When oil is displaced from a porous medium by a waterflood, a front advances through the reservoir, and across this front there is an abrupt change in the saturation profile (i.e., there is a discontinuity in $\partial S_w/\partial x$). There may also be a discontinuity in the saturation, S_w, but this is not necessary.

Ahead of the front, oil is flowing without water through a reservoir with a saturation corresponding to the initial irreducible water saturation S_{wi}. Immediately behind the front, the flow of water is just sufficient to keep up with the moving front.

Figure 5.16 depicts the situation in the immediate vicinity of the front. In time dt the front advances by a distance dx_f. During this period the water that flows past the plane with an initial abcissa equal to x_f is just sufficient to provide the extra water that remains behind the front in the elemental distance dx_f. A material balance for this is given by equation 5.36.

WATER BALANCE[5]

$$q_t f_{wf}\, dt = A\phi(S_{wf} - S_{wi})\, dx_f \tag{5.36}$$

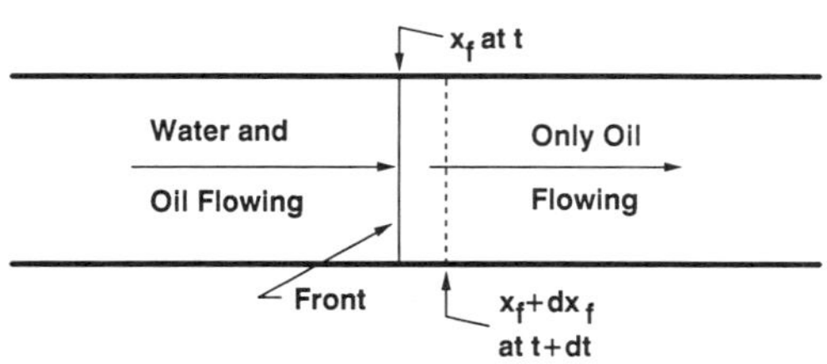

Figure 5.16 Conditions at the Shock Front

[5]Later on, it is shown that the transition zone behind the front may be terminated by a "trailing" front, which is accompanied by a second discontinuity, in $\partial S_w/\partial x$. The velocity at which this trailing front moves can be found in an analogous manner to that developed here. At the trailing front, the upstream fractional water flow is equal to that of the injected flooding fluid—usually unity.

This may be rearranged to give the front velocity dx_f/dt as a function of the fractional water flow behind the front and the saturation behind the front.

FRONT VELOCITY

$$\frac{dx_f}{dt} = \frac{q_t}{A\phi}\left(\frac{f_{wf}}{S_{wf} - S_{wi}}\right) \tag{5.37}$$

The term in brackets on the right-hand side of 5.37 is a function of S_{wf} and is known if the fractional flow curve is known. It is the slope of the straight line joining the point corresponding to the front conditions to the point $(S_{wi}, 0)$ on a graph of f_w versus S_w.

There is a maximum value to this slope, which may be found by drawing a tangent, as shown in Figure 5.17. The slope of this line when substituted into equation 5.37 yields the maximum velocity at which a shock front can move, and this represents the conditions for the shock front that forms in practice. A shock front at this condition will overrun any front having a different saturation.

For segregated flow, the maximum slope occurs when $f_w = 0$ (see Figure 5.15) and the water saturation at the shock front is S_{wi}. For this condition, both the numerator and the denominator of the term in brackets in equation 5.37 become zero. However, there is still a discontinuity in the water-saturation gradient at the front. In this case the front velocity can be found from the limiting form of equation 5.37 using L'Hospital's theorem.

$$\frac{dx_f}{dt} = \frac{q_t}{A\phi}\left(\frac{df_w}{dS_w}\right)_{S_w=S_{wi}} \tag{5.38}$$

The Saturation Behind the Front

Behind the front, the water saturation falls from $1 - S_{or}$ right at the point of injection down to the saturation at the shock front, as found by the construction in Figure 5.17.

Consider the change in saturation in the differential reservoir element shown in Figure 5.18. The saturation within this stationary element will, in general, tend to change with time because the concentration of the stream flowing from it is not the same as that of the stream that is entering; it is being depleted of oil. This relationship is represented by the continuity equation 5.39.

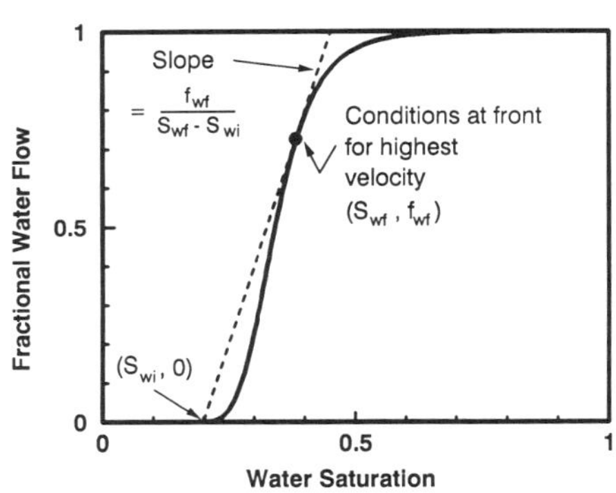

Figure 5.17 Shock Front Conditions

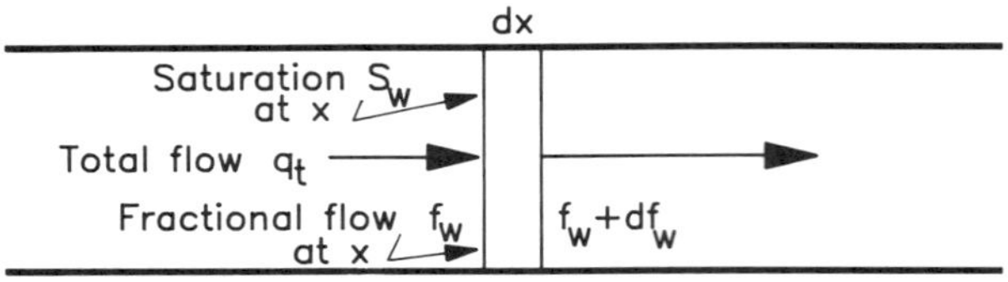

Figure 5.18 Conditions behind the Front. S_w Varies Continuously

$$\left(\frac{\partial S_w}{\partial t}\right)_x = \frac{-q_t}{\phi A}\left(\frac{\partial f_w}{\partial x}\right)_t \qquad \text{where } f_w = \frac{q_w}{q_t} \tag{5.39}$$

Behind the front, the water saturation S_w is a continuous function of x and t, and a general differential (equation 5.40) may be written.

$$dS_w = \left(\frac{\partial S_w}{\partial x}\right)_t dx + \left(\frac{\partial S_w}{\partial t}\right)_x dt \tag{5.40}$$

From this, the partial differential $(\partial x/\partial t)_{S_w}$ is obtained as equation 5.41. This may be combined with 5.39, as shown in the second part of 5.41.

$$\left(\frac{\partial x}{\partial t}\right)_{S_w} = -\frac{\left(\frac{\partial S_w}{\partial t}\right)_x}{\left(\frac{\partial S_w}{\partial x}\right)_t} = \frac{q_t}{\phi A}\frac{\left(\frac{\partial f_w}{\partial x}\right)_t}{\left(\frac{\partial S_w}{\partial x}\right)_t} \tag{5.41}$$

The term $(\partial f_w/\partial x)_t/(\partial S_w/\partial x)_t$ is simplified to df_w/dS_w. This is written as an ordinary differential, since f_w is assumed to be a function of S_w alone. This substitution results in equation (5.42), which is known as the Buckley-Leverett equation. It shows that the velocity at which a plane of a fixed saturation advances is proportional to the average fluid velocity multiplied by a function of saturation.

$$\left(\frac{x}{t}\right)_{S_w} = \left(\frac{\partial x}{\partial t}\right)_{S_w} = \frac{q_t}{\phi A}\cdot\frac{df_w}{dS_w} \tag{5.42}$$

or

$$N = \frac{q_t t}{\phi A x} = \frac{1}{f'_w}$$

$$\left(\text{Note } \frac{df_w}{dS_w} \text{ is written as } f'_w\right)$$

where N is the number of pore volumes of injected water (based on the volume of pores up to the point x) that are required to bring the water saturation at position x up to the level corresponding to f'_w. S_w must be greater than S_{wf}. If it is not, the assumption that S_w is a differentiable function of x and t, which is implied by the use of equation 5.40, is not correct.

The differential coefficient in the right-hand side of equation 5.42 is the slope of the fractional flow curve. This has been plotted for a typical example in Figure 5.19. Also shown is the tangent that determines the conditions for the shock front. The conditions behind the shock front correspond to that portion of the frac-

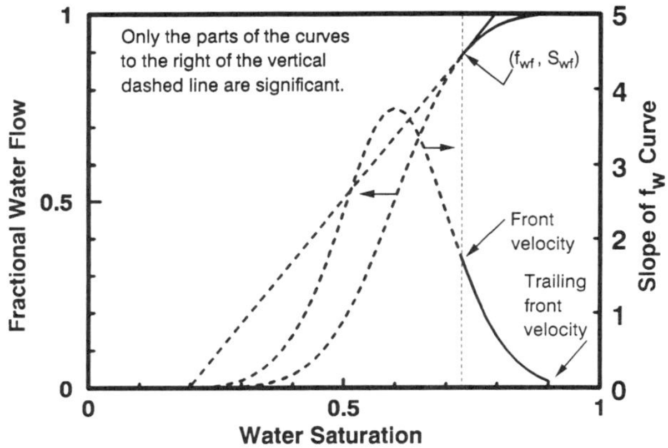

Figure 5.19 Slope of Fractional-Flow Curve

tional flow curve for which $S_w > S_{wf}$, i.e., to the upper right-hand part of the saturation curve. At the shock front, the saturation drops rapidly from S_{wf} to S_{wi}. For stratified flow, the whole range of effective water saturations occurs from S_{wi} to $1 - S_{or}$ because of the shape of the fractional flow curve.

The Upper Shock Front

In Figure 5.19, the curve for the slope does not fall completely to zero at the right-hand limit—i.e., at $S_w = (1 - S_{or}) = 0.9$. This is because the assumed fractional flow curve has a small slope at its upper end. This is a common occurrence. The result is that the saturation curves in Figure 5.22 have a small horizontal part where the water enters the reservoir.

The residual oil saturation extends a finite distance into the reservoir. The performance predicted by the Buckley-Leverett theory is extremely sensitive to the form of the fractional water flow curve at high water saturations.

It is of interest to consider the situation where the flood stream, which is introduced at the start of the reservoir, contains oil. While this circumstance is not of importance in normal waterflooding, it is significant in considering the conditions downstream of the condensation front in a steamflood. In this case, steam condensate and oil are forced continuously through the condensation front.

If the flood stream contains oil, then the fractional water flow at the entrance is fixed, and this flowing-stream condition persists up to a shock front, which moves along at a velocity corresponding to the velocity for the fractional water flow that is introduced. In the case shown in Figure 5.19, this fractional flow composition is 1.0.

In Figures 5.20 and 5.21 the diagram of Figure 5.19 is redrawn for flooding stream compositions of 0.95 and 0.5. In Figure 5.20 the flooding stream composition, $f_w = 0.95$, corresponds to a water saturation of $S_w = 0.78$. This saturation persists downstream up to a shock front, which is moving with a velocity corresponding to the slope of f_w at $S_w = 0.78$. Beyond this front is the intermingled, Buckley-Leverett zone in which the water saturation falls to the main shock front saturation of 0.74.

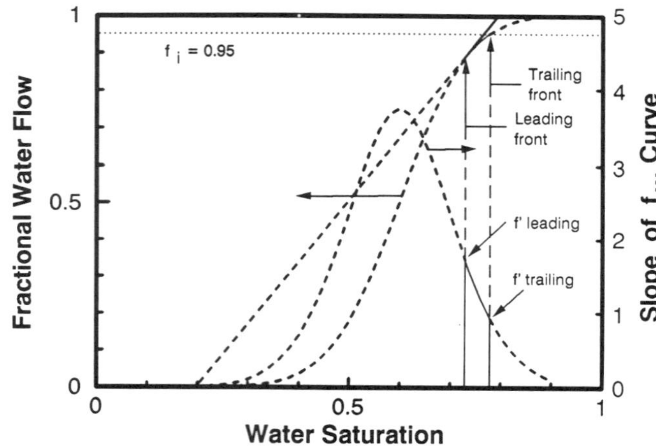

Figure 5.20 Diagram for Flooding with $f_i = 0.95$

In the situation shown in Figure 5.21, the saturation of the upper shock front falls below that of the lower front. In this circumstance, there is only one shock front, and the flowing-stream composition switches abruptly from $f_w = 0.5$ to $f_w = 0$ at the front. Note that in this case, the front velocity is higher than that which would correspond to the normal front velocitv determined by the tangency condition.

Figure 5.22 shows the saturation distribution for the example of Figure 5.19. The upper curved parts of the lines are dx/dt (from equation 5.24) multiplied by the appropriate time from the start of the flood. At the point where the water saturation falls to the shock front saturation S_{wf}, the saturation curve switches to the vertical line with the water saturation falling abruptly to the connate water saturation S_{wi}.

It should be noted that identical curves could have been obtained by using equation 5.42 over the whole saturation range and using the combined fractional flow curve of Figure 5.19, which is drawn by using the experimental fractional flow curve to the saturation front conditions and then the tangent line. This tangent has

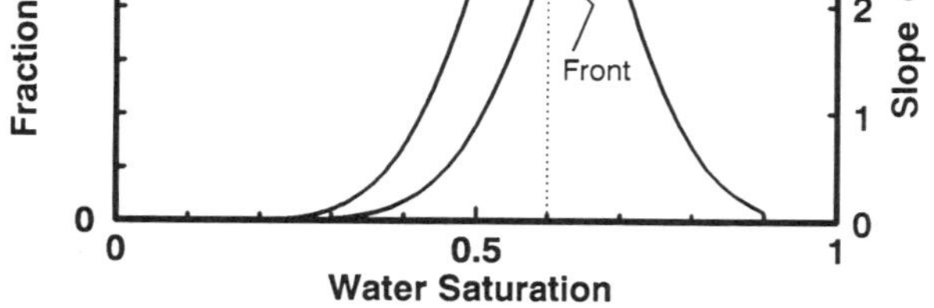

Figure 5.21 Diagram for Flooding with $f_i = 0.5$

a constant slope over the range of saturations that exist across the shock front, and the slope of this tangent gives the shock front velocity when it is inserted in 5.42 (compare 5.42 and 5.37).

Conditions at Breakthrough

Breakthrough occurs when the shock front reaches the limit of the reservoir—i.e., when $x_f = L$. Up until this time only oil is displaced from the reservoir. The quantity of water that has been injected can be found as $q_T t$ from equation 5.43, which is a rearranged form of 5.42.

$$q_t t = \frac{\phi AL}{(df_w/dS_w)_f}$$

or (5.43)

$$\text{No. of PV to breakthrough} = \frac{1}{(df_w/dS_w)_f}$$

Recovery at and after Breakthrough

At some time after breakthrough, the saturation profile will be as in Figure 5.23, and the average water saturation over the length of the reservoir L will be given by equation 5.44. This can be integrated by parts as shown.

$$\overline{S}_w = \frac{1}{L}\int_0^L S_w\,dx = \frac{1}{L}\left\{|xS_w|_0^L - \int_{1-S_{or}}^{S_{wL}} x\,dS_w\right\} \tag{5.44}$$

Substituting $x\,dS_w = (q_t t/\phi A)\,df_w$ from equation 5.42 in 5.44 and integrating gives

$$\overline{S}_w = S_{wL} + \frac{q_t t}{\phi AL}(1 - f_L) \tag{5.45}$$

Substituting L for x in equation 5.42 and combining the result with equation 5.45 results in equation 5.46, which, when rearranged, gives the remarkable equation 5.47. This was first published by Welge (1952).

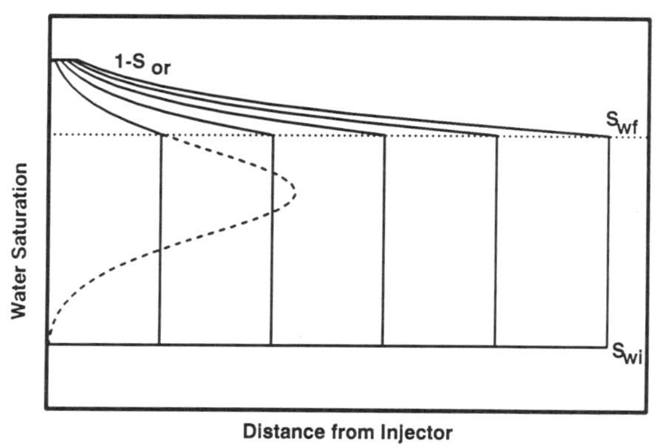

Figure 5.22 Distribution of Water Saturation

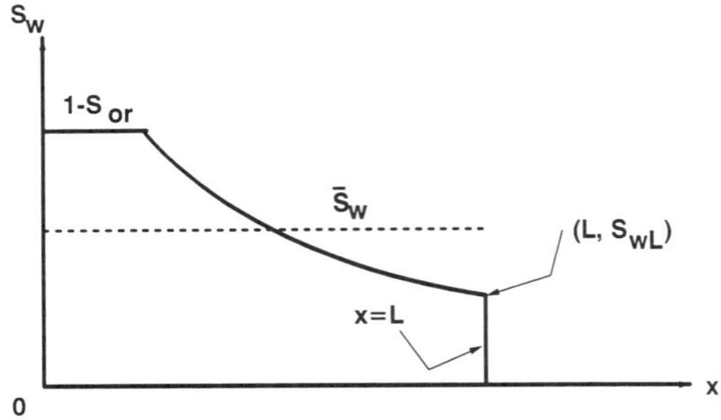

Figure 5.23 Average Water Saturation after Breakthrough

$$\bar{S}_w = S_{wL} + \frac{1 - f_L}{f'_L} \tag{5.46}$$

$$f'_L = \frac{1 - f_L}{\bar{S}_w - S_{wL}} \tag{5.47}$$

In the more general case where the flooding fluid already contains oil, the lower limit of the integral in equation 5.44 should be the saturation corresponding to the floodstream water fraction, f_i. Equation 5.47 then becomes

$$f'_L = \frac{f_i - f_L}{\bar{S}_w - S_{wL}}$$

The geometric significance of this equation may be seen from the construction shown in Figure 5.24.

For any point on the fractional flow curve, which lies at or above the point corresponding to the shock front (S_{wf}, f_{wf}), a tangent drawn upward intersects the line $f_w = f_i$ at a water saturation corresponding to the average water saturation in the reservoir. This is indeed a remarkably simple answer—an almost magical result.

The Buckley-Leverett-Welge method is employed as follows:

1. Draw the fractional flow curve.
2. Draw the tangent from the foot of the curve and determine the conditions at the shock front (S_{wf}, f_{wf}) and the average water saturation at breakthrough, $\bar{S}_{wf}$, from the intersection with the line $f_w = f_i$ (f_i is usually 1).
3. Calculate the oil recovery at breakthrough from

$$\text{Pore volumes of oil recovered at breakthrough} = \bar{S}_{wf} - S_{wi} \tag{5.48}$$

4. Calculate the time of breakthrough from the total injection volume $q_t t$ from 5.43 and the injection rate.[6]
5. For various arbitrary values of S_{wL}, draw the tangents and calculate the corresponding recoveries and injection volumes.

[6]Although it has been assumed throughout this discussion of the Buckley-Leverett theory that the injection rate is constant, this is not necessary for cases where f_w is assumed to be independent of rate, i.e., of q_t/A. For a given injection volume, the same conditions will prevail even though the injection rate varies.

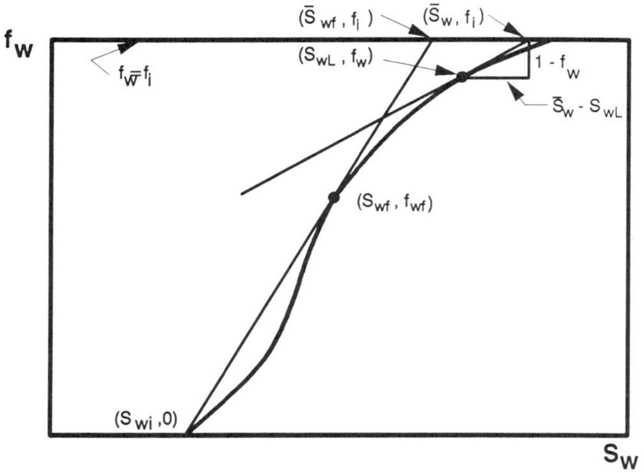

Figure 5.24 Relation between Outlet Conditions and Average Water Saturation

Effect of Viscosity Ratio

Figure 5.25 shows the series of fractional flow curves that was drawn earlier with the tangents corresponding to the conditions at breakthrough drawn in. In this example the water saturation at breakthrough increases as the viscosity of the oil decreases.

The breakthrough conditions listed in Table 5.3 were obtained from the curves in Figure 5.25.

As the viscosity of the oil is decreased, the fraction of a pore volume of water that can be injected before breakthrough and the fractional flow of water at breakthrough both increase. It should also be noted that with very viscous oils, the water saturation at the front is only slightly higher than the irreducible saturation. Only a slight increase in water saturation is required to allow much water to percolate through the largely oil-saturated reservoir. Only very low oil recoveries are obtained at breakthrough when viscous oils are displaced with water.

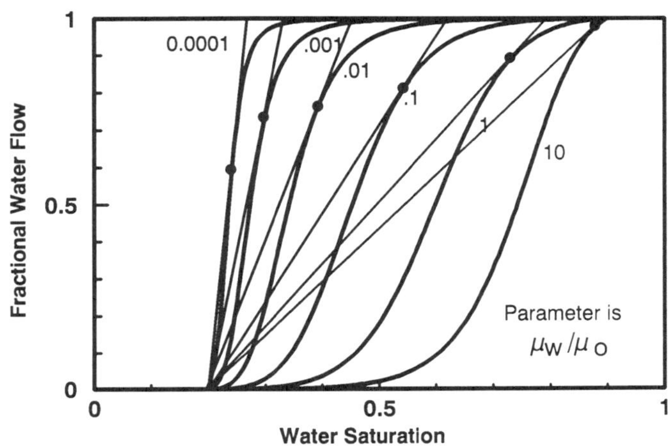

Figure 5.25 Effect of Viscosity Ratio on Breakthrough Conditions

TABLE 5.3 Conditions at Breakthrough for Various Ratios μ_w/μ_o

μ_w/μ_o	f_{wf}	S_{wf}	$\bar{S}_w$	PV	Recovery[1]	Pressure Gradient Ratio[2]
0	—	0.2	0.2	0	0	—
0.0001	0.60	0.24	0.26	0.06	0.09	2.64
0.001	0.73	0.29	0.33	0.13	0.19	2.28
0.01	0.75	0.39	0.48	0.28	0.40	1.70
0.1	0.81	0.54	0.61	0.41	0.59	0.97
1	0.90	0.73	0.78	0.58	0.83	0.33
10	0.98	0.88	0.89	0.69	0.99	—
∞	1.0	0.90	0.90	0.7	1.0	—

(1)Fraction of movable oil recovered.
(2)Pressure gradient in oil bank/pressure gradient just behind front.

Table 5.4 shows the viscosity of Cold Lake crude and water as a function of temperature. It is instructive to compare the viscosity ratios for this system with those in Table 5.3 above. Obviously displacement with water would be much more effective at higher temperatures. Similar viscosity ratio data are shown in Chapter 4, Figure 4.7.

Pressure Gradients during Displacement

The ratio of pressure gradients before and after the shock front are shown in Table 5.3 for the series of oils discussed previously. The pressure gradients upstream and downstream of the shock front were calculated by Darcy's equation (equations 5.49).

$$\begin{aligned} \text{Upstream:} \quad \left(\frac{\partial P_1}{\partial x}\right) &= -\frac{q_t f_{wf} \mu_w}{Akk_{rwf}} \\ \text{Downstream:} \quad \left(\frac{\partial P_2}{\partial x}\right) &= -\frac{q_t \mu_o}{Akk'_{ro}} \end{aligned} \tag{5.49}$$

If the absolute pressure gradient upstream of the front is smaller than that downstream, then the system will be unstable, since any small penetration of the front will be able to grow. The condition for this to be so is given by equation 5.50, which may be developed as shown to give the condition for stability of the front (which is

TABLE 5.4 Viscosity Ratio for Cold Lake Crude as a Function of Temperature

TEMPERATURE °F	°C	μ_o (cp)	μ_w (cp)	μ_w/μ_o	μ_o/μ_w
100	38	15,300	0.70	0.000046	22,000
200	93	181	0.30	0.0017	603
300	149	20.9	0.19	0.009	110
400	204	6.3	0.14	0.022	45

shown as equation 5.51). This may be rewritten as equation 5.52, which was given by Hagoort (1974).

FOR STABILITY

$$-\left(\frac{\partial P_1}{\partial x}\right) > -\left(\frac{\partial P_2}{\partial x}\right) \tag{5.50}$$

or

$$\frac{\mu_o k_{rwf}}{f_{wf}\mu_w k'_{ro}} < 1$$

where

$$f_{wf} = \frac{1}{1 + \dfrac{\mu_w}{\mu_o}\dfrac{k_{rof}}{k_{rwf}}} \tag{5.51}$$

$$M_s = \frac{\dfrac{k_{rof}}{\mu_o} + \dfrac{k_{rwf}}{\mu_w}}{\dfrac{k'_{ro}}{\mu_o}} < 1 \tag{5.52}$$

Graphs of dimensionless pressure gradients within reservoirs being flooded with water for the various ratios of oil viscosity to water viscosity are plotted against dimensionless distance in Figure 5.26. The steps in the curves correspond to the position of the shock front. In this figure the dimensionless quantities are

$$\text{Dimensionless pressure gradient} = \frac{Ak\left(\dfrac{\partial P}{\partial x}\right)}{q_t \mu_w}$$

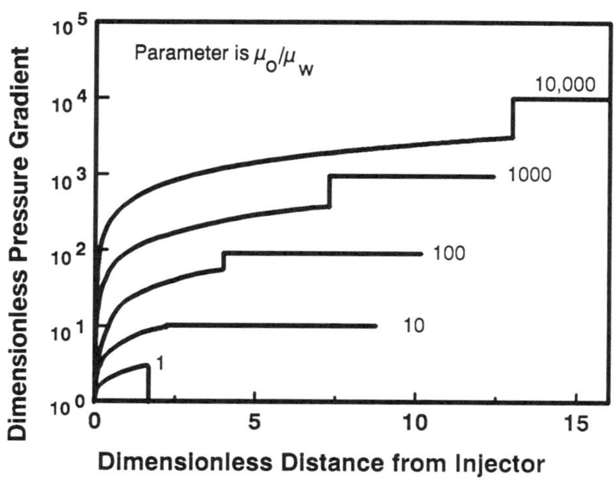

Figure 5.26 Change of Pressure Gradient at Front

and

$$\text{Dimensionless distance} = \frac{A\phi x}{q_t t}$$

For high viscosity ratios, there is an abrupt increase in pressure gradient at the shock front. This promotes instability, and fingering of the floodwater into the oil bank may be expected. For the particular system shown, the front becomes stable when the oil viscosity is ten times that of the water. For lower oil-to-water viscosity ratios, there is a drop in pressure gradient at the shock front.

Numerical Problem on Buckley-Leverett Theory

The relative permeabilities of oil and water in a particular core are given by the following equations:

$$k_{ro} = \left(\frac{0.85 - S_w}{0.55}\right)^3$$

$$k_{rw} = \left(\frac{S_w - 0.3}{0.55}\right)^3$$

where

$$1 - S_{or} = 0.85$$

$$S_{wi} = 0.3$$

Plot the fractional flow of water against the water saturation for two oil viscosities: 5 cp and 150 cp. Assume that the viscosity of the water is 1.0 cp, that the core is horizontal, and that capillary pressure can be neglected.

For each case, using Buckley-Leverett theory, calculate the average water saturation of the core during a waterflood when water first breaks through.

What is the composition of the produced fluid immediately after breakthrough?

How many pore volumes of water have been injected and how many pore volumes of oil have been produced at the breakthrough points?

Each of the floods is carried on to the point where the oil-water ratio in the effluent is 0.025. At this point, for each case calculate the fraction of the original oil in place (OOIP) that has been recovered.

How many pore volumes of water have been injected in each of the cases?

Solution Calculate fractional water flows for each oil viscosity using equation 5.26 (see Table 5.5):

$$f_w = \frac{1}{1 + \dfrac{\mu_w}{k_{rw}} \cdot \dfrac{k_{ro}}{\mu_o}}$$

Plot fractional flows against water saturation. See Figure 5.27. Many more points were used to define the curve than are given in Table 5.5.

Draw tangents from the origin and obtain the water saturation and fractional flows at the tangency points. Obtain the average water saturation at breakthrough

TABLE 5.5 Calculated Fractional Water Flows

S_w	k_{ro}	k_{rw}	Fractional Flow 5 cp	Fractional Flow 150 cp
0.30	1.000	0.000	0.0000	0.0000
0.35	0.751	0.001	0.0050	0.1304
0.40	0.548	0.006	0.0520	0.6221
0.45	0.385	0.020	0.2087	0.8878
0.50	0.258	0.048	0.4827	0.9655
0.55	0.162	0.094	0.7432	0.9886
0.60	0.094	0.162	0.8963	0.9962
0.65	0.048	0.258	0.9640	0.9988
0.70	0.020	0.385	0.9896	0.9996
0.75	0.006	0.548	0.9978	0.9999
0.80	0.001	0.751	0.9998	1.0000
0.85	0.000	1.000	1.0000	1.0000

from the intercepts with $f_w = 1$. The results are shown in Figure 5.27 and on an enlarged scale in Figure 5.28.

	S_{wf}	f_{wf}	$\bar{S}_w$	$\bar{S}_w - S_{wi}$
5-cp oil viscosity	0.575	0.833	0.630	0.33
150-cp viscosity	0.415	0.737	0.456	0.156

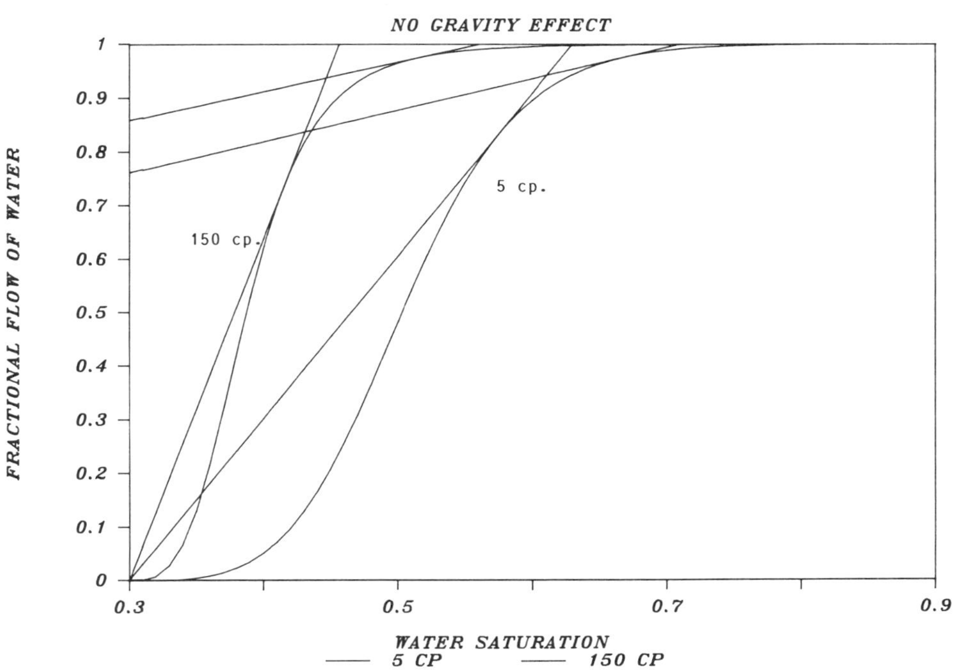

Figure 5.27 Effect of Oil Viscosity

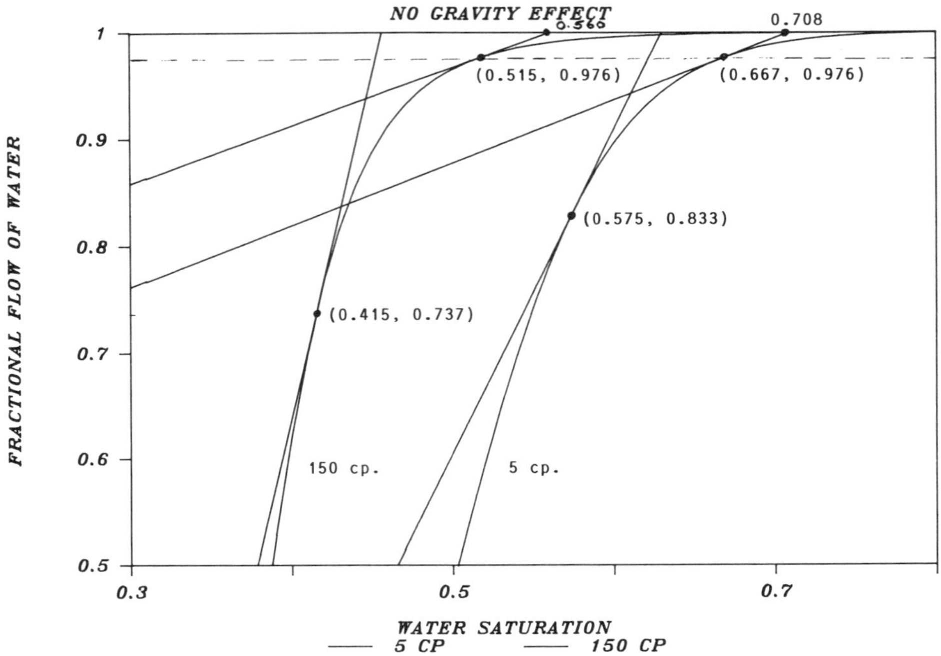

Figure 5.28 Effect of Oil Viscosity (enlarged scale)

$$\text{At breakthrough, volumes of oil produced} = \bar{S}_w - S_{wi}$$

The volumes of water injected up to breakthrough are equal to the volumes of the oil displaced.

Draw tangents at intersections with $f_w = 1.000/1.025 = 0.976$.

Read intercepts of tangents with $f_w = 1.0$ to obtain average water saturations and measure slopes of tangents.

	At $f_w = 0.976$		
	$\bar{S}_w$	f'_w	$1/f'_w$
5-cp oil viscosity	0.708	0.584	1.71
150-cp viscosity	0.560	0.545	1.84

$$\text{Total pore volumes of water injected} = 1/f'_w \quad \text{at} \quad f_w = 0.976$$

$$\%\ \text{OOIP recovered} = \begin{cases} 100(0.708 - 0.3)/(1 - 0.3) = 58.3\% & \text{for 5-cp oil} \\ 100(0.560 - 0.3)/(1 - 0.3) = 37.1\% & \text{for 150-cp oil} \end{cases}$$

More oil was recovered using less water for the less viscous oil.

Comparison of Displacement with Diffuse and Segregated Flows

The numerical example of the Buckley-Leverett theory is extended in this section to the case where the flow is considered to be completely segregated.

The effective relative permeabilities corresponding to segregated flow for the system considered are:

$$k_{ro} = \frac{0.85 - S_w}{0.55} \quad \text{and} \quad k_{rw} = \frac{S_w - 0.3}{0.55}$$

It is assumed that the porous medium is saturated with oils having a viscosity of either 5 cp or 150 cp and that the water saturation is at the irreducible level initially. As before, the viscosity of the water is taken as 1 cp.

Buckley-Leverett diagrams for the segregated flow cases are shown in Figure 5.29.

Conditions at Breakthrough

In the segregated case, only 0.11 pore volume of the 5-cp oil and a mere 0.004 pore volume of the 150-cp oil are recovered at the breakthrough point, but in each case the composition of the effluent stream is still nearly all oil immediately after breakthrough. In the diffuse case, 0.33 and 0.156 volumes of oil are recovered at breakthrough, and the compositions of the effluents jump to 83% and 74% water for the 5- and 150-cp oils, respectively.

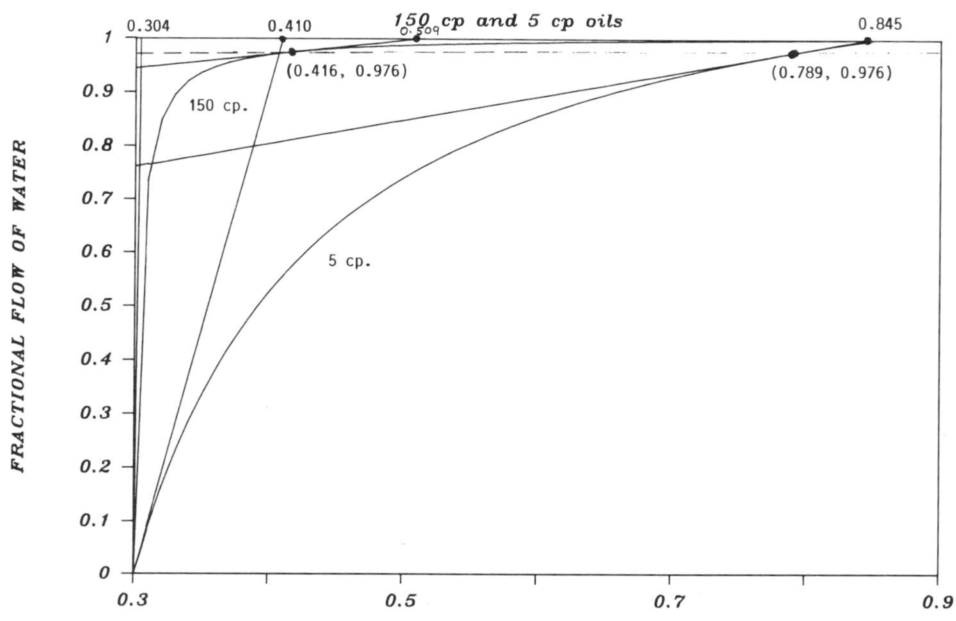

Figure 5.29 Effect of Oil Viscosity—Segregated Flow

	Conditions at Breakthrough			
	S_{wf}	f_{wf}	$\overline{S}_{wf}$	$\overline{S}_{wf} - S_{wi}$
5-cp oil				
Diffuse	0.575	0.833	0.630	0.330
Segregated	0.3	0	0.410	0.110
150-cp oil				
Diffuse	0.415	0.737	0.456	0.156
Segregated	0.3	0	0.304	0.004

Conditions when Oil-Water Ratio Falls to 0.025

The conditions when the oil-water ratio falls to 0.025 are shown by the upper points of tangency in Figure 5.29. The results are summarized next.

	Conditions at O/W = 0.025			
	S_{wf}	f_{wf}	$\overline{S}_{wf}$	$\overline{S}_{wf} - S_{wi}$
5-cp oil				
Diffuse	0.667	0.976	0.708	0.408
Segregated		0.976	0.845	0.545
150-cp oil				
Diffuse	0.515	0.976	0.560	0.260
Segregated		0.976	0.509	0.209

At this point, the performance for the segregated flow has surpassed that for the diffuse flow for the 5-cp oil; 0.545 PV of oil has been recovered compared to 0.408 for the diffuse flow. The reason for this difference is that the relative permeability for oil is significantly higher at high water saturations for the segregated flow than it is for the diffuse flow.

For the 150-cp oil, the recovery is much better than might have been expected from the poor results at breakthrough; 0.209 PV of oil has been recovered, as compared to 0.260 for the diffuse flow case. As may be seen by the data in the next section, the segregated case eventually results in better performance if the process is continued long enough.

Comparison of Oil Recoveries

The fractional recoveries of mobile oil for the four cases are compared in Figure 5.30 as functions of the pore volumes of water injected.

Water Saturation Profiles

Water saturation is plotted as a function of distance for the two cases involving 5-cp oil in Figure 5.31. A similar diagram could be drawn for the more viscous oil. The abcissa in Figure 5.31 is $\phi Ax/q_t t$, or f'_w. Several features should be noticed:

1. The abrupt change in saturation at the front for the diffuse flow.
2. The much greater distance that the segregated front has reached.
3. The position of the trailing shock front of mobile oil in the segregated case.

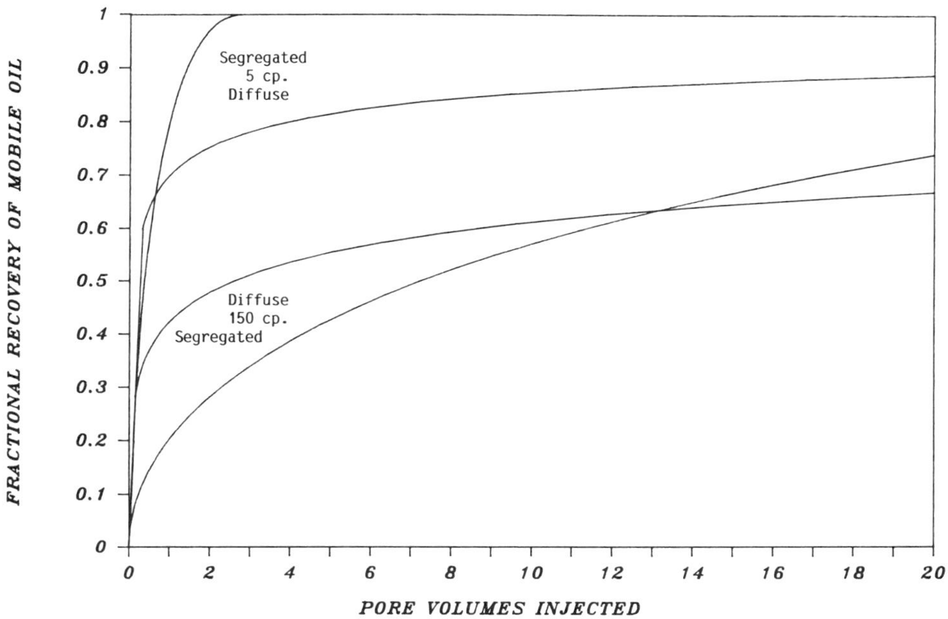

Figure 5.30 Fractional Recovery of Oil—Segregated and Diffuse Flow; 5 and 150 cp

The distance that the trailing edge of mobile oil has advanced is proportional to the slope of the fractional flow curve at $f_w = 1$. This is a positive value, $(\mu_o/\mu_w)(1 - S_{or} - S_{wi})$, for the segregated flow case but is zero for the diffuse flow

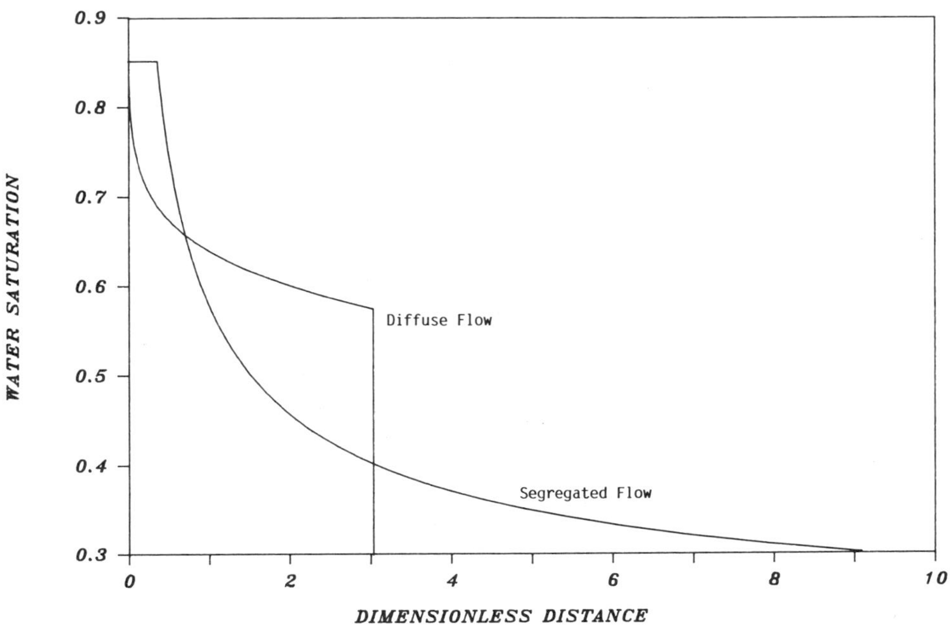

Figure 5.31 Water Saturation—Segregated and Diffuse Flow; 5 cp

because of the tangency of the oil relative permeability curve to the water saturation axis for this particular problem.

It is clear from this hypothetical example that the effectiveness of a waterflooding operation is critically dependent upon the shape of the fractional flow curve—particularly the shape at high water saturations.

C.W. NUTT'S CAPILLARY BUNDLE MODEL

The approach to the displacement of oil from a porous medium discussed in the last section depends upon the concept of both oil and water flowing in the same region, with the relative fluxes determined by the saturations of the phases. This concept allowed the development of relative permeability and fractional flow curves. The mathematical extension of the idea that was developed by Buckley and Leverett and added to by Welge results in a realistic model of the flooding process.

A paper by C.W. Nutt (1982) explains the observed phenomena from a different viewpoint. He shows that most of the phenomena of waterflooding, including the effect of viscosity ratio, can be explained by considering that the porous solid behaves as a simple bundle of capillaries of varying radii.

He introduces his idea by considering a situation in which the porous solid is represented by means of two capillaries connected in parallel, as shown in Figure 5.32.

It is assumed that the flow resistance in the feed and discharge lines can be neglected and that the flow in each capillary can be calculated by Poiseuille's equation. Also, the effects of capillary forces caused by the oil-water interface are neglected at first.

Consider the case where the capillaries are first full of an oil that is more viscous than water. Water is injected at a constant rate. The expected performance is shown in Figure 5.33, where it will be seen that the interface moves to the outlet

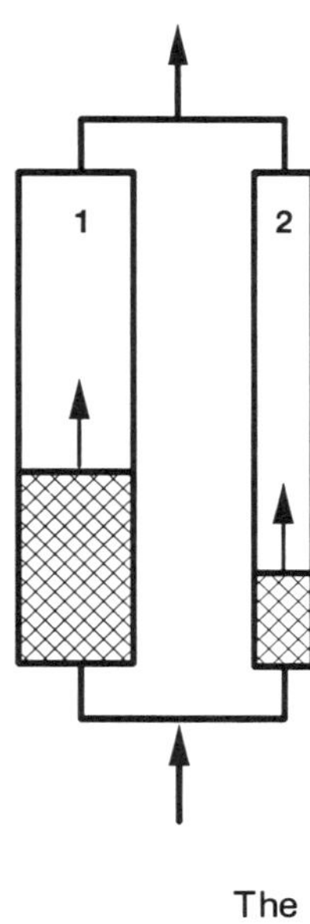

Figure 5.32 Displacement for Two Capillaries of Different Diameters (after Nutt 1982)

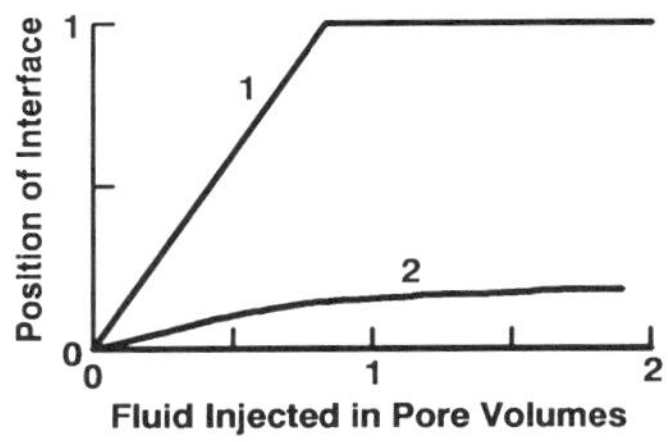

Figure 5.33 Position of Interfaces during Displacement (after Nutt 1982)

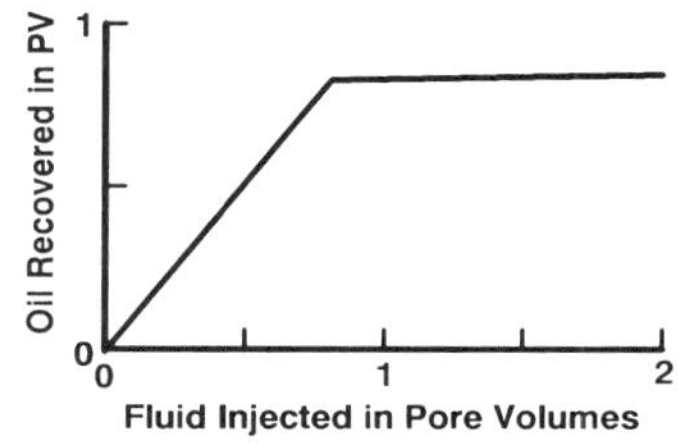

Figure 5.34 Cumulative Oil Production (after Nutt 1982)

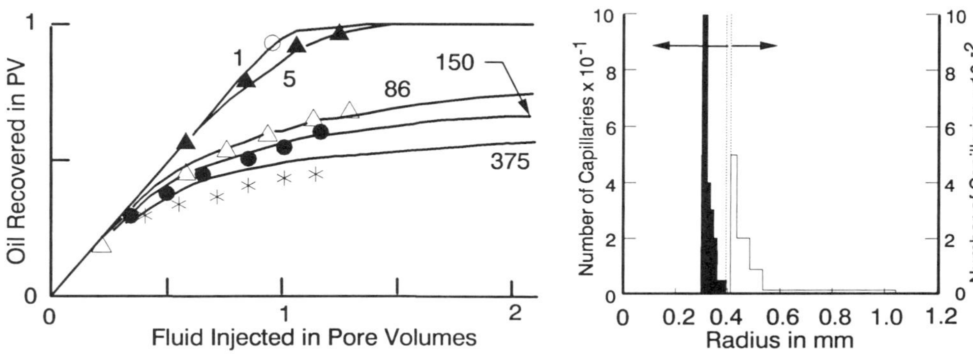

Figure 5.35 Comparison with Experimental Data from Sandpack Waterfloods. Parameter in left half of Figure is μ_o/μ_w (after Nutt 1982)

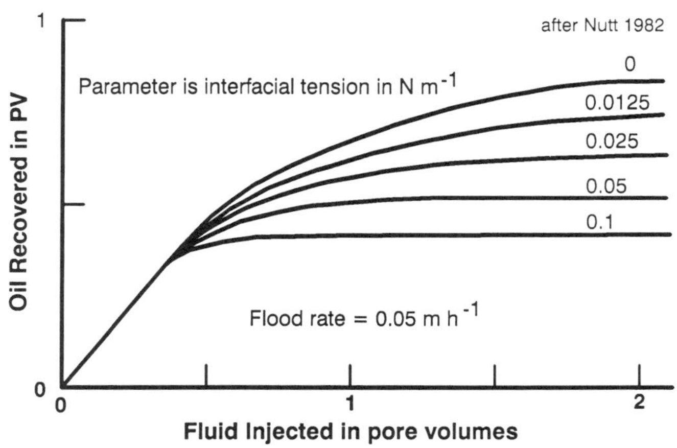

Figure 5.36 Effect of Interfacial Tension on Recovery for an Oil-Wet System

end more rapidly in the larger capillary. This occurs because, for a given pressure drop, the average flow velocity is greater in a larger capillary. As the displacement proceeds, the ratio of velocities increases because the larger capillary contains rela-

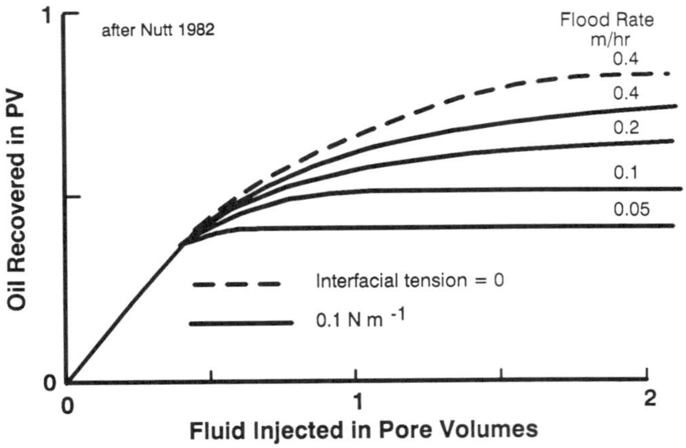

Figure 5.37 Effect of Flood Rate for Oil-Wet System

tively more water; this is less viscous, and, as a result, the pressure difference between the inlet and outlet of both tubes decreases.

When the water reaches the end of the large capillary, breakthrough occurs; after that, relatively little oil is produced because of the low pressure gradient. Most of the injected water flows through the large capillary. Figure 5.34 shows the qualitative effect on oil recovery that would be expected.

Nutt expands this idea by considering a whole range of capillaries of varying diameters in parallel and adjusting the distribution of sizes to match observed results. One of the more interesting comparisons is shown in Figure 5.35.

In the left part of this figure, the points show experimental displacement results reported by Blackwell for a series of flooding experiments using a single sandpack and various oil-water viscosity ratios. The curves in the diagram were calculated for the experimental conditions using the pore-size distribution shown in

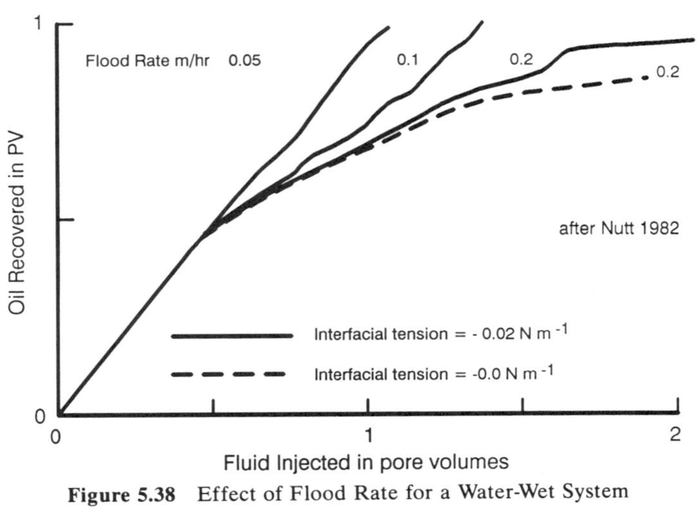

Figure 5.38 Effect of Flood Rate for a Water-Wet System

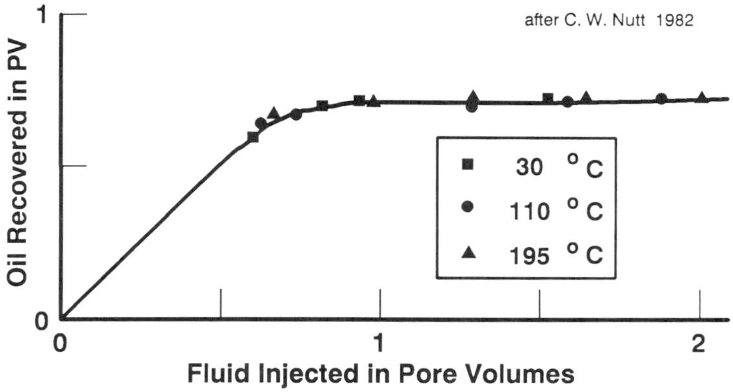

Figure 5.39 An Effect of Temperature. Recovery of a Mixture of *n*-decane, *n*-hexadecane, and Squalane from a Core. Points are Experimental Data from Sudibjo et al. (1978); The Curve is a Calculated One for all Three Temperatures Using a Selected Capillary Size Distribution

the right half. Nutt's simple theory gives a good prediction of the effect of the viscosity ratio on the displacement efficiency. As with the Buckley-Leverett theory, it predicts poorer displacement efficiencies and earlier breakthrough for the more viscous oils.

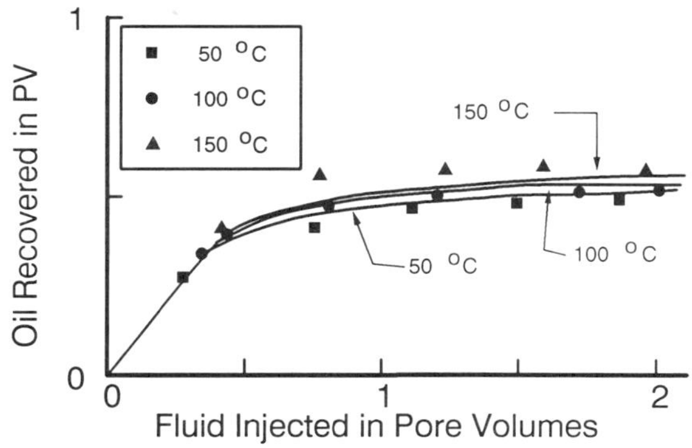

Figure 5.40 Another Effect of Temperature (after Nutt 1982)

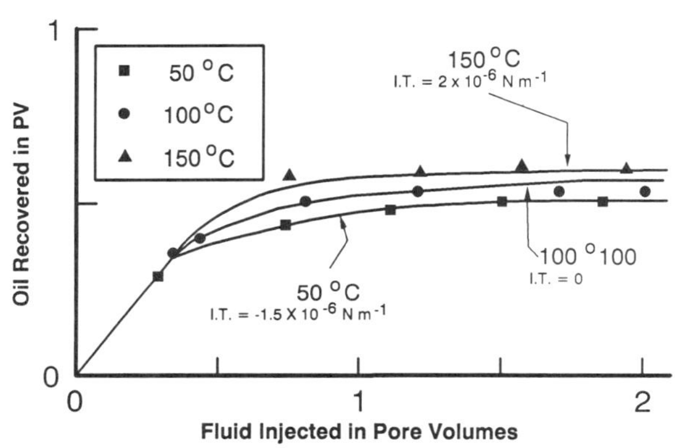

Figure 5.41 Effect of Temperature Combined with Change in Wettability (after Nutt 1982)

The effect of interfacial tension is also included by Nutt in his theory. The interfacial tension slows down the displacement of the oil if the solid is oil-wet and vice versa; in either case, the smaller the radius of the capillary, the greater the effect. Thus the interfacial tension effect tends to make the displacement for capillaries of different radii more uniform in water-wet systems and less uniform in oil-wet ones.

Nutt shows how a high oil-water interfacial tension may be expected to have little effect on the recovery at breakthrough for oil-wet systems but that it will have a significant effect on the ultimate recovery—high interfacial tension gives lower recovery. This is shown by Figure 5.36; the breakthrough points occur where the recovery curves deviate from the initial straight line.

Nutt also predicts that in an oil-wet system with a finite interfacial tension, a higher recovery will be obtained with a higher flood velocity. Predicted results are shown in Figure 5.37.

For water-wet systems Nutt's theory predicts that lower flooding velocities should give better recoveries because the capillary force tends to improve the displacement in the smaller capillaries. The effect is shown in Figure 5.38.

Another effect that Nutt considers is that of temperature. He shows that for some porous solids (i.e., for some pore-size distributions), the effect of temperature should be negligible—as it has been shown to be by Sudibjo et al. for one core. See Figure 5.39.

Conversely, Nutt predicts that for other pore-size distributions, there should be a temperature effect (Figure 5.40). By also assuming that the wettability changes with temperature, an effect very similar to the experimental data can be predicted (Figure 5.41).

Although Nutt's capillary model theory is able to explain many of the observed phenomena, it is not a complete explanation. The classical Buckley-Leverett theory is also able to explain the same phenomena. What is clear, however, is that the processes occurring in the displacement of oils—and particularly heavy oils—by water from porous solids are complex and are affected by a multitude of variables.

It is also apparent that attempts to extrapolate experimental data widely by theoretical concepts may lead to incorrect predictions because of the incorrect representation or even omission of important phenomena. It appears that the recent tendency to replace experimental studies with complicated mathematical models based on simplistic concepts is still very premature. Simple predictions based on realistic experimental information are much to be desired. We need to keep our laboratories for a while yet!

ANALYSIS OF STEAMFLOOD USING THE BUCKLEY-LEVERETT THEORY

In this section, the behavior of a one-dimensional, adiabatic steamflood is analyzed. The process that is considered is shown in Figure 5.42.

This diagram shows the flooding process at an intermediate stage. Steam is introduced at the left and sweeps through the steam-swept region, in which the temperature is constant at T_S. The small drop in temperature due to the pressure drop in the steam zone is ignored. At the condensation front, all the steam condenses abruptly and the liberated heat of condensation together with the sensible heat of the condensate is absorbed in raising the reservoir and its contents from the initial temperature T_R to T_S. The same assumptions are made as for the case of the steam displacement of a water-saturated reservoir (see Figure 5.6). It is assumed that there are no lateral heat losses, that the fluid velocities are uniform across the section, that the temperature of the solid is equal to that of the fluids at each point, and that thermal conduction can be ignored.

In the steam-swept region, steam flows at a relatively high velocity and moves the oil forward at the saturation temperature. The displacement of oil in this region can be described quantitatively using the Buckley-Leverett theory (Shutler and Boberg 1972; Boberg 1987) in the same manner as has been described for waterflooding. No condensation of steam occurs within the steam-swept region, since it is at constant temperature and it is assumed that there are no heat losses. The water saturation in the region corresponds to the irreducible value, S_{wi}. If the steam is wet, or if there are heat losses, a higher water saturation is required in order to move the liquid water to the interface; this is discussed later. For the present case, it is assumed that the steam is introduced dry and saturated and that there are no losses; even if water is flowing, the increase in water saturation required to provide water mobility will usually be small. The oil saturation increases from left to right in the steam-swept region in Figure 5.42.

The position of the condensation front is determined by a heat balance. The heat liberated by the condensation of the steam and by the cooling of the condensate from the start of the process to the time corresponding to the situation in Figure 5.42 is conserved as sensible heat within the steam-swept region. The position of the condensation front can be found from this heat balance. The volume of the steam-swept region increases proportionally to the quantity of injected steam.

At the condensation front, steam saturation drops to zero, the water saturation increases somewhat because there is need for water to flow in the oil-bank region beyond, and the oil saturation rises abruptly. The oil which is swept from the steam region, together with the condensate, flows forwards from the condensation front into the waterflood region.

In the waterflood region, some oil is displaced by the flowing water. The oil in the flooding stream from the steam chamber also flows through the waterflood region. Because of the high proportion of oil in the flooding stream leaving the condensation front, the conditions are similar to those which were discussed previously in connection with Figure 5.21. Immediately beyond the condensation front, the satu-

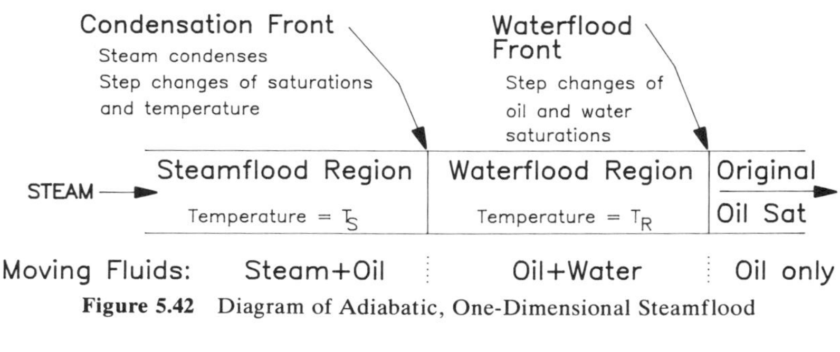

Figure 5.42 Diagram of Adiabatic, One-Dimensional Steamflood

ration conditions correspond to the fractional flow of oil and water that is leaving the front. If the fractional water-flow curve is concave to the left, as in Figure 5.21, then these conditions remain constant in the reservoir up to the waterflood front, where the saturation changes abruptly to correspond to those in the initial reservoir. The waterflood front advances much more rapidly than does the condensation front, and water breaks through long before the arrival of the condensation front.

Buckley-Leverett Theory Applied to the Steam Chamber

The general distribution of saturations that occur within the steam chamber are shown in Figure 5.43. The abscissa in this diagram is the dimensionless distance, which is equal to df_s/dS_s. The region with constant oil saturation, S_{or} occurs only if the curve of f_s is not tangent to the saturation axis at its upper terminal point.

The water saturation is constant at S_{wi} throughout because there is no water flow within the steamflood region. As has been mentioned previously, if the injected steam is wet, then the water saturation would be somewhat higher than S_{wi}.

The curve of steam saturation versus distance in Figure 5.43 is obtained directly from relative permeability data for the flow of steam and oil in the reservoir matrix with an irreducible saturation of water. In Figure 5.43, the steam saturation is plotted downward.

Calculation of Volume of Steam Within the Reservoir

Consider the position within the steamflood region that corresponds to the vertical dotted line shown in Figure 5.43; this represents some point behind the condensation front. The number of pore volumes of steam (measured as vapor and based on the total pore volume of the reservoir between the point of injection and the point corresponding to the vertical dotted line) that has been injected is equal to $1/f_s'$ (see equation 5.42).

Another way of looking at the horizontal scale in Figure 5.43 is to regard it as the distance along the reservoir expressed in pore volumes of reservoir per volume of injected steam. At any particular time, some of the injected steam remains within the steamflood region to the left as steam, and the remainder has passed beyond. The volume of remaining steam per volume of injected steam is given by the

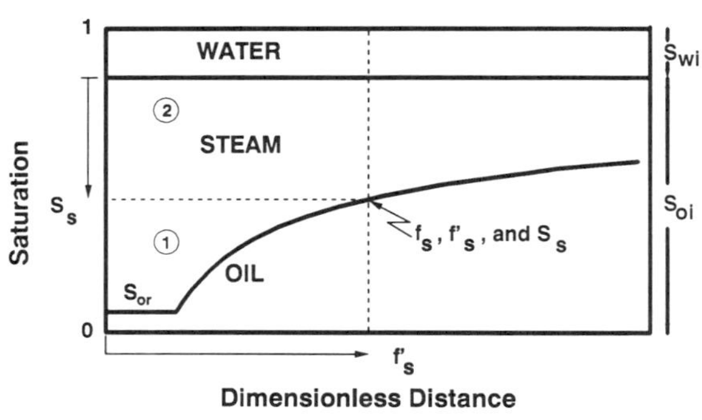

Figure 5.43 Saturations in the Steam Chamber

area marked STEAM in the figure; it is convenient to express this as the sum of the areas marked 1 and 2. This sum is given by

$$\frac{V_s}{V_{\text{inj}}} = \text{area 1} + \text{area 2} = \int_{S_s}^{1-S_{or}-S_{wi}} f_s' \, dS_s + f_s' \, S_s = 1 - f_s + f_s' \, S_s \tag{5.53}$$

The volume of water is given by the area marked WATER, which is equal to

$$\frac{V_w}{V_{\text{inj}}} = S_{wi} f_s' \tag{5.54}$$

The quantity of oil remaining is calculated by subtracting area 1 from the rectangle in the lower left corner, i.e.,

$$\frac{V_o}{V_{\text{inj}}} = f_s' \, S_o - 1 + f_s \tag{5.55}$$

The corresponding volume of rock matrix is given by multiplying the total pore volume by the ratio $(1 - \phi)/\phi$.

$$\frac{V_r}{V_{\text{inj}}} = \frac{1 - \phi}{\phi} f_s' \tag{5.56}$$

The fractional recovery of oil is equal to the pore volume occupied by steam divided by the pore volume initially filled with oil, i.e., by

$$\text{Recovery} = \frac{1 - f_s + f_s' \, S_s}{f_s'(1 - S_{wi})} \tag{5.57}$$

Heat Balance

At any particular time, the limit of the steam chamber is determined by a heat balance. At the condensation front, the injected steam is condensing and heating additional cold reservoir material. The balance is determined by equating the total heat within the contents of the chamber to the heat supplied by the steam when it condenses; the condensate cools to the reservoir temperature. This is shown by the following equation:

$$\begin{aligned}(\text{Steam volume supplied}) \cdot \mathrm{H}_s = {} & (\text{steam volume in chamber}) \cdot \mathrm{H}_s \\ & + (\text{water volume in chamber}) \cdot \mathrm{H}_w \\ & + (\text{oil volume in chamber}) \cdot \mathrm{H}_o \\ & + (\text{rock volume in chamber}) \cdot \mathrm{H}_r\end{aligned}$$

This balance, based on a unit volume of injected steam, is expressed in equation 5.58.

$$\mathrm{H}_s = (1 - f_s + f_s' \, S_s)\mathrm{H}_s + S_{wi} f_s' \, \mathrm{H}_w + (f_s' \, S_o - 1 + f_s)\mathrm{H}_o + \frac{1 - \phi}{\phi} f_s' \, \mathrm{H}_r \tag{5.58}$$

The heat content terms H_i are all measured above the reservoir temperature T_R. H_s is the heat in a volume of steam vapor measured above liquid water at T_R. In any particular numerical example, all the terms on the right-hand side of this equation will be either constants or known functions of f_s'. It is thus possible to plot the right-hand side against f_s' and thus determine the value of f_s' which balances the equation.

Numerical Example

The following numerical example illustrates the use of the Buckley-Leverett method to analyze a steamflood.

Problem A one-dimensional, adiabatic steamflood is carried out using dry saturated steam at 3.5 MPa (243°C).

The steam is injected into the core at a rate of 10 kg m^{-2} h^{-1}. The core has the following properties:

Initial temperature	25°C	
Length	50 cm	
Porosity	0.35	
Initial oil saturation	0.83	
Irreducible water saturation	0.17	both initially and for steamflooded core
Residual oil saturation	0.06	after exhaustive steamflood based on oil at T_R
Permeability to steam at S_{wi} and S_{or}	0.9 μm^2	
Permeability to oil at S_{wi} and $S_s = 0$	1.0 μm^2	

The materials involved have the following properties:

Physical Properties

	Oil	Rock	Water	Steam
Density at 25°C kg/cu m	966	2,600	1,000	17.54
Viscosity at 25°C cp	2,000		0.9	
Viscosity at T_S cp	4			0.018
Mean heat capacity kJ/kg °C	2.1	0.84	4.2	
Enthalpy at T_S above T_R				
kJ/kg	458	183	945	2,697
kJ/cu m H_o, H_r, H_w, H_s	442,235	476,112	944,700	47,293

Assume that the flow in the steam chamber is segregated—i.e., that the relative permeabilities are linear functions of S_o.

a. Calculate the time required for steam to breakthrough, the number of pore volumes of steam (measured as liquid water) that have been injected at this time, and the corresponding percent recovery of the original oil in place.

b. Plot a diagram of saturation (water, steam, and oil) versus fractional distance along the core at steam breakthrough.

c. Calculate the recovery as a function of pore volumes of steam injected measured as liquid and plot a curve.

Heat Balance, Saturations, and Recovery

The problem was solved by means of a tabular calculation using a spreadsheet program. The following table shows the calculated relative permeabilities, fractional flow of steam, and f' as a function of the oil saturation. The sixth column shows the computed heat content of the chamber.

Linear Steamflood Numerical Example

Oil Saturation	Permeability μm^2 Steam	Oil	f Steam	f' Steam	Chamber Heat[1]	Recovery % OOIP
0.060	0.900	0.000	1.0000	0.0000	0	92.77
0.060	0.900	0.000	1.0000	0.0065	7,193	92.77
0.095	0.860	0.050	0.9998	0.0071	7,895	92.58
0.130	0.820	0.090	0.9995	0.0078	8,715	92.01
0.165	0.780	0.140	0.9992	0.0087	9,678	91.05
0.200	0.740	0.180	0.9989	0.0097	10,818	89.72
0.235	0.700	0.230	0.9985	0.0108	12,181	88.00
0.270	0.650	0.270	0.9981	0.0122	13,824	85.91
0.305	0.610	0.320	0.9977	0.0139	15,827	83.43
0.340	0.570	0.360	0.9972	0.0159	18,300	80.57
0.375	0.530	0.410	0.9966	0.0185	21,394	77.32
0.410	0.490	0.450	0.9959	0.0216	25,330	73.70
0.445	0.450	0.500	0.9950	0.0257	30,433	69.69
0.480	0.410	0.550	0.9940	0.0311	37,197	65.31
0.515	0.370	0.590	0.9928	0.0382	46,408	60.54
0.518	0.360	0.590	0.9927	0.0389	47,271	60.20
0.550	0.330	0.640	0.9913	0.0483	59,373	55.39
0.585	0.290	0.680	0.9894	0.0628	78,385	49.86

[1] In kilojoules per cubic meter of injected steam.

Figure 5.44 shows the heat required to heat the steam chamber as a function of the reservoir pore volumes that have been heated per volume of injected steam (i.e., f'). It will be seen that the heat requirements are equal to the available heat in the steam for a reservoir pore volume of 0.0389. This is the pore volume in the steam chamber per volume of injected steam, measured as vapor.

Alternatively, we can say that $1/0.0389 = 25.7$ volumes of steam, as vapor, are required to heat and sweep one pore volume of reservoir. The steam volumes can be converted to the more conventional water equivalent by multiplying by the ratio of the densities, i.e., by 17.54/1000.

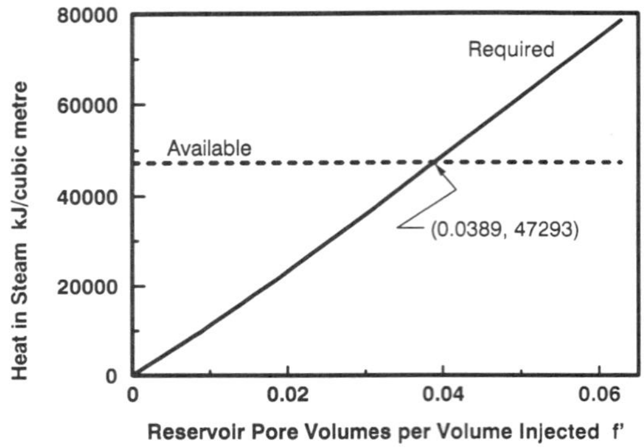

Figure 5.44 Heat Balance for Steam Chamber

$$\text{Pore volume of core} = 50 \times 0.35 \text{ cm}^3/\text{cm}^2$$
$$= 0.175 \text{ m}^3/\text{m}^2$$
$$\text{Steam required for breakthrough} = 0.175 \times 25.7 \times 17.54$$
$$= 78.9 \text{ kg/m}^2$$
$$\text{Time required for steam breakthrough} = \frac{78.9}{10} = 7.89 \text{ h}$$

The saturations are plotted as a function of f' in Figure 5.45, and the position of the condensate front is also shown. The water and oil saturations beyond the front are discussed later.

The oil recovery has been calculated in the seventh column of the preceding table using equation 5.57. The percent oil recovery is plotted as a function of the steam injected in Figure 5.46. The points for beyond the steam breakthrough come from the previous table; up to breakthrough, the percent recovery varies linearly with the steam injection.[7] It should be noted, however, that in general, the recovery is not volumetrically equal to the steam injection.

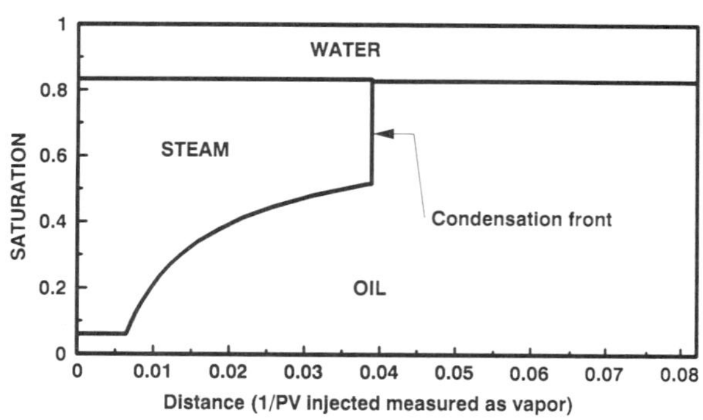

Figure 5.45 Saturations during Steamflood

[7]The straight line in Figure 5.46 should really be two straight lines, as in Figure 5.49. The change in slope occurs when water condensate breaks through. However, with the linear relative permeability relations assumed here, the oil displaced ahead of the condensation front is very small and can be neglected.

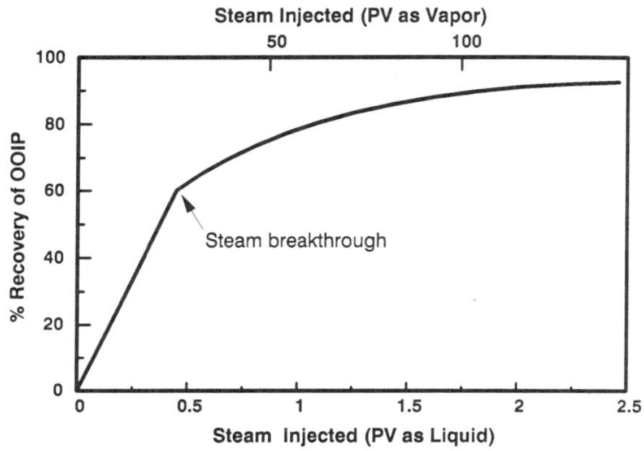

Figure 5.46 Oil Recovery as a Function of the Quantity of Steam Injected

Displacement of Oil ahead of the Condensation Front

The fluid mixture flowing from the steam chamber consists of viscous cold oil and water. Its composition may be obtained from a material balance for the steam chamber; this is shown in the following table.

Material Balance for Steam at Breakthrough

Basis: One volume of injected steam	Steam measured as vapor	Steam measured as liquid
Steam Chamber Contents:		
Connate water	0.0066	0.377
Remaining oil	0.0129	0.735
Steam	0.0194	0.019
Total pore volume of chamber	0.0389	2.218
Injected steam	1 (vapor)	1 (as liquid)
Effluent water	0.0172	0.981
Displaced oil	0.0194	1.108
f_w in effluent	0.47	0.47

The material balance is based on one volume of injected steam measured as vapor for the left-hand column and liquid for the right. One volume of injected steam (measured as liquid) produces a steam chamber having a total pore volume of 2.218. About 2% of the injected steam remains behind to replace the volume of the displaced oil. The steam condensate or effluent water measures 0.981 volumes, accompanied by 1.108 volumes of displaced oil. The fraction of water in the effluent from the steam chamber is 0.47. The reservoir saturation downstream from the condensation front adjusts itself to correspond to this fractional flow; the situation is similar to that shown in Figure 5.21, although in this case the water fractional flow curve is much steeper. With the relative permeability curves assumed here and the much lower viscosity of water compared to that of oil, only a slight increase in water saturation is required to accommodate the water flow. It is calculated that the water saturation will rise from the irreducible value of 0.17 to about 0.171—a very small change. A waterflood shock front races ahead of the condensation front, and it is characterized by this very slight increase in water saturation. In this example, the waterflood effect downstream of the condensation front is very small. The main effect is that of the displaced oil moving through the reservoir as it leaves the condensation front; the accompanying water condensate is able to flow with very little change in water saturation. The calculated saturations are shown by the solid lines as functions of the distance along the core of Figure 5.47.

EFFECT OF SHAPE OF RELATIVE PERMEABILITY CURVES

In the proceeding numerical calculation, it was assumed that the relative permeabilities were linear functions of saturation. The calculation has been repeated in this section assuming that the relative permeabilities vary with the cube of the mobile saturation. The results are tabulated in the following table. The calculated saturations are shown as dotted lines in Figure 5.47.

Flood Numerical Example with Relative Permeabilities Assumed to Vary with the Cube of the Mobile Saturation

Oil Saturation	Permeabilities μm^2 Steam	Oil	f Steam	f' Steam	Chamber Heat[(1)]	Recovery % OOIP
0.060	0.90	0.00	1.0000	0.0000	0	92.77
0.095	0.78	0.00	1.0000	0.0000	49	90.05
0.130	0.68	0.00	1.0000	0.0002	239	87.18
0.165	0.58	0.00	1.0000	0.0006	666	84.17
0.200	0.49	0.01	0.9999	0.0013	1,482	81.02
0.235	0.42	0.01	0.9999	0.0025	2,938	77.72
0.270	0.35	0.02	0.9997	0.0047	5,443	74.29
0.305	0.29	0.03	0.9995	0.0082	9,682	70.71
0.340	0.23	0.05	0.9991	0.0141	16,823	66.99
0.375	0.19	0.07	0.9983	0.0240	28,901	63.14
0.407	0.15	0.09	0.9972	0.0389	47,311	59.49
0.410	0.15	0.09	0.9971	0.0407	49,556	59.15
0.445	0.11	0.13	0.9950	0.0694	85,508	55.02
0.480	0.08	0.16	0.9914	0.1201	149,593	50.76
0.515	0.06	0.21	0.9852	0.2121	267,253	46.38
0.550	0.04	0.26	0.9739	0.3851	490,864	41.90
0.585	0.03	0.32	0.9531	0.7224	931,479	37.34

[(1)] In kilojoules per cubic meter of injected steam.

It is interesting to note that the quantity of oil remaining in the steam chamber is almost the same for the cubic relative permeability curves as for the linear ones. However, it is distributed differently. More oil remains near the injection

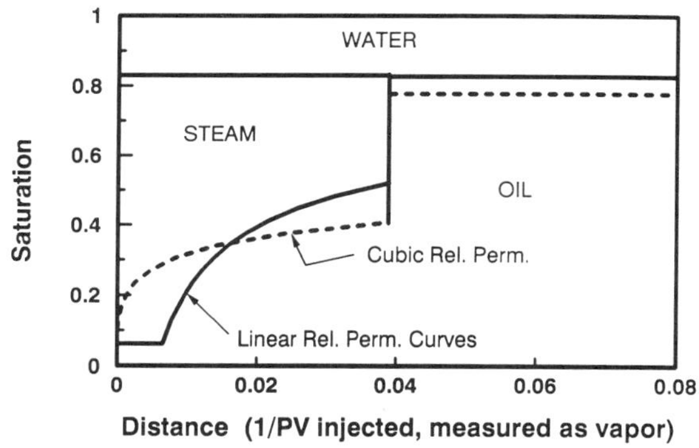

Figure 5.47 Saturations during Steamflood. Effect of Relative Permeability Curves

point and less near the front for the cubic curves. There is also a considerable difference in the waterflooding zone ahead of the steam front. With the cubic curves, a considerably higher water saturation is required to provide oil mobility. In this case the water is able to sweep additional oil ahead of the steam front, and the initial recovery is better. The Buckley-Leverett diagram in Figure 5.48 shows the conditions at the water front.

The calculated oil recovery is plotted against the volume of steam injected in Figure 5.49.

The following features should be noted.

1. In the initial stages, additional oil is recovered in the example with the cubic relative permeability curves as compared to that with the linear ones. This is because of the additional oil displaced by water downstream of the condensation front.
2. At steam breakthrough, the recoveries are almost exactly the same for the two examples.
3. After steam breakthrough, the system with the linear relative permeability relationships shows increasingly better recoveries. The reason for this is that at low oil saturations, the oil is more mobile in this system.

PRESSURE DROP FOR STEAMFLOODING

The pressure drop required to force steam into the reservoir is of great importance, particularly with viscous oils. The rate at which steam can be introduced into the reservoir is frequently controlled by the pressure drop.

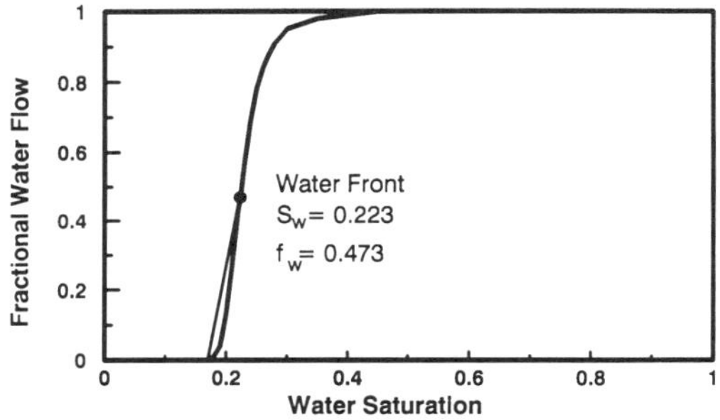

Figure 5.48 Calculated Conditions at Water Front for Steamflood with Assumed Cubic Relative Permeability Curves

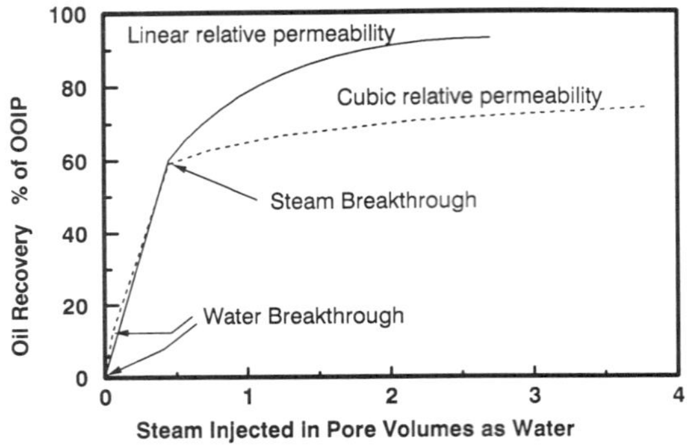

Figure 5.49 Calculated Oil Recovery as a Function of Steam Injection

Consider, for example, the laboratory-scale steamflood used in the numerical example described previously. Results of the calculation pertaining to the pressure drop are given in the following table.

Pressure Gradients in Steamflood Numerical Example

	Steam Zone		Zone Ahead
	Inlet	At front	of Condensation Front
Composition of flowing stream			
Oil %	0	0.73	53
Water	0	0	47
Steam	100	99.27	0
Volumetric Flow Rates			
Oil m^3/m^2 h	0	0.0042	0.0111
Water	0	0	0.0098
Steam	0.5701	0.5660	0
Permeability			
Oil μm^2	0	0.590	0.9996
Water	0	0	0.0004
Steam	0.900	0.360	0
Pressure Gradient			
kPa/m	3.17	7.86	6160
psi/ft	0.14	0.35	272

This table shows the conditions at the inlet to the core, just before the condensation front and just beyond the condensation front. The flowing stream consists almost entirely of steam even immediately before the condensation front (99.27 vol% steam). The very high ratio of steam to oil is the reason for the high effectiveness of the steam in driving the oil forward. Immediately beyond the front, where the steam has shrunk in volume because of its condensation to water, the ratio of oil volume to water volume in the flowing stream is much smaller; also, in this region, the temperature has fallen to the reservoir temperature and the oil is much more viscous. At the bottom of the table, the pressure gradients in the two zones are shown. The pressure gradient within the steam zone is only about 0.1% of that in the cold zone beyond the condensation front. There is little restriction to flow of the steam and oil within the steam zone, but as soon as steam condenses and the temperature falls, there is a very great resistance. In the particular numerical example shown, the pressure gradient within the cold flow zone is over 6 MPa/m. For the 500-cm core of the numerical example, the pressure drop is about 3 MPa, or over 400 psi.

Steam has little problem in sweeping the oil from the steam-saturated zone, but as it forces the oil through the condensation front, the resistance to flow increases enormously. The flows within the steam zone and beyond are coupled. Steam can flow only if it can condense at the perimeter of the steam zone. To condense, it must contact fresh cold reservoir. Fresh cold reservoir can become exposed only if there is sufficient pressure to force the oil through it and thus allow it to become exposed to the hot steam. One phase of the process cannot occur without the other.

Imagine a steamflooding process in which the mechanisms are balanced. Steam is being introduced continuously and is condensed by the cooling provided by the continuously uncovered reservoir. The swept reservoir is heated to steam temperature and the oil within it is forced downstream.

The steam chamber pressure reaches an equilibrium value that balances the process. If the rate of injection of steam is increased, more oil is supplied to the condensation front, more pressure is required to force this oil through the cold reservoir beyond, the required pressure differential increases, and the steam chamber pressure goes up accordingly.

A problem arises if the oil is very viscous and the injection pressure rises to a very high value. In this situation the pressure can reach a level at which reservoir fracture occurs. This is the situation in many, indeed most, virgin bitumen reservoirs when economic steam-injection rates are achieved. Under these conditions the steamflooding mechanism, which has been described previously, fails. Instead, the injected fluids flow into the opening fracture beyond, and because there is little pressure drop involved in their transport, the fracture extends.

Under these conditions, the heating occurs as a finger extending along the direction of the fracture. The amount of oil displaced by the process is drastically reduced. The pressure is dissipated not in advancing a broad condensation front but in parting the reservoir matrix. The situation is shown in Figure 5.50. The fracture may be vertical or horizontal depending upon the initial in situ compressive stresses within the reservoir. This is discussed further in Chapter 6. The in situ stresses may change as the result of stresses introduced by neighboring thermal recovery operations; it has been found at least in one project that reservoirs in which fractures originally were vertical change so that subsequent fractures become horizontal (Denbina, Boberg, and Rotter 1987).

It is interesting to note that although the fracture may advance rapidly and the condensate may be carried away along its length, the areal extent of the heated region—either vertical or horizontal, depending upon the fracture orientation)—does not increase much faster than it would if a broad condensation front were being

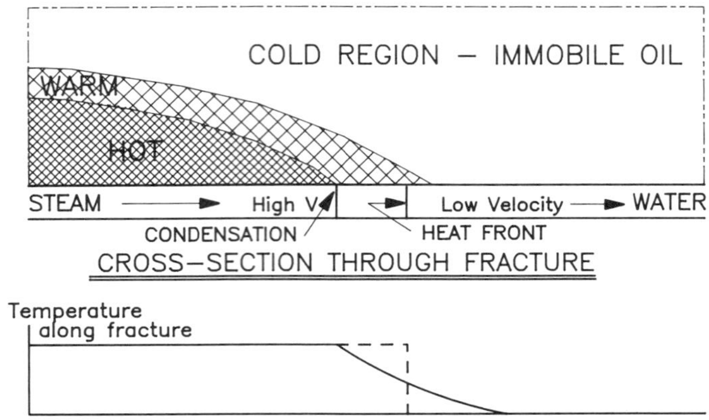

Figure 5.50 Steam Flow and Condensation within Fracture

carried forward by the same steam injection rate in a reservoir containing oil of a lower viscosity. The reader should refer to the results of the comparison of the heating of a reservoir from a fracture with those of the Marx-Langenheim frontal advance that were given in Figure 3.12. In that example, for instance, the heated area for the fracture was about double that for the frontal advance in a reservoir 100 ft thick.

When steam advances into the fracture, heat is still transferred to the oil sand, and the oil becomes fluid. The volume of oil heated by a given amount of steam is approximately the same. However, the pressure available is largely spent in disrupting the reservoir matrix[8] rather than in moving the oil. In Figure 5.50, the oil is moved forward somewhat by the pressure gradient along the fracture, but this is generally inadequate to achieve much movement. The hot oil remains beside the fracture, and the steam and water run through it. Eventually the condensate reaches the pressure sink, and steam breaks through without having displaced nearly as much oil as is possible in a nonfracturing system.

STEAM OVERRIDE

In a lateral steamflood in which steam is injected at a pressure below that required for fracturing, with the purpose of pushing oil horizontally toward one or more production wells, there is a tendency for the condensation front to become tilted so that steam runs over the top of the colder oil below. This is shown in Figure 5.51. As the steam front advances, the volume of the steam chamber increases and oil is displaced. This oil tends to flow downward and sideways towards the production well, and it is accompanied by condensate from the steam.

The effectiveness of this displacement is generally similar to that described previously. However, because the steam is advancing rapidly under the colder over-

[8]The energy is expended in carrying out work against the stresses in the reservoir matrix. This work, even for a vertical fracture, results in a lift of the surface of the ground above the reservoir. In cyclic steam projects, some of this energy becomes available, during the production cycle, to provide compaction drive to move the reservoir fluids (Denbina, Boberg, and Rotter 1987).

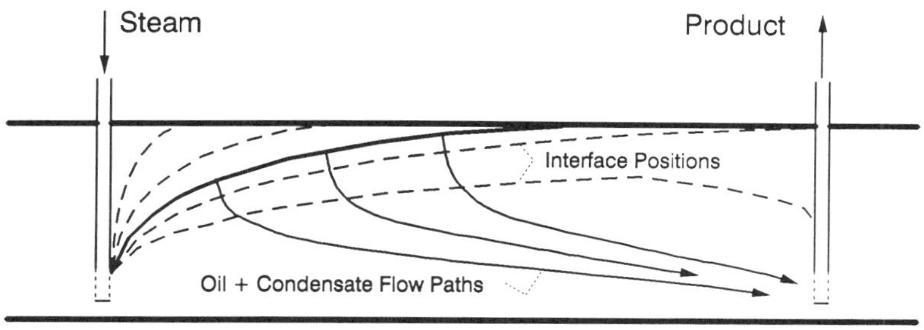

Figure 5.51 Steam Override during Steamflooding of Mobile Oil

burden, the heat losses are greater than for a linear flood because of the greater area that must be heated. The advancing steam chamber tends to produce a downward and sideways drive over a larger area than would be achieved with a simple horizontal displacement flow. The enlarged condensation front area tends to allow for an easier displacement of the oil.

Another factor which becomes important with highly overriding steamfronts is that the conductive transfer of heat beyond the condensation interface becomes increasingly significant as the surface of the condensation front grows. The sharp step-like temperature gradient which occurs initially at the interface becomes gradual. In Chapter 2 it was shown that the quantity of heat which is moved ahead of an advancing front is, in the steady state, inversely proportional to the velocity of advance (see equation 2.45 et seq.). The methods described in Chapter 2 can be used to estimate the heat which is ahead of the advancing front.

Eventually the steam chamber reaches the production well, and at this time the drive is mostly downward, so the movement of oil is assisted somewhat by gravity. Means for calculating the thermal efficiency of processes such as this have been described by Vogel (1984) and were described in Chapters 3 and 4. In these situations Vogel points out that the injection of excess steam results largely in bypass to the production well. The rate of injection should be controlled so as to minimize steam bypass; however, in such situations the rate of production may be below that which is economic.

As the point is approached where steam can break through to the production well, the mechanism by which oil is moved gradually changes from one where the oil is driven away from the advancing interface through the colder oil beyond (i.e., in a direction away from the interface) to one where the movement is more or less parallel to the interface with the driving force being provided by gravity combined with the drag of the steam flowing within the steam zone. This last effect is relatively small unless excessive steam bypass is permitted. The flow to the production well is hindered by the nature of the radial flow to it and the limited contact of the well with the reservoir.

As is mentioned in Chapter 7, the process can be improved by using extended horizontal production wells, which increase the collection capacity for draining oil, and by making use of the steam-assisted gravity drainage principle. A related approach is to drill in-fill wells to recover some of the remaining oil left between the injection and production wells; this is discussed in Chapter 4.

The use of hot water or low steam quality is also described in Chapter 4 as a means for recovering further oil from mature steam floods where steam has broken through to the producers. The density of the water allows it to fall to the bottom of the steam chamber and thus invade the region occupied by the remaining oil. Another approach is to employ low quality steam throughout the drive; this approach is discussed in Chapter 4 in conjunction with the correlations developed by Gomaa.

EFFECT OF STEAM QUALITY

In the analysis of the adiabatic, one-dimensional steamflood shown in Figure 5.42, it was assumed that the steam was dry and saturated. In practical processes, wet steam is employed. Qualitatively, the effect of water in the steam is to increase the water saturation in the steamflood region, since liquid water as well as steam has to flow. The heat in the liquid water that reaches the condensation front is also transferred and this contributes to the advance of the front. On a weight basis, the total heat of wet steam is less than that of dry steam; however, on a volume basis, with the volume measured as that of the wet steam mixture, the heat content is larger. Water at its boiling point has a higher heat content per unit volume than does steam. For a given quantity of injected heat, the condensation front moves forward a slightly smaller distance than it does for dry steam injection. The reason for this is that the flooded steam chamber contains slightly more heat because the water saturation and oil saturation are both slightly higher. However, the difference is relatively small, as will be seen.

The conditions within the steamflooded region can be approximated by assuming that the flowing ratio of steam vapor to liquid water remains constant within the steamflood region. This assumption is reasonable except in extreme cases because the amount of water left behind to provide the increased water saturation is only a small fraction of the total steam and water flow.

The fractional steam quality, f_s, is defined by the following equation.

$$f_s = \frac{q_s \rho_s}{q_w \rho_w + q_s \rho_s} \tag{5.59}$$

This may be rearranged as follows to give the ratio of water flow to steam flow (on a volumetric basis):

$$\frac{q_w}{q_s} = \frac{\rho_s}{\rho_w}\left[\frac{1 - f_s}{f_s}\right] = R \tag{5.60}$$

The flows of the individual components are given by the Darcy equations:

$$\left.\begin{aligned} q_o &= -\frac{k_o A}{\mu_o}\frac{\partial P}{\partial x} \\ q_w &= -\frac{k_w A}{\mu_w}\frac{\partial P}{\partial x} \\ q_s &= -\frac{k_s A}{\mu_s}\frac{\partial P}{\partial x} \end{aligned}\right\} \tag{5.61}$$

In these three equations, the relative permeabilities and absolute permeability have been combined as single terms. For the purpose of this analysis, it will be assumed that the three permeabilities can be expressed by equations of the following form:

$$\left.\begin{aligned} k_o &= C_o(S_o - S_{or})^n \\ k_w &= C_w(S_w - S_{wi})^n \\ k_s &= C_s(S_s)^n \end{aligned}\right\} \tag{5.62}$$

where n, C_o, C_w, and C_s are constants.

From the expressions for q_w and q_s in equation 5.61, a second expression for the ratio R can be obtained.

$$\frac{q_w}{q_s} = \frac{k_w \mu_s}{k_s \mu_w} = R \tag{5.63}$$

Equating the right-hand sides of equations 5.60 and 5.63 and rearranging leads to

$$\frac{k_w}{k_s} = \frac{\mu_w}{\mu_s} R = \frac{C_w}{C_s}\left(\frac{S_w - S_{wi}}{S_s}\right)^n \tag{5.64}$$

This may be rearranged to give the following expression for the water saturation in terms of the steam saturation:

$$S_w = S_{wi} + \beta S_s$$

where

$$\beta = \left(\frac{C_s}{C_w}\frac{\mu_w}{\mu_s}R\right)^{1/n} \tag{5.65}$$

The sum of the saturations of the three individual phases must be unity.

$$S_w + S_s + S_o = 1 \tag{5.66}$$

Eliminating S_w from 5.65 and 5.66 leads to

$$S_{wi} + (1 + \beta)S_s + S_o = 1 \tag{5.67}$$

This may be rearranged to give the following expression for the steam saturation as a function of the oil saturation:

$$S_s = \frac{(1 - S_o - S_{wi})}{(1 + \beta)} \tag{5.68}$$

The water saturation is obtained by difference.

$$S_w = 1 - S_s - S_o \tag{5.69}$$

Equations 5.68 and 5.69 allow the steam and water saturations to be calculated for any given oil saturation. The Buckley-Leverett method may now be applied as before; the combined water plus steam stream is treated as a single component.

We define the fractional flow of water plus steam as

$$f_{ws} = \frac{q_w + q_s}{q_o + q_w + q_s} = \frac{(1 + R)q_s}{q_o + (1 + R)q_s} = \frac{1}{1 + \dfrac{q_o}{q_s}\left(\dfrac{1}{1 + R}\right)} \tag{5.70}$$

The ratio of oil flow to steam (vapor) is given by

$$\frac{q_o}{q_s} = \frac{K_o}{K_s}\frac{\mu_s}{\mu_o} = \frac{\mu_s}{\mu_o}\frac{C_o}{C_s}\left(\frac{S_o - S_{or}}{S_s}\right)^n \tag{5.71}$$

Combining equations 5.70 and 5.71 leads to the following expression for f_{ws}.

$$\begin{aligned} f_{ws} &= \frac{1}{1 + \dfrac{\mu_s}{\mu_o}\dfrac{C_o}{C_s}\left(\dfrac{1}{1 + R}\right)\left(\dfrac{S_o - S_{or}}{S_s}\right)^n} \\ &= \frac{1}{1 + \gamma\left(\dfrac{S_o - S_{or}}{S_s}\right)^n} \end{aligned} \tag{5.72}$$

where

$$\gamma = \frac{\mu_s}{\mu_o}\frac{C_o}{C_s}\left(\frac{1}{1 + R}\right)$$

This may be differentiated to give an expression for f'_{ws}.

$$f'_{ws} = \frac{df_{ws}}{d(S_s + S_w)} = -\frac{df_{ws}}{dS_o} = \gamma n f_{ws}^2\left(\frac{S_o - S_{or}}{S_s}\right)^{n-1}\frac{1}{S_s}\left[1 + \frac{S_o - S_{or}}{S_s(1 + \beta)}\right] \tag{5.73}$$

The equations described here can be used to predict the saturations and recovery from a steamflood by means of a tabular calculation similar to that described previously. The position of the condensate front is determined by the same type of heat balance; it is necessary to treat the wet steam as a single component and to use the volume of the combined vapor and liquid in the calculation. The heat available in a cubic meter of the wet steam (measured as mixture) will be higher than that in a cubic meter of vapor.

The results of a series of calculations of this type are shown in Figures 5.52 and 5.53. These numerical examples were calculated using the data from the previous numerical example. The only difference in the input data is the steam quality.

Figure 5.52 shows the oil, steam, and, by difference, water saturation in the steam chamber for cases where the steam quality is 100%, 50%, and 25%. As before, the horizontal axis represents the distance from the injector measured in pore volumes per pore volume of injected steam, the steam being measured as the volume of the mixture of vapor and water. Decreasing the quality of the steam has the effect of increasing the heat supply per pore volume and the distance to which the condensation front advances.

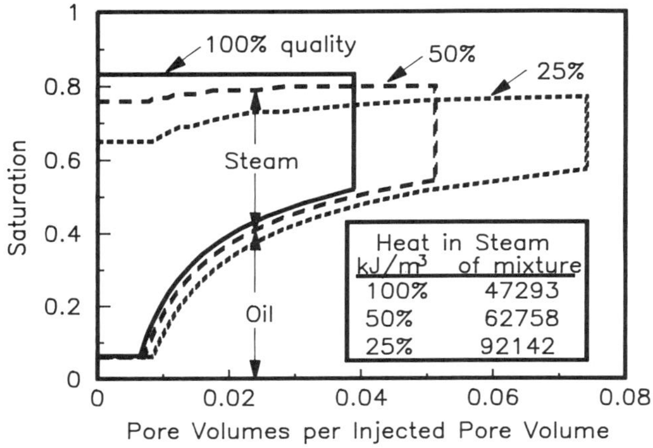

Figure 5.52 Effect of Steam Quality on Saturations. Linear Relative Permeability Functions

The calculated saturations are also shown in Figure 5.53. Here the scale for the distance from the injector has been expressed as cubic meters of pore volume per Gigajoule (GJ) of injected heat. These curves show that for a given quantity of heat, the frontal advance is nearly the same for each of the three injected steam qualities. The effect of lowering the steam quality is to reduce, slightly, the advance of the steam front because of the higher heat capacity of the chamber. The smaller steam saturations and the larger oil and water saturations of the chamber can also be seen. The water saturation is higher because of the need for water to flow through the chamber (i.e., saturations above S_{wi} are required). The oil saturations remaining behind the front are also higher because of the lower volume of the injected steam. Less steam flow is available to drag the oil to the condensation interface.

The calculated percent recoveries of the OOIP are tabulated in the box in Figure 5.53. For a given injection of heat, the dry steam is predicted to remove 60% of the oil at breakthrough, whereas, for the same heat injection, 25% quality steam would be able to remove about 51% from a slightly smaller steamed volume.

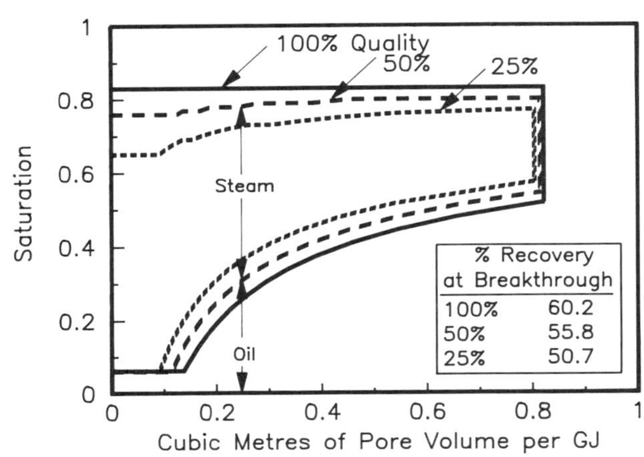

Figure 5.53 Effect of Heat Input on Saturations. Linear Relative Permeability Functions

EFFECT OF VERTICAL HEAT LOSSES

In the analysis just described, it is assumed that there are no vertical heat losses from the expanding steam chamber. Consider the situation shown in Figure 5.42, in which heat is lost by vertical conduction, both upwards and downwards from the steamflood region. Such losses have the effect of reducing the heat available to advance the front and also of reducing the quality of the steam that is flowing. The effect will vary with time. Initially in the flood, there will be little area for heat to be conducted away, and the effect on the process will be small. As the flood progresses, the rate of heat loss increases, and a smaller and smaller fraction of the injected heat is available to advance the front. Heat balances for this effect were discussed in Chapter 3 for several situations. These may also be applied to the present situation using the average steam chamber heat capacity, as predicted before. It is suggested that for practical purposes, the magnitude of the effect upon the recovery of oil can be estimated by including the losses in the calculation of the heat available to extend the front and by using an average quality of the steam in order to estimate its effect upon recovery.

More elaborate procedures for carrying out such calculations, which involve the approximation of the steam chamber as a number of discrete volumes, have been proposed (Shutler and Boberg 1972; Boberg 1987). These methods are rather involved and complicated; in many cases, the accuracy of the input data would probably not justify the complexity. The methods described here are relatively simple and can be readily carried out using tabular calculations with a spreadsheet-type microcomputer program.

EFFECT OF INCREASING STEAM VISCOSITY

If the steam had a higher viscosity it would be more effective in displacing the oil from the steam chamber. As an example of this effect, the numerical example which was described starting on page 224 was repeated assuming that the viscosity of the steam was 0.05 cp instead of the true value of 0.018 cp. This change had the result of decreasing, substantially, the oil saturations which remained in the steam chamber and improving the recoveries. The oil recoveries are plotted against the volume of steam injected in Figure 5.54. There was also a slight reduction in the quantity of steam required to reach breakthrough. This difference resulted from the lower heat capacity of the steam chamber.

This calculation indicates that additives such as foam producing materials should, because of their effect in increasing the apparent viscosity of the steam, result in a lower residual oil saturation in the steam-swept regions as well as increase the steam-drag effect.

GENERAL CONCLUSIONS ON DISPLACEMENT

The following qualitative conclusions on the nature of displacement processes for heavy oils can be drawn:

1. When heavy oils are displaced by water, the water tends to break through very quickly. This would be so even if there were no frontal instability.

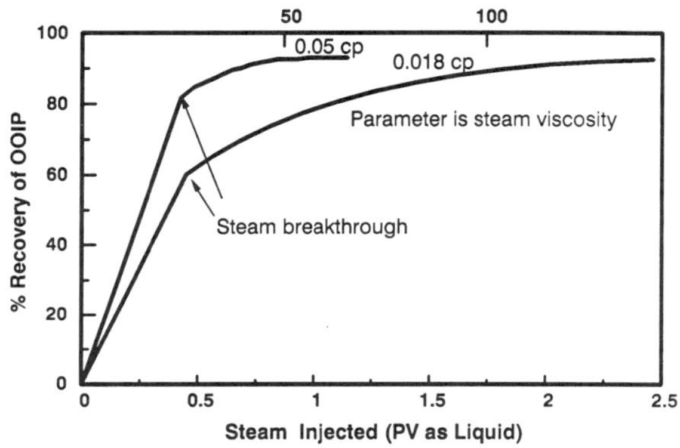

Figure 5.54 Effect of Higher Steam Viscosity upon Calculated Oil Recoveries. Steam Injected (*PV* as Vapor)

2. In the production of heavy oil by waterflooding, the water tends to run through the oil and drag some oil with it. The effectiveness of this process is greater the lower the oil viscosity. Higher temperatures give better oil-water production ratios because of the better viscosity ratio.
3. Because of the stabilization resulting from the shrinkage on condensation, steam has a much lower tendency to finger than does water. In many steamfloods the steam condensation front is more or less stable, although there is a strong tendency for the steam to override the liquids.
4. In a steamflood, the water condensate tends to drain through the oil either as fingers or in diffuse flow. This is often not undesirable, since in order to have a successful heating process using steam it is necessary to remove the condensate as it is formed. As was discussed in the section on convective heat transfer, the heat in the condensate can be effectively transferred to the front (steam front) as long as uncondensed steam remains at the front. As has been shown by Miller, there is also a tendency for the front to be stabilized by the temperature gradient. Steamflooding fronts can be relatively stable and advance in a regular manner through the reservoir with the steam overriding. Until all the latent heat of the steam is lost supplying vertical losses, most of the sensible heat of the condensate is given up at the steam interface as well. With hot waterflooding, the front is much more unstable and there is much more fingering of heat into the reservoir.
5. With very viscous oils, reservoir fracturing can occur if steam is injected at a practical rate. This results in the heating of the reservoir adjacent to the fracture as the steam flows into the fracture and condenses. The condensate flows along the fracture. Although large volumes of oil can be heated adjacent to the fracture, there is little driving force available to move the heated oil to the production well, and oil production is small.

BIBLIOGRAPHY

Armento, M. E., and Miller, C. A., "Stability of Moving Combustion Fronts in Porous Media," SPEJ, 423–430, December 1977. © 1977 SPE.

BAKER, P. E., "Effect of Pressure and Rate on Steam Zone Development in Steamflooding," SPEJ, 274–284, October 1973.

BOBERG, R. C., "Thermal Methods of Oil Recovery," J. Wiley, New York, 1987.

BUCKLEY, S. E., and LEVERETT, M. C., "Mechanism of Fluid Displacement in Sands," Trans. AIME, 146, 107–116, (1942).

CHUOKE, R. L., VAN MEURS, P., and VAN DER POEL, C., "The Instability of Slow, Immiscible, Viscous Liquid-Liquid Displacements in Permeable Media," Pet. Trans. AIME, 216, 188–194, (1959). © 1959 SPE.

DAKE, L. P., "Fundamentals of Reservoir Engineering Chapter 10," Elsevier Scientific Publishing Co., New York, 1978.

DENBINA, E. S., BOBERG, T. C., and ROTTER, M. B., "Evaluation of Key Reservoir Drive Mechanisms in the Early Cycles of Steam Stimulation at Cold Lake," SPE 16737, Dallas, 1987.

HAGOORT, J., "Displacement Stability of Water Drives in Water-Wet Connate-Water-Bearing Reservoirs," SPEJ, 63–74, February 1974.

LEVERETT, M. C., "Flow of Oil-Water Mixtures Through Unconsolidated Sands," Trans. AIME, 132, 149–171, (1939). © 1939 SPE.

MILLER, C. A., "Stability of Moving Surfaces in Fluid Systems with Heat and Mass Transport, III. Stability of Displacement Fronts in Porous Media," AIChE Journal, 21, 474–478, (May 1975).

MUSKAT, M., "Flow of Homogeneous Fluids," McGraw Hill, New York, 1937.

NUTT, C.W., "The Physical Basis of the Displacement of Oil from Porous Media by other Fluids: a Capillary Bundle Model," Proc. Roy. Soc. Lond., A382, 155–178, (1982).

PETERS, E. J., and FLOCK, D. L., "The Onset of Instability During Two-Phase Immiscible Displacement in Porous Media," SPEJ, 249–258, April 1981.

SAFFMAN, P. G., and TAYLOR, G. I., "The Penetration of a Fluid into a Porous Medium or Hele-Shaw Cell Containing a More Viscous Liquid," Proc. Roy. Soc., A245, 312–329, (1958).

SHUTLER, N. D., and BOBERG, T. C., "A One-Dimensional, Analytic Technique for Predicting Oil Recovery by Hot Water or Steamflooding," SPEJ, 489–498, Dec. 1972.

SUDIBJO, R., and BARDON, C., In *European Symposium on Enhanced Oil Recovery* (ed. J. Brown), Edinburgh: Institute of Offshore Engineering, Heriot-Watt University, 303–334, (1978).

VAN MEURS, P., and VAN DER POEL, C., "A Theoretical Description of Water-Drive Processes Involving Viscous Fingering," Pet. Trans., AIME, 213, 103–112, (1958).

VOGEL, J. H., "Simplified Heat Calculations for Steamfloods," JPT, 1127–1136, July 1984.

WELGE, H. J., "Simplified Method For Computing Oil Recovery By Gas Or Water Drive," Trans. AIME, 195, 91–98, (1952).

6

Cyclic Steam Stimulation

INTRODUCTION

The use of cyclic injection of steam to increase the flow of oil from reservoirs has proven to be an effective technique. It is useful over a wide range of reservoir conditions, and the mechanism by which it works varies.

A significant feature of steam stimulation is that the injected heat is concentrated near to the well bore where the streamlines converge and the pressure gradients are highest. Steam stimulation tends, inherently, to put the heat where it will do the most good. A major difference between cyclic steam stimulation and conventional steamflooding is that in stimulation, the displaced oil becomes and remains heated as it flows to the production well whereas in conventional flooding the oil must pass through cooler reservoir until the flood becomes mature.

At one end of the scale is the cyclic injection of steam into reservoirs containing oil that is so viscous that it may be considered as almost solid. The role of steam here is to "melt the solid" and thus allow it to flow through the reservoir. The steam requirements for this mode of operation are related to the quantity of steam required to raise the reservoir to steam temperature after an allowance of heat losses.

The other extreme case is where the oil within the reservoir already has appreciable mobility and conventional production is possible but at a low rate. In this circumstance the role of steam injection is to decrease the near–well bore resistance; this can be looked on as a true stimulation of production. The steam requirement is related to the heat required for the near–well bore region; normally this will be much lower than that required for general reservoir heating.

In the first case, the role of steam is to heat oil throughout the reservoir to make it mobile. In the second, the role of steam is to increase the production rate by reducing the near–well bore flow resistance. In both cases, the effect of steam injection decreases as the heated region cools, and it becomes necessary to repeat the injection cycle. Also, in both cases, subsequent cycles become less effective. In general reservoir heating it is necessary for successive cycles to heat the reservoir which is more and more remote from the production well. For the near–well bore stimulation mechanism, the effect of subsequent cycles deteriorates as the reservoir pressure (or other driving mechanism) becomes dissipated.

At any point in the spectrum of applications of the cyclic steam stimulation process, there must be an effective means to force the oil to the production well. If the oil already has substantial mobility and can be produced by conventional means without steam stimulation at appreciable rates, then the same driving force, the reservoir pressure, can transport the oil to the well. The flow is faster than in conventional production because of the reduction in the near–well bore resistance; this is discussed later.

Reservoir pressure is inadequate to move the oil at a practical rate to the production well when the cold oil is initially immobile or nearly so. In this case, other driving forces are required.

In some reservoirs, compaction drive results from the consolidation of the reservoir sand, with an accompanying decrease in average porosity as the pore pressure falls; this can provide drive pressure to transport the oil. The oil is squeezed from the porous rock as it compacts when the pore pressure is lowered. This mechanism has been important in the production of oil from the Bolivar Coast of Lake Maracaibo in Venezuela.

Another form of compaction recognized as being important to the production of oil in the early cycles of steam stimulation in the bitumen reservoir of Cold Lake is the compaction that follows reservoir expansion as the result of steam injection at fracturing pressure. In this reservoir, injection at fracturing pressure is the only means by which steam can be injected at practical rates. Steam injection causes an increase in the pore volume of the reservoir, which is reflected by an increase in the elevation of the ground surface above. Some of the energy used to inject steam into the reservoir is stored as potential energy by lifting the ground. When the well pressure is lowered, fluids can be squeezed towards the well by the settling of the lifted ground. The effect is not reversible, since movement of the sand grains in the vicinity of the fracture will prevent them from shifting back to their initial position: there is hysteresis.

A very important source of drive to move oil to the well in steam stimulation projects, particularly those producing bitumen, is gravity. This can only have a significant effect if there is a low-density phase to replace the oil as it drains downwards. Steam can fill this role. As oil is drained from the reservoir, an existing steam chamber can expand to replace it.

The cyclic steam stimulation process is also known as huff and puff, as steam soaking, and as steam stimulation; these are all acceptable descriptions.

THE STIMULATION OF WELLS WITH APPRECIABLE COLD FLOW

Steam injected into reservoirs, which are saturated with relatively mobile oil, flows into the formation by displacing reservoir fluids away from the well. At the same time, heat is transferred to the reservoir matrix and condensation of steam occurs. The condensate from the steam is cooled as it flows into the reservoir and more heat is transferred. Heat is also lost to the overburden and underburden.

The effect of steam stimulation can be visualized in an approximate manner by considering the flow to the well as being controlled by the steady-state flow through three concentric cylindrical regions, as shown in Figure 6.1.

Well Bore Skin

There may be special restrictions to flow in the immediate well bore region that are not characteristic of flow through the reservoir as a whole. Resistance in the region may occur as the result of damage due to mud invasion, inadequate or blocked perforations, chemical damage such as that caused by clay swelling, and damage due to migrating reservoir fines blocking the pore structure of the matrix. The resistance to flow is represented by the effect of a *skin factor, S,* in the formula

$$\frac{\Delta P_s}{q} = \frac{\mu S}{2\pi k h} \tag{6.1}$$

Even if the skin factor remains unchanged during stimulation, the well bore resistance, $\Delta P_s/q$, decreases with steam stimulation because the increased temperature reduces the viscosity, μ. In addition to this, steam may effect further improvement in the skin effect by cleaning the pores in the well bore region; i.e., it may reduce the factor S. This is particularly important where the skin effect is caused by wax or other deposits which can be removed by steam. The reduction of well bore skin resistance by heating can have a very dramatic effect if much of the flow resistance lies in the well bore skin. This effect is largely independent of the quantity of heat needed to heat the bulk of the reservoir.

Near–Well Bore Region

The steady-state resistance to flow in the cylinder of radius R_h that is heated around the well bore is given by

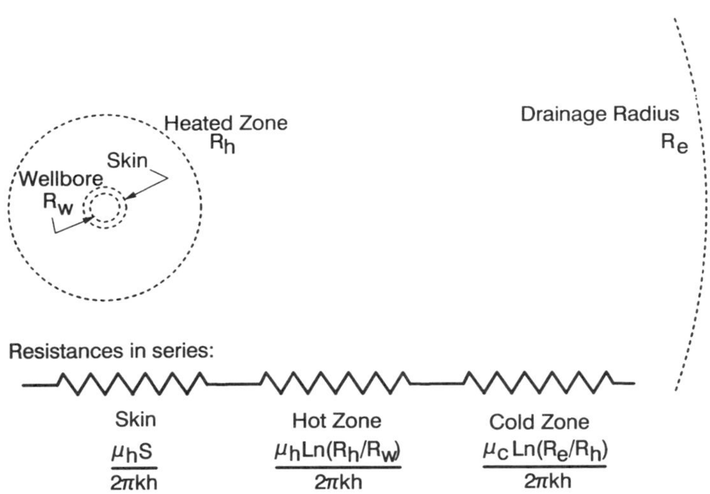

Figure 6.1 Steady-State Radial Flow to a Steam-Stimulated Well

$$\frac{\Delta P_h}{q} = \frac{\mu \ln(R_h/R_w)}{2\pi kh} \tag{6.2}$$

This resistance to flow is reduced if the region is heated because of the effect of temperature in changing the viscosity, μ.

Far–Well Bore Region

Beyond the heat front of radius R_h, the resistance to flow is given by

$$\frac{\Delta P_c}{q} = \frac{\mu \ln(R_e/R_h)}{2\pi kh} \tag{6.3}$$

If the sum of the resistances to flow for the cold situation is divided by the similar sum for the hot, the result is

$$\frac{\left[\sum \Delta P/q\right]_c}{\left[\sum \Delta P/q\right]_h} = \frac{\mu_c S_c + \mu_c \ln(R_e/R_w)}{\mu_h S_h + \mu_h \ln(R_h/R_w) + \mu_c \ln(R_e/R_h)} \tag{6.4}$$

If the total pressure drop, $\Sigma \Delta P$, is the same for both cases, this becomes

$$\frac{q_h}{q_c} = \frac{\mu_c S_c + \mu_c \ln(R_e/R_w)}{\mu_h S_h + \mu_h \ln(R_h/R_w) + \mu_c \ln(R_e/R_h)} \tag{6.5}$$

Two extreme cases of this are of interest.

1. If μ_h is negligible compared to μ_c, then

$$\frac{q_h}{q_c} = \frac{S_c + \ln(R_e/R_w)}{\ln(R_e/R_h)} \tag{6.6}$$

2. If $R_h = R_e$—i.e., the whole reservoir is heated—and $S_h = S_c$, then

$$\frac{q_h}{q_c} = \frac{\mu_c}{\mu_h} \tag{6.7}$$

This assumes that the skin factor remains constant. If S_h is less than S_c, then this too will tend to improve the flow and the ratio could, in favorable circumstances, be larger than that given by equation 6.7.

BOBERG AND LANTZ'S MODEL

A quantitative analysis of the process outlined in the previous section is described in a classical paper by Boberg and Lantz (1966); their method predicts the performance for isolated steam-stimulated wells in reservoirs containing relatively low viscosity oil. The basic idea involved is shown in Figure 6.2.

It is assumed that steam flow is radial and that the heated zone is a cylinder centered on the well. The reservoir may, if it is appropriate, be represented as shown by a number of thin sands divided by horizontal, impermeable shale barriers.

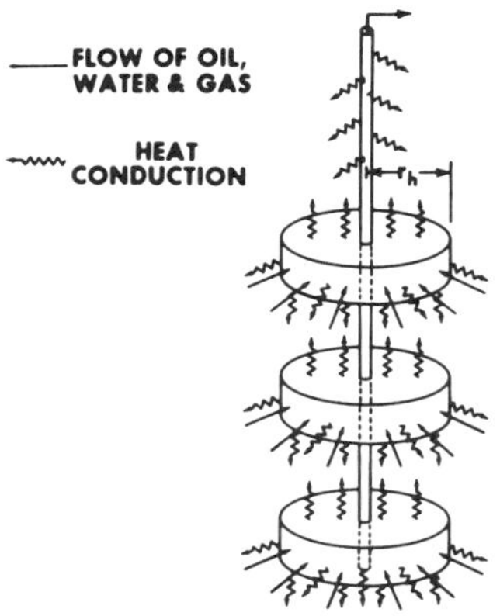

Figure 6.2 Boberg and Lantz's Steam-stimulation Model (from Boberg and Lantz 1966)

Allowance is made for the heat loss from the well bore during injection using methods such as that described in Chapter 2.

The radius of the steam heated zone is calculated using the Marx-Langenheim equation described previously.[1]

$$r_h^2 = \frac{hH_o f(t_D)}{4K\pi(T_S - T_R)}$$

$$t_D = \frac{4K_2\rho_2 C_2 t}{h^2(\rho_1 C_1)^2} \cong \frac{4\alpha_2 t}{h^2} \tag{6.8}$$

$$f(t_D) = e^{t_D}\operatorname{erfc}(\sqrt{t_D}) + 2\sqrt{\frac{t_D}{\pi}} - 1$$

After injection stops, the temperature of the heated zone falls as it loses heat vertically and also horizontally to the colder surroundings. These heat losses are estimated as a function of time from solutions to the conductivity equation. In addition, allowance is made for the heat removed from the heated region by the produced fluids.

The calculation is carried out in a stepwise manner in time using the following equation to estimate the average temperature of the heated cylinder at each step:

$$T_{\text{avg}} = T_R + (T_S - T_R)[\bar{v}_R\bar{v}_Z(1 - \delta) - \delta] \tag{6.9}$$

In this equation, δ is a term that accounts for the energy removed by the water and oil production, and $\bar{v}_R$ and $\bar{v}_Z$ are dimensionless factors that allow for the radial and vertical heat losses from the heated cylindrical volume. δ is obtained from the stepwise integration of the heat balance equation,

$$\delta = \int_{t_i}^{t} \frac{H_f\,dt}{\pi r_h^2 h(\rho C)_1(T_S - T_R)} \tag{6.10}$$

[1]In their paper, Boberg and Lanz consider a number of equal-sized, separate sand members, as shown in Figure 6.2, with *thick* shale members between. In this case, H_o is the heat injection per member and h is the thickness of each individual member, both in equation 6.4 and also in the calculation of t_D. If the shale members are *thin*, then their effect should be allowed for by considering the reservoir as a single entity.

where H_f is the rate at which heat is withdrawn with the products. At each step, it is changed using the quantities of the products, their heat capacities, and the production temperature. There is a decline in the production rate and temperature with time.

The factors $\bar{\nu}_R$ and $\bar{\nu}_Z$ are obtained from Figure 6.3. These were obtained from approximate theoretical solutions to the conduction equation ($\bar{\nu}_R$ can be obtained from equation 2.85).

The production rate at each time is calculated using the idea that the flow is through two concentric reservoir cylinders and the skin, with allowance being made for the changing resistance of the skin and heated cylinder as the hot-oil viscosity falls.

At the end of the production cycle, Boberg and Lantz add the heat remaining in the reservoir to the heat injected with the steam in the next injection period in order to calculate the total heat injected. This is used to estimate the new heated radius. This approximation is conservative, since it neglects the heat stored in the overburden and underburden at the end of the cycle. The method has been shown to produce good prediction for reservoirs having oil with an in situ viscosity of a few hundred centipoises.

For the assumed mechanism to be effective, it is necessary that there be an effective reservoir pressure that is able to move the oil to the well bore in the cold condition. The increased production of the stimulated well comes from the increased flow that this natural drive can produce when the effective radius of the well bore is increased by heating the reservoir around it.

Effect of Process Variables

Boberg and Lantz describe a study of the theoretical effect of various process variables and draw a number of interesting conclusions:

1. The method is intended for reservoirs with substantial drive and cold mobility; their method is not suitable for tar sands.
2. Wells having a high skin factor respond most favorably to stimulation even if no cleanup (i.e. reduction of S) is achieved. Figure 6.4 shows the calculated effect of steam stimulation on the production for a well assuming several skin

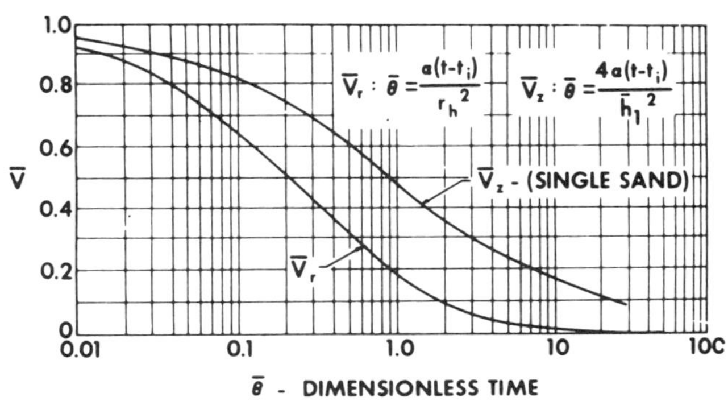

Figure 6.3 Chart for Estimating $\bar{\nu}_R$ and $\bar{\nu}_Z$ in Equation 6.6. (from Boberg and Lantz 1966)

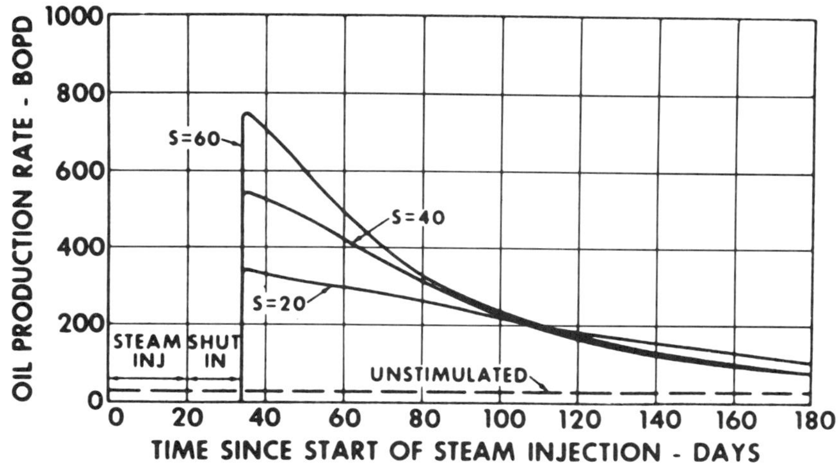

Figure 6.4 Effect of Skin Damage on Steam Stimulation Response (from Boberg and Lantz 1966)

factors. Steam stimulation allows oil to flow more easily through the "bottleneck" created by a high skin factor.

The skin factor is a variable that represents the additional (or, if negative, reduced) pressure drop around the well bore caused by local irregularities such as plugging (positive) or cleaning treatments such as acidizing (negative, it is hoped). It is defined by equation 6.1 (see Matthews and Russell 1967). The effect of the skin in radial fluid flow to a well is analogous to the effect of insulation on the flow of heat that was discussed in the latter part of Chapter 2. In some cases an additional effect of steam stimulation is to reduce the skin factor S by cleaning out deposits around the well bore.

3. The factors which tend to give higher incremental oil to steam ratios are:
   - High oil saturation and high oil sand to shale ratios. These reduce the heat that is required per unit volume of reservoir oil.
   - Low produced water-oil ratio (WOR). Water has a heat capacity that is approximately twice that of oil and thus water production tends to accelerate greatly the drainage of heat from the stimulated reservoir. The predicted effect of produced WOR ratio can be seen in Figure 6.5 which shows results of a process variable study made by Boberg and Lantz. Figure 6.5 also shows the beneficial effect which is predicted for using higher injection rates and also larger steam injection quantities. These improvements stem from the lower fraction of the heat which is lost.
   - Low produced gas to oil ratio. This is beneficial since produced gas is saturated with water vapor and the heat in this, particularly the latent heat, can be a serious drain on the heat pool.
   - High oil viscosity. The viscosity of very viscous oils drops more rapidly with increasing temperature than does that of less viscous ones and, as a result,

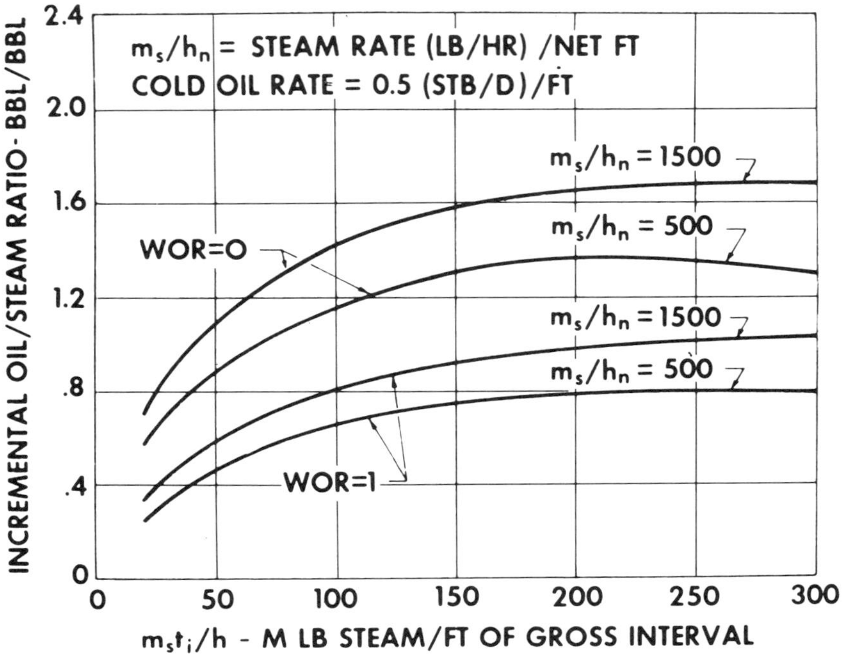

Figure 6.5 Theoretical Prediction of Incremental Oil-Steam Ratio versus Steam Injected (from Boberg and Lantz 1966)

the effect of steam stimulation can be larger with very viscous oils. Figure 6.6 shows the predicted effect for a particular set of conditions.

- Large sand thickness. This improves the OSR because of the reduced fraction of the injected heat which is lost.

4. Back-pressuring the well during the early part of the production cycle can be beneficial by reducing heated zone cooling caused by the flashing of water.

Boberg and Lantz's method has been extended and coupled with calculations for gas-lifting wells using the Orkiszewski correlations for calculating the two-phase pressure drop in a vertical pipe (Boberg, Penberthy, and Hagedorn 1973).

SCALING OF THERMAL MODELS

Physical laboratory models are usually scaled to the field situation by employing dimensional analysis. The most common scheme employed for doing this is to make the physical model geometrically similar to the field situation and to use the same fluids in the model, i.e., oil, water and steam. Scaling is usually carried out by making the Fourier number

$$\text{Fo} = \frac{\alpha t}{L^2}$$

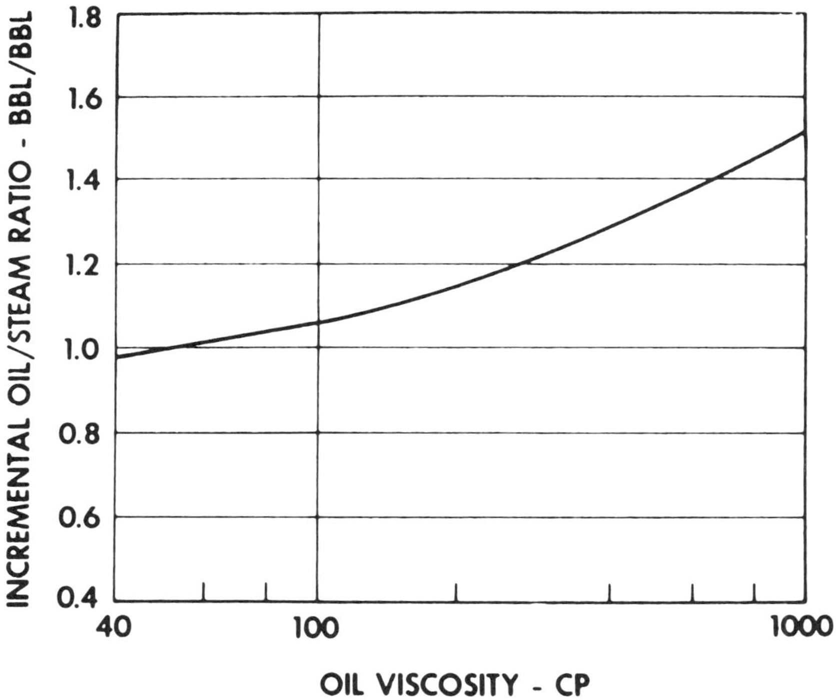

Figure 6.6 Effect of Viscosity on Incremental Oil-Steam Ratio (from Boberg and Lantz 1966)

equal for the model and the field at corresponding times. This means that the degree of heat penetration by conduction will be the same for each. It also means that the time scale will be shortened because the corresponding lengths in the model will be smaller.

$$\frac{t_{\text{model}}}{t_{\text{field}}} = \frac{\alpha_{\text{model}}}{\alpha_{\text{field}}}\left(\frac{L_{\text{model}}}{L_{\text{field}}}\right)^2$$

or if $\alpha_{\text{model}} = \alpha_{\text{field}}$,

$$\frac{t_{\text{model}}}{t_{\text{field}}} = \left(\frac{L_{\text{model}}}{L_{\text{field}}}\right)^2 = R^2$$

where R is the geometric scaling factor.

Thus, for example, if 1 cm in the model equals 1 m in the field, $R = 0.01$ and 1 h in the model will be equal to 10^4 h = 1.14 y in the field.

The other criterion used is to make the pressure gradients due to oil flow bear the same ratio to the potential gradient due to gravity in the model and in the field; i.e.,

$$\left(\frac{\Delta\rho g}{V\mu_o/k}\right)_{\text{model}} = \left(\frac{\Delta\rho g}{V\mu_o/k}\right)_{\text{field}}$$

The velocity V in each case is proportional to L/t. If the same fluids are used in the model as in the field, then $\Delta\rho$ and μ_0 will be the same in each, so

$$\left(\frac{tk}{L}\right)_{\text{model}} = \left(\frac{tk}{L}\right)_{\text{field}}$$

and

$$\frac{k_{\text{model}}}{k_{\text{field}}} = R\frac{t_{\text{field}}}{t_{\text{model}}} = \frac{1}{R}$$

Because of the need to shorten the time scale by the factor R^2 in the model, it is necessary to increase the permeability of the model by a factor of $1/R$ in order to maintain the viscous drag forces proportional to the gravity forces. This scaling procedure was described by Pujol and Boberg (1972).

More elaborate scaling procedures which allow models to be operated at pressures much lower than those in the reservoir are discussed by Stegemeier, Laumbach, and Volek (1980). The approach of operating scaled physical models under low pressure conditions using oils different from those in the reservoir and with steam of a different quality has been used by Shell and others for the physical model simulation of oil recovery processes using steam.

NIKO AND TROOST'S CYCLIC STEAM STIMULATION MODEL EXPERIMENTS

Niko and Troost (1971) carried out an interesting series of low-pressure, scaled-model studies of the steam stimulation process. Their physical model represented the near-well region of a reservoir in which there was adequate cold-oil mobility to provide drive. In such a model, it is necessary to represent the ability of the reservoir beyond the model to supply and to receive cold oil during the production and stimulation cycles.

Niko and Troost overcame this problem by using a series of resistance and capacitor tubes connected to the end of the sandpack. This arrangement is shown in Figure 6.7. The series of tubes and capillaries provided a volume into which liquid from the sandpack could be squeezed. The conditions were arranged so that the heat remained within the sandpack.

A number of process variable studies were carried out. The following conclusions are expressed in terms of the full-scale field that was modeled. The field data which were represented were for a typical Venezuelan Bolivar coast field.

Effect of Process Variables

Injection rate

A steam-injection rate in the range of 19 to 60 t/d into a 9-m layer of reservoir had little effect on subsequent performance for a given total quantity of injected steam.

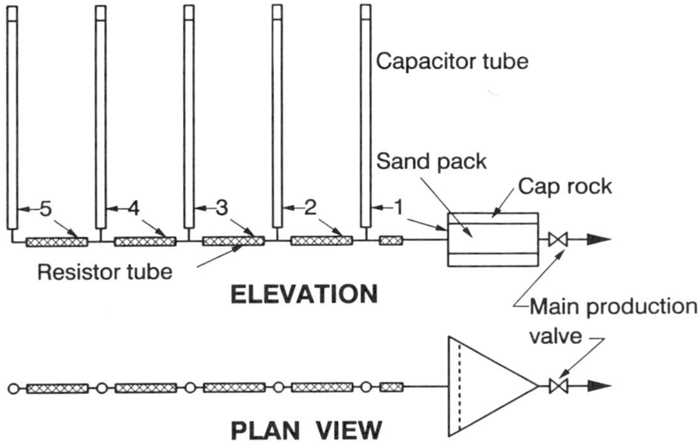

Figure 6.7 Niko and Troost's Steam Stimulation Apparatus (after Niko and Troost 1971)

Soak time

Soak time was not a significant variable in the range of 1 to 160 d. This means that hot-oil storage within the reservoir can be looked upon as a cushion to meet future lifting demands. This has also been found in field experience (Borregales 1977).

Differential pressure

The production rate was found to be proportional to the difference between the reservoir pressure and the well pressure.

Oil viscosity

The productivity index improvement (the ratio of the productivity index after stimulation to that before) was greater for more viscous oils because heating had a larger proportional effect on their viscosity (Table 6.1).

Reservoir thickness

Thicker layers did not respond as well as might be expected because the steam did not penetrate to the bottom. For a fixed injection of steam per unit thickness of reservoir, the steam was found to penetrate farther horizontally for thicker layers and per unit of original oil in place. For a 980-cp oil, a 1133-t steam slug injected

TABLE 6.1 Effect of Initial Oil Viscosity on Productivity Ratio

Viscosity in centipoise	980	4000	8000
Productivity ratio	6	13	20

into a 9-m layer gave an increased productivity ratio of 4.5; three times this quantity of steam injected into a 27-m layer gave a ratio of 3.8.

Steam slug size

The effect of steam slug size was found to be rather complex. A given quantity of steam injected as a number of small treatments rather than as fewer larger treatments (see Figures 6.8 and 6.9) gave a higher initial oil-to-steam ratio (Figure 6.10) and higher cumulative production initially (Figure 6.11).

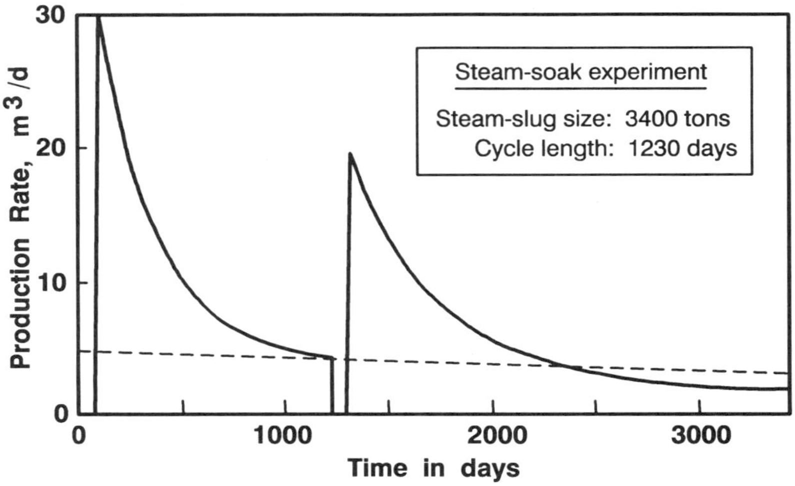

Figure 6.8 Large Steam-Slug Size (after Niko and Troost 1971)

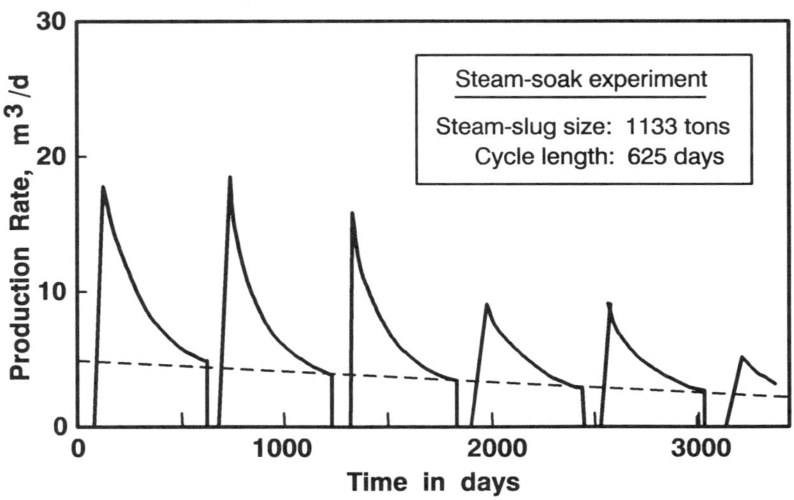

Figure 6.9 Smaller Steam-Slugs (after Niko and Troost 1971)

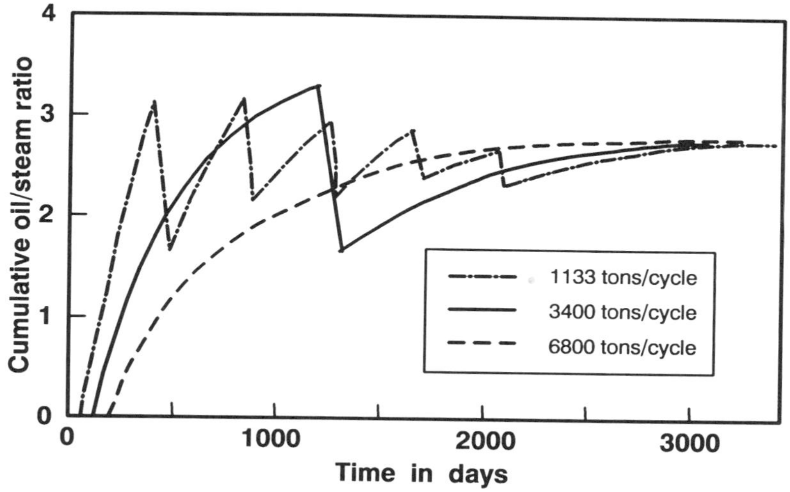

Figure 6.10 Cumulative Oil-Steam Ratio (after Niko and Troost 1971)

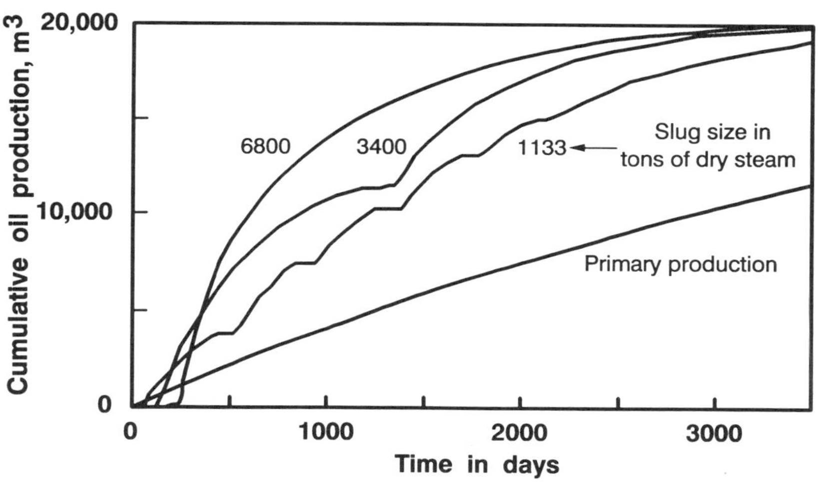

Figure 6.11 Cumulative Production (after Niko and Troost 1971)

After a year or so the cumulative production from the large treatment surpassed that from the smaller treatments (Figure 6.11) and the cumulative oil-to-steam ratios approached the same level (Figure 6.10).

Cycle length

The effect of cycle length was also studied for a fixed injection quantity of steam to each cycle. It was found that decreasing the cycle length (i.e., cutting off production earlier in each cycle) increased the cumulative oil production (Figure 6.12) but reduced the cumulative oil to steam ratio (Figure 6.13).

STEAM STIMULATION PRODUCTION MECHANISM

In the Boberg-Lantz analysis of the steam stimulation process and also in the experiments of Niko and Troost discussed in the previous section, the production mechanism for the steam-stimulated well was assumed to be basically the same as that for cold production. A reservoir pressure pushed the oil to the well; the effect of the steam was to make the oil flow more easily by heating the reservoir adjacent to the well bore.

The pressure that moved the oil was the reservoir pressure existing before the operation started. In the example of the Quirequire field used by Boberg and Lantz as a field example, this view of the mechanism is a reasonable one. In this case the cold-flow rate was already 135 B/d, and this was increased to 350 B/d by steaming.

Factors ignored in the Boberg-Lantz method include the following:

1. The movement of the oil from around the well by displacement with steam during the injection cycle and the refilling of the steam-saturated region during the production are ignored. The Boberg-Lantz theory assumes that the steam heats the near–well bore region but does not move oil away from the well bore. This is inconsistent with the ideas described in the previous chapter.

During the production cycle, the oil must first build an oil bank as it flows to the production well. Also as the pressure around the well bore is decreased during the onset of production, there will be vaporization of water and the generation of steam. Eventually, however, the steam will be displaced from the system and liquids will flow

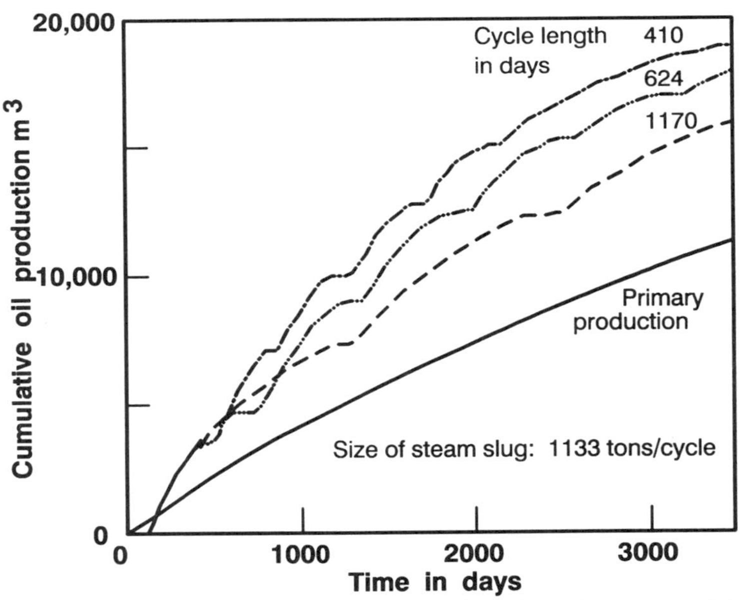

Figure 6.12 Effect of Cycle Length on Cumulative Oil Production (after Niko and Troost 1971)

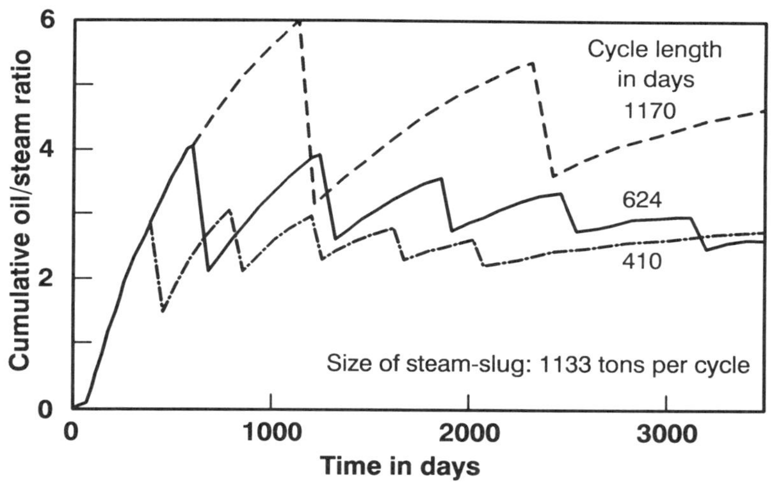

Figure 6.13 Effect of Cycle Length on Cumulative Oil-Steam Ratio (after Niko and Troost 1971)

2. The theory assumes that the near–well bore region remains at a uniform, but falling, temperature. In practice, much of the cooling comes from the flow of cool fluids into the perimeter, and a temperature gradient will be established in the cylindrical region. The heated cylinder is assumed to cool by thermal conduction. In the radial direction, this conduction will be offset, at least partially, by the flowing fluids carrying heat back by convection. The radial cooling will be less than estimated. The other mechanism of cooling that is considered is the vertical conduction of heat to the overburden and underburden. This estimation assumes that the overburden and underburden are initially at the reservoir temperature. In practice they will have already been heated somewhat during the spreading of the heat chamber, and the heat losses, calculated from the Marx-Langenheim formula, are already present in the adjacent reservoir boundaries. This, too, will tend to make the estimated heat loss high.
3. As Boberg and Lantz point out, the method does not allow for any depletion of oil within the heated zone—i.e., replacement of oil saturation by steam saturation. They recognize that in many cases the major part of the oil is produced in this manner. These cases are those in which the cold oil (or usually bitumen) is so viscous that it cannot flow at a practical rate to the heated boundry even if that boundary is very large in radius. These cases are also those where, even though the reservoir is thick, there is a relatively low oil-to-steam ratio because it is necessary to heat the entire reservoir volume from which the oil is produced.

This last point has been discussed by Burns (1969) who points out that in California fields, the increase in oil rate found in steam stimulation is often much larger than that which would be expected from the concept of an increase in the effective well bore radius.

Assuming steady-state conditions and neglecting any contribution of skin factor, the ratio of the production rate for a well surrounded by a heated region of radius R_h to that for an unheated well should be given by equation 6.11 (compare with equation 6.5):

$$\frac{q_h}{q_c} = \frac{\dfrac{\mu_c}{\mu_h} \ln \dfrac{R_e}{R_w}}{\ln \dfrac{R_h}{R_w} + \dfrac{\mu_c}{\mu_h} \ln \dfrac{R_e}{R_h}} \tag{6.11}$$

If the ratio of viscosities is very large, then this equation may be reduced to equation 6.12; i.e., the effective well bore radius is increased by heating from R_w to R_h (compare with equation 6.6):

$$\frac{q_h}{q_c} = \frac{\ln \dfrac{R_e}{R_w}}{\ln \dfrac{R_e}{R_h}} \tag{6.12}$$

As Burns points out, for reasonable values of R_h, the production ratio from equation 6.12 is limited to relatively low values. For example, if $R_w = 0.25$ ft and $R_e = 1000$ ft, the following values for the production ratio may be calculated:

Heated zone radius, feet:	50	100	200
Production ratio:	2.8	3.6	5.2

However, as may be seen from Table 6.2 (taken from Burns' paper), the ratios found in the field are usually much larger; the average value of q_h/q_c in this table is 12.8 or 9.0 if the extremely high value for the third row of data is left out.

Although large skin factors and/or reductions in the value of the skin factor could cause theoretical increases in the production ratio that are as large as those shown in Table 6.2, this is not consistent with the values found for the oil-steam ratio. When the improvement in production rate arises largely from its effect upon the skin factor, then it would be expected that a relatively small amount of steam would be sufficient, i.e., that there would be a relatively high oil-steam ratio. In Table 4.7 it was shown that for general reservoir heating to be achieved, steam-to-oil ratios of at least 0.5 to 2 would be required. These values are for production without heat loss. If allowance is made for losses and for incomplete displacement, then considerably more—probably several times more—steam would be required. Assuming that at least twice as much steam would be required then the SOR would be expected to be at least 1 to 4 (i.e. OSR would be no more than 0.25 to 1) for general reservoir heating. In Table 6.2, only the first three or possibly four of the projects shown display oil-to-steam ratios sufficiently high for skin and near–wellbore stimulation to be the main cause of the improved production rate. In the other cases, the quantity of steam that was required would be expected to be sufficient for there to have been extensive heating of the reservoir well beyond the well bore zone.

TABLE 6.2 Well Performance in Californian Steam Soak Projects

		Production (B/d)					Steam Injected			Oil Recovered		
Field	Zone	Before Steam (Q_{cold})	After* Steam (Q_{hot})	$\frac{Q_{hot}}{Q_{cold}}$	End	Net Sand Open (ft)	Total (B)	B/ft of Sand	Cycle Period (months)	Total (B)	Per B of Steam	Additional Oil Recovered per B of Steam
Huntington Beach	TM	15	160	11	25	40	4,500	112	15	29,000	6.5	5.0
San Ardo	Lombardy	25	360	14	35	220	14,000	64	18	50,000	3.6	2.8
Kern River	China	3	140	47	15	22	4,400	200	6	11,600	2.62	2.5
Midway Sunset	Potter (A)	10	110	11	25	250	6,000	24	5	9,240	1.54	1.29
Kern River	Kern River	14	65	4.6	20	220	6,500	30	5	4,730	0.73	0.43
Coalinga	Tremblor	3	52	17.3	15	107	9,000	84	5	4,300	0.48	0.40
Midway-Sunset	Tulare	5	56	11	10	240	12,000	50	6	4,640	0.38	0.31
Midway-Sunset	Potter (B)	5	35	7	10	250	7,700	31	4	3,000	0.39	0.29
White Wolf	Reef Ridge	30	85	2.4	30	75	14,000	187	4	6,750	0.48	0.23
Poso Creek	Etchegoin	7	20	3	10	80	6,700	84	6	2,660	0.40	0.21

*Average first 30 d.

(Burns 1969)

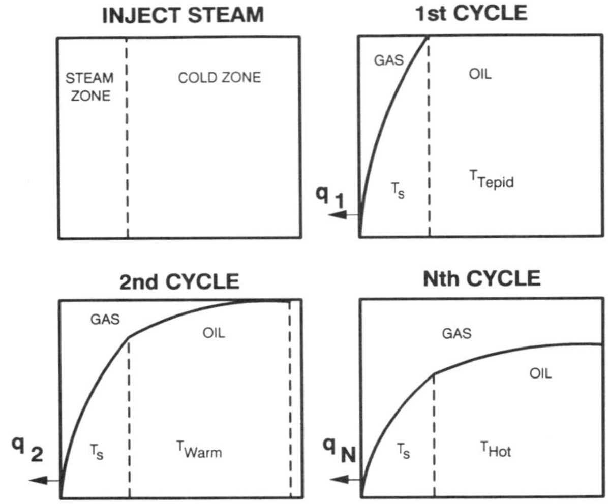

Figure 6.14 Gravity Drainage Effect (after Doscher 1966)

The Boberg-Lantz explanation does not allow for the displacement of oil from the steam-heated region by the effect of an expanding, overriding steam zone combined with gravity drainage to the well bore.[2] This mechanism was described by Doscher in 1966 (see Figure 6.14).

This mechanism is basically different from that involved in the cold flow process. It can, particularly with thick reservoirs containing very viscous oils, become the dominant production mechanism, particularly in the later cycles. In this mechanism, the heated region around the well bore is not just a conduit for the transportation of the heated oil but, as it becomes depleted, is the source of the produced oil.

For this mechanism to be dominant, it is necessary for the cold, in situ viscosity of the oil to be high enough to contain the injected steam in the vicinity of the well and thus to allow the steam chamber to be an expandable source of vapor to replace the oil and condensate as they drain. The compression and expansion of the steam within this chamber provides the cushion to accommodate the changes in liquid volume that are required during the steaming and production cycles. In the Boberg and Lantz model and in Niko and Troost's experiments with mobile oil, it is the movement of the fluid in the cold reservoir that provides this flexibility.

Gravity drainage can occur only if there are two fluids of different densities: in this case heavy oil and water (which have almost the same densities) and lighter steam.

[2]Even if the condensate front is assumed to remain vertical, the volume occupied by the injected steam is not accounted for in the Boberg-Lantz theory. During injection, steam is expected to sweep oil away from the well bore, as described in the previous chapter. This steam-driven oil is pushed through the condensation front and into the oil bank beyond. During the production cycle, this heated oil must flow backward toward the well. The steam will either flow to the production well and be produced or rise above the oil, forming an overriding chamber. As the oil flows through the hot rock matrix, it becomes heated and highly mobile.

The steam tends to rise and move away from the well, and the heavier oil and water tend to fall to the well. As the liquid drains away, heat continues to be conducted into the cooler reservoir adjacent to and below the growing steam chamber.

Production is accompanied by a falling pressure and temperature in the growing steam chamber, and, as the whole system cools, another steam injection cycle becomes necessary. During the period of falling pressure, sensible heat in the rocks in the depleted steam-chamber is transferred to the residual water and is removed by evaporation of the water to steam. At the same time, steam continues to condense at the boundaries of the chamber. With successive cycles, the steam chamber continues to grow, and the vertical heat losses grow larger. This reduces the oil-to-steam ratio.

For the process to be effective, it is desirable that the reservoir be thick and also that it be continuous. For example much better results were obtained in the San Ardo field in California, which contains 220 ft of more or less continuous sand (25 B/d cold to 360 B/d after steaming) than in the Coalinga field, which, although it contains 107 ft of pay, is split into layers less than 30 ft thick (3 B/d cold to 52 B/d after steaming); see Table 6.2.

Lennon (1976) has reviewed the geological factors which have been found important in steam-soak projects in the west side of the San Joaquin Basin in California, including the Coalinga and Midway-Sunset fields. He shows that much of the difference in the performance of projects in this area can be related to the depositional environment (particularly the presence of marine shales which tend to divide the reservoir), to the reservoir structure and type of trap (e.g. the dip of the reservoir layers in the Midway-Sunset field), and to the nature of the reservoir rocks themselves (porosity and permeability).

SIMPLIFIED ANALYSIS OF PRODUCTION RATE DECLINE DURING RESERVOIR COOLING

Insight into the nature of the cyclic steaming process can be obtained from the following simplified analysis of the production rate decline. The method can also be used to obtain descriptive parameters for the comparison of projects.

Suppose that a quantity of heat, Q_i, is injected rapidly into a reservoir. During the production cycle, heat is removed continually from this stored "heat bank." We will assume that the rate of withdrawal of heat from the bank is proportional to the production rate of the oil. If the heat in the produced oil were the only heat loss, then the rate of loss would be given by the heat capacity of the product stream multiplied by the difference between its temperature and that of the reservoir. In order to allow for the other sources of loss, particularly the heat in the water that is removed from the reservoir and also the heat loss to the reservoir surroundings, we will include a factor, η, and write.

$$\frac{dQ}{dt} = -\eta q \rho C(T - T_r) \tag{6.13}$$

where Q is the heat stored in the bank
t is the time
η is a dimensionless factor
ρC is the volumetric heat capacity of the oil
T is the temperature of the heat bank
T_R is the reservoir temperature

As heat is withdrawn from the initial bank, the temperature of the bank declines. If it is assumed that the mass and composition of the material that constitutes the heat bank remains constant or approximately constant and that it has a constant heat capacity, then

$$\frac{Q}{Q_i} = \frac{T - T_R}{T_S - T_R} = T^* \tag{6.14}$$

where T^* is the dimensionless temperature of the bank.

This equation may be differentiated to give

$$\frac{dQ}{dt} = Q_i \frac{dT^*}{dt} \tag{6.15}$$

Combining equations 6.13 and 6.15 leads to

$$Q_i \frac{dT^*}{dt} = -\eta q \rho C (T_S - T_R) T^* \tag{6.16}$$

As the temperature of the heat bank falls, the production rate q will decrease with time because of an increase in the viscosity of the produced oil. For simplicity, we will assume that the rate q is proportional to T^{*s}, where s is constant. If, for example, we assumed that q was proportional to the reciprocal of the oil viscosity, then s would be equal to the parameter m for a viscosity equation of the form

$$\frac{1}{\mu} = \frac{1}{\mu_s} T^{*m} \tag{6.17}$$

This type of equation is used in Chapter 7 in the material on gravity drainage, where it is shown that the parameter m has values of about 3 for typical heavy oils in thermal recovery situations. If, as happens for steam-assisted gravity drainage to horizontal wells, the rate is proportional to the square root of the reciprocal viscosity, then s might be expected to have a value of about $m/2$. In general, it would be reasonable to expect that s should be of the order of $m/2$ to m. The production rate is given as a function of the reduced temperature by the equation

$$\frac{q}{q_i} = T^{*s} \tag{6.18}$$

In this equation, q_i is the oil production rate extrapolated to zero time.[3] Eliminating q from 6.18 and 6.16 leads to

[3]In practical situations, the production rate normally grows with time during the initial stages of the production cycle. The reason for this is the need to purge steam from the steam chamber and also, frequently, the throttling of the production well to control excessive steam velocities. The initial rate used in equation 6.18 is the initial rate found by extrapolating the production curve after the initial period back to the start.

$$\frac{dT^*}{dt} = -\frac{\eta q_i \rho C(T_S - T_R)}{Q_i} T^{*s+1} \tag{6.19}$$

This may be integrated by separating the variables to give

$$\frac{1}{T^{*s}} = \frac{\eta q_i \rho C(T_S - T_R) st}{Q_i} + C^1 \tag{6.20}$$

where C is an integration constant.

Eliminating T^* from 6.18 and 6.20 leads to

$$\frac{1}{q} = \frac{\eta \rho C(T_S - T_R) s}{Q_i} t + \frac{1}{q_i} \tag{6.21}$$

This equation indicates that a plot of reciprocal oil production rate (e.g., in days/barrel) should give a straight line when it is plotted against time.[4] The slope of the straight line is a measure of the rate at which the production declines. The theory indicates that this decline is expected to be less for large heat injections, Q_i, and to be greater for higher steam temperatures and particularly for higher values of the factor η. It will be recalled that this factor is equal to the total heat-loss rate divided by the rate of sensible heat removal in the oil. Thus, for example, if large quantities of water were produced with the oil—as, for example, might happen if the oil lies above a water leg—then η would be higher and the production rate decline would be greater. Similarly, in thin reservoirs with higher vertical heat losses, η would be expected to be greater, and the rate of decline would also be greater. The fraction of the condensate from the injected steam that returns with the oil can have a strong impact (see, for example, Martin 1967). In many projects, it is found that the produced water is much less in quantity than the injected steam—particularly in the early cycles and in mobile oil reservoirs.

A factor not considered here is that the temperature within the heat bank is not uniform. There is a tendency for the near–well bore region to remain hotter as the heat bank cools. Since the flow restriction is greater, because of the nature of radial flow in the near–well bore region and also in the skin, the production rate will not fall quite as fast as expected; i.e., it is possible for η to be less than 1.

Although the preceding analysis is crude and seems to, perhaps fortunately, contain compensating errors, it does represent the type of decline rate found in published cases, and it is also possible to compare values of the factor, η, and to draw reasonable conclusions. In the following, it is assumed that $s = m$, and any deviations from this are included in the factor η; m is obtained using the methods of Chapter 7.

Burns (1969) gives data for the decline in the oil-production rate for two cycles of steam injection in the Potter Sand in the Midway Sunset field in southern California. The data are shown is Figure 6.15.

[4]Decline curves of this type are often referred to as *harmonic decline curves.* Seba and Perry (1969) mention the use of such curves for the successful correlation of production rate data for cyclic steam stimulation in the Yorba Linda field. They also indicate that harmonic curves fit their theoretical curves very closely—maximum deviation 5% and mostly within 1%.

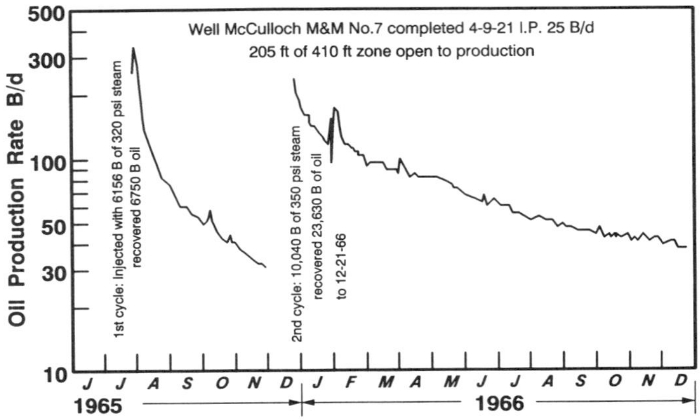

Figure 6.15 Typical Response to Cyclic Steam Injection in the Midway Sunset Field (after Burns 1969)

These data are replotted as the reciprocal of the oil-production rate against time in Figure 6.16; the data are represented well by straight lines.

The slopes of these straight lines can be used to calculate values of the apparent loss factor η (Table 6.3).

TABLE 6.3 Data for Well M & M No. 7

Cycle Number	1	2
Steam pressure, psig	320	350
Steam temperature, °F	428	436
Parameter $m = s$	2.5	2.5
Steam injection, B	6,156	10,040
Heat per barrel of steam assuming 70% quality	314,000	315,000
ρC Btu/B °F	175	175
Slope of graph, B^{-1}	0.00023	0.000059
Factor η	3.1	1.27

(Burns 1969).

The factors indicate that there was considerably less loss in the second cycle than in the first one. Possibly much of the heat injected in the first cycle was conserved, and this reduced the loss in the second.

Figure 6.17 shows data for the Quiriquire field studied by Boberg and Lantz plotted in the same format. Apart from the first points for each cycle that reflect the throttling of the well during the startup, these data also fall on straight lines. However, in this case, the second cycle (η = 2.53) showed a more rapid, rather than a slower, decline in the production with time. Possibly this reflects the declining reservoir pressure. Data are shown in Table 6.4.

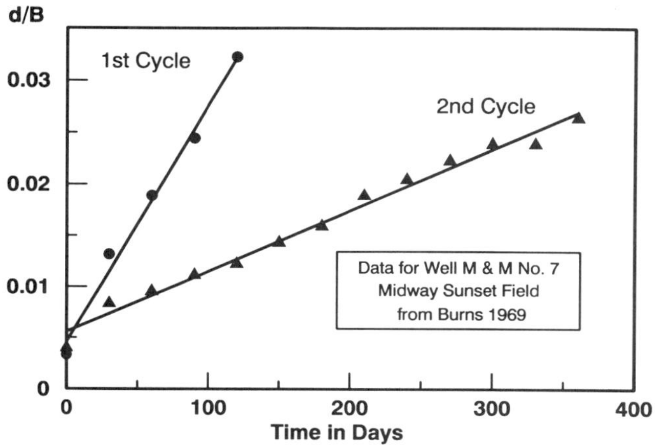

Figure 6.16 Reciprocal Production Rate for Cyclic Steam Stimulated Well in Midway Sunset Field

TABLE 6.4 Data for Cyclic Stimulation in the Quiriquire Field in Venezuela

Cycle Number	1	2
Steam pressure, psig	770	800
Steam temperature, °F	516	520
Parameter $m = s$	2.4	2.4
Steam injection, B	51,714	54,857
ρC Btu/B °F	175	175
Slope of graph, B^{-1}	0.0000187	0.000025
Factor η	1.8	2.5

(Boberg and Lantz 1966).

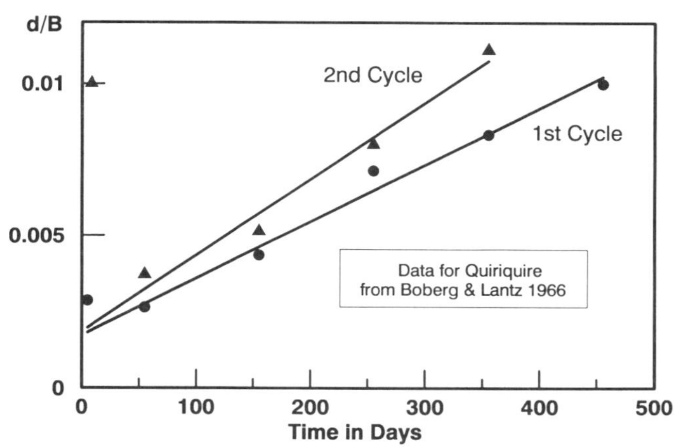

Figure 6.17 Reciprocal Production Rate for Cyclic Stimulation in the Quiriquire Field in Venezuela

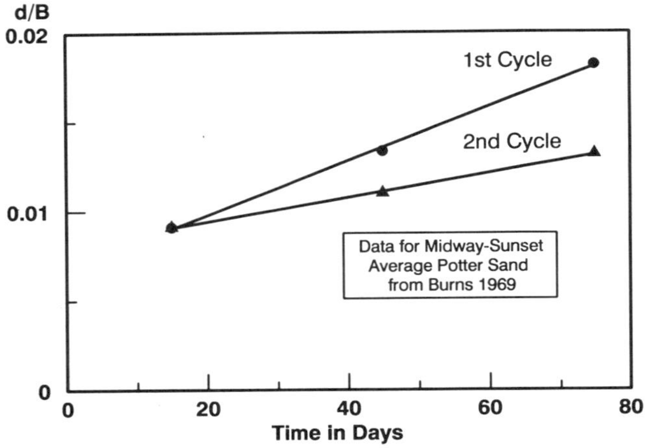

Figure 6.18 Reciprocal Production Rate for Cyclic Stimulation of Average Potter Sand

Data for other wells are shown in Figures 6.18, 6.19 and 6.20.

The data in Figure 6.20 are for a well that had a poor cement job and in which it was thought that steam was lost to nonproductive horizons. The lines are steeper than those in Figure 6.15, which is for the same reservoir—presumably the effective heat injection Q_i was less in well M & M 5A because of the steam loss.

Several papers describe theoretical methods for the analysis and prediction of steam stimulated production in which the Boberg and Lantz approach is extended to include gravity drainage (Towson and Boberg 1967; Seba and Perry 1969; Kuo, Shain, and Phocas 1970; and Jones 1977).

Each of these studies assumes the initial formation of a heated zone that is analyzed using the Marx Langenheim approach. Various approximations are made concerning the state of the region around the heated zone (e.g., it is cold; it is heated by conduction from the hot zone but is at a uniform temperature; it is heated by radial conduction and the temperature distribution varies accordingly).

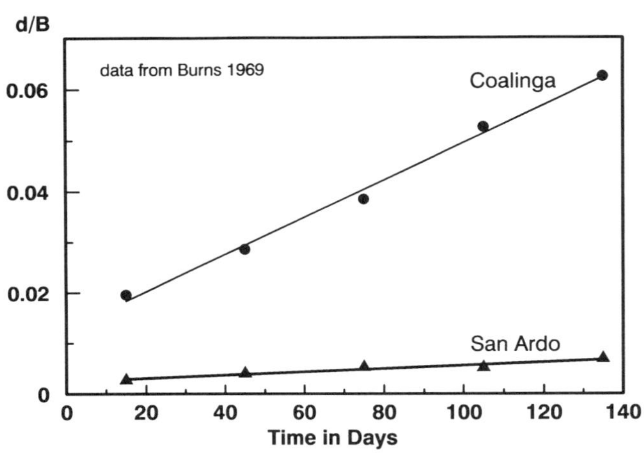

Figure 6.19 Reciprocal Production Rate for Cyclic Steam Stimulation of Coalinga and San Ardo Sands

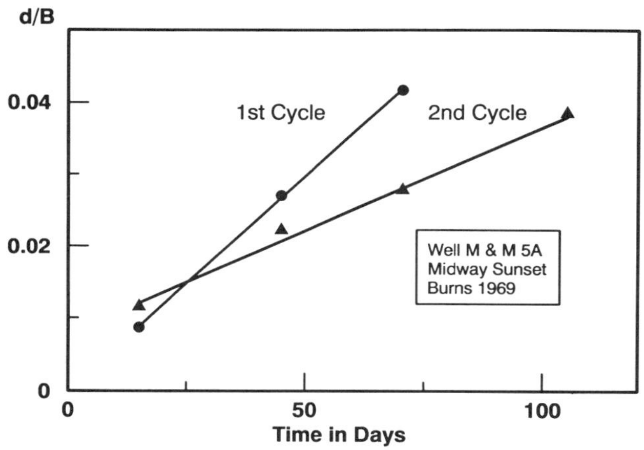

Figure 6.20 Reciprocal Production Rate for Cyclic Steam Stimulation of a Midway Sunset Well with Poor Cementing

The rate of drainage of the oil is calculated using the methods of Matthews and Lefkovits (1956), who developed an analytical method for predicting the effect of gravity drainage for conventional isothermal reservoirs.[5]

The theories predict that the interface between the liquids and the steam zone rise from a level at the bottom of the well in the manner envisaged by Doscher. As the liquid is drained away, the interface falls. The positions of the interface calculated for a specific example by Kuo, Shain, and Phocas are shown in Figure 6.21. This also shows the calculated temperature distributions.

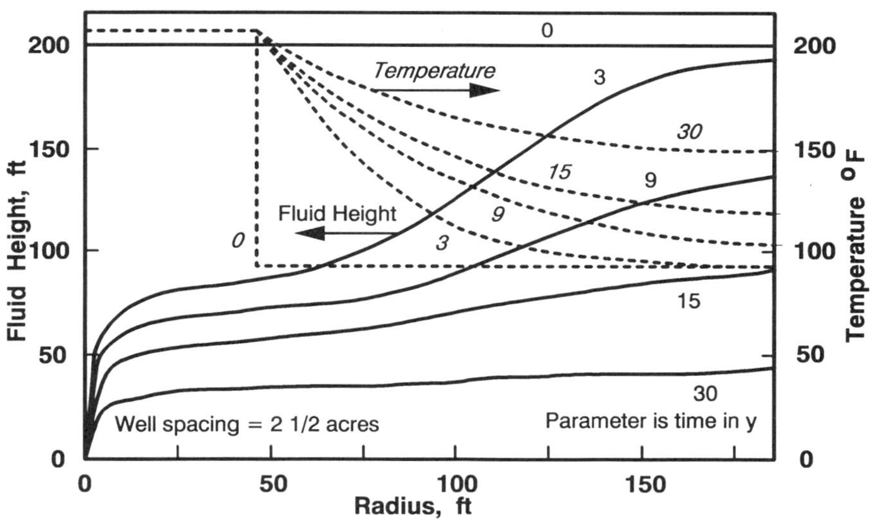

Figure 6.21 Conditions Around Steam-Stimulated Well (after Kuo, Shain, and Phocas 1970)

[5]This theory assumes that the viscosity of the oil is uniform. In practice, there is a sharp temperature gradient in the oil layer adjacent to a growing steam chamber, and there is a corresponding gradient in viscosity. It is largely the oil adjacent to the chamber that is flowing. This mechanism is discussed in Chapter 7.

THE PROBLEM OF THE FIRST CYCLE IN THE CYCLIC STEAM STIMULATION OF TAR SANDS

During one of the later cycles in the production of very viscous oils by steam stimulation, the steam flows into the low-pressure steam chamber left from the preceding cycle; as steam injection continues, the temperature and pressure increase. Steam condenses both within the chamber and at its perimeter. More oil is heated, and this drains to the lower part of the chamber together with the steam condensate.

In the first cycle the problem is different because there is no preformed low-pressure steam chamber that is able to accept the steam. If the in situ oil is reasonably mobile, a chamber can be formed by pushing oil away from the well. Most of the outwardly displaced oil, together with some of the condensate, will be returned later by the natural reservoir pressure. A similar result can be achieved if there is a high water-saturation layer in which there is fluid mobility. Another similar circumstance might occur with a layer of high gas saturation.

If the in situ oil is practically immobile and if the water saturation is too low for water to be mobile, as is often the case with tar sands, then it is necessary, in the first cycle, for the injection pressure to be high enough to disrupt the reservoir. The sand must be physically moved in order to make room for the steam. Some of the volume can come from the compression of the reservoir materials, but most comes by lifting the surface of the ground and creating new void space. Since the volume of the injected steam (measured as liquid) is larger than that of the oil it produces, it is apparent that the increase in liquid volume accommodated by the disruption of the reservoir must be very substantial.

For example, consider the injection of 50,000 B of steam into a reservoir with a sand thickness of 100 ft. Injections of this order are commonly carried out in the cyclic steam stimulation of the Cold Lake reservoir. If this injection resulted in a single fracture having the height of the reservoir and a total length of 800 ft (400 ft either side of the well), then, if it were of uniform width and contained the condensate from the steam, it would need to be 3.5 ft wide. In practice, of course, it is very unlikely that the increase in reservoir would be of the simple shape such as this. Nevertheless it is apparent that injections of this magnitude will cause significant changes to the reservoir matrix.

Whether this increase occurs through the creation of "cracks" or by a distributed increase in the pore volume over considerable volumes is still obscure, although the latter seems more probable in unconsolidated sands when one considers the large pore volume that must be generated.

In their original state, tar sands are often—and perhaps always—very closely packed (Dusseault 1977). They are often described as *locked*. By this it is meant that protuberances on individual sand grains tend to fit into hollows in their neighbors. This makes the sand much stronger than would be expected for a noncemented material. It is the locking of grains that is thought to explain the strength of the near-vertical oil sand cliffs found where the tar sands form outcrops along the bank of the Athabasca River. This is in distinct contrast to the low angle of repose found for typical loose sands.

When the tar sands in Athabasca are mined and separated, it is found that the separated sand occupies a volume that is about 25% greater than the original. It seems likely that fracturing oil sands in situ also creates an irreversible increase in pore volume. In any case, in the first cycle, it is the creation of additional volume within the reservoir by disruption from the entry of steam that provides the working volume.

As the steam enters the reservoir, fracturing occurs and additional pore volume is created; this allows the accommodation of the condensate from the steam. It is also necessary to accommodate the volume of the steam that is flowing through the heated reservoir on its way to the region where condensation occurs. The accumulation of condensate may be quite remote from the well. The condensate flows through open fractures to fill the volume behind the fracture front as it encroaches into the undisturbed reservoir. This fracture front is probably distributed rather than simply being the edge of an advancing crack. One concept of how such a distributed region might advance into the reservoir is shown in Figure 6.22.

In the steam-saturated region, the pressure is above the fracture pressure and the sand is loose and free to move. Steam flows through the disturbed region under the action of a small pressure gradient. At a surface, the condensation surface, the steam is converted to water, and this liquid water flows through multiple fractures into the tar sand beyond. As it flows, it cools rapidly.

In the region beyond and close to the condensation surface, heat is carried by conduction into the disturbed sand between the fractures. This heat causes the bitumen within the pores to try to expand (see Butler 1986). The increased pore pressure beyond the condensation surface causes a backflow of fluid toward the steam-saturated region. This increase of pressure beyond the condensation surface loosens the sand particles, and they join the loose sand within the disturbed region. The whole mechanism proceeds in a coupled fashion, and the disturbed region advances steadily. There is a dilation of the sand matrix; the porosity increases, and

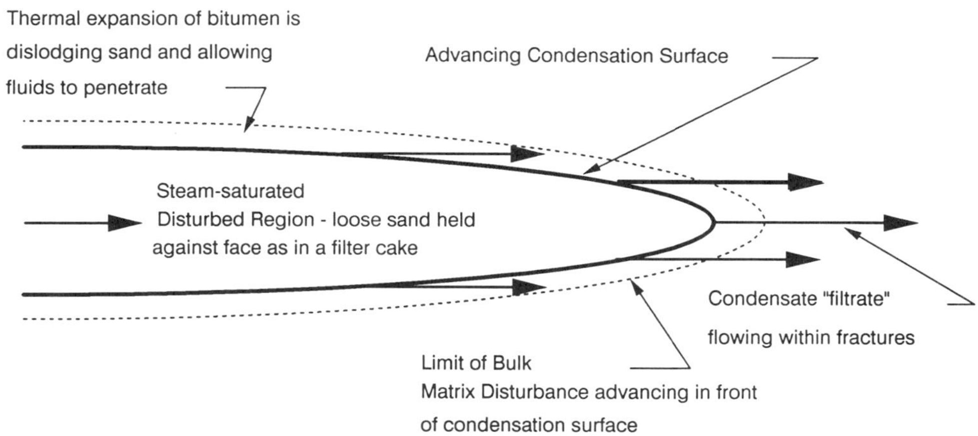

Figure 6.22 Possible Mechanism for the Disruption and Fracturing of a Tar Sand Reservoir with Steam

overall the increase in pore volume is achieved by raising the surface of the ground way beyond the disturbed region.

As the disturbance and fracturing continues, an increasingly large bank of disturbed, higher-porosity sand is created. Later on, when the pore pressure is released during the production cycle, this "fluffed-up" sand becomes compressed by the weight of the overburden sinking upon it and the porosity becomes reduced, although not to its original level. During the reduction of porosity, there is a squeezing of the fluids back to the production well.

As the fracture closes, the fluid pressure acts to move oil as well as the condensate to the well. However, much of the steam condensate is relatively remote from the well and is trapped. This trapping of the condensate beyond the oil explains why water production is less than might be expected during the first production cycles.

Although this mechanism is similar to that described previously for the steam stimulation of mobile heavy oils, there are important differences. With the mobile oils, the volume of the condensate from the steam was accommodated by the movement of oil away from the well bore and by the invasion of the condensate into the oil, probably either as fingers or stratified fingerlets. With essentially immobile bitumen contained in tar sands, the accommodation is by the dilation of the tar sands as the fracturing stress is exceeded. In the first case, the flow of fluids toward the well bore during the production cycle is driven by the reservoir pressure around the perimeter of the heated region. In the second case, much of the driven force comes from the compaction of the matrix squeezing fluids toward the production well. In later cycles, as has been described previously, the steam and its condensate are accommodated by the compression of steam within the existing chamber, and gravity provides much of the drive.

The similarity of the mechanisms is also seen by comparing the production curves for Cold Lake bitumen shown in Figure 6.23 to those found by Niko and Troost for mobile heavy oil in Figure 6.8.

Figure 6.24 shows the corresponding cumulative oil and water production for the same Cold Lake project. The water-to-oil ratio in the produced steam is only slightly larger than 1 after the first cycle; after four cycles it is about 2.

Denbina, Boberg, and Rotter (1987) have studied the production of oil in the early cycles of steam stimulation at Cold Lake using numerical simulation. They find that the field data can be closely simulated by incorporating a model for the response of the reservoir sand to shear failure during steam injection.[6] This provides increased pore volume during injection and allows a compaction drive during production. The pore volume decreases as the sand consolidates during the relaxation of the pore pressure. They also incorporate a relative permeability hysteresis into their model; this allows the penetraton of condensate into the dilated reservoir but restricts its back flow during the compaction drive. In their paper, Denbina,

[6]There is no shear stress within a static fluid in which the compressive stress is equal in all directions. In a solid however, it is possible and usual for the compressive stresses to be different in different directions. If this is the case, there must also be accompanying shear stress (see, for example, Holtz and Kovacs 1981). In many cases this can lead to the failure of the solid in shear.

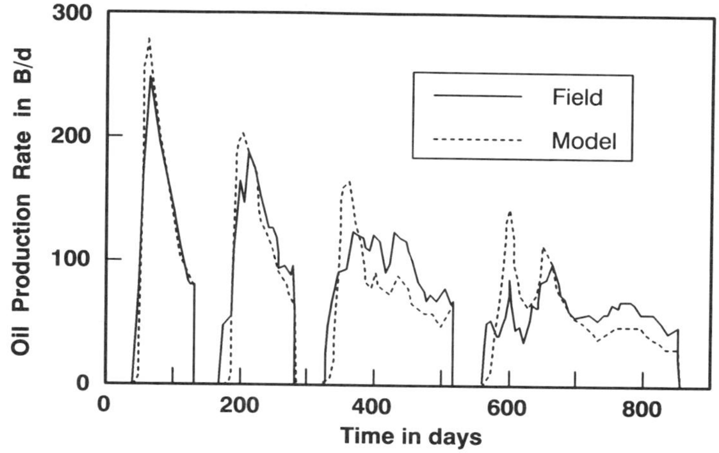

Figure 6.23 Oil-Production Rates from Esso's Cold Lake Steam-Stimulation Project. Data for "average well" (after Denbina, Boberg, and Rotter 1987)

Boberg, and Rotter conclude that formation compaction provides the main reservoir drive during early cycles. In later cycles, gravity becomes increasingly important. In the early cycles, much of the steam condensate flows into the reservoir but does not flow back because of the permeability hysteresis.

The need to use unusual relative permeability relations to match the observed performance of cyclic steaming projects by numerical simulation is a common problem. It has been discussed by Dietrich (1981) who finds that conventional thermal simulators can match the field performance if the relative permeability for water is assumed to be extremely small. The value of k_{rw} for water saturated sand with approximately 50% mobile water and 50% mobile oil from the curve recommended by Dietrich for cyclic steaming is compared to the corresponding value recommended

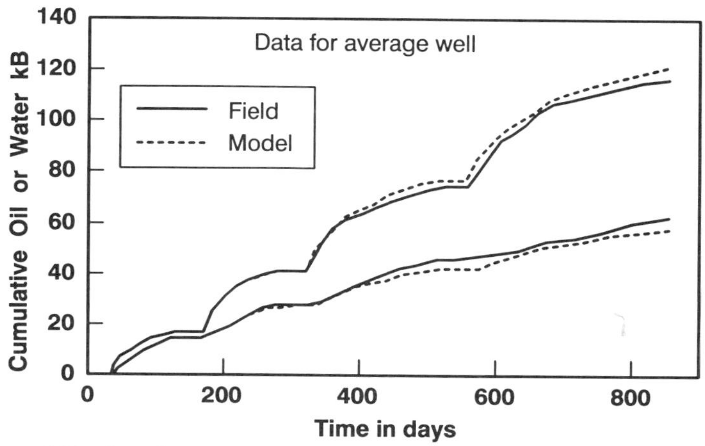

Figure 6.24 Cumulative Oil and Water Production from Esso's Cold Lake Stimulation Project (after Denbina, Boberg, and Rotter 1987)

by Gomaa (see Figure 4.32) for steamflooding, and to the range found in the conventional waterflooding of sand cores in the following table:

Cyclic stream, Dietrich (1981)	0.001
Steamflooding, Gomma (1980)	0.004
Conventional waterflooding	0.02–0.1

The difficulty is that if conventional curves are used for the relative permeability of water, then the water production predicted by the model is far greater than that found in practice. Water does not flow back to the production well as fast in the field as would be predicted by the simulator. Dietrich, like Denbina, Boberg, and Rotter, discusses the possibility of there being a hysteresis in which water could flow very easily into the formation but in which the relative permeability for backward flow would be far less.

A factor which could contribute to such hysteresis but, which seems to have been ignored, is that of the instability of the water-oil interface during injection. As has been described in Chapter 5, the flowing stream beyond the condensation interface is largely water but, because the viscosity of water is much lower than that of the oil, the water-oil interface is highly unstable and the water flows as fingers which can be described as rivulets through the oil. For the segregated flow, the saturation required for movement of the water is only very slightly larger than the irreducible water saturation and the water becomes rapidly dispersed deep into the oil. It seems possible that when the well pressure is lowered for the production cycle, the rivulets of water become unstable as the flow direction is reversed and they become broken by the oil into pockets. Then, as the oil proceeds to displace these pockets of water back toward the well, the relative permeabilities correspond to those for diffuse flow.

It is suggested that the flow away from the well is characterized, because of the abundant supply of water and the instability of broad oil-water interfaces, by the segregated flow of small swift water rivulets; this disperses the intruding water over considerable distances. The flow back to the well is with more normal flow because it is now the oil which is displacing the water.

This process would be expected to be modified by the override of the steam chamber. In this circumstance, heated oil tends to be bypassed below the steam, and water condensate will flow with the heated oil with both tending to be dragged forward by the pressure gradient. In this drag-layer, the temperature is considerably above that of the reservoir and the ratio of oil viscosity to water viscosity will be less than it would be at reservoir temperature (see Figure 4.7). As a result the water saturation within this mobile layer will be substantially higher and more water will be available to flow backwards when the flow is reversed. Thus, some of the steam condensate can be distributed far ahead of the advancing front, while some will remain with bypassed oil and will be much closer to the well.

For cyclic steaming of bitumen with fracturing, the fractures which were able to carry the water away from the condensation interface during steam injection, close, or at least partially close, during the production cycle. This effect may be expected to contribute substantially to the hysteresis effects.

CYCLIC STEAMING OF VACCA TAR, OXNARD, CALIFORNIA

This reservoir contains a 5° API crude oil at a depth of 1870 ft. The reservoir temperature is 100°F. The extrapolated viscosity of the crude oil at reservoir temperature is about 1 million centipoise; this is approximately the same as that of the bitumen in Athabasca. Although the Vacca tar is heavier and basically more viscous than Athabasca bitumen, the temperature is higher in the Vacca reservoir.

A project to recover this tar has been described by Bott (1967). His paper describes the results of four cycles of stimulation in well 702. The reservoir was about 450 ft thick, but the well was drilled into only the upper 217 ft of the pay. Casing was set into the upper 107 ft of this hole, leaving 110 ft of pay open below. A slotted linear was set in the open hole and packed with gravel.

As may be seen from Table 6.5, the project was very promising, and high production rates were achieved with excellent oil-to-steam ratios. Very high steam injection pressures were required in the first two cycles compared to those in cycles 3 and 4. It seems likely that fracturing of the reservoir occurred in cycles 1 and 2, but that in cycles 3 and 4, injection occurred into an existing steam chamber, as described previously. The lower oil-steam ratios found in cycles 1 and 2 are consistent with the idea of steam being injected into a fracture with much of the heated oil not being available during the subsequent production cycle. However, much of this heat is conserved and reduces the requirements in the subsequent cycles.

TABLE 6.5 Cyclic Stimulation of Vacca Tar (Well 702)

	5° API; h = 217 ft.; k = 5.5 D; T_R = 100°F				
Cycle	1	2	3	4	Total
Steam injection					
pressure	1,600	1,750	1,200	1,050	
B	22,839	14,756	7,500	10,671	55,766
Production					
Oil, B	5,153	13,192	11,497	13,125	42,967
Water, B	1,904	4,358	5,575	9,181	21,018
Oil B/SD average	115	131	107	75	99
OSR	0.23	0.89	1.53	1.23	0.77

(from Bott 1967).

It is interesting that the production data from the Vacca project plot as straight lines of reciprocal production rate against time. This is shown in Figure 6.25.

The data fall, in general, on straight lines for each cycle, and the slopes reflect the oil-steam ratio. It is interesting that these data, for an initially immobile oil, still show the harmonic decline discussed earlier. Also, the calculated values of η seem reasonable.

Bott also reports the water-to-oil ratio found for each cycle, and this makes it possible to calculate the average heat capacity of the products. This is compared

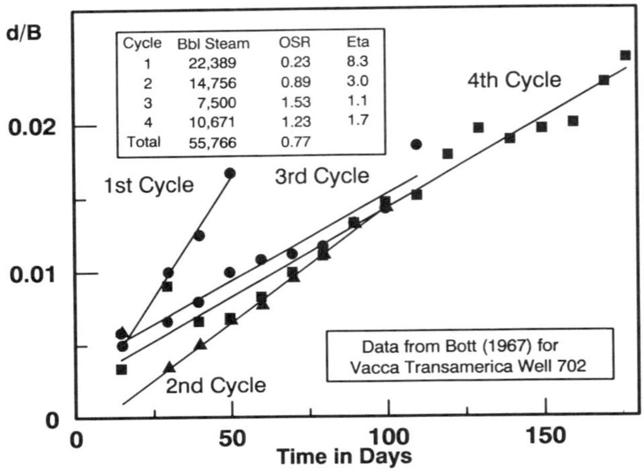

Figure 6.25 Cyclic Stimulation of Vacca Tar, Oxnard, California. 5°API, Reservoir Temperature 100°F.

to the slopes of the decline curves in Table 6.5. The comparison indicates that in cycles 1 and 2, more heat was lost from the heat bank than could be accounted for by the heat carried away with the products. This is shown by the ratio in the last line of the table being greater than 1. In cycles 3 and 4, the reverse was true. This indicates that heat lost in the first two cycles augmented the heat bank in the latter two. These comparisons were made assuming that the parameter s was equal to m; however, it is possible that with the mechanism of flow, s would be less than m and this, in turn, would make the calculated loss factors in the last row of Table 6.6 smaller than they should be.

TABLE 6.6 Comparison of Slope of Decline Curves for Vacca Project with Reported Water-Oil Ratios in Product.

Cycle	1	2	3	4
Slope of line in Figure 6.4 $B^{-1} \times 1000$	0.326	0.157	0.117	0.120
Calculated $\eta\rho C$, Btu/B °F	1459	519	200	297
Water-oil ratio in product	0.369	0.344	0.485	0.700
ρC for product, Btu/B_{oil} °F	304	295	345	420
$\eta\rho C/\rho C_{prod}$	4.8	1.8	0.58	0.71

COMPACTION DRIVE IN CONVENTIONAL HEAVY OIL RESERVOIRS

There has been extensive production of heavy oils both by conventional, nonthermal means and by cyclic steam stimulation in the Bolivar region of the east coast of Lake Maracaibo in Venezuela. The reservoirs in this area are very large, and enormous quantities of oil have been produced. The cold production from these reservoirs (the main ones are Tia Juana, Lagunillas and Bachquero) is very substantial, and the recovery, without thermal stimulation, is much higher than would be expected.

The reason for this higher production is compaction drive. As fluids are withdrawn from the reservoir, the pore pressure decreases and additional compressive load is transferred to the rock matrix. This matrix is sensitive to load, and rearrangement (particularly of plastically deformable clay constituents) occurs with a decrease in overall pore volume and a subsidence of the ground surface. The fluids are squeezed from the reservoir rather like water from a sponge.

This type of behavior probably occurs to some extent in most reservoirs but it is of greatest practical significance in reservoirs (both petroleum and water) with matrices which are sensitive to increased stress. It is thought that this is the case when the reservoir matrix has not been previously stressed in its geological history to the same extent as occurs when the pore pressure is relieved during fluid production. An extreme, well-documented case of compaction drive and the accompanying ground subsidence is that of oil production from the Wilmington field which lies partly under the city of Long Beach, California. In this case, the maximum subsidence of the ground surface was greater than 8 meters and very substantial damage was done to surface facilities such as buildings and bridges (Mayuga 1970).

If there were no compaction drive in the Bolivar reservoirs, the recovery would be limited to about 5% of the original oil in place (Figure 1.12). However, as the pore pressure is decreased, the reservoir compacts and additional oil is produced. In addition to compaction of the sand reservoir, there is also a contribution from water being squeezed from the pores of interbedded shale streaks; this water also tends to displace oil. The compaction in this area is well documented and evidenced by the settling of the ground surface. Measurements have been made of this, since the 1930s in some cases. Figure 6.26, which is taken from Borregales (1977), plots the cumulative subsidence of the surface of the ground against the cumulative production for a project in the Tia Juana field. In this diagram, a subsidence of 50 million barrels corresponds to a change in the elevation of the surface of about 2.3 m.

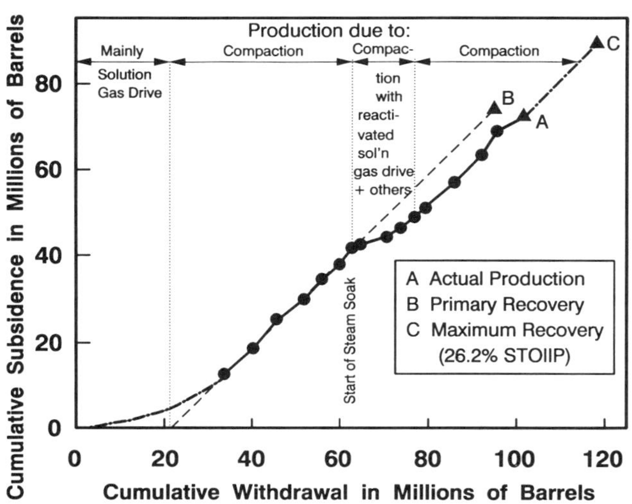

Figure 6.26 Drive Mechanisms in the D-2/E-2 Project in the Tia Juana Field (after Borregales 1977)

During the initial production, when the drive was largely due to the effect of solution gas, there was relatively little subsidence. Over the bulk of the production period shown, the surface subsidence tracks the cumulative production along a 45° line; the drive is essentially due to compaction alone. This would continue to the limit *B* if no steam were introduced.

Cyclic steam stimulation not only accelerates production, it also increases the recovery achievable. One reason for this is the additional gas drive that is initiated by the effect of the steam; oil is displaced by steam vapor remaining in the reservoir. In addition, the more rapid production rate makes more extensive depletion economic. The limit shown by the point *C* for cyclic steam stimulation corresponds to a recovery of 28% of the original oil in place. The projected recovery for other projects in the Bolivar Coast is shown in Table 6.7. It can be seen that the increase in oil recovery by steam soaking is between 5 and 15 % of the original oil. The very high oil-to-steam ratios indicate a stimulation mechanism rather than one depending on general reservoir heating. Figure 6.27 shows the average production behavior of all of the steam-soaked wells on the Bolivar Coast.

TABLE 6.7 Cyclic Steam Results—Bolivar Coast

Project Name	STOIIP $B \times 10^6$	Recovery as % STOIIP			Compaction % STOIIP	OSR
		Primary	Steam	Total		
D2/E2	454	19.8	8.2	28.0	18.9	2.94
J-7	713	10.6	12.2	22.8	14.6	4.94
H-7	98	8.2	13.2	21.4	16.1	5.03
D-6	604	4.4	13.1	17.5	14.1	7.37
T-6	305	7.5	5.6	13.1	23.4	4.83

(Data from Borregales 1977)

Borregales considers that, after the recovery of 20 to 30% of the oil in place, these projects should be converted to steam drives; presumably this would allow recoveries of the order of 50%. He points out that it is desirable to continue the cyclic steam process to the point where compaction drive is no longer operative. At this point, the reservoir is hot and the reservoir pressure is low. This allows the easier introduction of steam in the steam drive, which, at a lower pressure, is thermally more efficient. A particularly important consideration is that if compaction drive were still available when flooding steam is introduced, then the valuable compaction drive might be dissipated, moving compressible steam rather than the hard-to-push oil. The result would be a lower overall recovery and/or a higher overall steam requirement.

FRACTURING AND RESERVOIR EXPANSION DURING STEAM INJECTION

In order to obtain practical rates of steam injection in oil sand reservoirs, it is usual to use steam pressures high enough to fracture the reservoir and thereby allow injection. With lower pressures in a virgin reservoir, the injection rates are usually negligible.

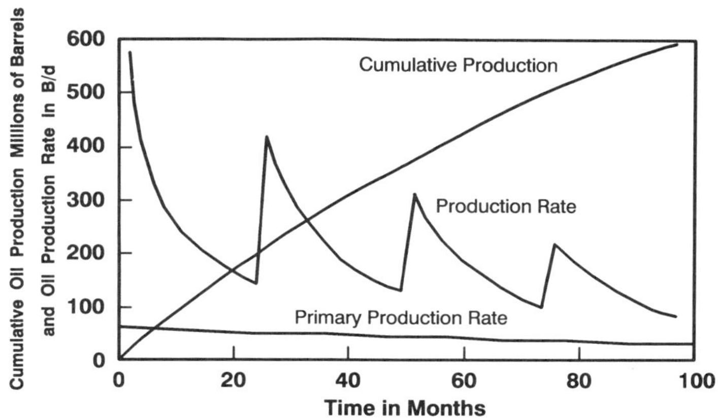

Figure 6.27 Average Production Behavior of Steam-Stimulated Wells in the Bolivar Coast Fields (after Borregales 1977)

When fluids are injected into porous cemented rocks, the fractures that form are usually thin cracks, and these tend to be approximately planar and either horizontal or vertical. With vertical fractures it is found that there is a preferred orientation. Even with horizontal fractures, there tends to be an azimuthal direction in which the fracture spreads preferentially (Aughenbaugh and Pullen 1966).

Settari, Kry, and Yee (1988) have reported the formation of asymmetrical fractures around steam-injected wells at Cold Lake in a region where horizontal fractures were expected from measurement of the initial reservoir in situ stresses.

It is accepted that the orientation of the fracture will be controlled by the in situ stresses that exist within the rock at the time of fracturing (see, for example, Howard and Fast 1970). In general these stresses may be resolved into three orthogonal, principal compressive stresses, and it is usually assumed that one of these is vertical and that the other two are horizontal.

Fracture theory predicts that the plane of a hydraulic fracture will lie normal to the minimum principal compressive stress. Thus, if the minimum principal stress happens to be vertical, then the fracture will be horizontal and vice versa. If the minimum principal stress is horizontal in a specific direction then the fracture will be vertical and will be at right angles to the direction of the minimum stress.

Stress Due to Gravity in a Semi-infinite Strain-Free Solid

Both fluids and solids can support compressive stresses. However, in a fluid, the stresses must be equal in all directions (the pressure), whereas a solid can support stresses which differ with direction. It is this property that allows a solid to maintain a shape.

As a simple example of the stresses[7] within a solid, consider a semi-infinite hypothetical solid that is stress- and strain-free and that is then in some imaginary manner exposed to a gravitational field normal to its surface. It might be useful for the reader to imagine a block of an elastic solid such as an eraser.

[7]This example resulted from a discussion with Dr. S. Bharatha of Esso Resources Canada Limited.

The gravity will compress the solid downward, and this will also tend to make it extend horizontally. We will assume that the horizontal growth of a large horizontal layer of the solid is constrained by some distant immovable barrier and that, as a result, the horizontal dimensions remain constant.

A small element in the solid is shown in Figure 6.28. It is being compressed downward by the vertical principal stress and horizontally by the two equal horizontal principal stresses.

The condition that the strain (i.e. the contraction per unit length) in direction 1 be zero is given by equation 6.22. This states that the sum of the contributions to the strain caused by the three principal stresses should be zero. From this is derived an expression for the horizontal stress as a function of the vertical stress.

$$\varepsilon_1 = \frac{\sigma_H}{E} - \frac{\nu\sigma_H}{E} - \frac{\nu\sigma_V}{E} = 0 \quad \text{or} \quad \sigma_H = \frac{\nu}{1-\nu}\sigma_V \tag{6.22}$$

where E is Young's modulus
σ_H is horizontal stress
σ_V is vertical stress
ν is Poisson's ratio

But $\nu < 0.5$; therefore, $\sigma_H < \sigma_V$.

Poisson's ratio must always be less than 0.5, since, if it were not, there would be an expansion in volume when a body is compressed.[8] For a fluid, Poisson's ratio is 0.5. If Poisson's ratio is less than 0.5, then equation 6.22 indicates that the horizontal stress will be less than the vertical. In such a situation only vertical fractures can form if the material is fractured by the injection of a fluid within its pores and if its tensile strength is either negligible or the same in each direction.

This conclusion applies only to the imaginary unstrained semi-infinite solid for the conditions just considered. A similar analysis applied to a large spherical body to which a gravity was suddenly applied (the gravity coming from its own mass) would lead to a result in which there was a residual hoop compressive stress at the surface, and, near the surface, horizontal fractures would be possible.

In Situ Reservoir Stresses

In actual practice, the stresses within the earth's crust are most complex and are related to events such as the movement of tectonic plates, mountain building, and the

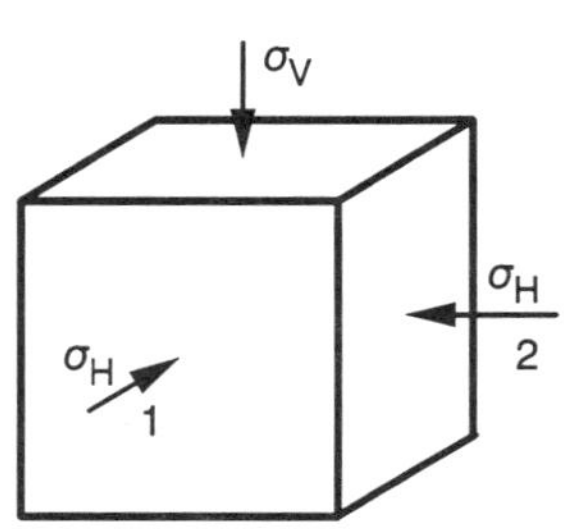

Figure 6.28 Principal Stresses Acting on Solid Element

[8]Notice that in equation 6.22, σ_H would be equal to σ_V and hence to the stress in any direction if $\nu = 0.5$. In this circumstance, the material would act like a fluid.

like. In general, it is found that in the real world, horizontal rather than vertical fractures tend to form from the hydraulic fracturing of porous rocks at relatively shallow depths (down to about 1200 ft.); below this, vertical fractures usually form. The direction of the horizontal principal stresses in regions where vertical fractures form can be inferred from measurements made of the out-of-roundness of well bores using a four-arm caliper. The direction corresponding to the smallest diameter is that of the larger horizontal principal stress and vice versa.

Using this technique it has been possible to plot maps showing the directions of the maximum and minimum horizontal stress directions. Figure 6.29 shows such a map for Alberta.

In the region shown, the maximum stress lies generally in the NE/SW direction with the minimum stress normal to it. Fractures at depths below about 1200 ft are thus expected to be vertical and to occur in the NE/SW direction. Chhina and Agar (1985) report the results of fracture tests at 12 locations in the Alberta tar sand reservoirs. They find that horizontal fractures normally occur at depths less than 250 m, and vertical ones occur at depths greater than 400 m. In between, vertical fractures are likely, although horizontal ones may occur. In one test in the Lower Grand Rapids formation at 342 m, a horizontal fracture started growing, but this changed to a vertical orientation as it grew.

Part of the data used in drawing Figure 6.29 was the observation that at Cold Lake, communications between steam-fractured wells tend to occur in this preferential direction. Gough and Bell also mention caliper measurements made on a well in the vicinity that indicated a direction of 41° east of north for the maximum stress.

Daneshy et al. (1986) describe measurements of in situ stresses at various levels in a gas well; these were obtained from microfracture tests during the drilling operation. Drilling mud was used as fracture fluid and oriented cores were obtained immediately after the microfracture tests. The minimum stresses were obtained from the pressure data and the directions of the vertical fractures were obtained from the position of the fractures within the retrieved cores.

In situ stress measurements in the Cold Lake field are described by Kry, Gronseth, and Morgenstern (1989).

Fracturing Pressure

The minimum pressure required to form a fracture is equal to the minimum principal stress. It is usual to take the vertical principal stress as being equal to the stress corresponding to the weight of the overburden, i.e., to the depth multiplied by the density and gravity. This gives values of about 1 psi per foot of depth, or 22.6 kPa/m. If vertical fractures tend to form, the fracturing pressure measured during a fracturing operation will be less than this value and can be as small as one-half of it (Howard and Fast 1970).

Ground Heave

There has been a lot of discussion without much data in regard to the nature of the fracture in unconsolidated tar sand. From this it seems likely that, rather than thin cracks forming as in the fracturing of consolidated rocks, relatively large volumes of

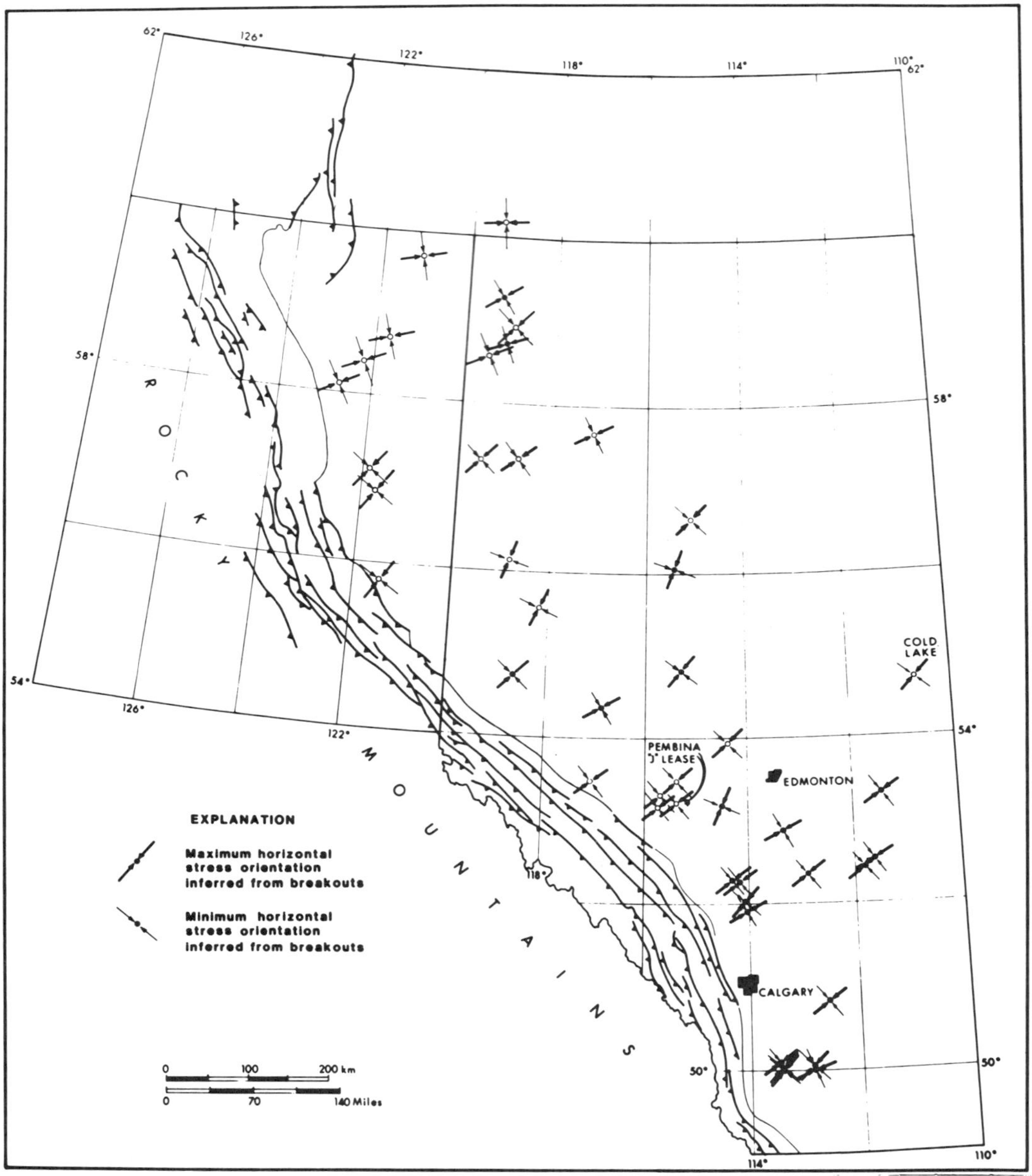

Figure 6.29 Principal Horizontal Stress Directions in Alberta (from Gough and Bell 1981)

tar sand become disrupted when high-pressure steam is injected. Evidence for this is contained in measurements of the surface disruption observed in a McMurray steam injection pilot, where it was found that substantial permanent ground heave resulted as a result of the steam injection (Agnew 1976). Figure 6.30 shows contours for constant elevation increase for one of the patterns at this pilot.

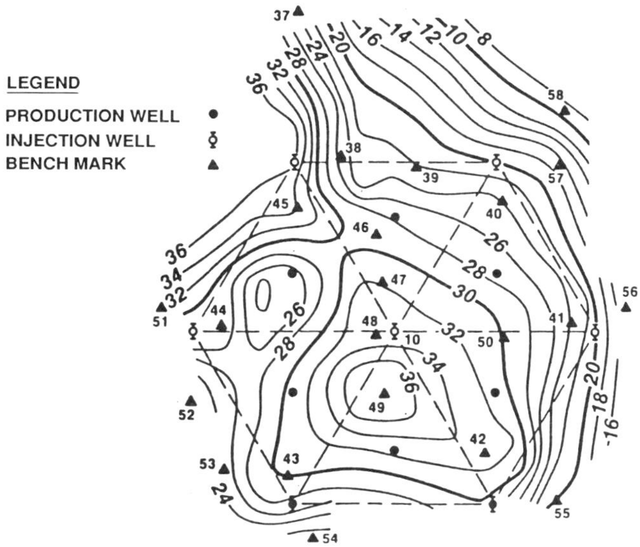

Figure 6.30 Changes in Ground Elevation at Texaco's In Situ Steam Flooding Pilot Near McMurray Alberta. Contours are for constant increases in ground elevation measured in hundredths of a foot. Data for May 12–Nov. 24, 1976 (from Agnew 1976)

The extent to which steam can penetrate a fracture before it condenses can be calculated using Hearn's equation, which was described previously (equation 3.57). It seems possible that this relation will find more application in the theory of steam stimulation.

EFFECT OF FRACTURE ORIENTATION ON PRODUCTIVITY FROM STIMULATION

If gravity drainage plays an important role in the production of extremely viscous crudes by steam stimulation, then the orientation of the fracture-disrupted reservoir zone that is allowing steam injection would seem to be important. In particular, one might speculate that vertically oriented fractures, because they allow use to be made of the vertical thickness of the reservoir to provide head for drainage of oil, might be much better than horizontal fractures.

There seems to be little data, either positive or negative, to support this view. However it might be noted that the Esso pilot at Cold Lake (Buckles 1979 and Shepherd 1979) and the subsequent commercial extensions are successful projects, operating with economic oil-to-steam ratios and with economic well productivities in spite of the very high in situ oil viscosity. Predominantly vertical fractures were reported in the early operations, although it has been found that with extensive steaming, horizontal fractures may be formed (Denbina, Boberg, and Rotter 1987).

Kry, Gronseth, and Morgenstern (1989) have shown that, in a test at Cold Lake, the maximum in situ stress,[9] measured by mini-fracture tests, increased almost immediately from 9.3 to 10.5 MPa as a result of steam injection into neighbor-

[9]Kry, Gronseth, and Morgenstern define the pressure at the point of inflection in the pressure time curve following the well shut-in after the mini-injection as being the in-situ stress or instantaneous shut-in pressure. "Judgement is required to infer the value of the in-situ stress from measured shut-in pressures."

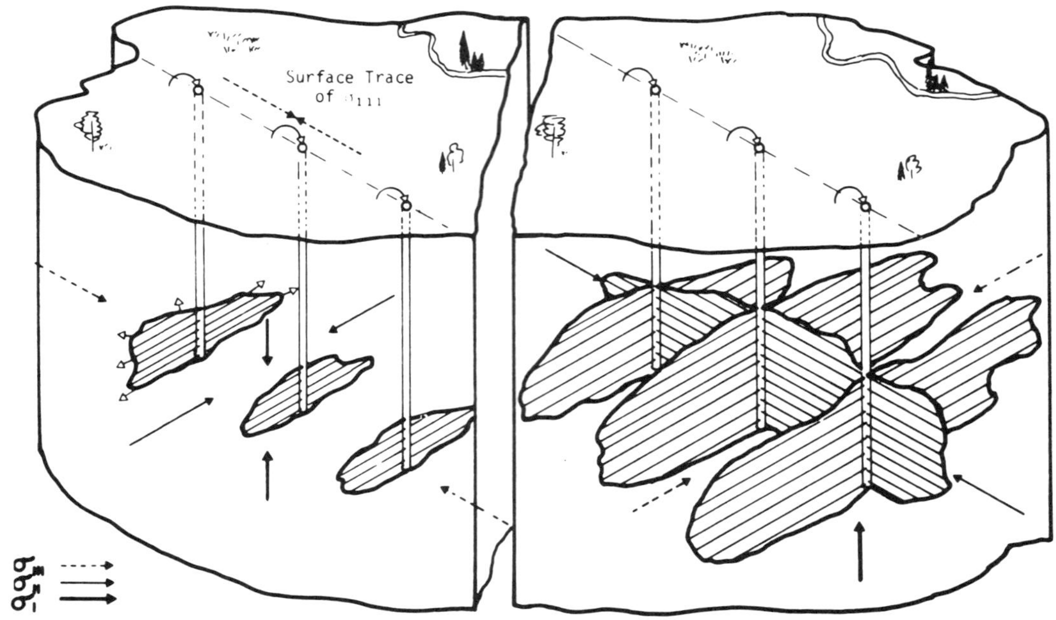

Figure 6.31 Stress Redistribution Could Result in Interwell Communication (from Dusseault 1977)

Stage 1: Initiation of multi-well simultaneous hydraulic fracture. Fracture propagation is normal to minor stress.

Stage 2: Increase in minor principal stress causes relocation and change in fracture orientation (case where σ_1 remains vertical)

ing wells. This was interpreted as an increase in the horizontal stress which had to be overcome to initiate a vertical fracture from the well bore. In succeeding minifracture tests, it was found that the instantaneous shut-in pressure (ISIP) fell in less than two hours to a stabilized value of 9.8 MPa. This is significantly higher than the initial ISIP of 9.3 MPa and approximately equal to the calculated vertical stress. This behavior was interpreted by Kry, Gronseth, and Morgenstern as indicating that the vertical fractures initiating from the well bore had turned to become predominantly horizontal.

Another factor is that vertical fractures may assist recovery of the oil by providing an improved means of passage of the oil past horizontal tight streaks. Horizontal movement of the fluids in the unfractured reservoir is usually less restricted than vertical movement. Vertical fractures should improve this situation.

POSSIBLE PRODUCTION OF ORTHOGONAL VERTICAL FRACTURES FROM THE FRACTURING OF A LINE OF WELLS

An interesting possibility that was described by Dusseault (1977), based on an earlier suggestion by Schuck and Advani, is that the fracturing of a row of wells may so modify the stress fields in the reservoir that fractures normal to the usual direction may also be formed. This possibility is shown in Figure 6.31.

In the Esso production area at Cold Lake, reservoir fracturing occurs extensively as the result of steam injection. Vertical fractures are normally found when steam is first injected into a new reservoir, and these generally have the expected NE/SW orientation (Mainland and Lo 1983). However it has been found that after an area of the field has undergone intensive steam injection, new fractures are generally horizontal (Denbina, Boberg, and Rotter 1987). This change in fracture direction is probably the result of changes in the tectonic stresses caused by thermal treatment. Heating increases the horizontal compressive stresses, leaving the vertical stress the minimum principal stress. The vertical stress can be relieved by the upward expansion of the ground. This is evidenced by the surface heave observed above thermal in situ operations.

BIBLIOGRAPHY

AGNEW, H., U.S. Patent 4,143,714 (1976).

AUGHENBAUGH, N. B. and PULLEN, M.W., "Directional Hydrofracturing: Fact or Fiction," 3d Symp. on Salt Proc. 2, 393–403 (1966).

BOBERG, T. C. and LANTZ, R. B., "Calculation of the Production of a Thermally Stimulated Well," *JPT,* 1613–1623 (December 1966). © 1966 SPE.

BOBERG, T. C., PENBERTHY, JR., W. L., and HAGEDORN, A. R., "Calculating the Steam-Lifted Performance of Gas-Lifted and Flowing Heavy-Oil Wells," *JPT,* 1207–1215 (October 1973).

BORREGALES, C. J., "Steam Soak on the Bolivar Coast," Canada-Venezuela Oil Sands Symposium, Edmonton, May 30–June 4, 1977 pub. as *The Oil Sands of Canada-Venezuela 1977,* CIM Special Volume 17, (1977).

BOTT, R. C., "Cyclic Steam Project in a Virgin Tar Reservoir," *JPT,* 585–591 (May 1967).

BUCKLES, R. S., "Steam Stimulation Heavy Oil Recovery at Cold Lake, Alberta," *SPE* 7994, (1979).

BURNS, J., "A Review of Steam Soak Operations in California," *JPT,* 25–34 (January 1969). © 1969 SPE.

BUTLER, R. M., "The Expansion of Tar Sands During Thermal Recovery," *JCPT,* 51–56 (September–October 1986).

CHHINA, H. S. and AGAR, J. G., "Potential Use of Fracture Technology for Recovery of Bitumen from Oil Sands," 3d UNITAR/UNDP International Conference on Heavy Oil and Tar Sands, Long Beach, Calif., 771–789, (July 1985) published by AOSTRA 1988.

DANESHY, A. A., SLUSHER, G. L., CHISHOLM, P.T., and MAGEE, D. A., "In-Situ Stress Measurements During Drilling," *JPT* 891–898, August 1986.

DE HAAN, H. J. and VAN LOOKEREN, J., "Early Results of the First Large-Scale Steam Soak Project in the Tia Juana Field, Western Venezuela," *JPT,* 101–110 (January 1969).

DENBINA, E. S., BOBERG, T. C., and ROTTER, M. B., "Evaluation of Key Reservoir Drive Mechanisms in the Early Cycles of Steam Stimulation at Cold Lake," *SPE* 16737, Dallas (1987). © 1987 SPE.

DIETRICH, J. K., "Relative Permeability During Cyclic Steam Stimulation of Heavy-Oil Reservoirs," *JPT,* 1987–1989 (October 1981).

DOSCHER, T. M., "Factors Influencing Success in Steam Soak Operations," Petroleum Industry Conference on Thermal Oil Recovery, Sponsored by Rockwell Mfg. Co., Los Angeles, Calif., 76–80, June 6, 1966. (reference from Burns 1969).

DUSSEAULT, M. B., "Stress State and Hydraulic Fracturing in the Athabasca Oil Sands," Oil Sands of Canada and Venezuela, 27–35, CIM Special Volume 17, (1977).

DUSSEAULT, M. B. and MORGENSTERN, N., "Sampling and Testing of Athabasca Oil Sands for Stability Studies," *Oil Sands of Canada and Venezuela,* 260–269, CIM Special Volume 17, (1977).

GOUGH, D. I. and BELL, J. S., "Stress Orientations from Oil-Well Fractures in Alberta and Texas," *Can. J. Earth Sci.,* 18: 638–645 (1981).

HOLTZ, R. D. and KOVACS, W. D., *An Introduction to Geotechnical Engineering,*" Englewood Cliffs, N.J.: Prentice Hall (1981).

HOLZHAUSEN, G. R., et al., "Results of Deformation Monitoring During Steam Stimulation," Conf. on Appl. Oilsands Geoscience, Edmonton, Alberta (June 11–13, 1980).

HOWARD, G. C. and FAST, C. R., "Hydraulic Fracturing", SPE Monograph 1970.

JONES, J., "Cyclic Steam Reservoir Model for Viscous Oil, Pressure Depleted, Gravity Drainage Reservoirs," *SPE* 6544 (1977).

KRY, P. R., GRONSETH, J. M., and MORGENSTERN, N. R., "Geotechnical Properties," In: *AOSTRA Technical Handbook on Oil Sands, Bitumens and Heavy Oils,* Hepler, L. G. and Hsi, C. (Editors), Alberta Oil Sands Technology and Research Authority, Edmonton, Alberta.

KUO, C. H., SHAIN, S. A. and PHOCAS, D. M., "Gravity Drainage Model for the Steam-Soak Process," *SPEJ,* 119–126 (June 1970). © 1970 SPE.

LEFKOVITS, H. C. and MATTHEWS, C. S., "Application of Decline Curves to Gravity-Drainage Reservoirs in the Stripper Stage," *Pet. Trans. AIME,* 213: 275–280 (1958).

LENNON, R. B., "Geological Factors in Steam-Soak Projects on the West Side of the San Joaquin Basin," *JPT,* 741–748, July 1976.

MAINLAND, G. G. and LO, H.Y., "Technological Basis for Commercial In-Situ Recovery of Cold Lake Crude," 11th World Pet. Cong., London, Session RDT3(1) (1983).

MARTIN, J. C., "A Theoretical Analysis of Steam Stimulation," *JPT,* 411–418 (March 1967).

MATTHEWS, C. S. and LEFKOVITS, H. C., "Gravity Drainage Performance of Depletion-Type Reservoirs in the Stripper Stage," *JPT,* 207: 265–274 (December 1956).

MATTHEWS, C. S. and RUSSELL, D. G., "Pressure Buildup and Flow Tests in Wells," *SPE Monograph* 1, (1967).

MAYUGA, M. N., "Geology and development of California's giant—Wilmington Oilfield," In: Halbouty, M.T. (Editor), "*Geology of giant petroleum fields,*" Mem. Am. Ass. Pet. Geol., 14, 158–186, (1970).

NIKO, H. and TROOST, P. J. P. M., "Experimental Investigation of the Steam-Soak Process in a Depletion-type Reservoir," *JPT* 251: 1006–1014 (August 1971). © 1971 SPE.

OWENS, W. D. and SUTER, V. E., "Steam Stimulation—Newest Form of Secondary Petroleum Recovery," *Oil Gas J.,* 82–90 (April 26, 1965).

PUJOL, L. and BOBERG, T. C., "Scaling Accuracy of Laboratory Steam Flooding Models," *SPE* 4191 (1972).

SEBA, R. D. and PERRY, G. E., "A Mathematical Model of Repeated Steam Soaks of Thick Gravity Drainage Reservoirs," *JPT,* 87–94 (January 1969).

SETTARI, A., KRY, P. R., and YEE, C.T., "Coupling of Fluid Flow and Soil Behaviour to Model Injection into Unconsolidated Oil Sands," 39th Annual Technical Meeting of the Pet. Soc. of CIM (June 12–16, 1988), Preprints paper 88-39-72.

SHEPHERD, D.W., "Predicting Bitumen Recovery from Steam Stimulation," *World Oil,* 68–72 (September 1979).

SHEPHERD, D.W., "Steam Stimulation Recovery of Cold Lake Bitumen," 1st Unitar Conference, Edmonton, Canada (June 4–12, 1979), reported in *The Future of Heavy Crude Oils and Tar Sands,* New York: McGraw-Hill, 349–360 (1981).

STEGEMEIER, G. L., LAUMBACH, D. D. and, VOLEK, C.W., "Representing Steam Processes with Vacuum Models," *SPEJ,* 151–174, June 1980.

SUN, R. J., "Theoretical Size of Hydraulically Induced Horizontal Fractures and Corresponding Surface Uplift in an Idealized Medium," *J. Geophysical Res.,* 74, no. 25: 5995–6011 (November 15, 1969).

TOWSON, D. E. and BOBERG, T. C., "Gravity Drainage in Thermally Stimulated Wells," *JCPT,* 130–135 (October–December 1967).

7

Steam-Assisted Gravity Drainage

INTRODUCTION

The recovery of heavy crudes using a special form of steamflooding that has become known as *steam-assisted gravity drainage* (SAGD) is discussed in this chapter. In this process the movement of oil to the production well is caused by gravity forces, and the geometry is such that the oil moves approximately parallel to the interface that forms the boundary of a growing, steam-saturated zone known as the *steam chamber.*

A theory that predicts the rate at which this process will occur has been developed by the Heavy Oil Research Division at Esso Resources in Calgary and has been described in a series of papers (Butler, McNab, and Lo 1981; Butler and Stephens 1981; Butler, Stephens and Weiss, 1980; Butler 1985). This theory, together with recent developments is summarized.

CONCEPT

The intention in developing the steam-assisted gravity drainage process was to devise a means whereby heavy oil or bitumen could be removed in a systematic manner in order to give a more complete recovery than is possible in conventional steamflooding processes, where the oil is moved by pushing it with injected fluids.

Gravity is already present throughout the reservoir, and by using it as the chief driving force to effect oil movement, it is possible to avoid the differential fingering that occurs when viscous oils are moved by pushing with a less viscous fluid.

It was reasoned that if steam were injected above but close to a production well that was completed at the base of the reservoir, the steam would tend to rise and the condensates, together with warmed oil, would fall. The oil and condensate would be removed continuously from the production well. It was thought that if these liquids were not removed too quickly, then the tendency of the steam to flow directly to the production well—and thus to bypass the reservoir—could be reduced or possibly even eliminated. This is analogous to the ability of a steam trap to allow the flow of condensate from the bottom of a steam-heated radiator without allowing significant bypassing of the steam.

An attraction of this concept is that, although the injection well and the production well can be very close, the mechanism will cause the steam chamber to expand gradually and eventually allow drainage from a very large area. The injector and producer do not have to span the drainage area. Another advantage is that the heated oil remains hot as it flows towards the production well. In conventional steamflooding, as was discussed in Chapter 5, the oil that is displaced from the steam chamber is cooled and is hard to push to the production well. In SAGD the oil remains heated as it flows around the steam chamber.[1]

Although the concept is attractive, it was not clear initially whether the mechanism could produce oil at practical rates. The first efforts were to analyze the process theoretically to determine the production rates that might be anticipated. Following this, a number of scaled recovery-model experiments were carried out in the laboratory to study the process, and the results were compared with theory.

RELATIONSHIP TO CONVENTIONAL STEAMFLOODING

In previous chapters it has been shown that steamflooding tends to produce a stable displacement without steam fingering. Condensate that leaves the steam zone flows through the oil zone either as fingers or, in exceptional cases, by diffuse Buckley-Leverett type flow.

A major problem with conventional lateral steamflooding is the tendency for steam to override the oil zone and break through at the production well. This effect can be reduced by completing the producer near to the base of the reservoir and by removing product at a controlled rate in order to allow gravity to keep the steam zone segregated. A considerable improvement can be achieved if the reservoir dips and if the steam is injected so that it flows downdip.

Operation in this fashion has much in common with the steam-assisted gravity drainage discussed in the following sections. However, with conventional wells the pressure gradients associated with radial flow to the production well make the maximum rate that is achievable without steam coning relatively low. In the steam-assisted gravity drainage process, horizontal wells are employed, and production rates of the order of 0.3 B/d per foot of horizontal well are indicated to be practicable without steam coning; with long horizontal wells this provides an economic drainage rate.

A major potential for the steam-assisted gravity drainage process lies in overcoming the shortcomings of conventional steamflooding that are created by the tendency of the steam to override.

[1]There is a similarity between the SAGD process and reverse in-situ combustion which is described in Chapter 9. From one point of view SAGD can be looked upon as reverse steamflooding.

GRAVITY DRAINAGE THEORY

In the development of the concept just described, the objective was to develop a means whereby steam could be introduced continuously and the condensate removed as it formed, together with the oil that has been heated. The concept is shown in Figure 7.1.

Steam is introduced near the bottom of the reservoir and tends to rise; the heavier condensate and heated oil tend to fall to the bottom. Means are provided for introducing steam quite near to the bottom if there is difficulty in displacing cold oil or higher up if the oil is mobile, and for removing liquids as they drain to a lower location. As the liquids are removed, the space that is left in the pores is filled with steam.

A horizontal well is placed at the bottom. A steam chamber is formed above the well, and steam is injected continuously into this chamber by means of another horizontal well placed close to and usually somewhat above the production well. In some applications the injection well may be a vertical well rather than a horizontal well (for long horizontal production wells, a number of vertical wells may be used). In reservoirs containing mobile oil, a horizontal injection well placed higher in the reservoir, rather than low down, can be used with advantage.

The steam pressure is usually maintained constant during much of the process. The chamber is surrounded by colder oil sand. Steam flows through the sand within the chamber to the interface and condenses. The liberated heat is conducted into the colder oil sand. This heats the oil near the condensation surface and allows it to drain by gravity to the production well. The condensate also drains. As the oil is removed, the steam chamber grows upwards and sideways. The pressure within the steam chamber remains essentially constant. Flow is caused by gravity. Oil and steam condensate drain downwards and steam rises.

In some respects the process is analogous to that of water draining from a bathtub. In the major mechanism of steam-assisted gravity drainage, steam does not

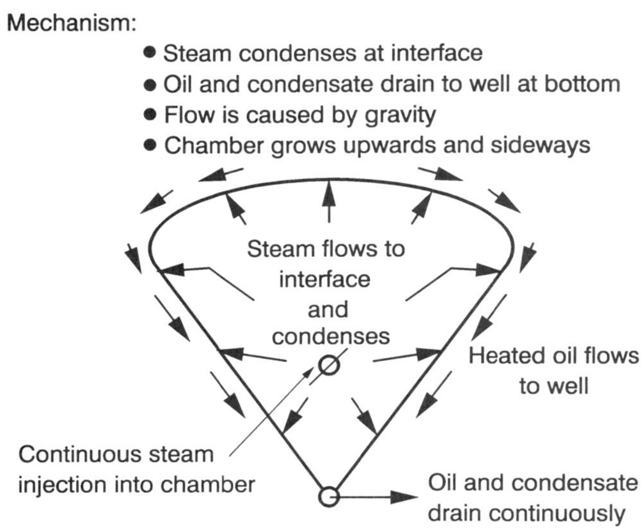

Figure 7.1 Steam-Assisted Gravity Drainage

push the oil out anymore than the air pushes water from the tub when it empties. Yet, eventually, the steam fills the volume originally occupied by oil just as air eventually fills the tub.

It has long been realized that the gravity drainage of conventional crude oil from below a gas cap can produce unusually high oil recoveries. Dykstra (1978) summarizes work in this area. Terwilliger et al (1951) showed that the recovery decreases and the drainage rate increases when pressure gradients are imposed upon the gravity drainage process. Increasing the rate by lowering the production well pressure tends to leave additional liquid behind in the gas-saturated region.

While the dynamic hold-up of oil in the gas-saturated steam chamber is significant, it is relatively small, since the viscosity of the oil within the chamber is very low compared with the average viscosity of the oil draining below and around it.

The average oil saturation remaining in the steam chamber can be estimated using the integrated form of an equation developed by Cardwell and Parsons (1949); this is shown in equation 7.1.

$$\bar{S}_{or} = \frac{(b-1)}{b}\left(\frac{\nu_S \phi Z}{bkgt}\right)^{1/(b-1)} \tag{7.1}$$

where $\bar{S}_{or}$ is the average residual oil saturation after time t
Z is the drainage height
k is the permeability
b is the exponent in Cardwell and Parson's equation for relative permeability, $k_r = S^b$
ν_S is the kinematic viscosity of the oil at the temperature of the steam

If b is set equal to a typical value of 3.5 and Z is set equal to the maximum possible value h, then the result is equation 7.2.

$$\bar{S}_{or} = 0.43\left(\frac{\nu_S \phi h}{kgt}\right)^{0.4} \tag{7.2}$$

Typical field values for a steam chamber temperature of 216°C and for Cold Lake crude are as follows.

$\nu_S = 0.452\ \text{m}^2/\text{d}$ (5.2 cs) $\quad \phi = 0.3; \quad h = 30.5$ m

$k = 0.987 \times 10^{-12}\ \text{m}^2$ (1 darcy) $\quad g = 7.32 \times 10^{10}\ \text{m/d}^2$

The average residual oil saturation has been calculated in Table 7.1 for various times t.

TABLE 7.1 S_{or} as a Function of Time

t (days)	100	200	1,000	2,000	10,000
S_{or}	0.34	0.26	0.14	0.10	0.06

For a practical case with a project life of 6 to 15 years, the average time that the chamber would drain might be of the order of 2000 days, and the corresponding residual oil saturation would be 0.1. Values calculated from (7.2) are of the same order as those given in Figure 4.9 in Chapter 4.

In the initial analysis of this concept, it appeared that the upward rate of growth of the steam chamber would be larger than that sideways. Eventually the upward growth is limited by the top of the reservoir, and the sideward growth then becomes critical. Figure 7.2 shows how steam chambers are expected to grow for a series of parallel horizontal production wells in a reservoir.

The horizontal well shown in the center of Figure 7.2 is flanked by similar wells on either side. The positions of the condensation surface are shown diagrammatically at successive periods of time. The steam chambers grow to the top of the reservoir and then spread sideways. After a period they intermingle and form a single steam layer above the oil. Heating continues and oil drains to the horizontal wells. This method allows almost complete coverage of the reservoir volume. The initial analysis focused on the rate at which steam chambers can grow sideways. This theory is discussed in the next sections.

Darcy's Law

Figure 7.3 shows a small part of the drainage interface. It is a vertical section. Oil heated by steam flows approximately parallel to the condensation surface down towards the production well somewhere below the left of the depicted region.

The steam is at a temperature T_S, and the reservoir is initially at T_R. Steam is condensing at the condensation surface, or the interface, which, in the region shown, is inclined at an angle θ to the horizontal. The interface is at temperature T_S. Heat is transferred into the cooler reservoir by conduction. Beyond the interface, into the reservoir, successive layers of material are cooler. At a distance ξ from the interface, where the viscosity of the oil is μ and the kinematic viscosity of the oil is ν, Darcy's law may be written, for a section with unity thickness measured into the paper, as:

$$dq = \frac{k(d\xi \times 1)(\rho_o - \rho_g)g \sin \theta}{\mu} = \frac{kg \sin \theta}{\nu} d\xi \tag{7.3}$$

Note that the potential gradient is $(\rho_0 - \rho_g)g \sin \theta$. ρ_g is neglected in comparison to ρ_0, and μ/ρ_0 is set equal to ν. The equation gives the rate of drainage of oil, dq, within the element $d\xi$. If the interface velocity measured normal to the interface is U and it is asssumed that the heat transfer is by conduction only, then the temperature ahead of the interface for a steady-state advance is given by equation 7.4; this is the same as equation 2.44.

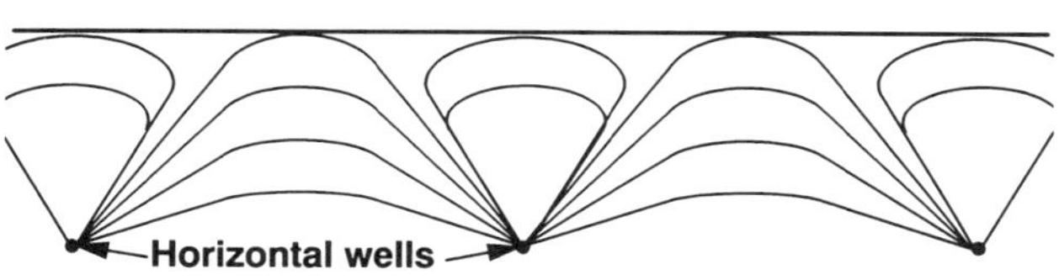

Figure 7.2 Growth of Steam Chambers Above Parallel Adjacent Horizontal Wells

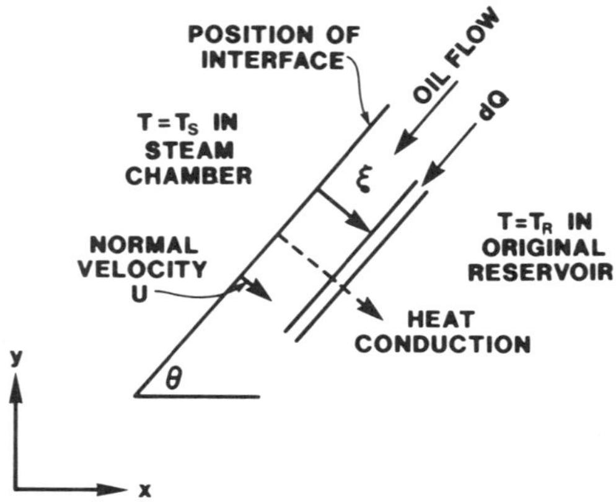

Figure 7.3 Small Vertical Section of Interface

$$\frac{T - T_R}{T_S - T_R} = e^{-U\xi/\alpha} \tag{7.4}$$

High values of U result in the temperature falling rapidly with distance, and low values give a slowly falling temperature.

If the reservoir were unheated, then the corresponding differential flow would be given by equation 7.5:

$$dq_r = \frac{kg \sin \theta}{\nu_R} d\xi \tag{7.5}$$

It is useful to subtract this flow from that given by equation 7.3 to give the increased flow due to heating.

$$dq - dq_r = kg \sin \theta \left(\frac{1}{\nu} - \frac{1}{\nu_R} \right) d\xi \tag{7.6}$$

We now redefine dq as $(dq - dq_r)$, as given in 7.7.

$$dq = kg \sin \theta \left(\frac{1}{\nu} - \frac{1}{\nu_R} \right) d\xi \tag{7.7}$$

This manipulation is done because otherwise the total flow that would be determined by integrating 7.3 would be infinite, since ν_R, although it is likely to be very large, must be finite. By making this change, the difficulty is overcome.[2] Integration of equation 7.7 results in 7.8.

$$q = kg \sin \theta \int_0^\infty \left(\frac{1}{\nu} - \frac{1}{\nu_R} \right) d\xi \tag{7.8}$$

To evaluate the integral it is necessary to know the viscosity of the oil as a function of distance from the interface. Since equation 7.4 gives the temperature as a

[2]An alternative means for circumventing the problem of the infinite cold flow is to neglect ν_R in equation 7.8 and make the upper limit of the integral some hypothetical finite value ξ_{max}. In principle, ξ_{max} is chosen to be such that T has fallen to a value at which the drainage is negligible.

function of distance, it is necessary to know the viscosity only as a function of temperature to evaluate q.

The variation of viscosity with temperature depends upon the properties of the particular oil in the reservoir. One arbitrary form of temperature function that corresponds reasonably well to the performance of actual oils over the range of interest is given by equation 7.9.

$$\frac{\nu_S}{\nu} = \left(\frac{T - T_R}{T_S - T_R}\right)^m \tag{7.9}$$

This function also has the attraction of being of a form that makes the evaluation of the integral of 7.8 particularly simple. Also, ν_R is infinite, i.e., $1/\nu_R = 0$.

In order to use equation 7.9 it is necessary to specify the viscosity at the steam temperature, T_S, and a value for the parameter m. For heavy crudes, it is found that the parameter m should have a value of about 3 to 4.

The integral of equation 7.8 may be evaluated with the result shown in equation 7.10.

$$\int_0^\infty \left(\frac{1}{\nu} - \frac{1}{\nu_R}\right) d\xi = \frac{\alpha}{U} \frac{1}{m\nu_S} \tag{7.10}$$

Integrated Flow

Eliminating the integral from equations 7.8 and 7.10 gives the expression shown in equation 7.11 for the flow q. By itself this equation is not too useful, since it involves the unknown variables U and $\sin\theta$.

$$q = \frac{kg\alpha \sin\theta}{m\nu_S U} \tag{7.11}$$

There is a trivial solution to 7.11 that should be noted. If both U and $\sin\theta$ are zero, then q can be any arbitrary value. The significance of this result is that if U is zero, then the steady-state temperature distribution that was assumed would provide an infinite amount of heated reservoir and only an infinitesimal slope would be needed to move any amount of oil. This is, of course, unrealistic.

Material Balance

A second relationship between the flow of oil q and the front velocity can be defined by considering the material balance at the interface. Consider again the small region depicted in Figure 7.3.

If the interface is advancing, then oil must be flowing out of the region at a faster rate than it is flowing in; it is the difference in the rates that determines the advance of the interface rather than the rate itself. A material balance about a thin vertical element results in equation 7.12.

$$\left(\frac{\partial q}{\partial x}\right)_t = \phi \Delta S_o \left(\frac{\partial y}{\partial t}\right)_x \tag{7.12}$$

Velocity of the Interface

The velocity of the interface U is related to the term $(\partial y/\partial t)$ in equation 7.12 and to the angle θ by equation 7.13.

$$U = -\cos\theta\left(\frac{\partial y}{\partial t}\right)_x \tag{7.13}$$

In this expression the term $(\partial y/\partial t)$ can be expected to be negative. U from equation 7.13 is substituted into equation 7.11, and this is simplified by setting $\sin\theta/\cos\theta = \tan\theta = (\partial y/\partial x)$. When this is done, the result is 7.14.

$$\begin{aligned} q &= -\frac{kg\alpha \sin\theta}{m\nu_S \cos\theta(\partial y/\partial t)} \\ &= -\frac{kg\alpha\left(\dfrac{\partial y}{\partial x}\right)}{m\nu_S\left(\dfrac{\partial y}{\partial t}\right)} \\ &= -\frac{kg\alpha\phi\,\Delta S_o}{m\nu_S}\left(\frac{\partial y}{\partial q}\right)_t \end{aligned} \tag{7.14}$$

Equation 7.14 may be rearranged and integrated by separating the variables, as in 7.15.

$$\begin{aligned} \int_0^q q\,dq &= \int_0^{h-y} \frac{\phi\,\Delta S_o kg\alpha}{m\nu_S}\,dy \\ q &= \sqrt{\frac{2\phi\,\Delta S_o kg\alpha(h - y)}{m\nu_S}} \end{aligned} \tag{7.15}$$

or, at the bottom of the steam chamber where $y = 0$,

$$q = \sqrt{\frac{2\phi\,\Delta S_o kg\alpha h}{m\nu_S}} \qquad \text{one side}^3 \tag{7.16}$$

This is a remarkable result, since it indicates that the rate of drainage is a function of the drainage height but is not dependent on the shape of the interface or on its horizontal extension. Extending the interface horizontally increases the area for heat transfer, but this effect is just counterbalanced by the effect of the longer and more sloping path in restricting the flow.

It is interesting to realize that all the variables in equation 7.16 and similar ones have equal weight. For example, changing any of them by a factor of 2 changes the predicted rate by a factor of $\sqrt{2}$.

[3]Equation 7.16 is marked *one side*. It gives the rate at which oil drains from one side of the steam chamber. For the usual field situation where oil is draining from both sides of the steam chamber, the rate must be doubled.

Position of the Interface

The horizontal velocity of the interface is given by equation 7.17.

$$\left(\frac{\partial x}{\partial t}\right)_y = \frac{-\left(\frac{\partial y}{\partial t}\right)_x}{\left(\frac{\partial y}{\partial x}\right)_t} \tag{7.17}$$

Multiplying equation 7.17 by equation 7.14 (as shown on the second line of 7.14) and substituting the value for q from equation 7.15 results, after rearrangement, in equation 7.18.

$$\left(\frac{\partial x}{\partial t}\right)_y = \sqrt{\frac{kg\alpha}{2\phi\,\Delta S_o m\nu_S(h - y)}} \tag{7.18}$$

Note that the horizontal velocity is a function of the vertical height but is independent of time. If it is assumed that the steam chamber is initially a vertical plane above the production well, then the horizontal displacement is given as a function of time t and height y by 7.19.

$$x = t\sqrt{\frac{kg\alpha}{2\phi\,\Delta S_o m\nu_S(h - y)}} \tag{7.19}$$

Equation 7.19 may be rearranged to give y as a function of x and t, as in equation 7.20.

$$y = h - \frac{kg\alpha}{2\phi\,\Delta S_o m\nu_S}\left(\frac{t}{x}\right)^2 \tag{7.20}$$

Equation 7.20 may also be written in the dimensionless form of 7.21. The dimensionless variables X, Y, and t' are defined by 7.22.

$$Y = 1 - \frac{1}{2}\left(\frac{t'}{X}\right)^2 \tag{7.21}$$

$$Y = \left(\frac{y}{h}\right)$$

$$X = \left(\frac{x}{h}\right) \tag{7.22}$$

$$t' = \frac{t}{h}\sqrt{\frac{kg\alpha}{\phi\,\Delta S_o m\nu_S h}}$$

Values of Y calculated from equation 7.21 are plotted against X in Figure 7.4.

A characteristic feature of this set of curves is that the lower part of the interface moves horizontally away from the production well. If the well were at the

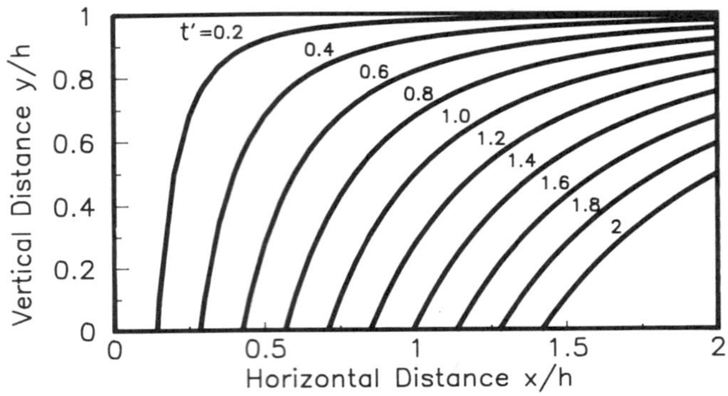

Figure 7.4 Calculated Interface Curves

origin of Figure 7.4, then it is apparent that the oil would have to move farther and farther horizontally along the base of the diagram as the bottom of the interface moved away.

In the discussion of equation 7.11 it was pointed out that, based on the assumptions made, oil could indeed move horizontally below a stationary horizontal interface. This was indicated to be possible because there is assumed to be an infinite thickness of oil, all at steam temperature below a stationary interface. With such conditions only an infinitesimal slope would be needed to cause oil flow. Although this is not realistic, the concept of two interfacial curves with oil sliding down one that is advancing and then moving along another almost stationary one with a much lower slope is conceivable and resembles some practical situations.[4]

Another way of looking at this result is to imagine that a pool of heated oil is maintained around and above the horizontal well with the rate of removal being controlled in order to maintain the level of the liquid constant within this sump. As the process proceeds, the width of the sump increases because of the receding interface.

Because of the problem of the recession of the interface from the production well, the rates calculated by equation 7.16 are recognized as being too high. This is discussed further in the discussion on TANDRAIN, which is given later.

THE EXPONENT *m*—AN EXTENDED DEFINITION

The property *m* was introduced with the empirical equation 7.9 to allow for the effect of temperature on viscosity. In the development of the theory, this form of equation is particularly attractive, since it allows the evaluation of the reciprocal viscosity-distance integral (equation 7.8) and provides a simple and useful result.

[4]One example of such behavior is given by the process shown in Figure 7.63. In this experiment oil from a steam chamber drained downwards to a horizontal shale barrier which was heated from below. The oil was able to flow along the surface of the barrier with an interface that was inclined only very slightly.

Equation 7.9 provides a particularly simple result because it results in an integral which is evaluated readily; it also makes the viscosity of the oil at reservoir temperature infinite and makes the manipulation to eliminate the cold flow used above unnecessary. If the temperature viscosity relationship is limited to equation 7.9, then it is not possible to allow for the effect of reservoir temperature.

In this section it will be shown how any realistic form of viscosity-temperature function can be employed and how the effect of reservoir temperature can be incorporated (Butler 1985).

Change of Variable of Integration

The dependence of temperature on distance from the interface ξ is given by equation 7.4. Using this relationship it is possible to change the variable of integration in 7.8 from distance to temperature. The expression for $d\xi$ given by equation 7.23 is obtained by differentiating equation 7.4 and combining the result with 7.4 by eliminating the exponential term.

$$d\xi = -\frac{\alpha}{U}\frac{dT}{(T - T_R)} \tag{7.23}$$

Substitution of $d\xi$ from 7.23 gives the following expression for the integral of equation 7.8.

$$\int_0^\infty \left(\frac{1}{\nu} - \frac{1}{\nu_R}\right) d\xi = \frac{\alpha}{U}\int_{T_R}^{T_S} \left(\frac{1}{\nu} - \frac{1}{\nu_R}\right)\frac{dT}{T - T_R} \tag{7.24}$$

The integral on the left-hand side of 7.24 was previously evaluated by equation 7.10, which contains within it the exponent m from the empirical viscosity-temperature relation. Equation 7.24 allows the evaluation of the integral for any specified dependence of viscosity ν on temperature T; furthermore, it allows the inclusion not only of the effect of the steam temperature T_S (equation 7.10 also allows this) but also of the effect of the reservoir temperature T_R.

In order to continue to use the expression developed previously, it is convenient to redefine m using equation 7.25. Combining 7.10 and 7.24 and solving for m results in expression 7.25.

$$m = \left[\nu_S \int_{T_R}^{T_S} \left(\frac{1}{\nu} - \frac{1}{\nu_R}\right)\frac{dT}{T - T_R}\right]^{-1} \tag{7.25}$$

This defines m as a function of the viscosity-temperature characteristics of the oil, the steam temperature, and the reservoir temperature.[5] It is a relatively simple matter to write a computer program that will calculate the integral of equation 7.24

[5]It is also possible to consider the term $m\nu_S$ as a property of the oil that is defined by the integral of 7.25 and is a function of T_R and T_S. In the development here, m is considered as a separate parameter. This has some advantage, since ν_S is a strong function of temperature and is visualized readily; m is a dimensionless number that does not vary rapidly with either T_R or T_S. In many applications it is adequate to consider m as a constant.

or the corresponding value of m for specific crudes and input parameters. Specific values of the parameter m are calculated in a later section using equation 7.25 and viscosity-temperature curves for specific oils.

ORIGINAL SCALED VISUAL MODEL

The earliest experiments carried out in Esso Canada's laboratory involved the use of glass-sided reservoir models operated at atmospheric pressure. The results of an experiment of this type are shown in Figure 7.5 as a series of positions of the observed interface.

A glass-sided model 29 cm long, 11 cm high and 2.5 cm thick was filled with glass beads and saturated with Cold Lake crude. A wire mesh along the left-hand vertical side of the model represented a fracture. Steam at atmospheric pressure was introduced into the top of the model, and liquids were allowed to drain from the bottom as shown. The permeability of the glass beads was chosen to make the model dimensionally similar to the field.

If the theory described previously were complete and accurate, the dimensional similarity between the model and the field could be achieved by making the dimensionless time as determined by equation 7.22 the same for the model as for the field. For a given value of t', equation 7.21 predicts a specific curve of Y versus X.

Examination of equation 7.22 indicates that with this restriction alone, it would be possible to compensate for a low permeability k by employing a matrix with a high thermal diffusivity α. The low mobility of the oil would be compensated for in this case by allowing a deeper penetration of the heat below the interface. Such a compensation is realistic only if there is, in fact, a sufficient depth for heat flow to occur; i.e., the assumption that the reservoir extends to infinity that is implicit in the integration of equation 7.8 may not be realistic.

An analysis of this problem has shown that a condition for dimensional similarity that overcomes this problem is that not only must t' (as given by equation 7.22) be the same for the model and the field, but also, a dimensionless number B_3 given by equation 7.26 must be made the same.

$$B_3 = \sqrt{\frac{kgh}{\alpha \phi \Delta S_o m \nu_s}} \tag{7.26}$$

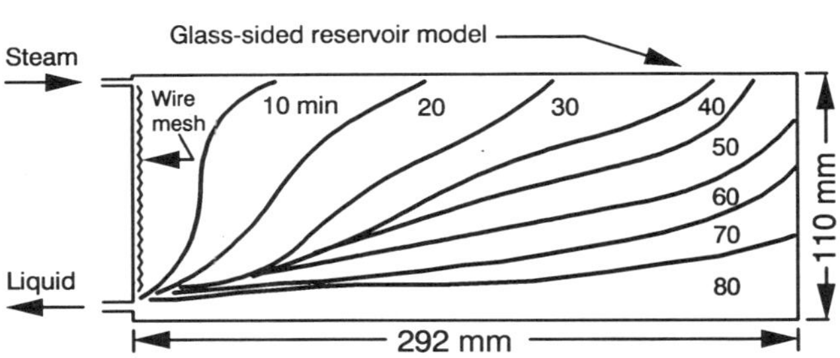

Figure 7.5 Glass-Sided Reservior Model (after Butler, McNab and Lo 1981)

DIMENSIONAL SIMILARITY

As discussed in Chapter 2, the extent of the rise of the temperature in a solid body being heated by conduction is determined by the dimensionless number

$$F_o = \frac{\alpha t}{h^2} \tag{7.27}$$

F_o is sometimes known as the Fourier number. It may be looked upon as the dimensionless time, which compares the depth of the penetration of isotherms into a body that is being heated by conduction to its physical dimensions. For dimensional similarity between a model and the field, both F_o and t' should be the same.

$$t' = \frac{t}{h}\sqrt{\frac{\alpha k g}{\phi\,\Delta S_o m \nu_S h}} \tag{7.28}$$

If F_o and t' are each equal in the field and in the model, then it follows that their quotient,

$$B_3 = \frac{t'}{F_o} = \sqrt{\frac{kgh}{\alpha\phi\,\Delta S_o m \nu_S}} \tag{7.29}$$

will also be equal. This is the condition that was mentioned in the previous section.

If there is dimensional similarity between the model and the field, then the velocity at which the fluid is running down the interface at any particular point in the field should be equal, in scaled terms, to the velocity in the model at the corresponding point.

The velocity at the interface is given by

$$\text{Velocity at interface} = \frac{1}{\phi\,\Delta S_o}\left(\frac{dq}{d\xi}\right)_{\xi=0} = \frac{kg \sin\theta}{\nu\phi\,\Delta S_o} \tag{7.30}$$

This equation recognizes that the flow is confined to a fraction $\phi\,\Delta S_o$ of the reservoir matrix. For there to be similarity at the point, the velocity of the interface must be proportional to h/t, and it follows, therefore, that $kgt \sin\theta/h\nu_S\phi\,\Delta S_o$ should be the same for both the model and the field. If these are dimensionally similar, then $\sin\theta$ will also be the same, and it can be dropped from the expression. A condition, then, that the fluids should be draining at similar rates is that the dimensionless time t'' should be the same in both.

$$t'' = \frac{kgt}{\phi\,\Delta S_o\,\nu_S h} \tag{7.31}$$

If equation 7.31 is divided by 7.28, there results another condition for dimensional similarity.

$$\sqrt{B_2} = \sqrt{\frac{kghm}{\alpha\phi\,\Delta S_o\,\nu_S}} \tag{7.32}$$

It will be noted that $\sqrt{B_2}$ is equal to mB_3.[6] The parameter m will be the same for the field and the model if they operate with the same oil, steam temperature, and reservoir temperature. Even if these conditions are not true, the values of m will not differ very much; as a result, although it will be impossible to satisfy both 7.31 and 7.32 simultaneously, the error is probably not very great.

If it is assumed that m has the same value in both the field and the reservoir, then the conditions 7.29 or 7.32 can both be replaced by

$$B_4 = \sqrt{\frac{kgh}{\alpha\phi\,\Delta S_o\,\nu_S}} \tag{7.33}$$

It is suggested that the condition involving B_3, equation 7.29, should be used, since this is the one that comes out of the improved theory as a dimensionless parameter; this is discussed later.

The dimensionless time defined by equation 7.31 is quite similar to the *drainage modulus* × *time* referred to in the paper on gravity drainage by H. Dykstra (1978).

For the model experiment shown in Figure 7.5, the corresponding model and field conditions are given in Table 7.2.

TABLE 7.2 Corresponding Model and Field Parameters

	MODEL	FIELD
$m^{(1)}$	3.9	3.9
kg m^3/d^2	$1079^{(3)}$	$0.072^{(4)}$
h m	0.105	30.5
α m^2/d	0.0557	0.0557
ν_S m^2/d	12.25 (98°C)	0.452 (215°C)
$\phi\,\Delta S_o$	0.4	0.21
B_3	10.3	10.3
$\alpha t/h^2$	$5.05t^{(2)}$	$6.00 \times 10^{-5}t^{(2)}$

(1) The data in the table come from the original paper on this subject. It was not realized then that m could be expressed as a function of T_R and T_S, and it was assumed that since the oil was the same, m would have the same value for both the model and the field.
(2) t in days.
(3) Corresponds to 15000 D.
(4) Corresponds to 1.0 D.

In order to obtain dimensional similarity, it is necessary to employ a much more permeable medium in the model than is present in the field. The time scale is very compressed in this example; 1 min for the model is equivalent to $(5.05/(6.00 \times 10^{-5}))$ min, or 0.16 y, in the field.

[6] B_2 has been defined in this way in order to make it consistent with the usage in the cited papers.

The position of the interface at 10-min intervals is shown in Figure 7.5. It will be noted that the oil drained in a systematic manner. The tendency of the interface lines to curve upward at the top is believed to be due to heat losses through the top of the model. These cause pressure gradients in the steam chamber near the top of the model that are not recognized in the preceding theory. Such pressure gradients also tend to cause the actual production to be somewhat lower than that estimated by the theory.

It also seems likely that noncondensable gas will tend to accumulate at the top of the reservoir and reduce the rate of heating along the top. Such an effect may be desirable in some cases, since it will limit the override of the steam and reduce heat loss—particularly in isolated well systems (Butler and Yee 1986a, 1986b).

As was shown previously, 10 min in this particular experiment is equivalent to about 1.6 y production in a full-scale field experiment corresponding to these conditions.

The recovery of oil from the model is plotted against time in Fig. 7.6, and the results are compared with the recovery predicted by equation 7.16.

The observed rate was of the same order as that predicted but slightly lower. The agreement is encouraging. The deviation of the observed rate is thought to be due to factors not recognized in the derivation of the equation. With time, the effective height becomes lower than h because of depletion and because some of the heat is used to cause the lateral transfer of the draining fluid to the fixed well (as will be shown later, this last effect can be compensated for approximately by substituting 1.5 for the constant 2 within the square root sign in equation 7.16. The line modified in this manner is also shown in Figure 7.6 where it is marked TanDrain; it is in better agreement.

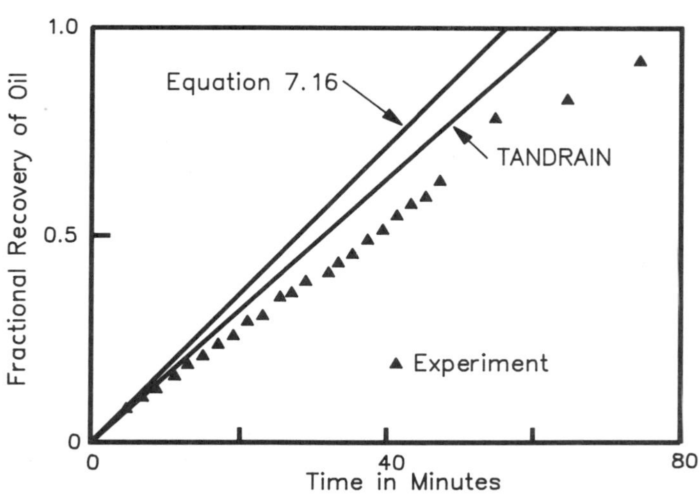

Figure 7.6 Recovery of Oil from Model

ORIGINAL SCALED, PRESSURIZED MODELS

Esso Canada has carried out gravity drainage experiments using a scaled, pressurized reservoir model in addition to atmospheric-pressure, visual models. Figure 7.7 shows a photograph of a horizontal section through one of these models after the oil had been partially produced by gravity drainage using steam at about 3 MPa. The central circular steam chamber can be seen clearly in contrast to the black oil-saturated region that surrounds it. The well arrangement that was used for this experiment is shown in Figure 7.8.

In this particular experiment the flow was radial. Steam was injected from a vertical well located almost immediately above the production well. The shape of the steam chamber formed is depicted by the curves in Figure 7.8; these were located by observations from thermocouples that were buried in the reservoir sand.

Production rates from these experiments were about the same as would be predicted by the gravity drainage theory with an allowance for radial flow.

CALCULATED DRAINAGE RATES FOR FIELD CONDITIONS

Figure 7.9 shows the expected rate of drainage to a horizontal well under a practical field condition. The curves predict the rates of drainage of Lloydminster, Cold Lake, and Athabasca crudes for the particular set of reservoir conditions shown as a function of the steam temperature employed.

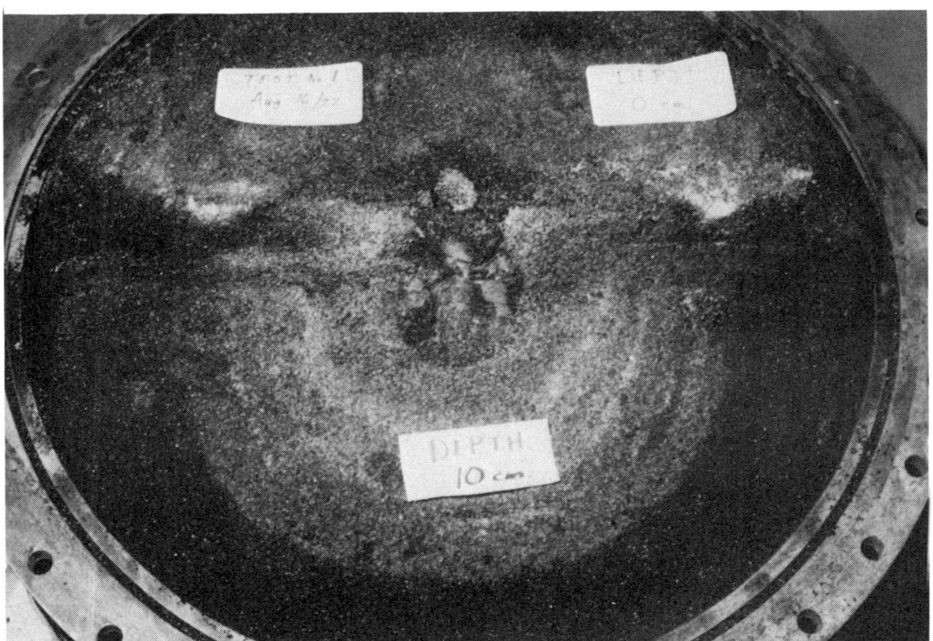

Figure 7.7 Photograph of High Pressure Gravity Drainage Model (from Butler, McNab and Lo 1981)

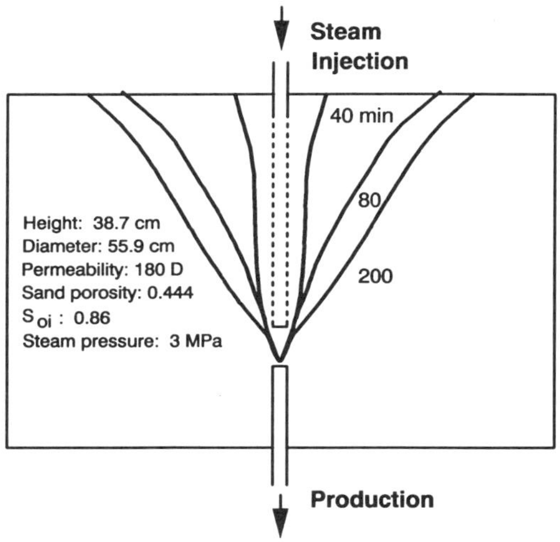

Figure 7.8 Positions of Interface from Thermocouple Measurements

The predicted rates increase with steam temperature because of the effect of temperature upon oil viscosity. Rates from about 0.05 to 0.8 m^3/d per meter of horizontal well are predicted (0.1 to 1.6 B/d ft). For example, at 200°C a horizontal well 500 m long would be expected to produce about 150 m^3 of Cold Lake oil per day. This is calculated from equation 7.16. In practice, maximum rates of about 80% of the values shown might be expected. Also, of course, depletion will lower the rate, and average rates that are only a fraction of the maximum will be found.

These calculations assume that the steam chamber starts out extending over the whole vertical height of the reservoir and along the whole length of the horizontal well. In practice it may take time for the chamber to grow to these limits. Esso researchers (Griffin and Trofimenkoff 1986) have suggested that the rate at which a

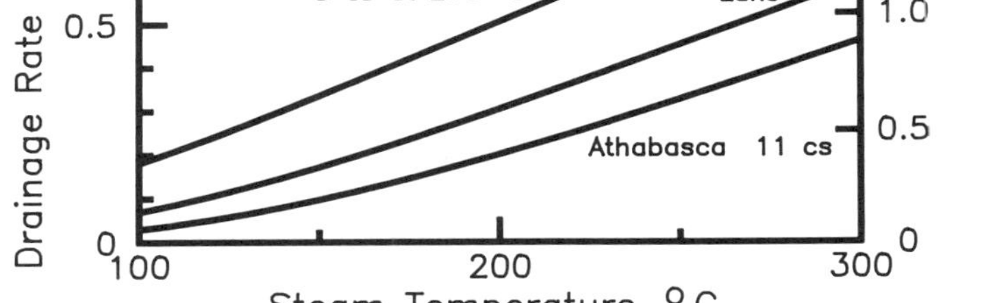

Figure 7.9 Predicted Drainage Rate for Various Crudes (based on Equation 7.16 for typical conditions; 2 sides)

steam chamber spreads along the horizontal well if the chamber starts from a vertical injector can be calculated from equation 7.19 by setting $y = 0$; they present laboratory results to support this. Rates about 50% faster than this were found when the horizontal well was heated.

TANDRAIN—AN EXTENSION TO THE ORIGINAL SAGD THEORY

The theory described in the previous sections has been modified and extended (Butler and Stephens 1981) in two ways:

1. The calculated interface curves of Figure 7.4 were modified so that they remain joined to the production well.
2. The theory was modified to allow for the confining effect of adjacent wells. Whereas in the previous treatment the interface spread horizontally to infinity, in this paper it spreads only to a vertical no-flow boundary located halfway to the next adjacent well.

A point of concern with the solution derived previously was that the oil draining down the interface curves would have to drain horizontally to the well after it reached the bottom. Some of the available head must be used to cause this lateral flow.

As a simple approximation, it is assumed that the lower parts of the interface curves of Figure 7.4 can be replaced by tangents drawn from the well to the curves. This is shown by the curves in Figure 7.10.

The name TANDRAIN was used at one time for a computer program which drew these curves.

Figure 7.11 shows the effect for a typical interface. The TANDRAIN assumption reduces the rate of drainage of oil to the value given by Equation 7.34.

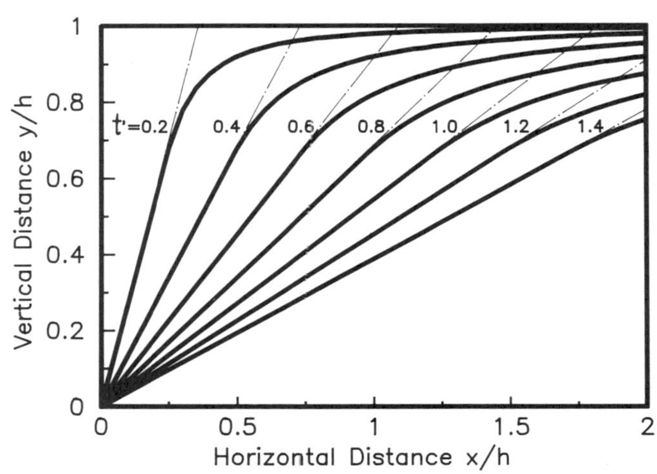

Figure 7.10 Calculated Interface Positions for an Infinite Reservoir Using the Tandrain Assumption

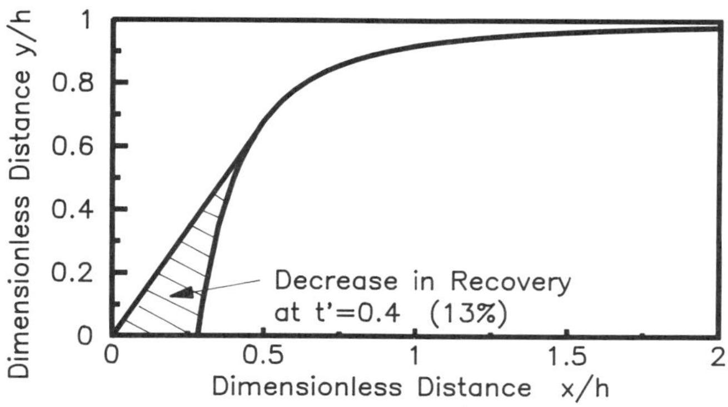

Figure 7.11 Effect of Tandrain Assumption on Predicted Recovery

$$q = \sqrt{\frac{1.5kg\alpha\phi\Delta S_o h}{m\nu_s}} \tag{7.34}$$

The rate is 87% of that calculated by 7.16, and—as may be seen from Figure 7.11—the recovery, for the same value of the dimensionless time, is reduced in the same proportion.[7]

Another way of comparing equations 7.19 and 7.34 is to say that the effective head causing vertical drainage has been reduced from h to 75% of h. The remainder of the head is used to cause horizontal movement of the draining oil.

Effect of No Flow Boundary

The theory in the TANDRAIN paper (Butler and Stephens 1981) leads to the set of interface curves shown in Figure 7.12 for a confined reservoir.

The dimensionless time and the abscissa used in this figure are defined differently from those of equation 7.25; w rather than h is used as the dimension in the expression for the dimensionless time; w is defined as half of the horizontal distance between wells.

The revised theory provides a relation that allows for the effect of depletion on the drainage rate. A numerical solution was developed in the paper, and it was found that it could be represented by the simple equation 7.35a.

$$Q^* = \sqrt{\tfrac{3}{2}} - t^{*2}\sqrt{\tfrac{2}{3}} \tag{7.35a}$$

The relationship between this and the previous expressions for the rate of drainage is shown in Figure 7.13.

[7]A variation of the above derivation is termed LINDRAIN. In this variation it is assumed that the interface remains straight right up to the top of the reservoir. As the steam chamber grows in size this straight interface becomes more inclined and longer. It is also assumed that, at a limiting intermediate point on the interface, the heat ahead of the interface is just equal to the steady state value. Above and below this critical point there is less heat ahead of the interface than would correspond to the steady state. The location of the critical point is chosen so that the drainage rate is a maximum; this leads to a position of $y/h = 1/\sqrt{3}$. With these assumptions it is found that a drainage rate equation similar to 7.34 is obtained except that the constant 1.5 is replaced by $0.75\sqrt{3} = 1.3$.

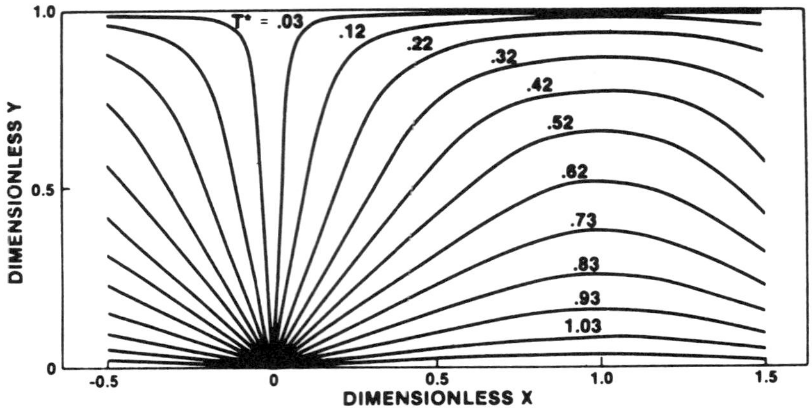

$$t^* = \frac{t}{w}\sqrt{\frac{kg\alpha}{\phi \Delta S_o m \upsilon_s h}}$$

Figure 7.12 Confined Horizontal Well Interfaces

The cumulative recovery of mobile oil may be obtained by integrating equation 7.35a with respect to dimensionless time; the result is

$$\text{Recovery} = \int_0^{t^*} Q^* dt^* = \sqrt{\tfrac{3}{2}}t^* - \tfrac{1}{3}\sqrt{\tfrac{2}{3}}t^{*3} \qquad (7.35b)$$

The value of t^* required to obtain a particular recovery, f, may be obtained by solving (7.35b) for t^*. The root of interest is given by

$$t^* = 2\sqrt{\frac{3}{2}}\cos\left(\frac{2\pi - \cos^{-1}(-f)}{3}\right)$$

For instance, if $f = 0.5$, then $t^* = 0.4253$.

The corresponding curves for cumulative recovery are shown in Figure 7.14.

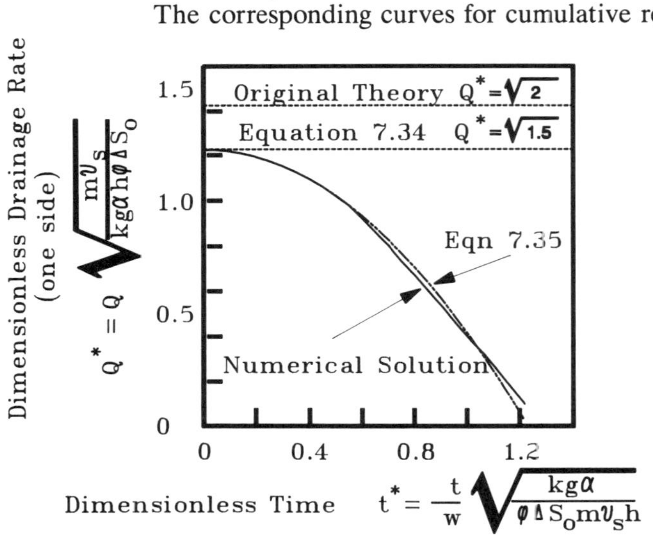

Figure 7.13 Comparison of Drainage Rates from Different SAGD Equations

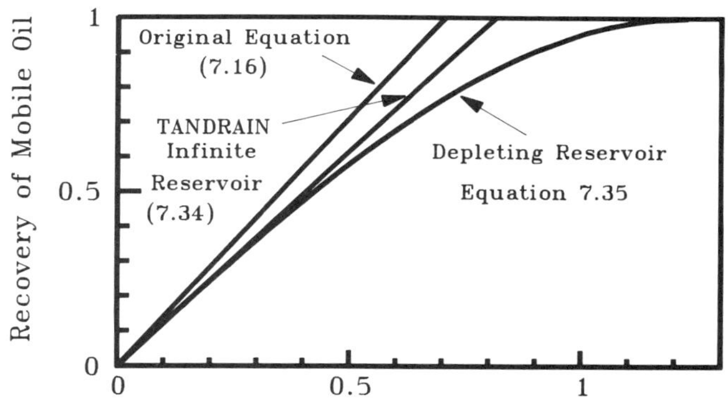

$$t^* = \frac{t}{w}\sqrt{\frac{kg\alpha}{\varphi \Delta S_o m \upsilon_s h}}$$

Figure 7.14 Cumulative Recovery for a Horizontal Well

Further Experimental Data

Esso Resources Canada has carried out a considerable number of scaled visual model experiments. One such experiment (Butler and Stephens 1981) involved a box having transparent front and rear walls that was 36 cm wide, 26 cm high, and 2.5 cm thick. The model was packed with a coarse-grained sand that was fully saturated with Cold Lake crude. The grain size of the sand was chosen so that the permeability of the packing would make B_2 the same for the model and for the field.

The photographs in Figures 7.15 show the stages in the development of a steam chamber due to the gravity drainage of oil during continuous flow of steam into the chamber. The steam chamber is the lighter-colored, growing region in the center of the photographs. The major part of the oil has drained from this region. The elapsed time is indicated at the bottom of each picture; the grid drawn on the front window of the model has a line spacing of 20 mm.

In this model the steam inlet and product outlet were both placed in the center of the bottom of the model, with the inlet 2 cm above the outlet; they were concentric tubes. Steam flowed in to replace the hot oil and the condensate as they flowed out. The residual oil saturation in the drained space of the steam chamber was typically 5%. Models of this type with a variety of different permeabilities, well configurations, and overall geometry were used.

Figure 7.16 compares the production rate from one of these models with the prediction made by the TANDRAIN theory. In this particular experiment the steam was injected into a central well, which was located slightly above a lower, horizontal production well.

The measured rate rises as the steam chamber grows toward the top of the reservoir and then decreases as the reservoir becomes depleted. The TANDRAIN theory does not predict the rising-rate period, because in its derivation it is assumed that the steam chamber extends for the whole vertical height at the start. However, the theory predicts a maximum production rate that is close to the measured value, and it represents the decline during depletion reasonably well.

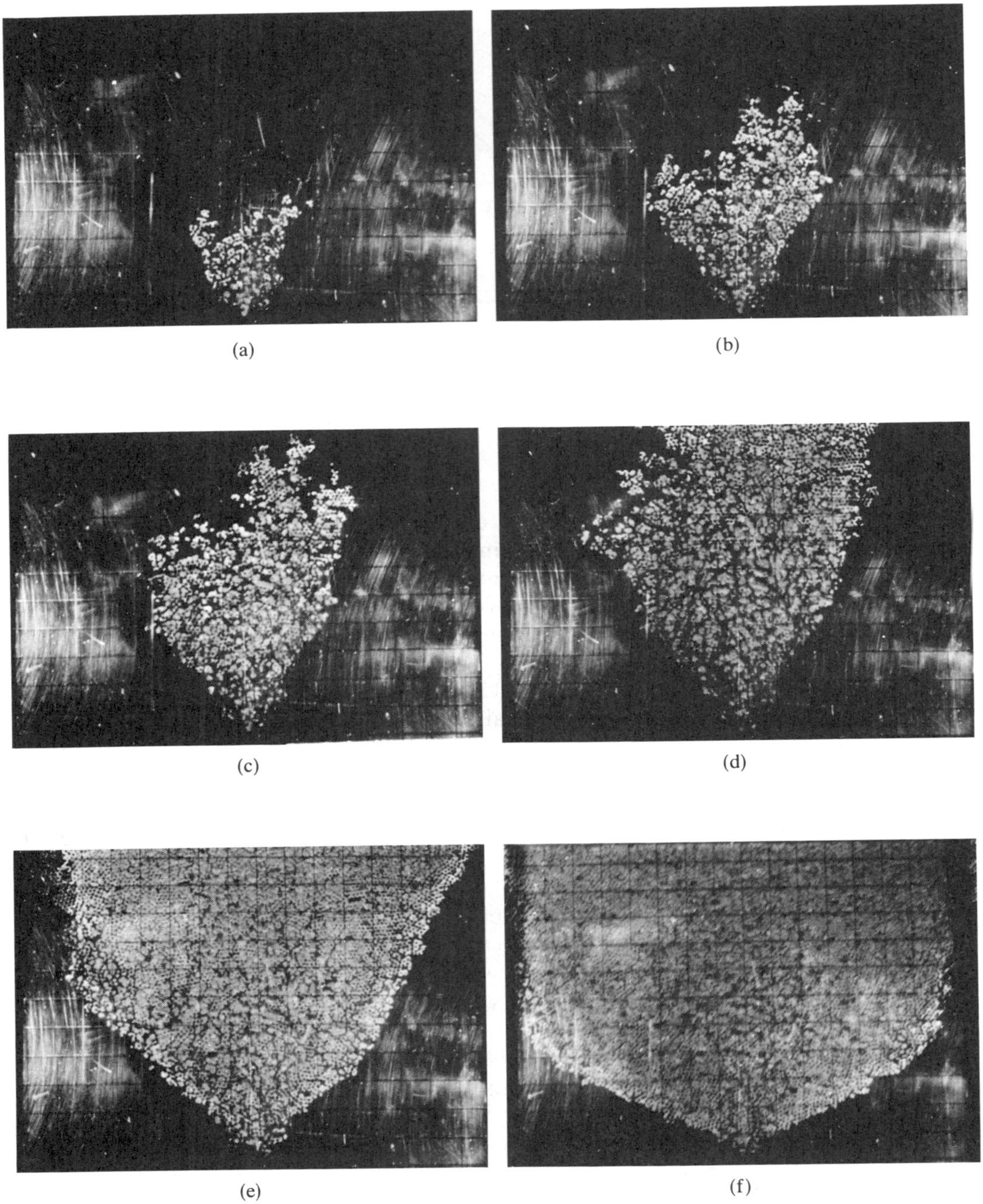

Figure 7.15 Development of the Steam Chamber During Gravity Drainage in Laboratory Model

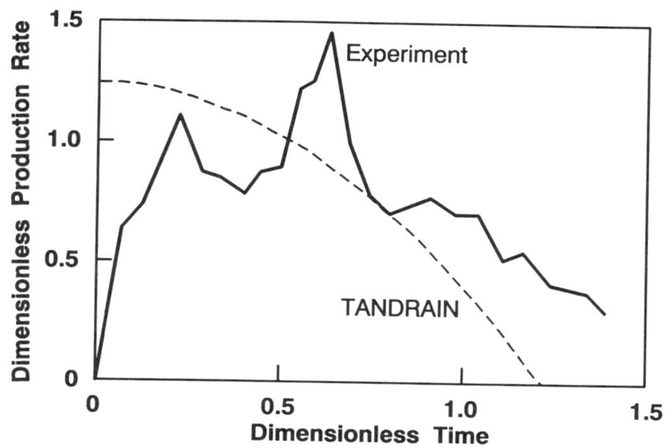

Figure 7.16 Oil-Production Rate from Low-Pressure Laboratory Model

Extrapolation of the Model Experiments to the Field

Using the theory and the scaling methods described, the model experiments were extrapolated to the field scale as shown by the data in Table 7.3.

THE RISING STEAM CHAMBER

In this section the vertical growth of a steam chamber above a horizontal well is considered, and equations are derived for the rate at which it rises and for the rate at which oil is produced.

The derivation that follows differs from that given by Butler, Stephens, and Weiss (1980), although the final results (i.e., equations 7.43 and 7.44) are very similar to those in the reference, and the same experimental data are used.

It is assumed that the problem is a two-dimensional one and that the shape of the steam chamber remains geometrically similar as it rises. This similarity of shape is, at least to a first approximation, found in experiments with visual models such as those shown by the earlier series of photographs.

In the previous sections on TANDRAIN, it was shown that the drainage rate that would be predicted from equation 7.16 overestimates the production because not all the head h is available; some of the head is required to move the oil horizontally to the well. For TANDRAIN, this is allowed for by changing the factor 2 that is within the square root sign to 1.5. This is equivalent to assuming that 75% of the head is available.

During the rising-chamber period, similar considerations apply, and it may be assumed that the available head is less than the total height of the chamber, h, at any particular time. We will allow for this by including an unspecified factor β in the equation, with the expectation that it will turn out to be less than the 2 in equation 7.16. This is shown in equation 7.36.

TABLE 7.3 Extrapolation of the Model Experiments to the Field

	Model Experiment	Extrapolation to Field		Theoretical Prediction, Equation 7.35a (Field)
Time (min)	Cumulative Production (mL/cm)	Time (y)[1]	Cumulative Production (m^3/m)	Cumulative Production (m^3/m)[2]
29	51	0.65	38	57
48	92	1.07	69	93
67	128	1.49	96	127
96	200	2.14	150	174
125	259	2.79	194	211

Physical Parameters:

	Model	Field
k (m^2)	3×10^{-9}	0.75×10^{-12}
α (m^2/day)	0.056	0.056
ϕS_o (·)	0.4	0.21
m (·)	4.2	3.4
h (m)	0.23	27.5
W (m)	0.18	21.5
T_S (°C)	97	200
ν_S (m^2/day)	11	0.52
B_2 (·)	850	850

(1) $$\frac{t_{\text{field}}}{t_{\text{model}}} = \frac{\left[W\sqrt{\dfrac{\phi S_o m \nu_S h}{kg\alpha}}\right]_{\text{Field}}}{\left[W\sqrt{\dfrac{\phi S_o m \nu_S h}{kg\alpha}}\right]_{\text{Model}}} = 11{,}700$$

(2) $$\frac{\text{prod}_{\text{field}}}{\text{prod}_{\text{model}}} = \frac{[Wh\phi S_o]_{\text{Field}}}{[Wh\phi S_o]_{\text{Model}}} = 7{,}505$$

The factor 2 outside the square root sign in equation 7.36 recognizes that oil is draining to the well from both sides, whereas 7.16 gave the quantity draining from only one side.

$$q = 2\sqrt{\frac{\beta k g \alpha \phi \, \Delta S_o h}{m \nu_S}} \qquad (7.36)$$

Since it is assumed that the steam chamber remains geometrically similar as it grows, the cumulative oil production will be proportional to the mobile oil per unit area multiplied by the square of the chamber height. This is shown by equation 7.37, where the constant γ is determined by the shape of the chamber; the area of the chamber is γh^2.

$$q_{\text{cum}} = \int_0^t q \, dt = \gamma \phi \, \Delta S_o h^2 \qquad (7.37)$$

Differentiating 7.37 with respect to time gives another expression for the production rate, equation 7.38.

$$q = 2\gamma \phi \, \Delta S_o h \frac{dh}{dt} \qquad (7.38)$$

Setting the right-hand side of 7.36 equal to the right-hand side of 7.38 results in equation 7.39, which may be integrated as in 7.40 to give an expression for the height of the chamber, h, as a function of time. This is equation 7.41.

$$2\gamma \phi \, \Delta S_o h \frac{dh}{dt} = 2\sqrt{\frac{\beta k g \alpha \phi \, \Delta S_o h}{m \nu_S}} \qquad (7.39)$$

$$\int_0^h h^{1/2} dh = \frac{1}{\gamma}\sqrt{\frac{\beta k g \alpha}{m \nu_S \phi \, \Delta S_o}} \int_0^t dt \qquad (7.40)$$

$$\begin{aligned} h &= \left(\frac{3}{2}\frac{t}{\gamma}\right)^{2/3}\left(\frac{\beta k g \alpha}{m \nu_S \phi \, \Delta S_o}\right)^{1/3} \\ &= \left(\frac{9}{4}\frac{\beta}{\gamma^2}\right)^{1/3}\left(\frac{k g \alpha}{m \nu_S \phi \, \Delta S_o}\right)^{1/3} t^{2/3} \end{aligned} \qquad (7.41)$$

Value of Proportionality Constant in Height Equation

Equation 7.41 shows that the chamber height h should be equal to a constant (the first brackets), multiplied by a factor involving the reservoir and oil properties (the second bracketed term), and multiplied by the time raised to the $\frac{2}{3}$ power.

The initial constant term in 7.41 has been evaluated from a scaled-model experiment similar to that shown in the earlier photographs. In this experiment the height of the chamber was determined from the behavior of a series of thermocouples implanted into the model in a vertical line above the production well. The measured temperatures are shown in Figure 7.17. As the steam chamber approached each thermocouple, the temperature increased abruptly from the initial level to the steam temperature.

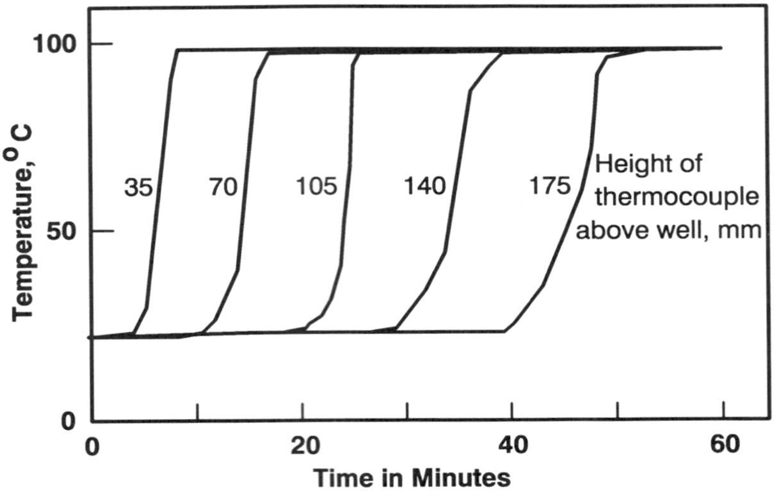

Figure 7.17 Measured Temperatures above Production Well

From these measurements it was possible to plot the height of the steam chamber against time, as in Figure 7.18. Also shown in Figure 7.18 is a theoretical line based on equation 7.41 with the constant set equal to 2; i.e., this theoretical curve is based on equation 7.42. The slope of the experimental curve is close to the theoretical value of $\frac{2}{3}$.

$$h = 2\left(\frac{kg\alpha}{m\nu_S\phi\,\Delta S_o}\right)^{1/3} t^{2/3} \tag{7.42}$$

The Oil-production Rate

When the value of h from equation 7.42 is substituted into equation 7.37, the result is equation 7.43. This has within it the *shape factor*, γ. By curve fitting the oil-rate-production data from the experiment, γ has been found to be about $\frac{9}{16}$.

$$q_{\text{cum}} = 4\gamma\left(\frac{kg\alpha}{m\nu_S}\right)^{2/3} (\phi\,\Delta S_o)^{1/3} t^{4/3} \tag{7.43}$$

From curve fitting,

$$4\gamma = 2.25 \quad \text{or} \quad \gamma = \tfrac{9}{16}$$

$$q_{\text{cum}} = 2.25\left(\frac{kg\alpha}{m\nu_S}\right)^{2/3} (\phi\,\Delta S_o)^{1/3} t^{4/3} \tag{7.44}$$

Using this value leads to equation 7.44 for the cumulative oil production, and differentiating this with respect to time results in equation 7.45 for the instantaneous production rate.

$$q = \frac{dq_{\text{cum}}}{dt} = 3\left(\frac{kg\alpha}{m\nu_S}\right)^{2/3} (\phi\,\Delta S_o)^{1/3} t^{1/3} \tag{7.45}$$

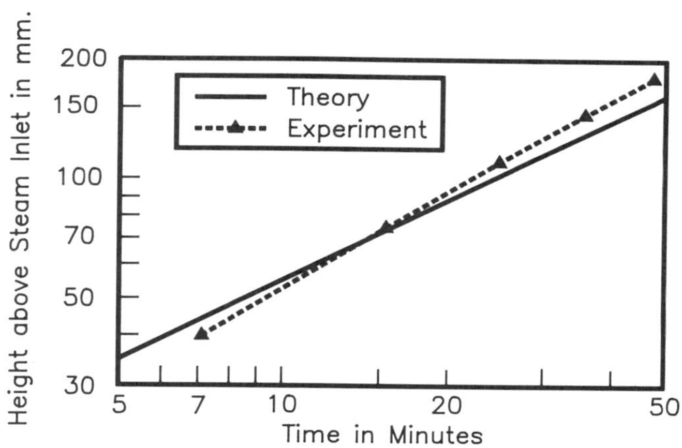

Figure 7.18 Steam Chamber Rise

Equation 7.44 is compared to the experimental data in Figure 7.19. The agreement is very satisfactory.

Shape of Steam Chamber

If it is assumed that the shape of the steam chamber is approximately a sector of a circle, then the angle subtended by the sides of the sector that gives $\frac{9}{16}$ as the value of gamma is 64°; this is shown in Figure 7.20. This shape is in approximate agreement with the observed shapes of rising steam chambers.

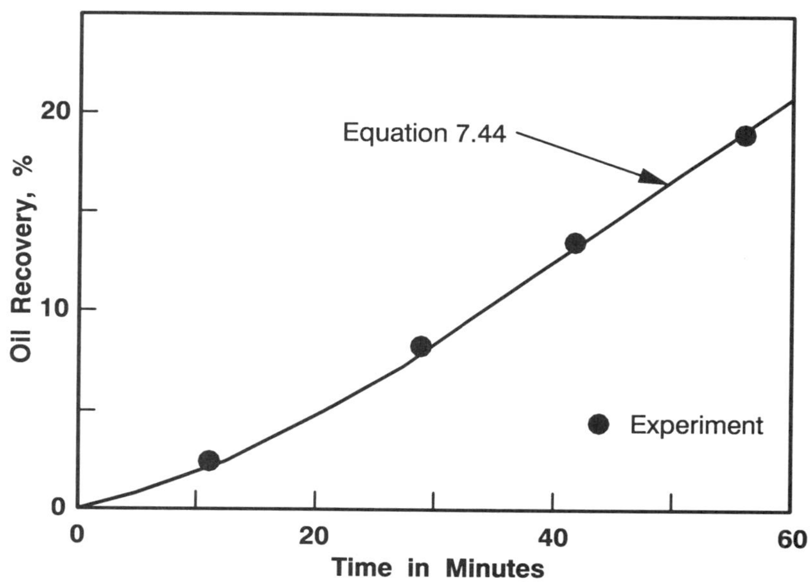

Figure 7.19 Oil Recovery During Chamber-Rise Period

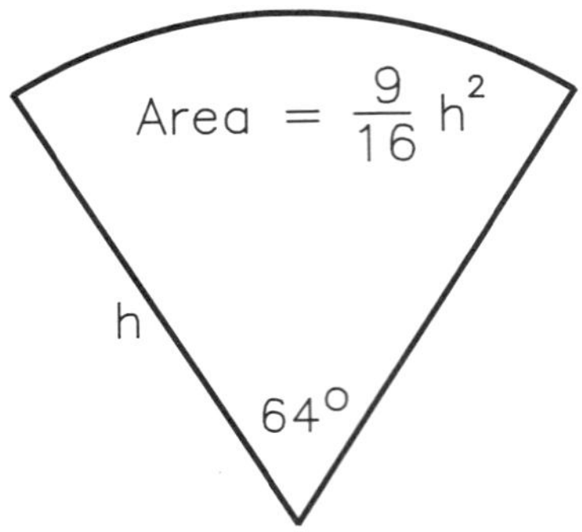

Figure 7.20 Approximate Shape of Rising Chamber Sector with $\gamma = 9/16$

Available Head

It is also possible to calculate the value of the head availability factor β. This is done in equation 7.46; the result—$\frac{9}{8}$, or 1.125—indicates that nearly half the available head is required to move the oil laterally.

$$\beta = 8 \times \frac{4\gamma^2}{9} = \frac{9}{8} \tag{7.46}$$

This value is less than the value of 1.5 found for the TANDRAIN equation. The photographs in Figure 7.15 indicate that during the chamber-rise period, the top of the steam chamber is ill-defined because of the instability of the rising front (but the sideways spreading front has a stable interface). Because of the ragged nature of the front, it is not surprising that less of the head is available to effect drainage to the production well. The photographs show clearly the effect of gravity in stabilizing and destabilizing steam fronts.

Equation 7.39 (with $\beta = 1.125$) predicts a lower drainage rate at the point where the chamber reaches the top of the reservoir than does the TANDRAIN equation for the same value of h. It is practical to continue to use equations 7.44 and 7.45 after the chamber reaches the top until the rate reaches that predicted by the TANDRAIN equation. After this point the TANDRAIN equation should be employed to follow the further depletion of the reservoir. This approach is shown in the numerical example which is developed starting on page 316.

The rate predicted by equation 7.45 for the rising chamber period and that by equation 7.35 for the depletion period can be plotted on a single chart, as shown in Figure 7.21.

FINGER RISE THEORY

Another approach to the prediction of the rate of rise of a steam chamber has been described by Butler (1987). This paper describes a theoretical approach in which the frictional drag for the falling oil around a rising finger is balanced with the frictional drag within the finger and the driving force provided by gravity.

The theory predicts a rate of rise for the fingers which is proportional to the permeability and inversely proportional to the viscosity of the oil at steam tempera-

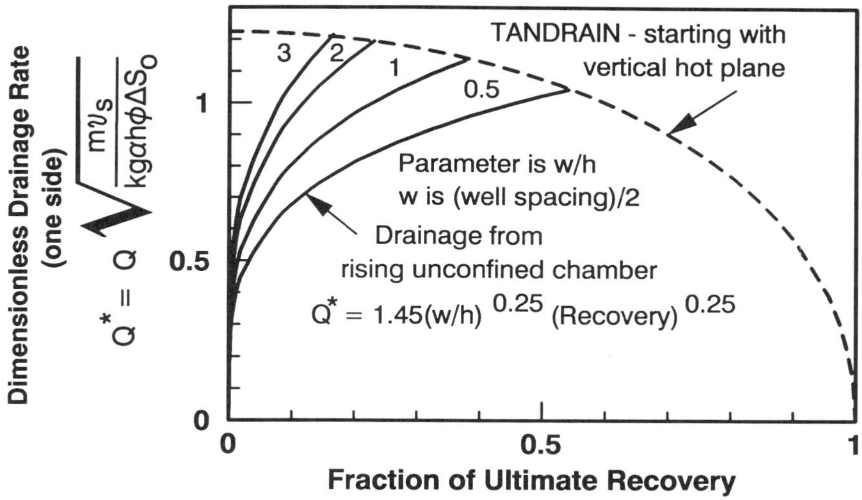

Figure 7.21 Calculated Drainage Rates for Rising and Depleting Chambers

ture. It also predicts the curvature of the top of the rising finger. Fingers rise beside each other and the oil flowing downwards between them falls as meandering fingers which interfere with each other and with the rising steam.

Typical predicted values of the rate of rise and of a characteristic dimension of the fingers are given in the following table:

Viscosity of reservoir oil at 100°C cs	30	100	300
Velocity of rise *m/d*			
Steam temperature 100°C	0.0163	0.0082	0.0043
Steam temperature 200°C	0.0822	0.0571	0.0415
Steam temperature 300°C	0.190	0.149	0.121
Finger dimension f_D m (see equation below[1])			
Steam temperature 100°C	8.97	4.57	2.65
Steam temperature 200°C	4.47	3.29	2.53
Steam temperature 300°C	2.56	2.22	1.94

[1]The top of the finger is parabolic according to the equation

$$(y/f_D) = (x/f_d)^2$$

EFFECT OF STEAM TEMPERATURE, RESERVOIR TEMPERATURE, AND OIL PROPERTIES ON DRAINAGE RATES

Steam and Reservoir Temperatures

The equation used for the general definition of m (7.25) allows the prediction of the effect of reservoir temperature on production rate (Butler 1981).

Figure 7.22 shows a plot of the integrand in the right-hand side of equation 7.25 as a function of temperature for a typical heavy crude having a viscosity of 129.1 mm^2/s (or cs) at 100°C and 6.8 mm^2/s at 200°C.

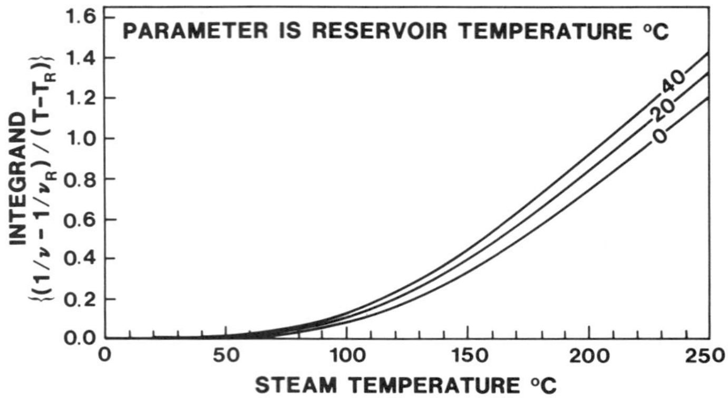

Figure 7.22 Effect of Temperature on Value of Integrand in Equation 7.25

Curves are shown corresponding to reservoir temperatures of 0, 20, and 40°C. Increasing the reservoir temperature has the effect of increasing the value of the integrand and, thereby, the corresponding value of the integral and the oil-production rate.

The curves in Figure 7.22 were integrated numerically and the corresponding values of m were calculated, with the results shown in Figure 7.23. It is found that m changes somewhat with steam temperature and decreases with increasing reservoir temperature. A lower value of m for a given set of conditions has the effect of increasing the drainage rate.

In general, it is found that the rate of drainage varies inversely with the square root of $m\nu_S$. This factor is plotted against steam temperature for the same crude oil in Figure 7.24. This figure shows the significant effect of the reservoir temperature and how it may be allowed for by the extended theory.

Oil Properties

The three curves in Figure 7.25 give calculated values of m as a function of steam temperature for three different hypothetical oils. The upper curve corresponds to a

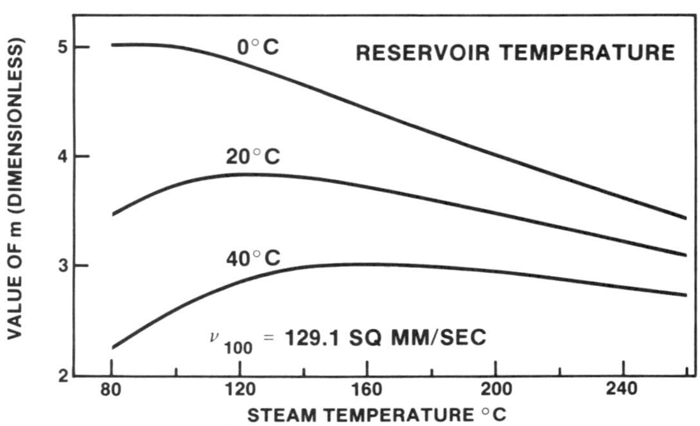

Figure 7.23 Effect of Reservoir and Steam Temperatures on m

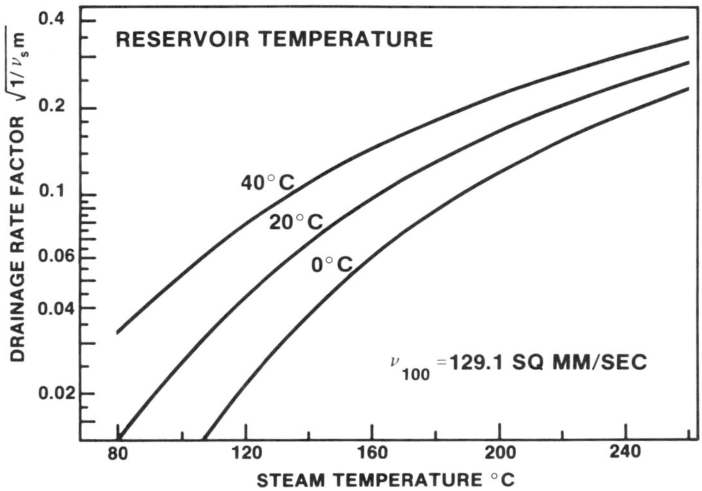

Figure 7.24 Effect of T_R and T_S on Rate Factor

very viscous oil similar to Athabasca bitumen. The middle curve is for a somewhat more fluid crude similar to Cold Lake crude. The lowest curve is for what might be termed a conventional heavy oil similar to the oils found in the Lloydminster area.

Lower values of m are calculated for the less viscous oils because ν changes more slowly with temperature; this has the effect of tending to give predicted drainage rates higher than would be expected from the viscosity of the oil at steam temperature alone.

The calculated drainage-rate factors for these three oils are plotted in Figure 7.26. The calculated effect of the oil properties on drainage rate is significant, but the overall effect is not overwhelming.

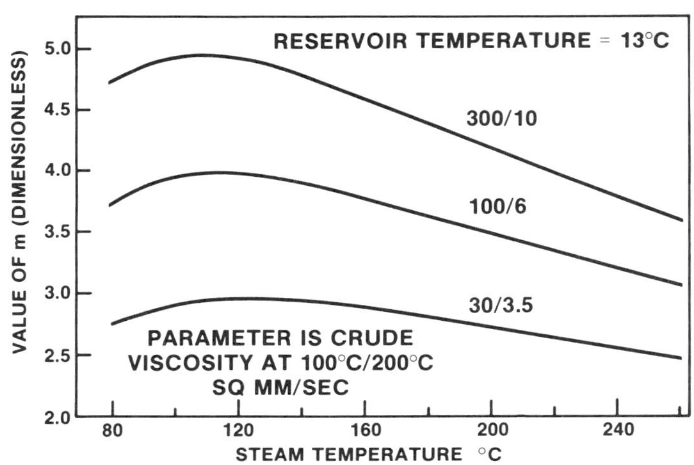

Figure 7.25 Values of m for Different Crude Oils

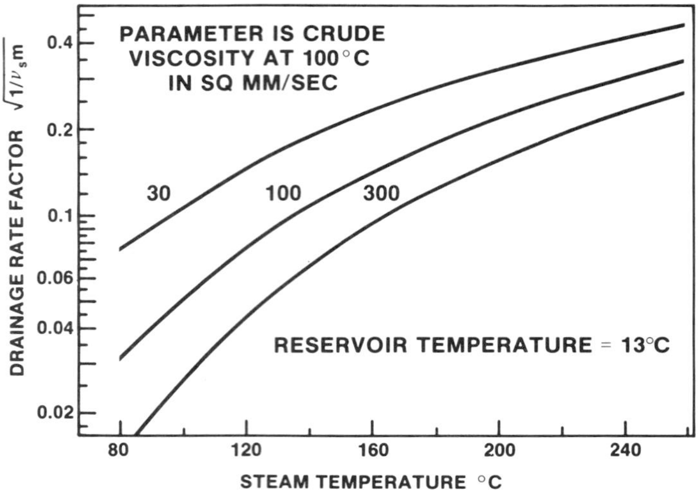

Figure 7.26 Drainage-Rate Factors for Three Heavy Crudes

There are, of course, other variables in equations such as 7.16 that are equally effective in modifying the drainage rate; these (e.g., thickness and permeability), as well as the oil properties, can vary significantly between reservoirs.

Numerical Problem on Steam-Assisted Gravity Drainage

A tar sand reservoir has the following properties:

Reservoir temperature	15°C
Oil viscosity at T_r	100,000 cs
Bitumen density	1.00 g/cc
Bitumen viscosity at 100°C	80 cs
Reservoir thickness	20 m
Thermal diffusivity	0.07 m^2/D
Porosity	0.33
Initial oil saturation	0.75
Residual oil saturation	0.13
Effective permeability for oil flow	0.4 d[8]

The field is to be drained by a series of parallel horizontal wells with a spacing of 75 m between wells. The wells will be located 2.5 m above the base of the reservoir. Steam will be injected from separate horizontal injection wells placed above the producers. Assume that initial thermal communication is achieved and that the system will be operated with a steam pressure of 1.2 MPa. Estimate the percent recovery of the original oil in place as a function of time for a period of 7 y.

It is planned to carry out a model experiment representing the field in the laboratory. The model will have a height of 35 cm and will operate using the same

[8]Assumed to correspond to an absolute permeability of 1.0 D.

bitumen and at the same pressure as the field. What should be the permeability of the sand used in the model? How many minutes in the model will be required to represent 1 y in the field?
(a) *Obtain value of parameter m:* Determine m using either equation 7.25 or by obtaining an approximate value by interpolation from Figure 7.25.

$$T_S = 188°C$$

$$T_R = 15°C$$

$$\nu_S = 7.8 \text{ cs (by interpolation on Figure A.5.2)}$$

$$= 0.078 \text{ stokes} \qquad (cm^2/s)$$

$$= 7.8 \times 10^{-6} \text{ m}^2/\text{s}$$

Estimating m from Figure 7.25 leads to

$$m = 3.4$$

(b) *Tabulate parameters for problem:*

$$\phi = 0.33$$

$$\Delta S_o = 0.75 - 0.13 = 0.62$$

$$k = 0.4 \times 10^{-12} \text{ m}^2$$

$$g = 9.81 \text{ m/s}^2$$

$$\alpha = 0.07/(24 \times 3600) \text{ m}^2/\text{s} = 8.10 \times 10^{-7} \text{ m}^2/\text{s}$$

$$h = 20 - 2.5 = 17.5 \text{ m}$$

$$m = 3.4$$

$$\nu_S = 7.8 \times 10^{-6} \text{ m}^2/\text{s}$$

$$w = 75/2 = 37.5 \text{ m}$$

(c) *Calculate factor to convert q to q*:*

$$\text{Factor} = \sqrt{\frac{m\nu_S}{kg\alpha h\phi\Delta S_o}} = 1.527 \times 10^6 \text{ s/m}^2 = 17.67 \text{ d/m}^2$$

(d) *Equation for t*:*

$$t^* = \frac{t}{w}\sqrt{\frac{kg\alpha}{\phi\Delta S_o m\nu_S h}} \qquad \text{(see Figure 7.12)}$$

$$= 4.215 \times 10^{-4} t_{\text{days}}$$

(e) *Calculate rates and recovery for depletion assuming chamber starts as vertical plane:*

TIME IN YEARS	t^*	q^* (1)	$\int q^* dt^*$ (2)	$q = 2q^*/17.67$
0	0	1.225	0	0.139 m^3/m d
1	0.154	1.205	0.188	0.136
2	0.308	1.147	0.369	0.130
3	0.462	1.050	0.539	0.119
4	0.615	0.916	0.690	0.104
5	0.769	0.742	0.818	0.084
6	0.923	0.529	0.916	0.060
7	1.077	0.278	0.979	0.031

(1) From equation 7.35a.
(2) Fractional recovery of mobile oil.

(f) *Rates and recovery during rising-chamber phase:*

$$q_{\text{cum}} = 2.25\left(\frac{kg\alpha}{m\nu_S}\right)^{2/3}(\phi\,\Delta S_o)^{1/3}t^{4/3}$$

$$= 32.11t^{4/3} \qquad \text{(with } t \text{ in years)}$$

Fractional recovery $= q_{\text{cum}}/(17.5 \times 75 \times 0.33 \times 0.62) = 0.120t^{4/3}$, with t in years.

$$q_{\text{cum}}\ (\text{in m}^3/\text{m}) = 32.11\left(\frac{t_{\text{days}}}{365}\right)^{4/3} = 0.0123t_{\text{days}}^{4/3}$$

Differentiating with respect to t gives q in m^3/m d:

$$q\ (\text{m}^3/\text{m}\ d) = (\tfrac{4}{3})0.0123t_{\text{day}}^{1/3} = 0.0164t_{\text{day}}^{1/3}$$

Rate and Recovery during Rising-Chamber Period

Time Years	q	Fractional Recovery
0	0	0
0.5	0.093	0.048
1.0	0.117	0.120
2.0	0.148	0.302

The recovery rates are plotted against the percent recovery in Figure 7.27. The two curves intersect at 21% recovery and a rate of 0.135 m^3/m d.

Assume that the rising-chamber curve applies up to the point of intersection and that the rate then follows the depletion curve. The changeover point occurs at 558 d, but this point corresponds to only 401 d for the depletion curve. It is neces-

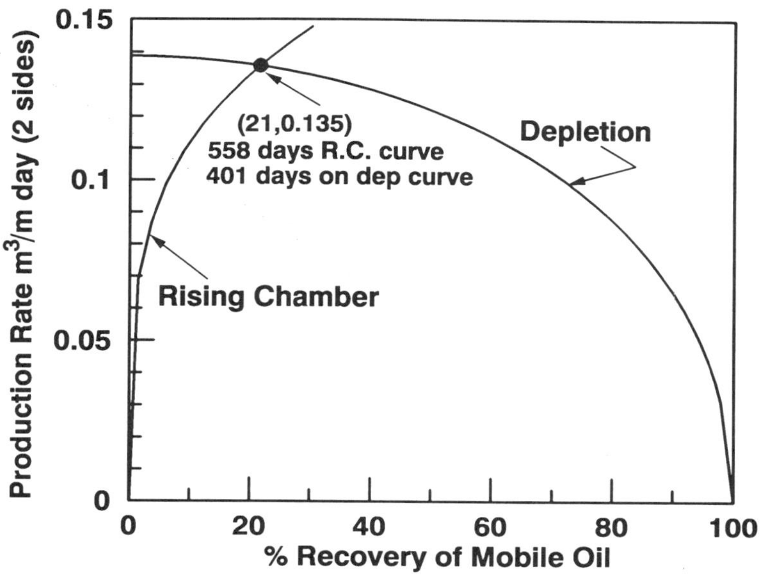

Figure 7.27 Oil-Production Rate versus Recovery

sary, then, to subtract 157 d from the time in order to calculate consistent rates from the depletion equation.

Summary Table for Rates and Recovery as a Function of Time

Years	q	t^*	q^*	Recovery[(1)]
0	0			0
0.5	0.093			0.048
1.0	0.117			0.120
1.53	0.135	0.169	1.202	0.212
2.0	0.133	0.242	1.177	0.293
3.0	0.124	0.395	1.097	0.467
4.0	0.111	0.549	0.979	0.627
5.0	0.093	0.703	0.821	0.766
6.0	0.071	0.857	0.625	0.878
7.0	0.044	1.011	0.390	0.957

[(1)]Fractional recovery of the movable oil above the production well.

These rates are plotted in Figure 7.28.

(g) *Scaling of the Model* The dimensionless numbers B_3 and t^* should be the same for the model as for the field; i.e.,

$$B_3 = \left(\sqrt{\frac{mkgh}{\phi \Delta S_o \alpha m \nu_S}}\right)_{\text{field}} = \left(\sqrt{\frac{mkgh}{\phi \Delta S_o \alpha m \nu_S}}\right)_{\text{model}}$$

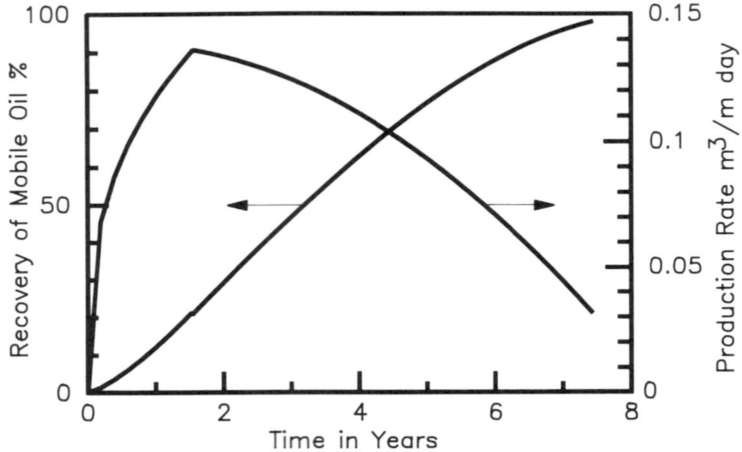

Figure 7.28 Production Rates and Percent Recovery

Since the temperatures and pressure will be the same in the model and the field, the only variables that will differ will be h and k (we will assume that the porosity, saturation, and thermal diffusivity will be the same). Then

$$(kh)_{\text{field}} = (kh)_{\text{model}}$$

and the permeability of the sand to be used in the model will be

$$k_{\text{model}} = \frac{h_{\text{field}}}{h_{\text{model}}} \cdot k_{\text{field}} = \frac{20}{0.35} \times 0.4 = 22.9 \text{ D effective permeability}$$

Using the same relative permeability of 0.4 assumed initially, this means that a porous medium having an absolute permeability of 22.9/0.4 = 55 D should be used.

If the dimensionless time for two corresponding stages in the model and the field are the same, then

$$t^* = \left(\frac{t}{w}\sqrt{\frac{kg\alpha}{\phi \Delta S_o m \nu_S}}\right)_{\text{field}} = \left(\frac{t}{w}\sqrt{\frac{kg\alpha}{\phi \Delta S_o m \nu_S}}\right)_{\text{model}}$$

Hence,

$$\frac{t_{\text{field}}}{t_{\text{model}}} = \left(\frac{w_{\text{field}}}{w_{\text{model}}}\right)\sqrt{\frac{k_{\text{model}}}{k_{\text{field}}}\frac{h_{\text{field}}}{h_{\text{model}}}}$$

$$t_{\text{model}} = \left(\frac{h_{\text{model}}}{h_{\text{field}}}\right)^2 t_{\text{field}} = \left(\frac{0.35}{20}\right)^2 t_{\text{field}}$$

If $t_{\text{field}} = 1$ y = 525,600 min, then $t_{\text{model}} = 161$ min.

STEAM-INJECTION WELLS

There are two general arrangements of steam-injection wells that have been proposed to be used with horizontal production wells in steam-assisted gravity drainage:

1. Horizontal injection wells, with one well positioned above each producer.
2. Vertical steam-injection wells located directly above the producers.

Horizontal Injection Wells

In this arrangement a horizontal injection well is placed directly above each producer. For example, in AOSTRA's Underground Test Facility project (Edmunds et al. 1987) horizontal wells are drilled upward from the mined tunnel, which is located just below the Athabasca tar sand reservoir, and then horizontally. The wells are drilled as injector-producer pairs. In this application, the wells are very close together, about 2 m apart, to allow interwell reservoir heating and communication. Communication can be achieved initially by heating both wells and pressuring them cyclically.

In this application, the steam chamber, as it forms, must grow upward, to the top of the reservoir, in order to achieve high vertical conformance.

If the reservoir contains oil having an appreciable mobility, then it is possible to locate the injection well higher up in the reservoir. The limiting factor is that the mobility of the oil within the reservoir should be high enough to allow the steam chamber to advance downwards from the injector to the producer in a reasonable length of time. During this advance there is a displacement of cold oil and steam condensate to the production well, and, with adequate oil mobility, this communication period can be highly productive.

The length of time required to achieve communication can be estimated approximately by calculating the breakthrough time from equation 4.23. In order to allow for the changing conditions during the displacement, the time calculated from this equation should be divided by 2; this has been done in equation 7.47.

$$t_{BT} = \frac{\phi \, \Delta S_o \, \mu_o s^2 \ln(s/R_W)}{6k \, \Delta P} \tag{7.47}$$

In this equation, s represents the interwell distance.

The allowable vertical separation between the injector and the producer that would give a one year breakthrough time has been plotted against the logarithm of the oil viscosity in Figure 7.29. Typical values were assumed for the other variables; these are shown in the figure.

Bitumens have viscosities ranging from 10^5 to 10^6 cp, and it will be seen that separations of 2 m or less are required. The lower curve in Figure 7.29 is pessimistic in this range, since it does not allow for the effect of well bore heating. The radius that is heated can be estimated using Figure 2.12. For typical tar sand, it is estimated that a radius of 3 to 5 m is heated appreciably in a period of 60 d. Also shown in Figure 7.29 is a curve for a well bore radius of 3 m.

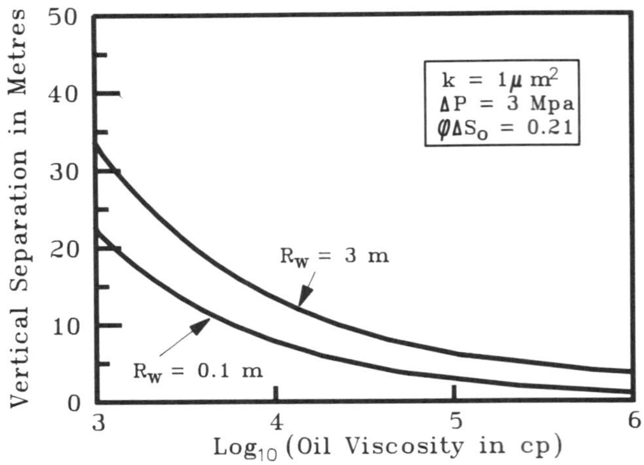

Figure 7.29 Allowable Vertical Separation for One Year Breakthrough Time

It is clear from Figure 7.29 that if the viscosity of the oil in the reservoir is below about 10,000 cp, then quite high vertical separations can be employed, particularly if the production well is preheated. A large separation is advantageous, since the pressure gradient resulting from the flow of steam assists the drainage of oil. In some situations, there may be an advantage in employing two horizontal injection wells, with one located close to the producer to initiate steam chamber formation and a second located higher in the reservoir to be used as the steam chamber grows to it (Butler 1984).

The quantity of oil displaced during the establishment of thermal communication can be estimated using the ideas of inter-well displacement discussed in the development of equation 4.24 in Chapter 4. This equation was written for two fully-penetrating vertical wells. It can be rewritten as follows, with a change of symbols to represent two parallel horizontal wells having lengths which are great enough for the problem to be considered as two-dimensional:

$$V_{bt} = (\pi/3)\,(s^2 \phi\, \Delta S_o) \tag{7.48}$$

where s represents the distance between parallel horizontal wells within a reservoir of infinite extent and V_{bt} is the production volume per unit length of well at breakthrough. Although this equation is also applicable to other arrangements we will consider the case where the injection well is vertically above the producer.

When oil is displaced by downward steamflooding the volume that is displaced is somewhat less than would be calculated by equation 7.48 because of the difference between the properties of the steam and the oil, because of the need for the condensate from the steam to flow with the oil, and because of the need to heat the reservoir.

The displacement of oil by downward steamflooding in two dimensions has been studied theoretically by Butler and Petela (1989). In their theoretical model it is assumed that the streamlines for the steamflooding phase remain the same as they would be for single fluid flow. As the oil is displaced from each streamtube the condensation interface advances along it as determined by a heat balance. The effect of this is that the velocity within shorter streamtubes increases more rapidly than within longer ones and breakthrough occurs relatively more rapidly (i.e. with a smaller displaced volume) than it would with single-fluid flow. Figure 7.30 shows the position of the interface for a particular set of conditions at times corresponding to fractions of 0.25, 0.5, 0.75, and 1.0 of the breakthrough time. Also shown in the figure is the broken line which corresponds to the position of the interface at the breakthrough point for single fluid flow. The volume of the displaced zone is smaller for the steamflooding case than for the single fluid case.

If the injection well is at the top of the reservoir just below the upper boundary and the production well is at the lower boundary then the volume of displaced oil at breakthrough is less than for the previous calculation. For reservoirs bounded by horizontal surfaces the volume of displaced oil for single fluid flow is given by:

$$V_{bt} = (2/\pi)(s^2 \phi \Delta S_o) \tag{7.49}$$

where the symbols are as before. This equation leads to a displaced volume which is 61% of that for the unbounded reservoir. Butler and Petela studied this geometry for steamflooding and calculated positions of the condensation front are shown in Figure 7.31. The broken line in this figure shows the position of the interface at breakthrough. Again, as for the unconfined case, the volume of the displaced zone

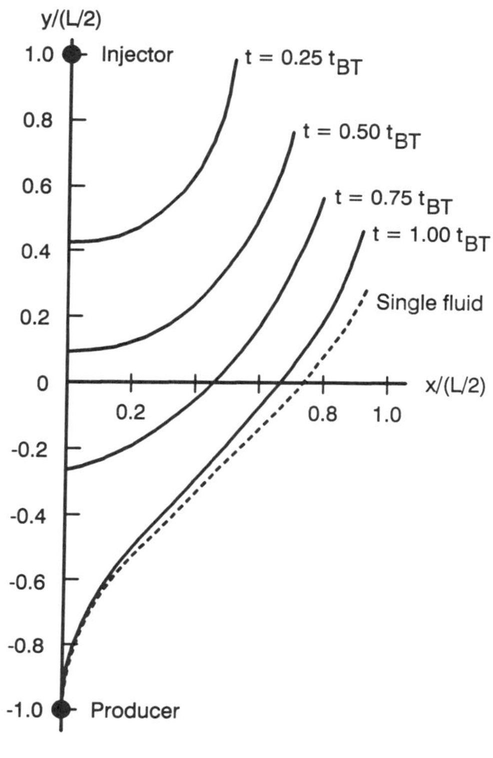

Figure 7.30 Positions of Steam Front for Times Corresponding to 0.25, 0.5, 0.75 and 1.0 Times the Breakthrough Time in an Unconfined Reservoir. The Broken Line is the Position of the Interface for Single Fluid Flow at Breakthrough (from Butler and Petela 1990)

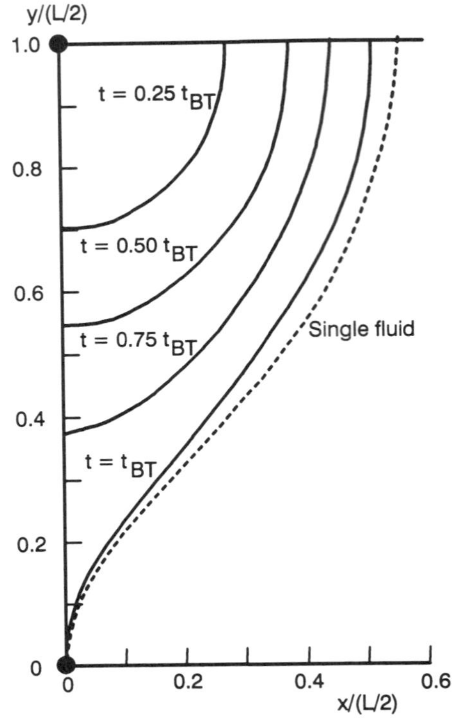

Figure 7.31 Positions of Steam Front for Times Corresponding to 0.25, 0.5, 0.75 and 1.0 Times the Breakthrough Time for a Reservoir with Horizontal Boundaries. The Broken Line is the Position of the Interface for Single Fluid Flow (from Butler and Petela 1990)

at breakthrough is smaller than for single fluid flow. For the particular example shown the volume of the displaced region is 91% of that for single fluid flow.[9]

The downward vertical steamflooding of a Lloydminster-type heavy oil has been studied by Sugianto and Butler (1989) using a scaled physical model. In their experiments they found that the time for steam breakthrough could be predicted reasonably accurately by the theory described above but that the shape of the interface was rather broader than that shown in Figure 7.31. The reason for this was that the injected steam spread beneath the overburden and, as a result, the steam source was more like a distributed plane than a line (a line is a point on a two dimensional diagram such as Figure 7.31).

Steamflooding calculations made by Butler and Petela show excellent agreement between the shape of the experimental interface at breakthrough and that which they predicted for steamflooding using the same method that is described

[9]In estimating the volume of displaced oil it is necessary to include the factor $\phi \Delta S_o$. The value of ΔS_o may be rather less than for the long term project because the drainage of oil may not be complete at the time of steam breakthrough. For example suppose that in a vertical flood project the vertical distance s is 20 m, ϕ is 0.35, S_o is 0.85, and S_{or} is 0.3 (this might be a reasonable value for a time of about 150 days; see Table 7.1), then the volume of displaced oil at steam breakthrough might be expected to be about 0.91 of the volume calculated from the above equation [i.e. $V_{bt} = 0.91(2/\pi)(s^2\phi \Delta S_o) = 0.91(2/\pi)(20^2 \times 0.35 \times 0.55) = 45\ \text{m}^3/\text{m}$]. For a 500 m well this would correspond to a production of 22500 m^3, or 140 kB, during the establishment of thermal communication. This production could have a very significant effect on the economics.

above but with a planar steam source. Figure 7.32 shows comparisons of the position of the experimental interface and those predicted by theory for both a linear source and a planar one.

The reason for the spreading of the steam below the overburden is the steam-assisted gravity drainage effect (see Figure 7.33). At the edge of the spreading chamber the heated oil falls away and allows the steam chamber to spread laterally. In other experiments it has been found that the degree of spreading can be controlled to some extent by adjusting the vertical position of the injector and the rate at which steam is injected.

Injecting the steam at an intermediate elevation results in a steam chamber which grows upwards because of the SAGD effect and downwards because of the pressure drive. The relative extent of these effects depends upon the injection rates. High rates force the chamber down to the production well rapidly, whereas with low rates chamber rise is the predominant effect. Under these conditions it is possible to operate with a steam chamber which is only in the upper part of the reservoir. Oil draining to the bottom of this chamber is forced through the intervening oil saturated region to the production well. This operation with a steam chamber detached by a considerable distance from the production well is undesirable because only a part of the potential drainage head is employed and the heat losses to the overburden are excessive. It is desirable that steam be injected at a rate sufficient to keep steam low down in the reservoir. There may, in some cases, be difficulties in doing this because of inadequate steam generation capacity.

Vertical Injectors

Vertical wells can also be used for steam injection above horizontal producers.

At Yarega in Russia, horizontal production wells have been drilled from underground chambers and steam injected above them from separate, near-vertical wells, which are drilled from separate underground workings.

The Hopco project in California involved eight horizontal production wells drilled horizontally from an underground chamber at the foot of a mine shaft. Steam was injected from separate vertical wells drilled from the surface. The project operated for about $1\frac{1}{2}$ y; although it was successful in producing oil, the oil-steam ratio was very low. However, the operators consider that this approach is promising,

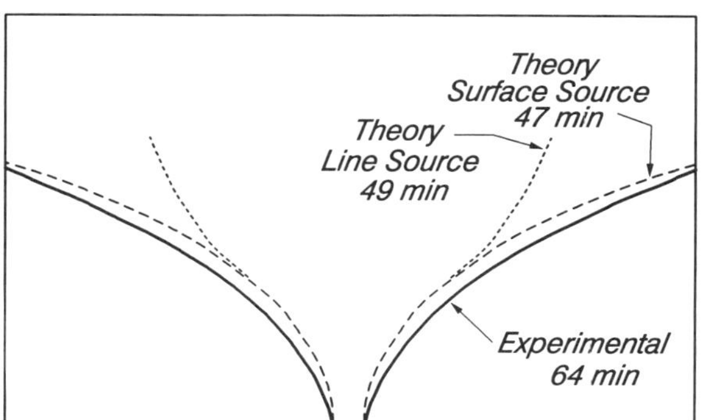

Figure 7.32 Comparison of the Experimental and Theoretical Positions of the Steam Front at Breakthrough (from Butler and Petela 1990)

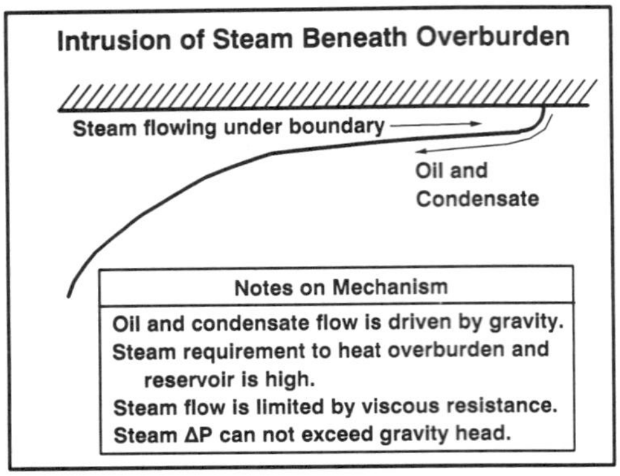

Figure 7.33 Mechanism of Spreading of Steam Chamber Just Beneath the Overburden

and they ascribe the poor results obtained to a pilot site that was unfavorable. They suggest that further projects should be considered (Dietrich 1988).

The Esso Resources first horizontal pilot at Cold Lake (Bezaire and Markiw 1979) involved a deviated production well drilled from the surface in order to pass horizontally near the base of the formation. A vertical injector is drilled from above. Although there have been no detailed results published from the Esso project, a paper at the 1983 Petroleum Congress indicated that it had produced more than 100,000 barrels of oil. It is believed still to be producing oil in 1990.

A second horizontal pilot has been drilled at Cold Lake by Esso, which has about 1 km of open horizontal length. The latest plans are to use this together with a number of vertical wells above. These vertical wells will produce first by cyclic steaming and then, presumably, some or all of them will be used for injection.

In the Sceptre oil project in the Tangleflags field in Saskatchewan, the horizontal producer is located toward the bottom of the oil zone. Steam is injected from existing vertical wells above. The project is considered to be successful and, during the downward flooding phase, very high production rates have been achieved.

There are several advantages to using vertical injection wells. They are cheaper and simpler to construct, and there is not the same need for drilling accuracy that would be required for horizontal wells. Also, it is possible to change the point of injection of the steam vertically as the project matures. In the initial stages it is desirable to have the steam injection close to the production well to facilitate communication. However, as the project continues, it becomes advantageous to raise the point of injection so that the motion of the steam through the chamber produces a favorable rather than an unfavorable pressure gradient.

A major disadvantage of vertical injection wells is that each injector covers only a relatively limited length of the producer. For very long producers, it is necessary to employ more than one injector. It has been shown by Griffin and Trofimenkoff (1986) that equation 7.19 can be used to estimate the half-length of the active well. In their paper they point out that with some production systems, the well will remain unheated from the end of the active portion in one direction, but because the hot fluids are flowing within it, it will be heated in the other direction.

In their laboratory experiments, these authors found that the active portion of the horizontal well spread more rapidly than would be calculated from equation 7.19 in the heated direction. They suggested that this effect could be included in the prediction by using a value of m in the equation for the hot end that corresponded to the temperature of the production fluids rather than the reservoir temperature. In their example, it was found that this effect increased the spreading rate by about 50% in the hot direction.

In a production system involving multiple injectors, it would presumably be only the remote end that would be cold. Even in this case, the difficulty might be overcome by employing a production tubing reaching to the end of the well.

Other interesting observations in the Griffin and Trofimenkoff paper include the following:

1. Experiments in both high-pressure and low-pressure models agree with the general theory presented here.
2. In systems with vertical injection wells, the production is much greater than would be calculated from the active length of the production well. The reason for this is the production from the ends of the steam chamber. The authors show that if allowance is made for this, good agreement is obtained between their laboratory experiments and the theoretical prediction.
3. The authors find, in their experiments, that the extent of steam override immediately under the overburden is much less than would be expected theoretically. They recommend that the width of the heated surface be estimated by assuming that the interface is a straight line rising from the production well to the top of the reservoir. While this assumption has little effect on the rate of oil production, it affects the estimates of thermal efficiency and oil-steam ratio by reducing the area for heat loss both to the overburden and to the reservoir below the interface.[10] The estimation of heat losses is discussed later in this chapter.

Joshi and Threlkeld (1985) of Phillips Petroleum compared horizontal and vertical injectors and found the horizontal injector to be the most effective and to provide significantly faster recovery. Joshi and Threlkeld also studied the effect of vertical fractures in the reservoir. They found that these are desirable in that they promote the early rapid production of oil. Multiple fractures are better than single ones.

A third area studied in this paper from Phillips involves the effect of the rate of product removal upon the efficiency of the process. It is found that if the product is removed too slowly, the thermal efficiency is low, but that it rises to an optimum as the removal rate is increased. At very high rates, steam is bypassed from the chamber, and the efficiency falls again. The rate of withdrawal of product should be controlled to prevent significant bypassing of steam. If the rate is too low, then drained liquid product builds up above the production well, and the process is needlessly prolonged. This adjustment is found to be straightforward in practice.

[10]This corresponds to the LINDRAIN assumptions described in the footnote to page 303.

AVOIDING THE STEADY-STATE HEAT-DISTRIBUTION ASSUMPTION

In the preceding theoretical material in this chapter, it has been assumed that the temperature distribution ahead of the advancing front is that corresponding to the steady state for the particular local advance velocity U; this is given by equation 7.4. It is this assumption that gives rise to the discrepancy of the front advancing from the production well; this was discussed earlier. Also, since the heat ahead of the front will generally be somewhat less than that corresponding to the steady state, the earlier equations tend to overestimate the production rate.

An approach that avoids the steady-state heat assumption has been described (Butler 1985). The interface is divided into small elements, and the heat storage ahead of each of these is calculated at successive time steps using an approximate differential equation. For each time step, the flow of oil behind the element is calculated and then the movement of the interface is obtained from material balance considerations similar to those involved in equation 7.12.

Figure 7.34 shows the positions of the interface calculated using this technique for a case involving an unconfined well with the parameter B_3 set equal to 8. In the new theory it is found that the results depend somewhat on B_3; this is defined by equation 7.29.

Although the curves for different values of B_3 are generally similar to those of Figure 7.10, they tend to become more S-shaped as B_3 is increased. This tendency toward an S shape was also displayed by the experimental curves shown in Figure 7.5. Calculated interface curves for various values of B_3 at a constant dimensionless time of $t^* = 0.5$ are shown in Figure 7.35.

Also plotted on Figure 7.35 is a dotted curve drawn from equation 7.21; this curve is based on the assumption that the heat ahead of the advancing interface corresponds to the steady state. As B_3 is increased, the curves from the improved theory become closer to the original theory. This is consistent with the calculations

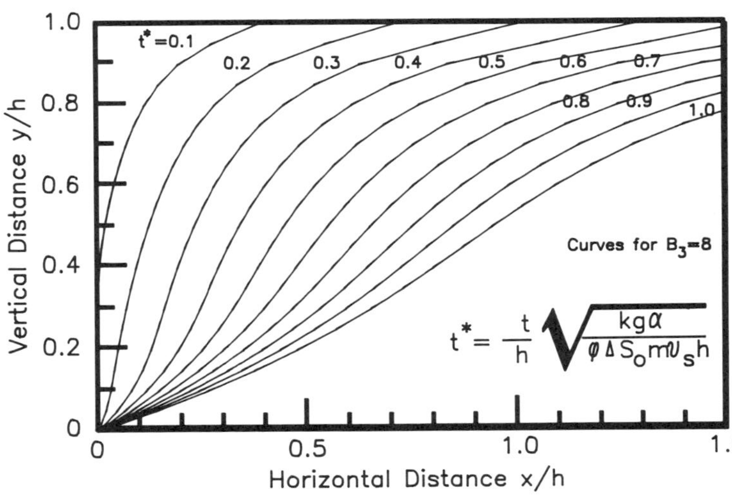

Figure 7.34 Position of Interface for an Unconfined Well

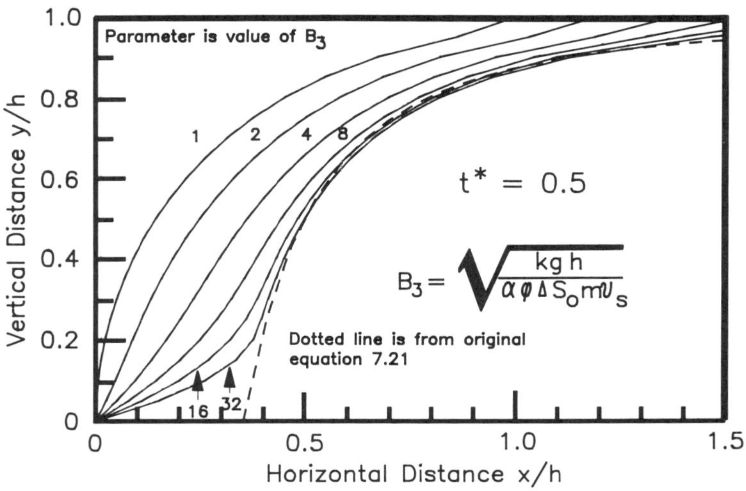

Figure 7.35 Effect of Parameter B_3 on Shape of Interface

for the transient temperature profile ahead of an advancing front that were made in Chapter 2. Figure 2.8 shows that the approach to the steady state increases for smaller values of α. Small values of α also give larger values of B_3 in Figure 7.35.

Figure 2.8 shows that the approach to the steady-state heat equilibrium ahead of an advancing front is a function of U^2t/α. For a given time, t, this approach is less for larger values of α; these larger values of α correspond to smaller values of B_3. Thus for a given value of t^* the degree of heat penetration (i.e. the fraction of the steady state value which is achieved) will tend to be less the larger the value of α or the smaller the value of B_3.

This explains the effect of B_3 on the position of the interface curves in Figure 7.35. The smaller the value of B_3 the farther away the heat penetration from the steady state value. As a result there is a lower drainage rate while the heat bank ahead of the front is building. Large values of B_3 correspond to the case where the permeability is large compared to the thermal diffusivity. In this circumstance a steady state heat distribution is achieved rapidly and only a very thin layer of heated oil drains, but it drains quickly because of the favorable permeability (and/or high gravity, high head, low viscosity, and so on). High values of B_3 result in relatively thin layers of rapidly-flowing mobile oil whereas low values of B_3 result in the flow of thick layers. Considerable time is required for a thick flowing layer to become established because of the need to build up the heat bank beyond the advancing front.

The buildup of the heat bank takes longer in the lower part of the reservoir because the frontal advance rate is slower there and for a given time U^2t/α is lower (Figure 2.8). As a result, for the calculation where the deviation from the steady state is included (Figure 7.34) compared to that where it is not (Figure 7.4), the advance of the interface is slower near the base of the reservoir than it is at the top. It will be noted in Figure 7.34 that initially the steam chamber does not advance into the reservoir at the bottom; it overrides at the top without steam penetrating to the

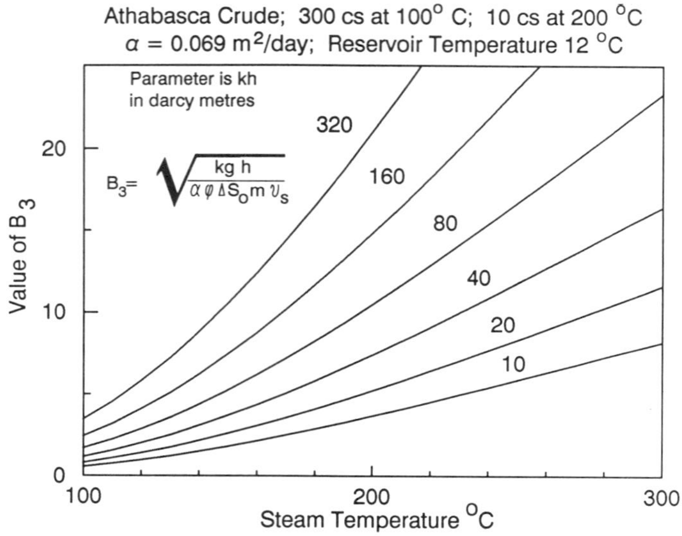

Figure 7.36 Effect of Steam Temperature and kh on B_3

production well.[11] This is reminiscent of the behavior predicted by van Lookeren's theory described in Chapter 4 (see Figures 4.29 and 4.30). The phenomenon shown for early times in Figure 7.34, and which would be even more evident if the value of B_3 had been smaller, occurs even though it is assumed that initially the entire vertical plane above the production well is at steam temperature. The flow of draining oil prevents the advance of the steam downwards.

Values of the Parameter B_3

Figures 7.36, 7.37, and 7.38 give values of B_3 as a function of steam temperature and the product kh for three different crude oils using typical values of α and T_R. Values of B_3 for a whole range of situations can be estimated from these diagrams by interpolation.

[11]The theory described does not allow for steam being introduced at a higher pressure than the draining oil. In practice the steam chamber could be forced to the production well more rapidly by either increasing the steam injection pressure or by lowering the pressure within the production well. Strategies such as this are desirable because they make the whole drainage head available more rapidly.

Also there is a need to allow for the resistance to radial flow in the immediate vicinity of the production well; this too requires an increase in the pressure difference between the injector and the producer. Another factor which requires an increase in the pressure difference is the resistance to the radial flow of steam around the injector. Nevertheless there remains the concept of the continuous steam-assisted gravity drainage with the production of oil controlled so that oil is withdrawn at a rate equal to that of the drainage around the perimeter of the chamber without allowing live steam to bypass in excessive quantities. This can also be looked upon as the production of oil from below an expanding steam (gas) cap without allowing the coning or, perhaps more accurately, cresting of steam.

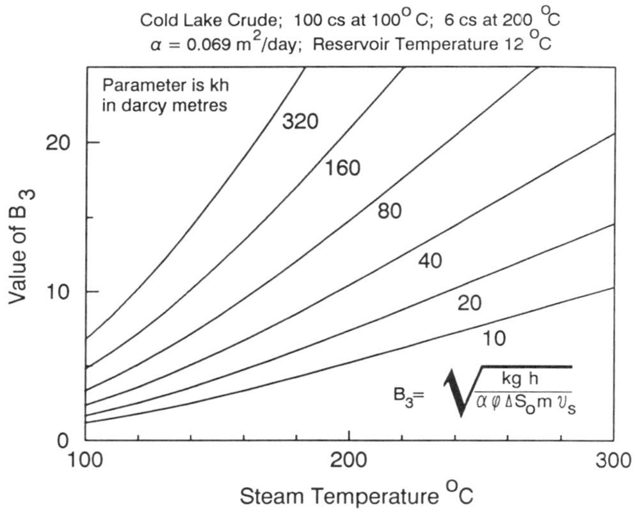

Figure 7.37 Effect of Steam Temperature and kh on B_3

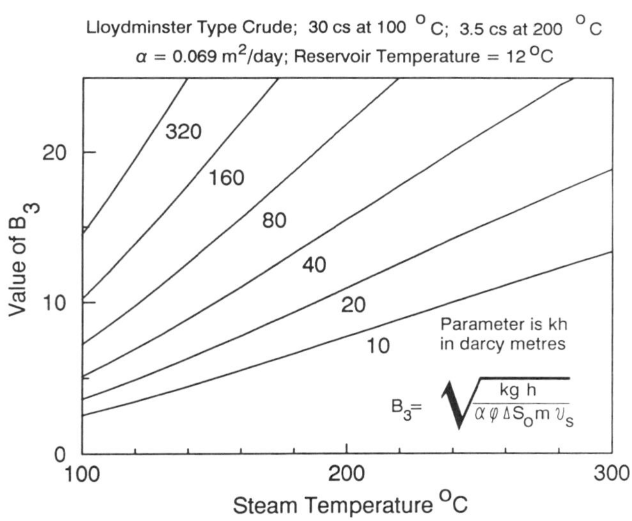

Figure 7.38 Effect of Steam Temperature and kh on B_3

Heat Penetration as a Function of Distance Along Interface

Figure 7.39 shows the interface positions calculated using the same approach for a case involving a confined well. Although the curves are similar to those of Figure 7.12, it will be found, on careful examination, that the production rate is somewhat lower, particularly initially.

The reason for the lower rate is that the heat penetration beyond the interface is lower than that corresponding to the steady state.[12] The fraction of the steady-

[12]This is particularly true near the start where it is assumed that the reservoir is cold beyond the initial vertical hot plane.

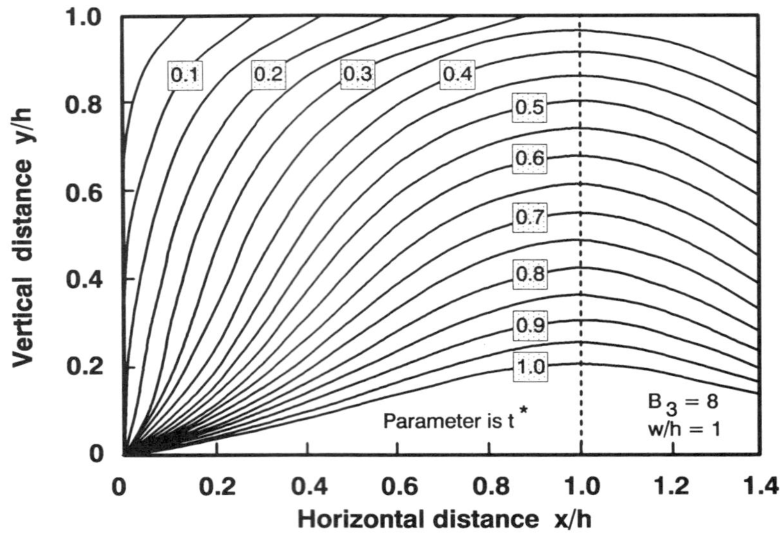

Figure 7.39 Position of Interface for a Confined Well

state heat penetration that is achieved is plotted against the vertical position on the interface for a dimensionless time of 0.3 and for $B_3 = 8$ in Figure 7.40.

The fraction of the steady-state heat penetration achieved is above 0.8 over most of the interface and is much lower near the well. It is this deviation from the steady-state heat penetration that makes the drainage rate somewhat less than that predicted by the earlier equations.

Figure 7.41 compares the drainage rates predicted from the earlier equations with those calculated from the curves in Figure 7.39. The three dotted curves in this diagram are the same as those in Figure 7.13. The solid curve shows the rate

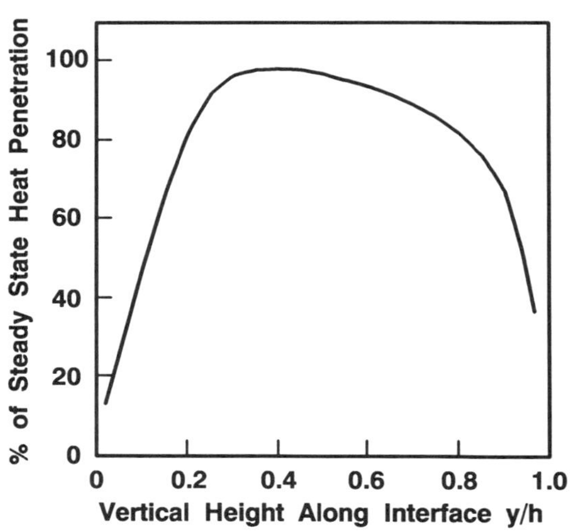

Figure 7.40 Heat Penetration Along the Interface as Percent of Steady State: $t^* = 0.3$; $B_3 = 8$

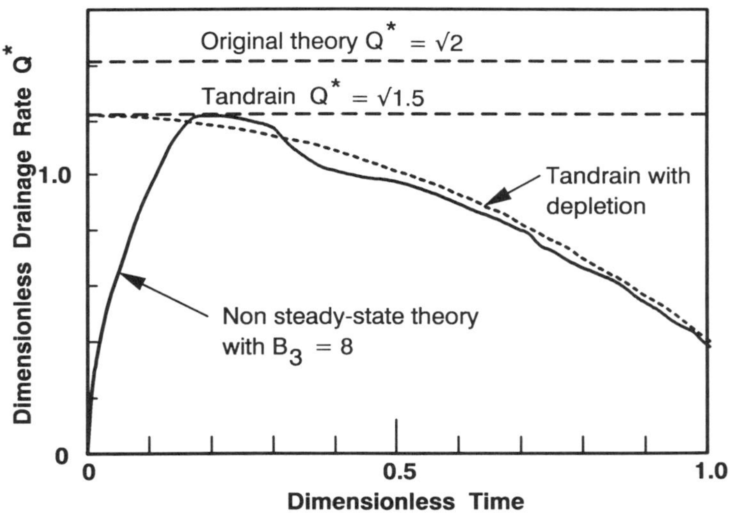

Figure 7.41 Comparison of Predicted Production Rates

from the newer theory. It starts much lower and then reaches the level predicted by 7.34. After this, it falls because of depletion. Although the solid curve in Figure 7.41 exceeds the rate predicted by equation 7.35a over part of the range, it should be realized that during this period, the reservoir is less depleted because of the earlier low drainage rate. If the curves had been plotted against the fractional recovery, the curve for the new theory would have fallen below the other throughout. The irregularities in the curve from the new theory are due to instabilities in the method of calculation.

Predicted Oil-Steam Ratios

The steam consumption for the processes described in this chapter may be estimated using the equations developed in Chapter 2.

In general, steam is required to provide the following:

1. The heat to raise the steam chamber from T_R to T_S.
2. The heat required to raise the produced oil from T_R to the production temperature. It may be assumed that this is the same as the steam temperature. This assumption is somewhat pessimistic because the oil leaving the system is actually below the steam temperature.[13]
3. The heat losses to the overburden above the steam chamber. These may be calculated using equation 2.28.
4. The heat in the reservoir beyond the advancing front. An approximate method of evaluating this is to assume that it is equal to the heat loss to the

[13]It is relatively simple to show that the mixing temperature of the draining oil is given by $(T_m - T_R)/(T_S - T_r) = m/(m + 1)$. It seems likely that the draining condensate will also intermingle with the oil beyond the interface and reach this same temperature.

overburden. Alternatively, equation 2.48 can be employed using an average value for U. A more accurate method involves the use of the heat-penetration function described by Butler (1984). It will be found that whichever of the
The heat to the reservoir increases rapidly at first as heat penetrates the side
approximately the same.

The heat losses for the unconfined well of Figure 7.34 are plotted against time in Figure 7.42.

In this diagram the cumulative heat requirements are converted to dimensionless values by means of the following equation:

$$q_c^* = \frac{q_c}{h^2 \rho C(T_S - T_R)} \tag{7.50}$$

In using this equation, the value of ρC should be for the reservoir or for the overburden, as appropriate. In Figure 7.39 the curve for the cumulative heat to the chamber gives the total heat to the chamber plus the produced oil if the value of ρC is for the fully saturated reservoir.

The curve for the chamber may also be looked on as a curve for the cumulative produced oil in appropriate units. It is the integral of the production-rate curve.

The heat to the reservoir increases rapidly at first as heat penetrates the side of the initial hot plane. The rate of heat supply to the reservoir decreases after this initial period and then grows as the extent of the heated interface increases.

The heat requirements for the confined well of Figure 7.39 are shown in Figure 7.43 as thinner lines; the thicker lines are for the unconfined well and are taken from Figure 7.42.

Lines for the confined well deviate starting at the point where the steam chamber meets the one growing from the neighboring well. At this point the rates of increase of the heat loss to the overburden and to the reservoir decrease very

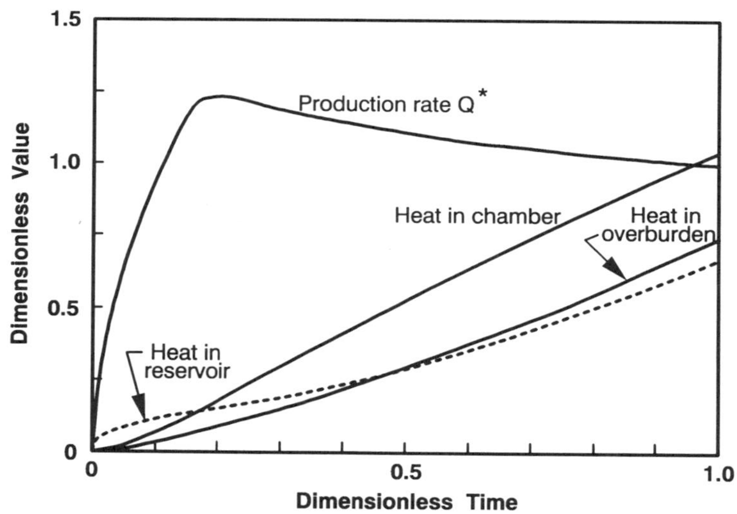

Figure 7.42 Production Rate and Cumulative Heat Distribution

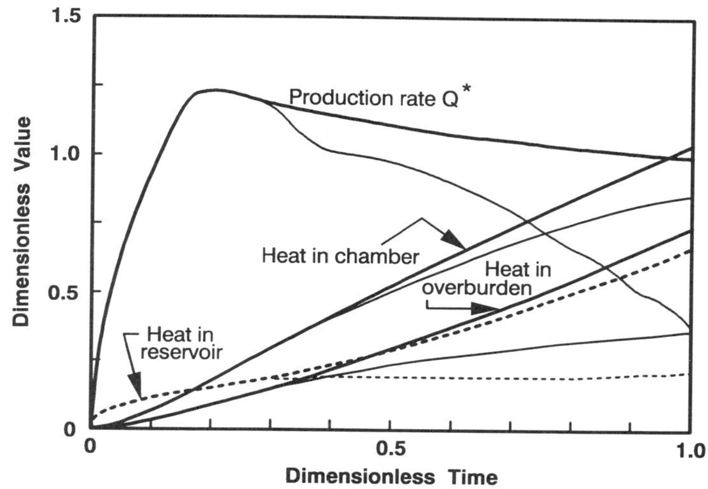

Figure 7.43 Effect of Well Confinement on Cumulative Heat Requirements Thicker Lines are for Unconfined System and Thinner Lines are for Confined System.

markedly and proportionately more so than the decrease in the growth of the chamber. The net result is that the thermal efficiency for the confined well case increases.

A comparison of the thermal efficiency for the two cases is shown in Figure 7.44.

In this figure the thermal efficiency represents the fraction of the steam heat that is used to heat the steam chamber and product. The efficiency is much larger for the confined case; the effect is very significant.

Effect of Steam Pressure

Table 7.4 shows the effect of varying the operating steam pressure on the predicted performance. In each of the three calculations shown, the reservoir was assumed to

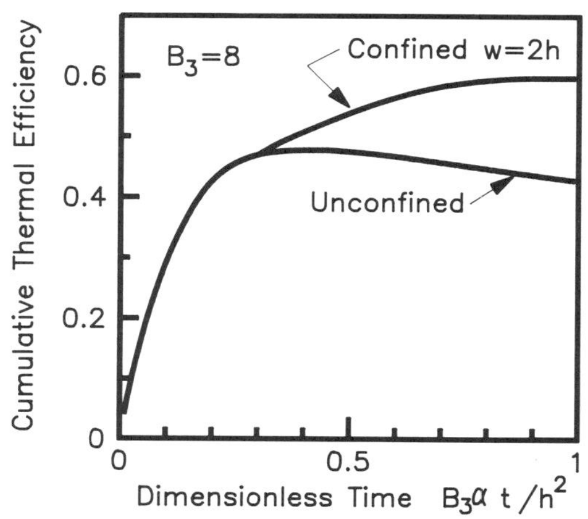

Figure 7.44 Effect of Well Confinement on Thermal Efficiency

TABLE 7.4 Effect of Steam Pressure on the Performance and Thermal Efficiency of an Unconfined Horizontal Well

Steam pressure, MPa	0.45	2.0	5.7
Steam temperature, °C	148	213	272
Years to produce 92 m^3/m	8.7	3.8	2.4
Average rate, m^3/m D	0.029	0.068	0.104
Thermal efficiency	0.37	0.49	0.56
Oil-steam ratio	0.37	0.36	0.34
B_3	4	8	12

(from Butler 1985)

have the properties shown at the bottom of the table; these properties correspond to a reservoir similar to that at Cold Lake.

The effect of operating at higher steam pressures is to raise the temperature of the steam chamber; this allows the oil to drain more rapidly. Because of the shorter time involved in the operation at the highest pressure, the thermal efficiency is also highest, i.e., a smaller fraction of the injected heat is lost to the overburden and reservoir beyond the steam chamber.

There is, however, a counteracting effect that offsets the improved efficiency at high pressure. This results from the increased heat needed to raise the system to steam temperature; this is higher simply because the steam temperature is higher. The net result is that the overall oil-steam ratios in Table 7.4 are almost independent of the steam pressure employed. There may, however, be significant economic advantages in operating at higher pressures because of the faster production that is obtained.

Calculations were based on following parameters:

$$\text{Steam quality} = 0.7 \qquad \rho_{\text{res}} = 2040 \text{ kg/m}^3$$

$$T_R = 6°\text{C} \qquad C_{\text{rock}} = 963 \text{ J/kg °C}$$

$$H = 22 \text{ m} \qquad C_{\text{oil}} = 2093 \text{ J/kg °C}$$

$$\phi = 0.35 \qquad K_{OB} = 1.73 \text{ J/s m °C}$$

$$\Delta S_o = 0.61 \qquad \rho_{OB} = 2400 \text{ kg/m}^3$$

$$k = 0.5 \times 10^{-12} \text{ m}^2 \qquad C_{OB} = 837 \text{ J/kg °C}$$

$$K_{\text{res}} = 1.3 \text{ J/s m °C} \qquad \nu = 100 \text{ cs at 99°C}$$

SAGD RESULTS FROM SCALED LABORATORY RESERVOIR MODELS OPERATING AT BOTH HIGH AND LOW PRESSURES

Chung and Butler (1989b) describe experimental studies of the Steam-Assisted Gravity Drainage Process which were carried out in two-dimensional vertical laboratory scaled models. Some of the experiments were carried out with steam pressures near to atmospheric and others, using a stronger apparatus, with a pressure of

3.55 MPa (507 psia). The results were compared with predictions from the non-steady state theory which was described previously (Butler 1985).

Cold Lake Bitumen was employed and the experimental conditions were chosen so that the value of B_3 would be the same in the model as in the field. This requirement means that a coarser packing is required in the model than that in the field. In order to model the same field conditions it is necessary to use higher permeability packing for low pressure experiments than for ones carried out at the field pressure and temperature.

In the first experiments a preheated vertical injection well was employed (see Figure 7.45) and steam was circulated within the well before production was allowed. During the production period steam was injected at the top of the reservoir. Figure 7.46 shows photographs of the steam chamber at various stages of development for one of the experiments and Figure 7.47 shows the position of isotherms within the reservoir for the same times. It is interesting to compare the photographs in Figure 7.46 for a steam chamber spreading from a central hot well to the initial photographs in Figure 7.15 which show a rising chamber. The interface in Figure 7.46 is stable and advances in a steady and systematic manner as contrasted to the fingering displacement which occurs above the rising chamber in Figure 7.15. The wavy nature of the isotherms in Figure 7.47 is caused by the computer program which drew the contours from the limited temperature measurement information and is not significant.

It was found necessary to modify the theory slightly to allow for the reservoir preheating which occurred during the initial steam circulation period. With this adjustment it was found that the theoretical predictions of both the oil production rate and the position of the interface were in excellent agreement with the experimental data.

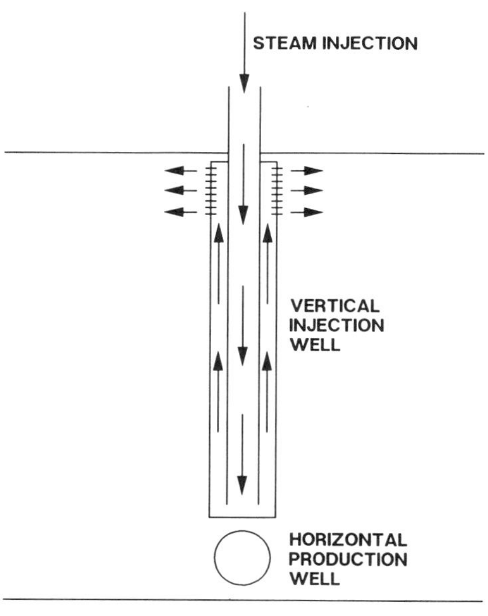

Figure 7.45 Preheated Vertical Steam Injector used in Model Experiments (from Chung and Butler 1989b)

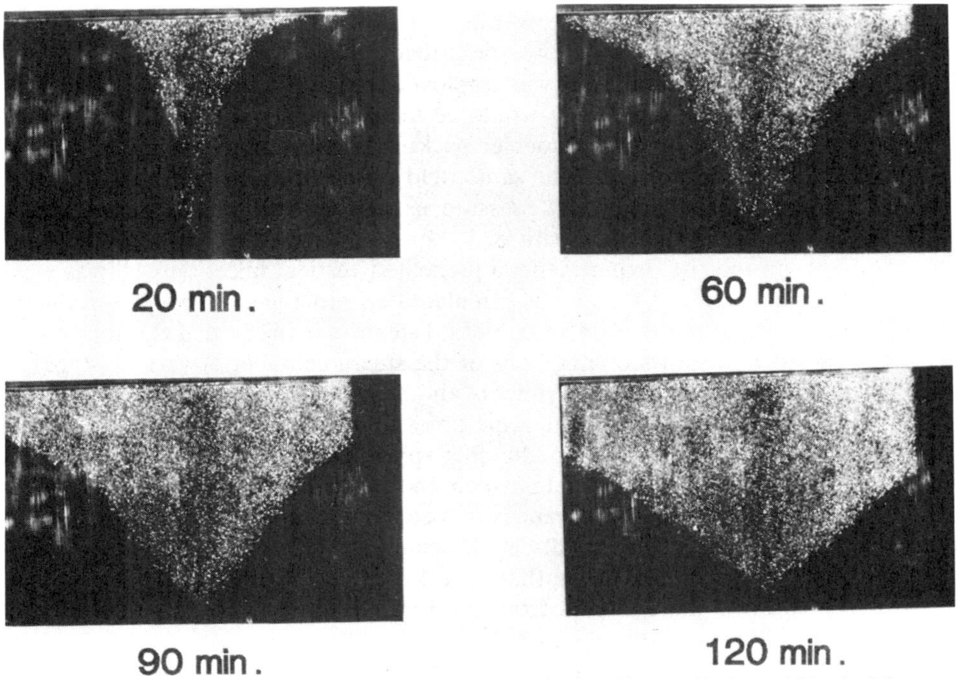

Figure 7.46 Photographs of Steam Chamber Developing About a Preheated Vertical Injection Well in a Low Pressure Experiment with a Transparent Cell Wall (from Chung and Butler 1989b)

The oil production rate is compared to the theoretical predictions in Figure 7.48. Two theoretical curves are shown. The dotted curve is the theoretical curve which is obtained if it is assumed that the reservoir is all at the initial temperature at the start of production. The solid curve makes an allowance for the preheating before production. Figure 7.49 shows, for the same experiment, a comparison of the position of the interface as determined from photographs with the predicted values. The agreement with the measured data is satisfactory.

In experiments of this type the oil-steam ratios are far lower than those which would be found in the field because of the excessive heat losses from the large vertical surfaces of the two-dimensional model. It is however quite practicable to predict the oil-steam ratios for the field situation and examples of this are given in the paper (Chung and Butler 1989b).

A similar agreement between theoretical predictions and experimental data was also found for the high pressure experiments. Figure 7.50 shows a comparison for a high pressure run.

Experiments were also conducted using multi-well models at both high and low pressures and again the results were similar. Figure 7.51 shows isotherms from a high pressure model experiment in which five parallel horizontal well pairs were modeled. The model represents a two-dimensional section through the reservoir. In these experiments the steam was injected from five injection wells each of which was located immediately above a producer.

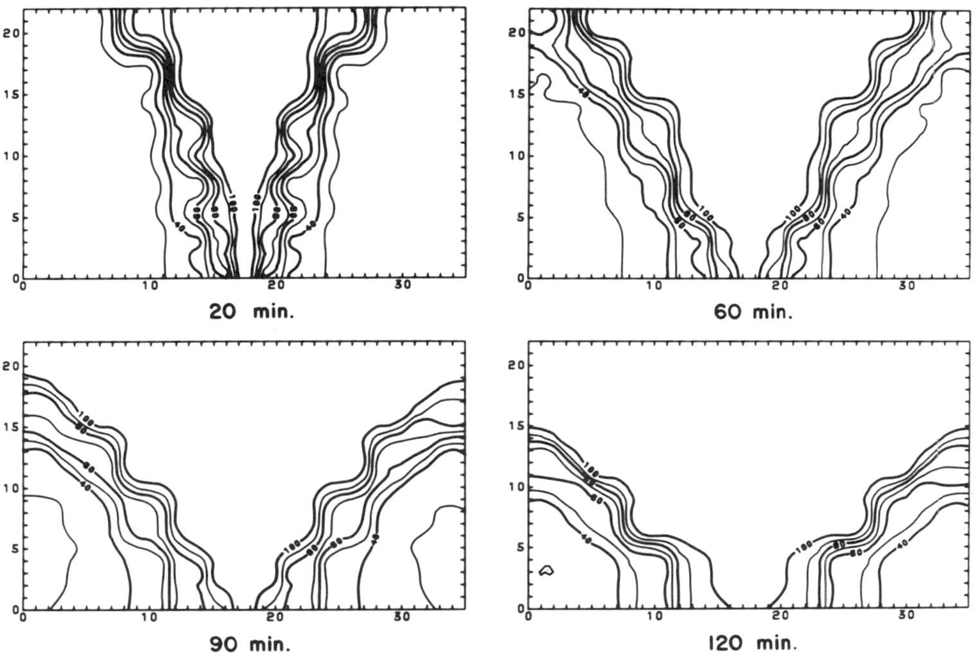

Figure 7.47 Temperature Distribution within the Reservoir for the Experiment of Figure 7.42. Isotherms are Labelled in °C. Cell Dimensions are Marked in cm (from Chung and Butler 1989b)

This geometric arrangement is similar to that employed in the steam chamber pilot at AOSTRA's Underground Test Facility (Edmunds et al. 1987). This test facility consists of tunnels immediately below a section of the Athabasca reservoir. In the steam chamber pilot three pairs of parallel deviated wells are drilled, upwards from the tunnels and then horizontally into the reservoir. The production well in each pair lies near to the base of the reservoir and its steam injection well is

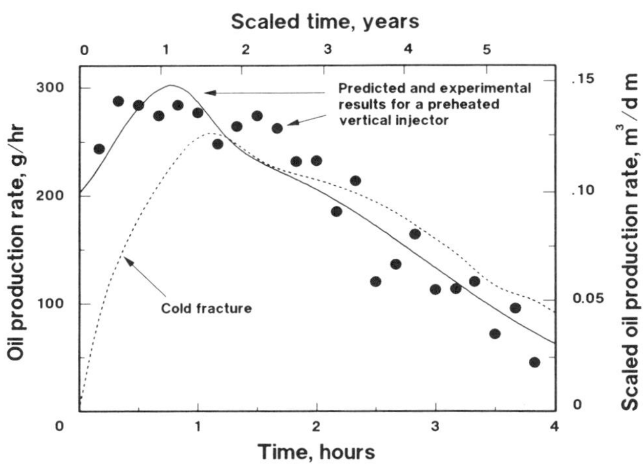

Figure 7.48 Predicted and Experimental Oil Production Rates for a Low Pressure Experiment (from Chung and Butler 1989b)

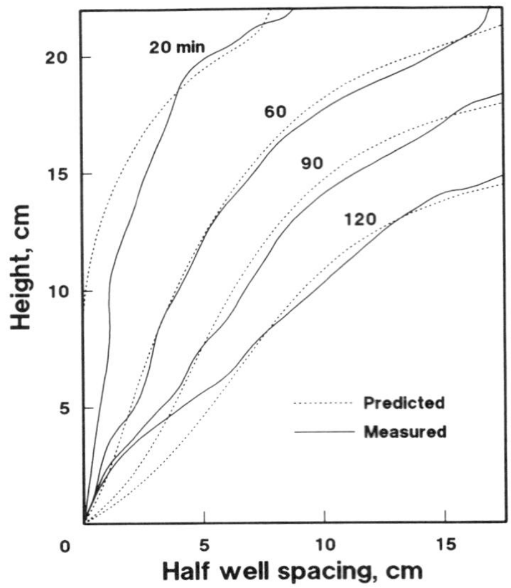

Figure 7.49 A Comparison of the Predicted and Experimental Positions of the Interface (from Chung and Butler 1989b)

parallel and slightly above. The project was adopted by an industry steering committee following a proposal made by the author to AOSTRA in 1984 (Butler 1984). The project has been constructed and operated and it has been very successful; there are plans to develop an expanded project.

One of the factors studied in the laboratory work was the effect of the steam chambers interfering with each other before they grew to the top of the reservoir. This can happen if the wells are located with a close horizontal spacing. The ability to drill closely spaced wells economically might possibly be the major justification for a commercial tunnel-based bitumen recovery project. Projects with large spacings would be more economical using wells drilled from the surface.[14]

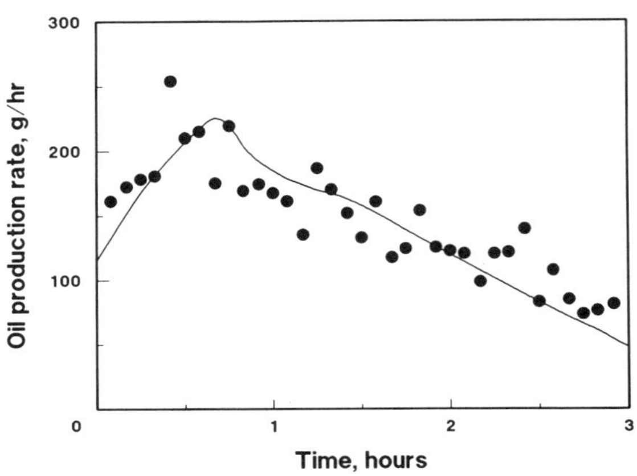

Figure 7.50 A Comparison of the Predicted and Experimental Oil Production Rates for a High Pressure Experiment (from Chung and Butler 1989b)

[14]There is doubt as to whether the cost of developing underground mine-workings as a base for the construction of horizontal wells is economically justifiable. A significant part of the mine cost is the provision of safe working conditions, escape routes and the like, for the underground facility in the event of a well failure or of a steam or hot oil leak. At the AOSTRA project there are

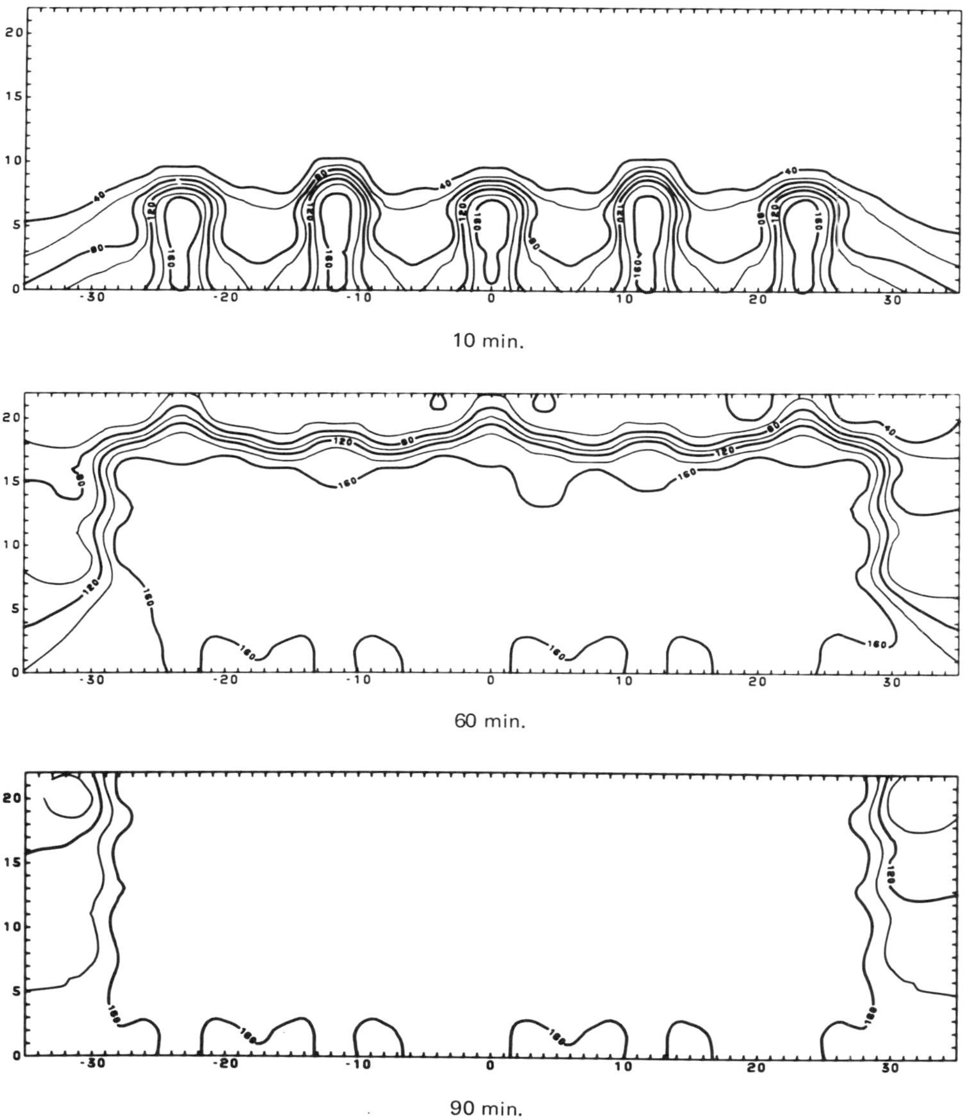

Figure 7.51 Temperature Distribution for a High Pressure 5 Well-Pair Experiment. Steam is Injected Just Above Each of the Five Production Wells Through Separate Injectors. The Upper Part of the Figure Shows Five Separate Rising Steam Chambers. At 60 Minutes These Have Coalesced To Form a Separate Chamber. By 90 Minutes Essentially All of the Model is Saturated with Steam Except for Colder Spots Between the Production Wells and the Ends of the Cell (from Chung and Butler 1989b).

two alternate escape routes throughout; these are based upon dual mine shafts. There is also a refuge room to be occupied by personnel trapped within the mine. A major factor that detracts from the mine approach is the substantial improvement in the cost of drilling horizontal wells from the surface which has occurred during recent years.

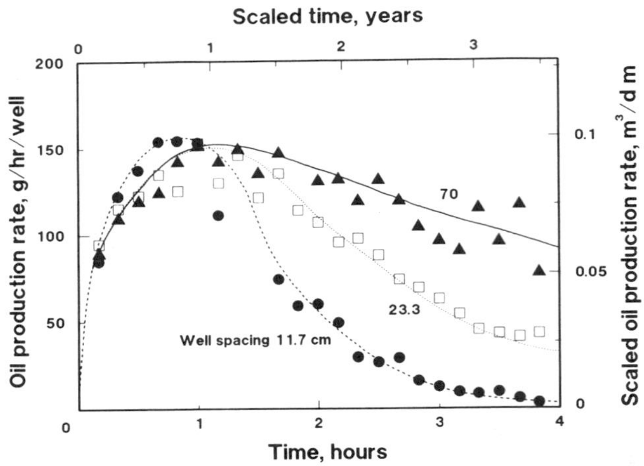

Figure 7.52 Oil Production Rates for High Pressure Experiments with Varied Horizontal Spacings Between Adjacent Wells. The Curve for 11.7 cm Spacing is for the Experiment which is Depicted in Figure 7.47 (from Chung and Butler 1989b)

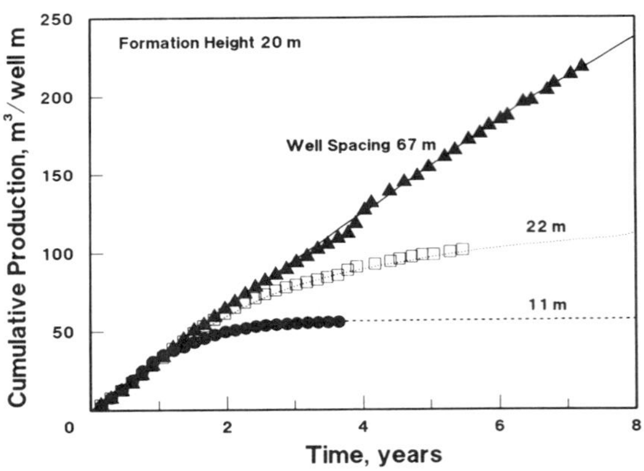

Figure 7.53 Predicted Cumulative Oil Production for the Field at AOSTRA's UTF Site for Various Horizontal Well Spacings; Formation Height 20 m (from Chung and Butler 1989b)

Coalescence of the steam chambers occurred in the experiment depicted in Figure 7.51. This coalescence did not result in a reduced rate of production per well as had been feared. In fact the experiment indicated a higher rate of production per well after the interference than from wells which were more widely separated for the same time in the experiment.

Figure 7.52 shows the rate of production as a function of time for various well spacings. In each experiment it took about one year of time (scaled to the Athabasca reservoir conditions) for the steam chamber to reach the top of the reservoir. However for the most closely spaced wells represented by the solid circles the rate was somewhat higher and the peak rate which corresponds to the chambers reaching the top of the reservoir was achieved earlier than in the other experiments.

The cumulative production for the Athabasca field conditions which would be predicted from these experiments is shown in Figure 7.53. This diagram corresponds to a steam temperature of 200°C (i.e. steam at 1.6 MPa or 232 psia).

OIL PRODUCTION AFTER STOPPING STEAM INJECTION

In the steam-assisted gravity drainage process, it is possible, toward the latter part of the productive life, to continue production of oil without steam injection. During this operation, the pressure in the steam chamber falls as the system cools. Heat in the rocks within the steam chamber is transferred to water in the pores, and further steam is produced. This heat is transferred from within the steam chamber to the boundaries, where it heats the oil and promotes gravity drainage.

The process has been studied by Ferguson and Butler (1988). They showed that in a confined development, the steam injection should be stopped at approximately the time that the adjacent steam chambers meet each other halfway between the adjacent wells. Under these circumstances, the production of oil obtained following the cessation of steam injection is approximately 50% of that produced before, although the rate drops off fairly rapidly. Figure 7.54 shows the predicted production rate as a function of time for a Cold Lake reservoir with half-spacings of 60, 90, and 120 m. The rate rises rapidly and then falls off with time. The three curves show the rate predicted if the steam is shut off at the time the steam chambers intermingle (the *confinement* time).

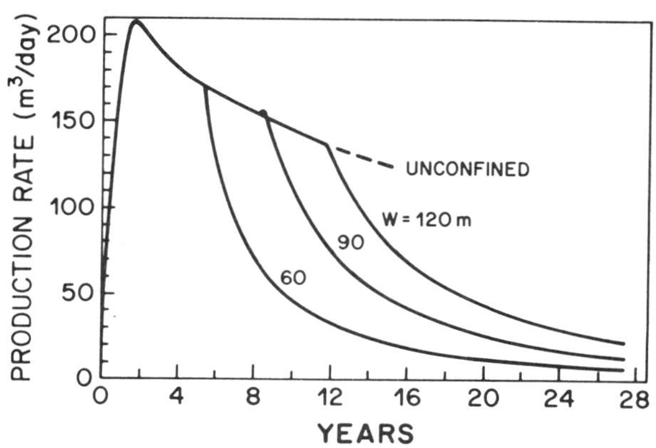

Figure 7.54 Predicted Production Rate for Parallel Horizontal Wells in Cold Lake Reservoir
Well Length 1000 m; Reservoir Height 30 m; Injection to Time of Confinement; Parameter W = Well Half-Spacing in Meters (from Ferguson and Butler 1988)

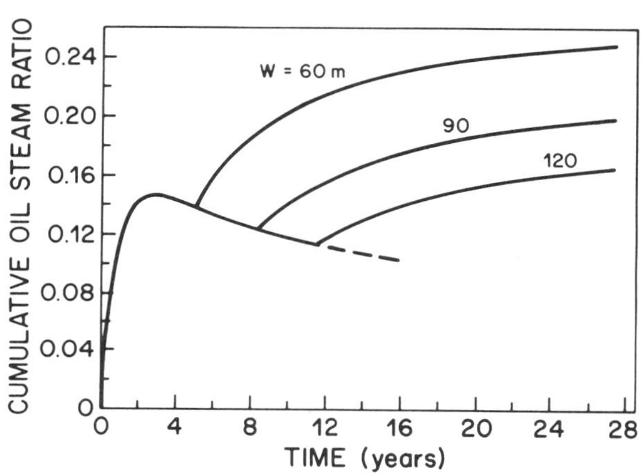

Figure 7.55 Predicted Oil-Steam Ratio for Parallel Horizontal Wells in Cold Lake Reservoir
Reservoir Height 30 m; Injection to Time of Confinement; Parameter W = Well Half-Spacing in Meters (from Ferguson and Butler 1988)

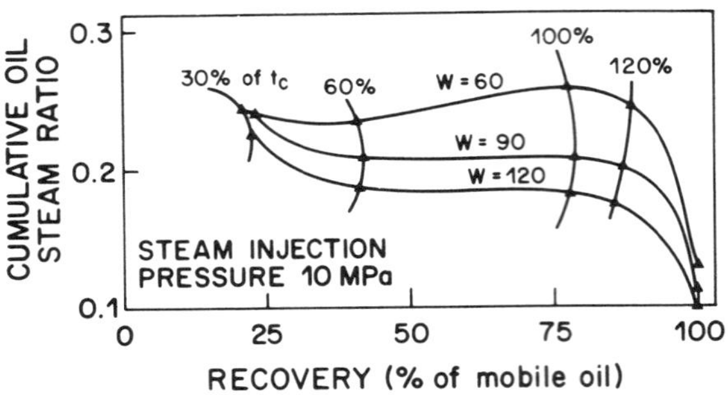

Figure 7.56 Cumulative Oil-Steam Ratio versus Recovery for Parallel Horizontal Wells in Cold Lake Reservoir Reservoir Height 30 m; Parameters: Injection Time as Percent of t_c, Half-well Spacing in Meters (from Ferguson and Butler 1988)

The cumulative oil-steam ratio is shown for the same three cases in Figure 7.55. In each case, the COSR increases rapidly after the steam is shut off; oil is produced during this period without the further consumption of steam. Higher oil-steam ratios are obtained with the closer well spacings because of the more rapid production and reduced time for heat loss. The generally low level of oil-to-steam ratios that are shown in Figure 7.55 and 7.56 reflect the choice of conditions used for the case—in particular, an extremely high steam-injection pressure (10 MPa) with its associated extreme temperature (311°C). It is thought that much more economic OSRs would have been found if a lower injection temperature had been assumed. Nevertheless, similar trends would be expected.

In Figure 7.56 the cumulative oil-steam ratio is plotted against the percent recovery of mobile oil (the mobile oil lying above the production wells). Curves are shown for the same three well spacings and also for the time at which the steam injection was stopped. This time is expressed as a percentage of the time of confinement. For each well spacing, the overall recovery increases as the time of steaming increases, and the cumulative oil-steam ratio rises to rather flat maxima in the vicinity of $t_c = 1$; at this point the recovery of the mobile oil is about 75%.

RECOVERY OF HEAVY OIL ABOVE WATER

The thicker conventional heavy oil reservoirs in Saskatchewan and Alberta frequently contain substantial aquifers at the base. This water tends to limit the applicability of conventional steam recovery techniques because of the extra heat load it creates. If steam is introduced above reservoir pressure it tends to flow into the water layer and excessive heat can be lost. Similarly cold water may be drawn towards the production well and consume valuable heat.

Conventional cold production in these reservoirs is usually not economical because coning of water to the production wells results rapidly in excessive water-oil ratios. The use of SAGD to produce these reservoirs can be attractive because of the possibility of controlling the pressure within the production well to be almost equal to that within the aquifer. Under these conditions the oil can be produced without drawing much water, other than the condensate from the steam, into the producing well. The process is shown schematically in Figure 7.57.

As is shown in the first diagram of Figure 7.57, (or as was discussed on pages 322 to 325 for an intermediate elevation) steam is injected into a well at the

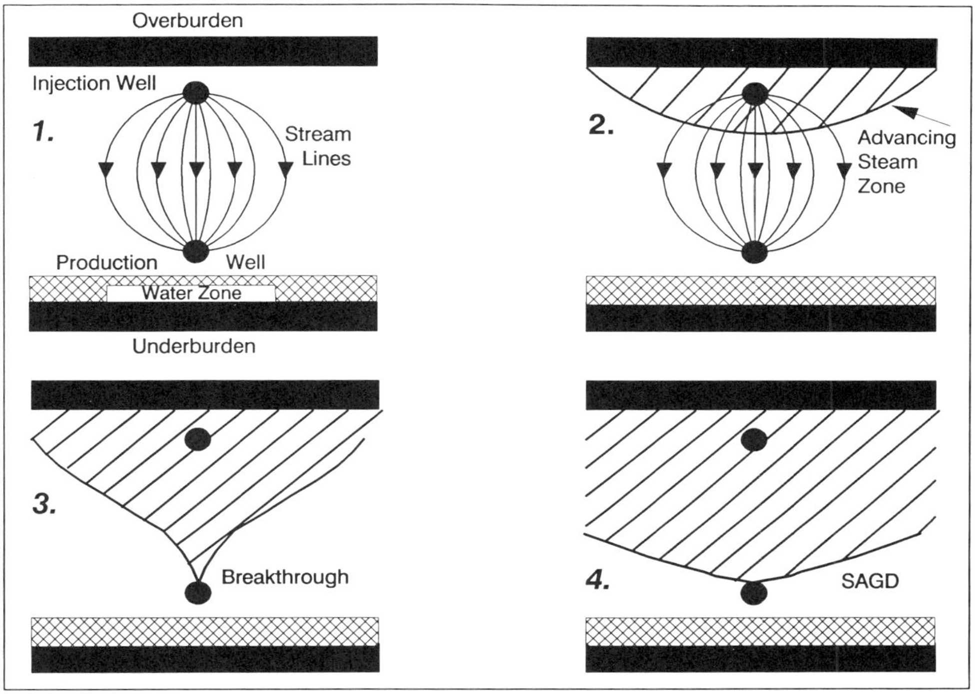

Figure 7.57 Diagram Showing Production of Conventional Heavy Oil which Lies above an Active Aquifer using Downward Steamflooding and Steam Assisted Gravity Drainage

top of the formation. In the first phase of the process oil is displaced downwards by a growing steam chamber. During this phase the steam pressure is substantially higher than the reservoir pressure and the production well pressure is controlled by throttling to prevent water from the aquifer flowing to the well.

The steam zone advances downwards as shown in the second and third diagrams in Figure 7.57. It advances more quickly along the central stream line, and after a time of perhaps one year steam arrives in the vicinity of the production well. Up until this the produced oil and the accompanying condensate from the steam chamber have been flowing through relatively cold reservoir and the produced fluids have been cold.

As the steam chamber approaches the production well the temperature of the produced fluid rises rapidly. At this point the production well should be throttled to allow oil and water to be produced without steam. The steam injection rate should be controlled so that the pressure within the steam chamber balances that of the aquifer. Under these conditions, the injection well pressure will be somewhat above the general steam chamber pressure because of the near well bore pressure drop around the injection well, and the production well pressure will be lower because of its radial flow pressure drop. The SAGD process phase is continued, using this control strategy, as shown in the fourth diagram of the figure, until the OSR is no longer economic.

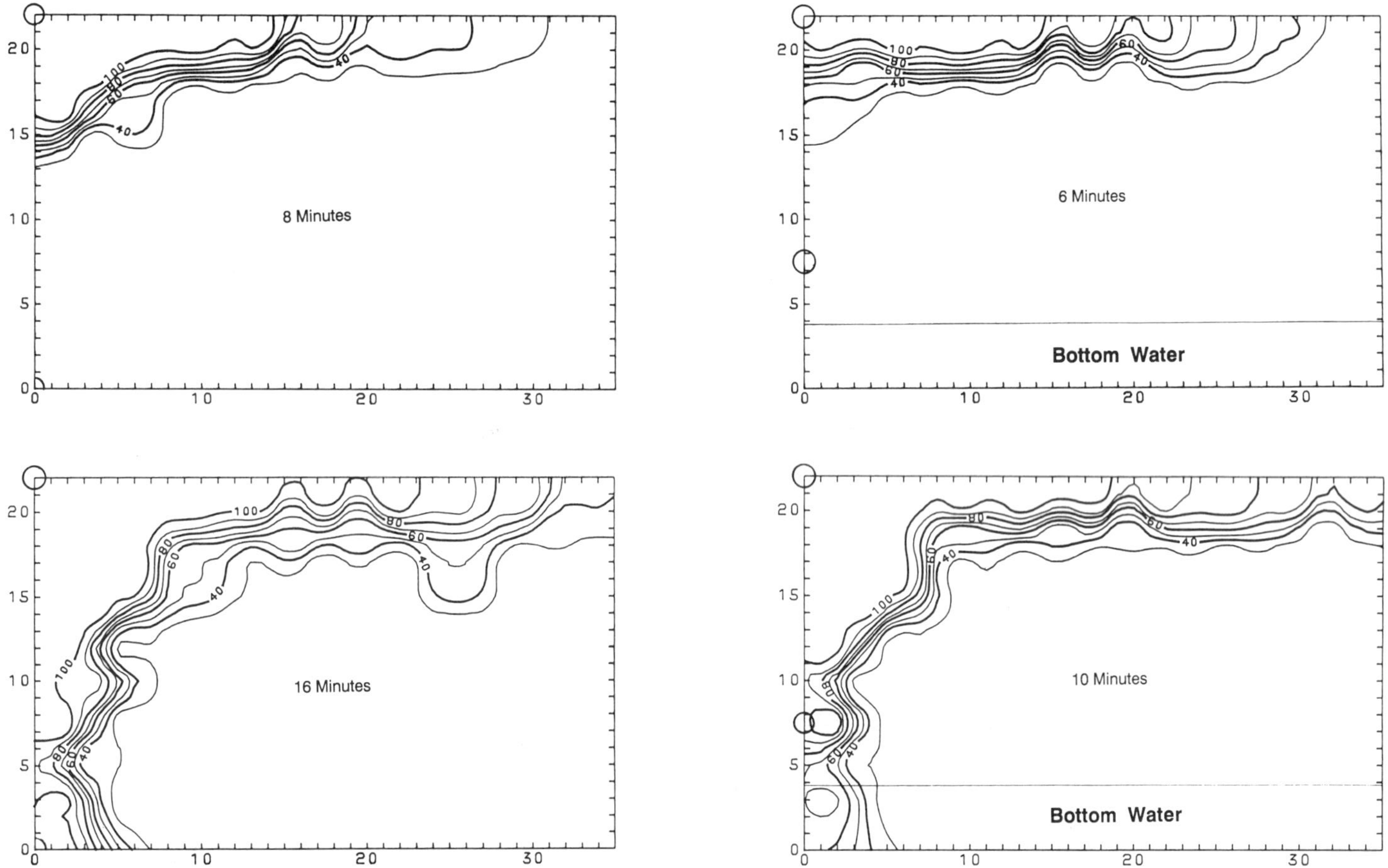

Figure 7.58 Temperature Distribution in Scaled Model Experiments Before and at Steam Breakthrough. Upper Diagrams Are for an Experiment without Bottom Water and Lower Diagrams Are for an Experiment with Bottom Water (from Sugianto and Butler 1989)

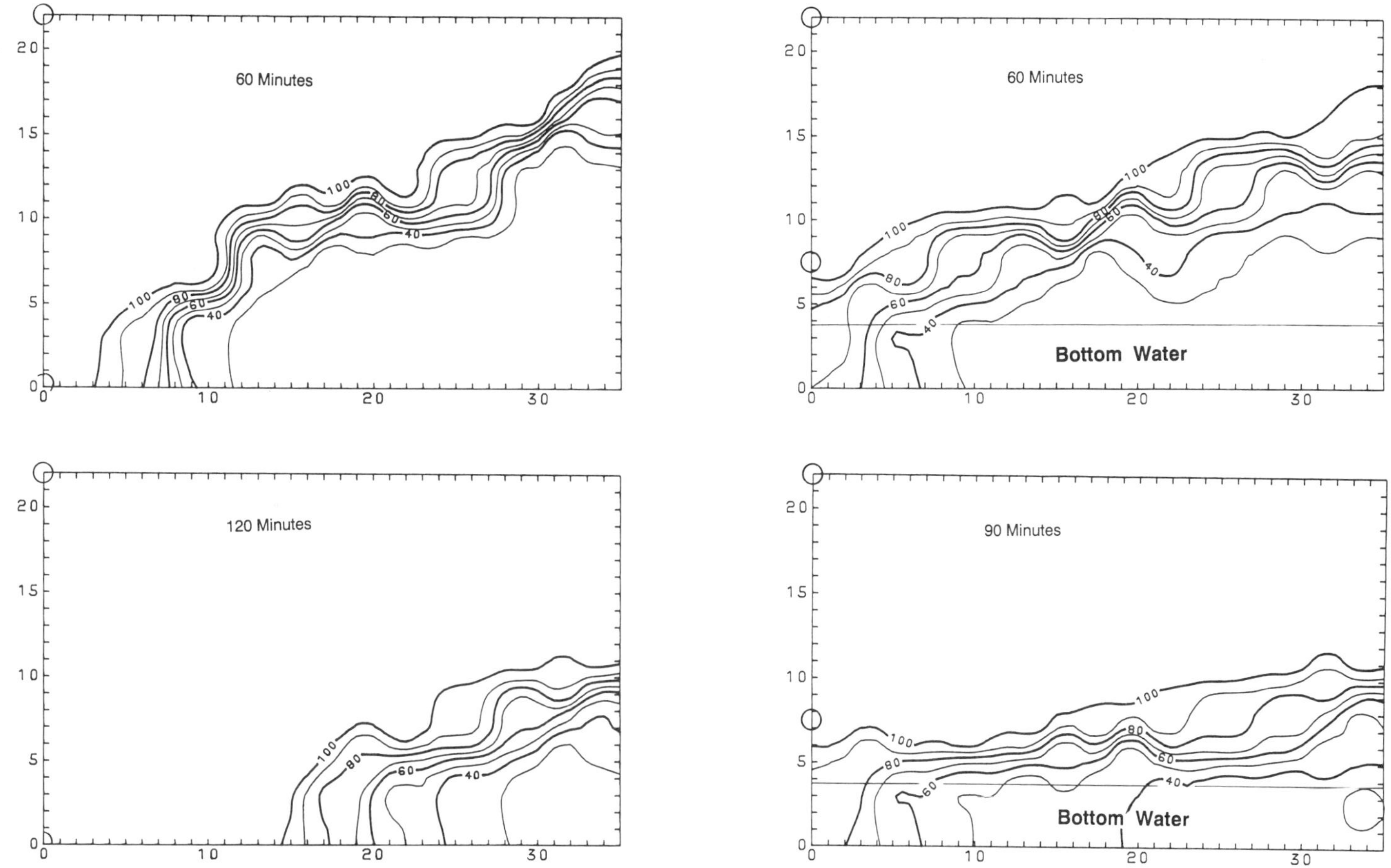

Figure 7.59 Temperature Distribution in Scaled Model Experiments After Steam Breakthrough. Upper Diagrams Are for an Experiment without Bottom Water and Lower Diagrams Are for an Experiment with Bottom Water (from Sugianto and Butler 1989)

The process has been studied in a series of 12 scaled laboratory model experiments by Sugianto and Butler (1989) and encouraging results were obtained. High recoveries of oil with little water production other than that of the steam condensate were achieved.

The extrapolated results indicated that a typical Lloydminster formation (14 m thick) should produce over 160 m^3/d (1000 B/d) from a horizontal well 500 m in length.

Studies of the effect of varying the vertical elevation of the horizontal production well showed that satisfactory operation could be achieved with the well located somewhat above the water interface. However when this strategy is adopted the oil located between the well and the aquifer can not be produced because it cannot drain by gravity. In some experiments the well was located deliberately at the bottom of the formation and thus at the base of the aquifer. This arrangement was found to give good results with little production from the aquifer if the initial production was throttled to allow displacement of water from around the production well as oil was forced downwards by the advancing steam front. In this mode of operation there is little heating of the aquifer by injected steam.

Figures 7.58 and 7.59 show the temperature distributions measured during two companion experiments, one without a bottom water layer (see the left-hand diagrams in each of the figures) and one with an active water level (see the right-hand diagrams). In the experiments with bottom water, water could flow freely into and out of the bottom of the model through tubes connected to an external constant pressure vessel containing cold water. By adjusting the injection pressure in the reservoir it was possible to operate with little flow of water to or from the model.

It is apparent from the position of the isotherms in the two figures that the operation with bottom water was conducted with very little heat passing into the bottom water zone. An oil recovery of 87% OOIP was the highest which was achieved in experiments without bottom water. The recovery achievable with a water zone present was somewhat less than that obtainable with a water free system but satisfactory; the highest recovery obtained with a bottom water layer was 79%. The reason for the lower oil recoveries in the experiments with bottom water was that some oil was forced into the water layer.

EFFECTS OF RESERVOIR HETEROGENEITIES

Scaled model studies of the effect of some reservoir heterogeneities on the SAGD process have been reported by Yang and Butler (1989). They investigated the effects of reservoirs containing horizontal layers of material of different permeabilities and also the effect of horizontal shale barriers.

When a higher permeability layer lies near the base of the reservoir there is a tendency for the steam chamber to undermine the layer above as by the photographs in Figure 7.60. However heat is transferred upwards into the lower permeability layer as the process proceeds and this tend to limit the degree of undermining.

When a layer of lower permeability lies at the base of the reservoir then the interface curves become steeper in the lower permeability section; this steeper

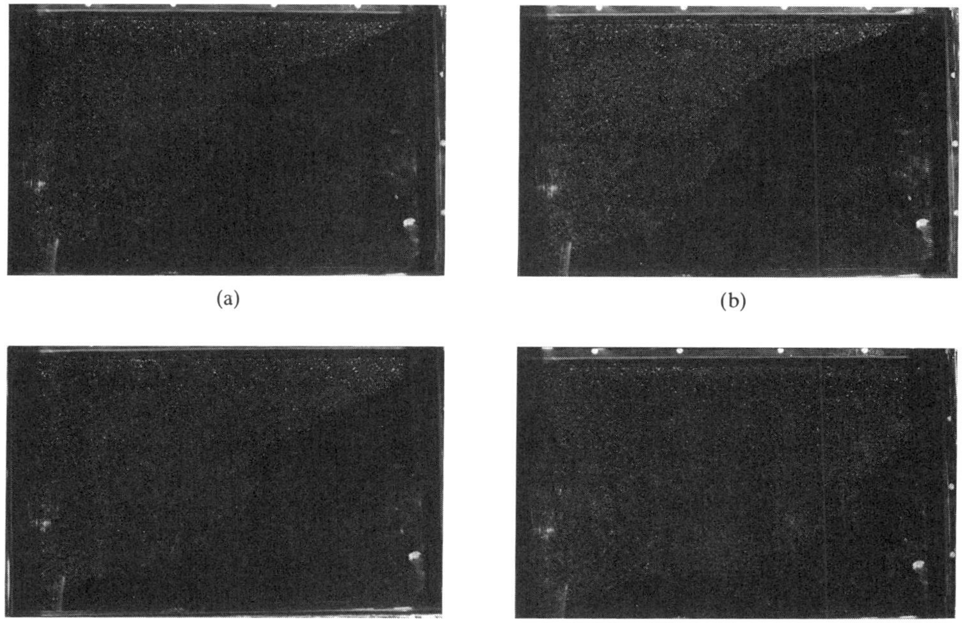

Figure 7.60 Positions of Interface During Scaled Visual Model Experiment with Upper Half of Model Packed with 2 mm Glass Beads and Lower Half with 3 mm Beads (from Yang and Butler 1989)

slope is needed in order to accommodate the higher flow from the upper part of the reservoir. Positions of the interface are shown in Figure 7.61 for an experiment of this type.

The effect of horizontal barriers in the reservoir was found by Yang and Butler to depend upon the geometry of the particular situation. Short barriers such as that shown in the photographs in Figure 7.62 had relatively small effects. As the

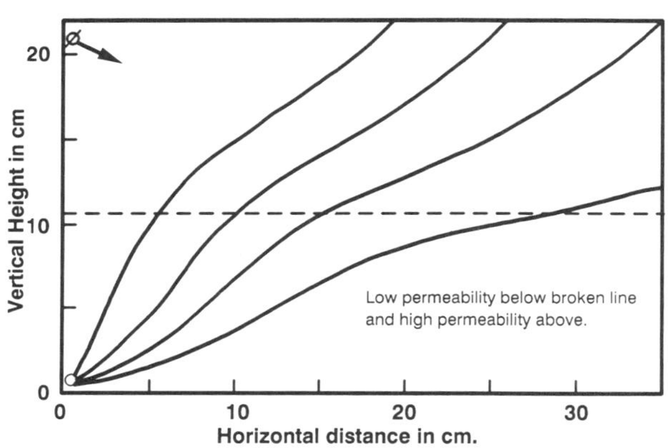

Figure 7.61 Positions of Interface During Scaled Visual Model Experiment with Upper Half of Model Packed with 3 mm Glass Beads and Lower Half with 2 mm Beads (from Yang and Butler 1989)

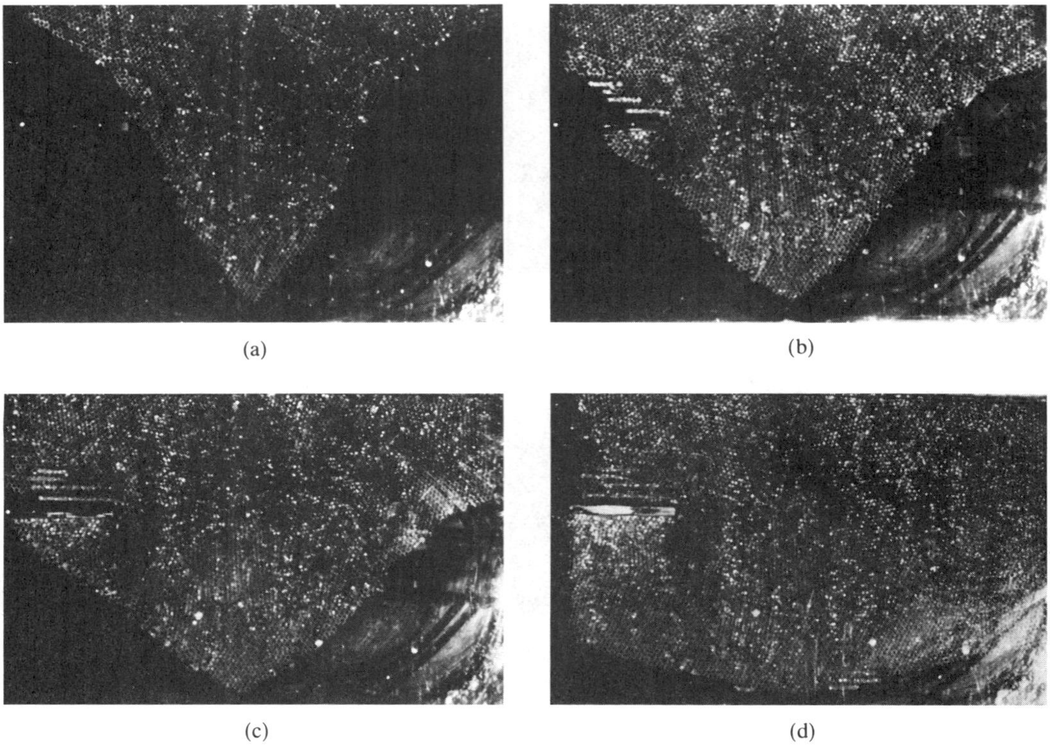

Figure 7.62 Positions of Interface during Scaled Model Experiment with a Short Horizontal Barrier in the Left-Hand Side of the Reservoir (from Yang and Butler 1989)

steam chamber reached the barrier it spread underneath, but heat was also transferred through the barrier to the reservoir above; because of this transferred heat the oil above could continue to drain and the barrier had no long term effect.

Long barriers in some locations, such as that shown in the photographs in Figure 7.63, had only a limited effect. In this experiment both the oil above the barrier and that below were produced. An important effect was that heat was transferred through the baffle and as a result only a small inclination of the interface above the baffle was sufficient to allow production from above. Nevertheless, as may be seen from Figure 7.64, the overall production was retarded as compared to that in a companion run without a barrier.

Different results were obtained with the geometry shown in Figure 7.65 and 7.66. In this experiment a long barrier extended across most of the width of the model above the injection and production wells. A normal steam chamber rose to the underside of the barrier and then spread sideways as if the barrier were the top of the reservoir. When it reached the end of the barrier however the drainage process stalled and no steam chamber formed in the upper half of the model. The oil above, although it gradually became heated, could not fall downwards because

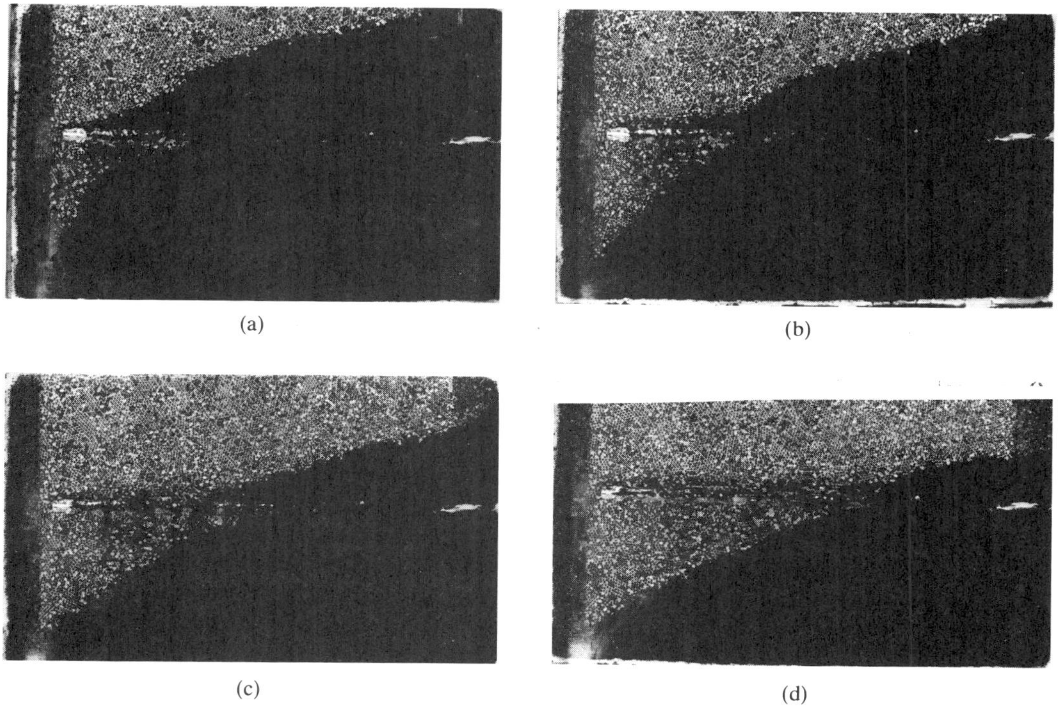

Figure 7.63 Positions of Interface during Scaled Model Experiment with a Long Horizontal Barrier in the Right-Hand Side of the Reservoir (from Yang and Butler 1989)

there was no continuous steam chamber and no density difference to allow a gravity drainage process. The oil above the barrier could be produced by injecting steam from a new well located in the upper zone at the top of the reservoir directly above the producer. With this arrangement the injected steam could sweep the heated oil downwards relatively easily.

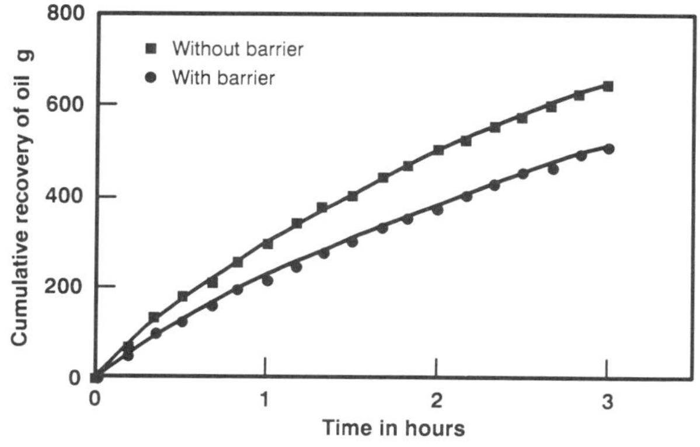

Figure 7.64 Cumulative Recovery of Oil from the Experiment of Figure 7.63 Compared to that from an Experiment without a Barrier (from Yang and Butler 1989)

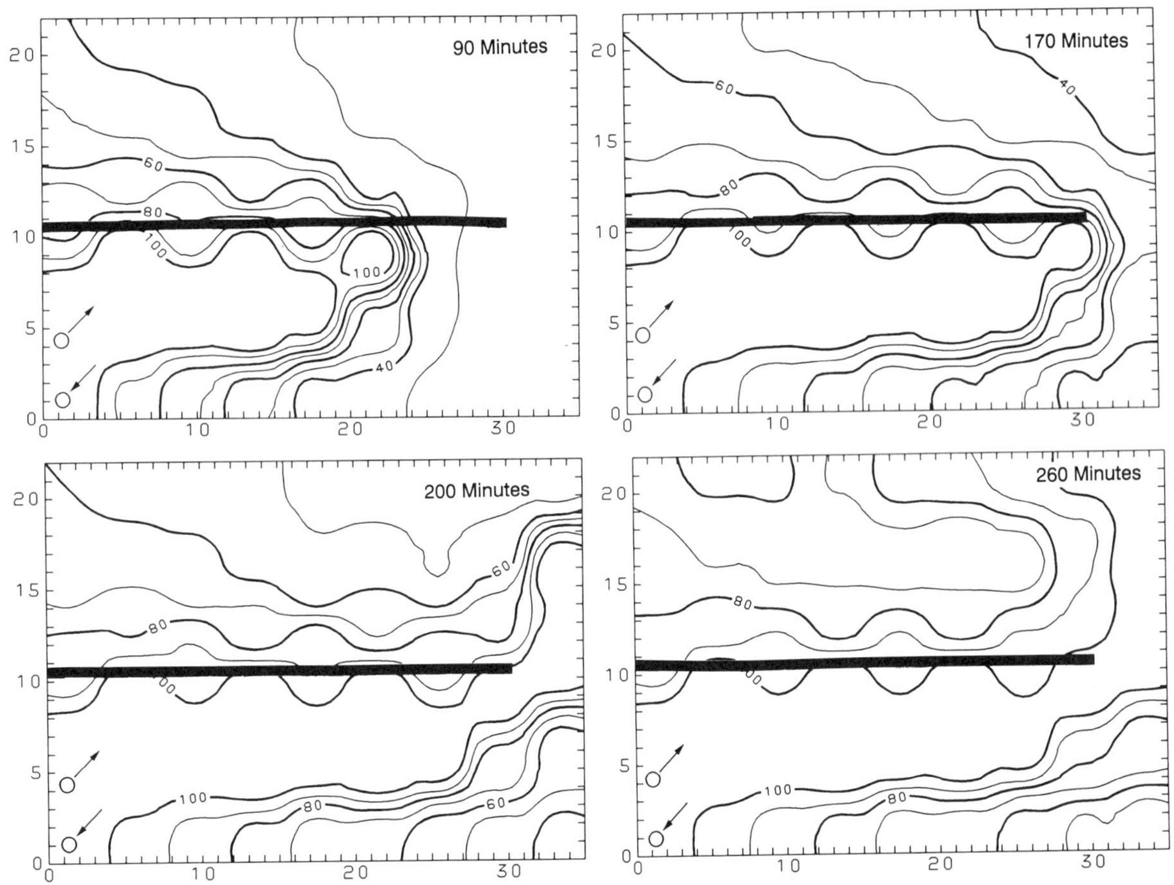

Figure 7.65 Positions of Interface during Scaled Model Experiment with a Long Horizontal Barrier above the Injection and Production Wells (from Yang and Butler 1989)

FORMATION OF W/O EMULSIONS WITHIN THE RESERVOIR

The formation of water-in-oil emulsions is very common in thermal recovery operations. The mechanism of formation of emulsions within the reservoir has been discussed by Jamaluddin and Butler (1988). They consider that the main cause of emulsion formation within the reservoir during recovery processes which involve steam is the condensation of steam on cooler bitumen surfaces. The tendency of bitumen to spread on water surfaces causes small droplets of water which are created by the condensation of steam to become buried within the bulk of the bitumen.

For small water droplets to form it is necessary, because of the effect of the small radius of curvature of the droplet on vapour pressure, for the steam to be supersaturated (i.e. the partial pressure of the water vapor needs to be somewhat above the vapor pressure of liquid water at the temperature of the condensate). It is this supersaturation which provides the driving force for the emulsification. The degree of supersaturation which can be achieved depends upon how easy it is for water to condense elsewhere. In particular if the reservoir rock is water-wet then there is considerable water available with a flat surface on which steam can condense without droplet formation. From this reasoning it would be expected that emulsion formation would be greatest in circumstances where steam can contact cooler bitumen surfaces without contacting relatively flat water surfaces. Jamaluddin and Butler show from a thermodynamic argument that the work required to disperse water within circular capillaries is less if the capillaries are oil wet than if they are water-wet.

Experimental data which are in support of the above ideas have been reported by Chung and Butler (1988 and 1989a) as well as by Jamaluddin and Butler (1988). Jamaluddin and Butler showed that when oil is displaced by steam from a strongly oil-wet packed bed (Teflon beads or toluene-washed, dried sand) higher ratios of emulsified water to oil were found in the product than in similar experiments which employed water-wet sand (sand washed prior to run with detergent). The measured ratios for these experiments are shown in Figure 7.67.

Measurements of the emulsified-water oil ratio were reported by Chung and Butler (1988 and 1989a) for the products from scaled SAGD experiments carried out at both low and high pressures. It was found that more emulsification was found when there was a rising steam chamber than when the steam chamber was spreading. This is consistent with the theoretical ideas which were described previously since there is more opportunity for steam to contact bitumen as the steam fingers rise into the cold reservoir.

Figure 7.68 shows a comparison of the emulsified-water oil ratios for the products from two companion experiments; one was carried out with a steam pressure of 153 kPa (22 psia) and the other with a steam pressure of 790 kPa (115 psia). In both experiments the model was saturated initially with bitumen and there was no connate water. It was filled by upwards flooding of the dry packing and it is likely that the packing was oil-wet initially. The results from the high pressure and low pressure runs were very similar. In both, the emulsified water-oil ratio was relatively high during the period when the steam chamber was rising and then it fell as the steam chamber spread sidewards. The low initial values of the ratio are the re-

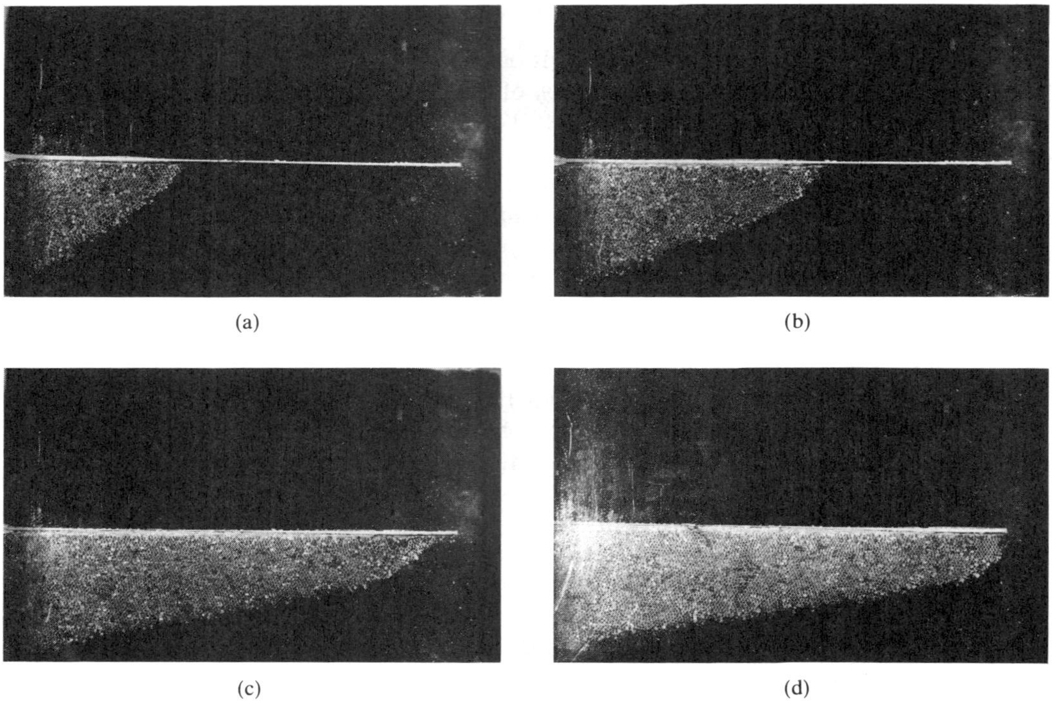

Figure 7.66 Photographs of Experiment Shown in Figure 7.65 (from Yang and Butler 1989)

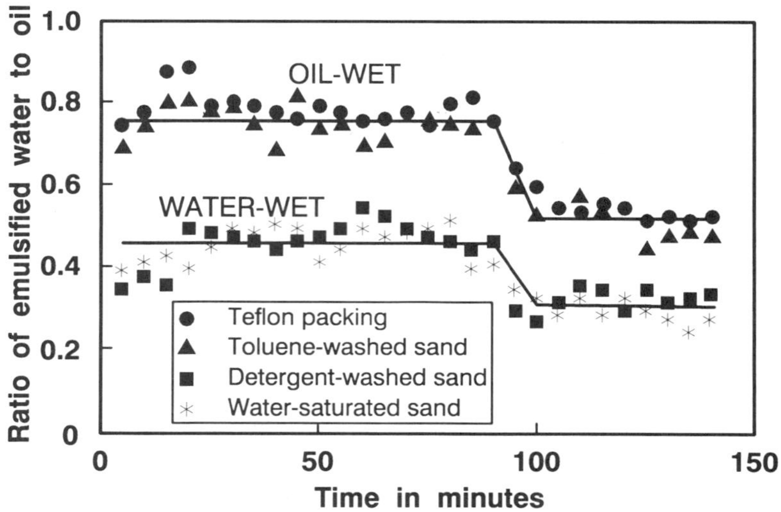

Figure 7.67 The Effect of Sand Pretreatment on the Ratio of Emulsified Water to Oil in the Product from SAGD Experiments with Cold Lake Crude Bitumen (from Jamaluddin and Butler 1988)

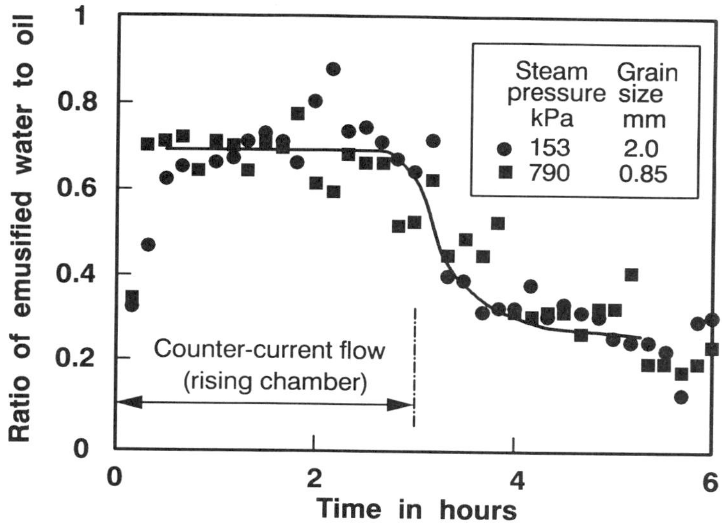

Figure 7.68 Ratio of Emulsified Water to Oil in the Product from SAGD Experiments with Steam Injected Just above the Production Well. Reservoir Contained No Water at the Start of the Experiments (from Chung and Butler 1989a)

sult of the production of the bitumen which was within the well at the start of the experiments.

The results from run 1 of Figure 7.68 are compared with those from a similar run in which the packing had a saturation of 12.5% of connate water at the start

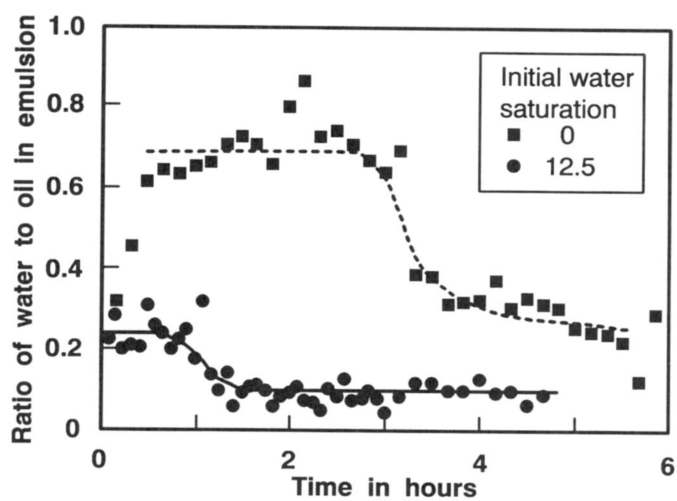

Figure 7.69 Ratio of Emulsified Water to Oil in the Product from SAGD Experiments with Steam Injected Just Above the Production Well. In One Experiment the Reservoir was Dry Initially and in the Other It Contained 12.5% Water Saturation (from Chung and Butler 1989a)

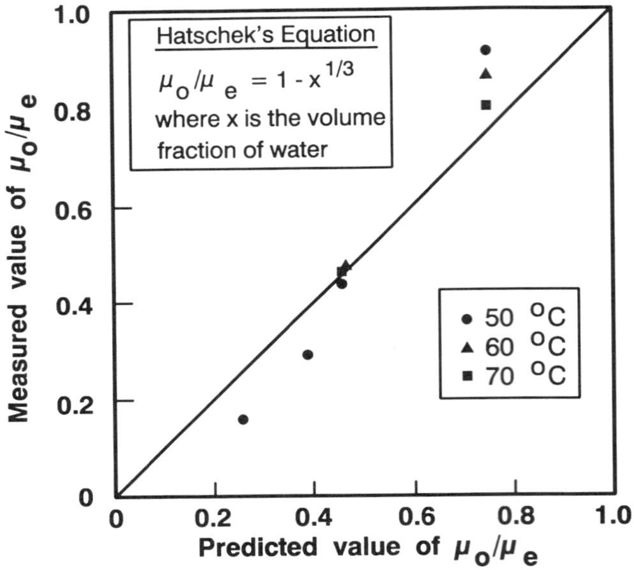

Figure 7.70 Comparison of Viscosities of Cold Lake Emulsions Predicted using Hatschek's Equation with Measured Values (from Jamaluddin and Butler 1988)

(see Figure 7.69). Although the results of this second experiment show the same trends the connate water had a large effect and much lower levels of emulsification were observed. This is in agreement with the theoretical ideas which were discussed previously.[15]

The viscosities of water-in-oil emulsions are higher than those of the base oil. A convenient, approximate equation to predict the viscosity is that of Hatschek (1911); this is given below,

$$\mu_e = \mu_o/(1 - x_w^{1/3}) \tag{7.51}$$

where μ_e is the viscosity of the emulsion
μ_o is the viscosity of the pure oil at the same temperature
and
x is the volume fraction of water in the emulsion.

Emulsion viscosities predicted from this equation are compared to measured values for Cold Lake crude emulsions in Figure 7.70. In Chapter 8, Figure 8.25, which is taken from Chung and Butler 1989, shows measured values of viscosities of Cold Lake crude emulsions as a function of temperature.

WELL BORE RESISTANCE

Although in the analysis in this chapter it is assumed that the pressure within the horizontal production well is constant there is a need to consider this in practical

[15] In the experiment with a high permeability reservoir layer below a lower permeability layer that was described on page 348 and illustrated in Figure 7.60, more water emulsification and a lower drainage rate were found than in the experiment with the high permeability layer at the top. It is thought that this difference was caused by the greater contact of steam with bitumen which occurred because of the undermining effect.

applications. An analysis of the effect of pressure drop along the horizontal well bore has been described by Ong and Butler (1989). They considered three processes as occurring in series:

i) Gravity drainage around the steam chamber. The rate at which this occurs has been discussed previously.
ii) Flow of oil from below the chamber to the production well. The resistance here is that due to the radial converging flow.
iii) The pressure drop along the length of the well bore. The pressure gradient to achieve this increases from zero at one end to a maximum at the outlet of the well.

Ong and Butler show that the effect of the well bore pressure drop is to cause a slope in the bottom of the steam chamber along the well. This slope reflects the pressure difference along the well. In practical field situations the effect is relatively small unless the oil viscosity within the well is high because it is cold. If the well is cold then there is an advantage in heating it by circulating steam or otherwise.

An interesting finding in their paper is that it is particularly important to consider the well bore pressure drop in three-dimensional scaled laboratory models having relatively long horizontal wells. A well scaled to have the same relative pressure gradient (measured as the slope of the bottom of the steam chamber) in the laboratory model should have a diameter larger than that which would come from simple geometric scaling.

CONCLUSIONS

In this chapter the Steam Assisted Gravity Drainage Process has been described. This process involves the use of one or more horizontal production wells located near to the bottom of the reservoir with steam introduced above from separate injection wells. It has been shown that such an arrangement can lead to satisfactory production rates with good recovery and oil-to-steam ratios.

The process is a logical extension of conventional steamflooding but, because of the improved contact with the reservoir which is achieved by the use of horizontal wells, much higher rates per production well can be obtained. It is possible to operate at satisfactory rates without steam-coning. Because of this, better conformance and hence recovery can be obtained thus with conventional steamflooding.

The process can be used for the production of bitumen or conventional heavy oils. Although extensive field demonstration data are not yet available there are promising indications of success which are in line with expectations. Recently two promising field demonstrations, one in Athabasca and one in the Lloydminster area, have joined the long-standing Esso pilots which are at Cold Lake.

In Athabasca AOSTRA has been testing the process at their UTF site and it is believed that the results are successful. An expansion is said to be in the planning stages. The AOSTRA demonstration is believed to be the most promising field steamflooding operation that has been conducted yet in the Athabasca field.

The Sceptre Resources project in the Tangleflags field near Lloydminster is also very promising. Very high production rates have been obtained with reasonable water to oil ratios and steam requirements. The results are notable not only because of the very high production rates (up to 1000 B/d or more of oil from a producer which is 420 m long) but because they are obtained in a field which, with conventional production, is uneconomic because of excessive water production from the underlying aquifer.

BIBLIOGRAPHY

BEZAIRE, G. E. and MARKIW, I. A., "Esso Resources Horizontal Hole Project at Cold Lake," CIM 79-30-10, 30th Annual Technical Meeting of the Petroleum Society of CIM (1979).

BUTLER, R. M.: "New Interpretation of the Meaning of the Exponent "m" in the Gravity Drainage Theory for Continuously Steamed Wells," *AOSTRA J. of Research,* 2, 67–71 (1985).

BUTLER, R. M., "A Steam Chamber Pilot for AOSTRA's Underground Test Facility," presented at AOSTRA's UTF-Industry Meeting in the Glenbow Museum Auditorium, Calgary (May 8, 1984).

BUTLER, R. M., "A New Approach to the Modelling of Steam-Assisted Gravity Drainage," *JCPT,* 42–51 (May–June 1985).

BUTLER, R. M., "Rise of Interfering Steam Chambers," *JCPT,* Vol. 26, No. 3, 70–75 (May–June 1987).

BUTLER, R. M. and PETELA, G., "Theoretical Estimation of Breakthrough Time and Instantaneous Shape of Steam Front During Vertical Steamflooding," *AOSTRA J. of Research,* Vol. 5, No. 4 (1989), pp. 359–382.

BUTLER, R. M., MCNAB, G. S., and LO, H.Y., "Theoretical Studies on the Gravity Drainage of Heavy Oil During Steam Heating," *Can. J. Chem. Eng.,* 59: 455–460 (August 1981).

BUTLER, R. M. and STEPHENS, D. J., "The Gravity Drainage of Steam-Heated Heavy Oil to Parallel Horizontal Wells," *JCPT,* 90–96 (April–June 1981).

BUTLER, R. M., STEPHENS, D. J., and WEISS, M., "The Vertical Growth of Steam Chambers in the In-Situ Thermal Recovery of Heavy Oils," *Proc. 30th. Can. Chem. Eng. Conf.,* 4: 1152–1160, (October 19–22, 1980).

BUTLER, R. M. and YEE, C.T., "A Theoretical Study of Steam Condensation in the Presence of Non-Condensable Gases in Porous Solids," *AOSTRA J. of Research,* 3, no. 1: 1–14 (September 1986).

BUTLER, R. M. and YEE, C.T., "An Experimental Study of Steam Condensation in the Pressure of Non-Condensable Gases in Porous Solids," *AOSTRA J. of Research,* 3, no. 1: 15–24 (September 1986).

CARDWELL, W.T. and PARSONS, R. L., "Gravity Drainage Theory," *Trans. AIME* 179, 199–211 (1949).

CHUNG, K. H. and BUTLER, R. M., "Geometrical Effect of Steam Injection on the Formation of Emulsions in the Steam-Assisted Gravity Drainage Process," *JCPT,* Vol. 27, No. 1 (January–February 1988).

CHUNG, K. H. and BUTLER, R. M., "In-Situ Emulsification by the Condensation of Steam in Contact with Bitumen," *JCPT,* Vol. 28, No. 1 (January–February (1989).

CHUNG, K. H. and BUTLER, R. M., "A Theoretical and Experimental Study of Steam-Assisted Gravity Drainage Process," in R. F. Meyers and E. J. Wiggins (Editors), The Fourth UNITAR/UNDP International Conference on Heavy Crude and Tar Sands, Vol. 4: In-Situ Recovery, AOSTRA, Edmonton, (1989b), pp. 191–210.

DIETRICH, J. K., "The Kern River Horizontal Well Steam Pilot," *SPE Reservoir Engineering,* 935–944 (August 1988).

DYKSTRA, H., "The Prediction of Oil Recovery by Gravity Drainage," *JPT,* 818–830 (May 1978).

EDMUNDS, N. R., WONG, A., MCCORMACK, M. E. and SUGGETT, J. C., "Fourth Annual Heavy Oil and Oil Sands Technical Symposium," University of Calgary, February 18, 1987, Calgary (1987).

FERGUSON, F. R. S. and BUTLER, R. M., "Steam-Assisted Gravity Drainage Model Incorporating Energy Recovery from a Cooling Steam Chamber," *JCPT,* Vol. 27, No. 5, 75–83, Sept.–Oct., 1988.

GRIFFIN, P. J. and TROFIMENKOFF, P. N., "Laboratory Studies of the Steam-Assisted Gravity Drainage Process," *AOSTRA J. of Research,* 2, no. 4: 197–203 (1986).

HATSCHEK, E., *Kolloid-Z.,* 8, 34 (1911).

JAMALUDDIN, A. K. M. and BUTLER, R. M., "Factors Affecting the Formation of Water-in-Oil Emulsions During Thermal Recovery," *AOSTRA J. of Research* (May, 1988).

JOSHI, S. D. and THRELKELD, C. B., "Laboratory Studies of Thermally Aided Gravity Drainage Using Horizontal Wells," *AOSTRA J. of Research,* 2, no. 1: 11–19 (1985).

ONG, TEE, S. and BUTLER, R. M., "Wellbore Flow Resistance in Steam-Assisted Gravity Drainage," *JCPT,* Vol. 29, No. 2 (March–April 1990).

PRATS, M., "A Current Appraisal of Thermal Recovery," *JPT,* 1129–1136 (August 1978).

SUGIANTO, S. and BUTLER, R. M., "The Production of Conventional Heavy Oil Reservoirs with Bottom Water Using Steam-Assisted Gravity Drainage," *JCPT,* Vol. 29, No. 2 (March-April 1990).

TERWILLIGER, P. L., WILSEY, L. E., HALL, H. N., BRIDGES, P. M., and MOISE, R. A., "An Experimental and Theoretical Investigation of Gravity Drainage Performance," *Trans. AIME* 146, 28–53 (1951).

YANG, GUIHUA, and BUTLER, R. M., "Effects of Reservoir Heterogeneities on Heavy Oil Recovery by Steam-Assisted Gravity Drainage," Paper No. 89-40-72, presented at the 40th Ann. Tech. Mtg. of the Petroleum Society of CIM (May 28–31 1989).

8

Steam Recovery Equipment and Facilities

INTRODUCTION

In this chapter the equipment and surface facilities needed for thermal recovery are discussed—the equipment for steam generation and steam distribution, the wells, the facilities for treating the crude, and, finally, the processes involved in treating produced water to make it suitable for recycling to steam generators.

In a typical steam recovery operation, the volume of produced water may be about three to five times larger than the volume of produced oil. As a result, there are very significant capital and operating costs for the water-treating and water-handling facilities. Since the performance and service factor of the whole project depends upon the satisfactory handling of water and generation of steam, it is vital to develop good designs and operating procedures. For these reasons, a good understanding of the water reuse process is essential if oil recovery is to be successful technically and economically.

STEAM GENERATION

The main roots of the industrial revolution are to be found in the discovery and development of practical steam power in England at the end of the seventeenth century.

Early steam boilers such as the one shown in Figure 8.1 were of simple design and were limited in capacity and pressure by the small size of the steel plates available. There were many explosions and accidents as a result of improper operating practices and design.

It soon became apparent to boiler designers that it was necessary to provide extended heat-transfer surface area in order to build boilers of increased capacity. In general, two approaches were followed:

1. Fire tubes, in which combustion occurs within the inside of one or more tubes that are submerged below the surface of the water in the boiler vessel. Air and fuel are introduced into one end of these tubes, and the combustion products flow through the remainder of the tube; this provides additional heat transfer. In some cases, the convective heat transfer is extended by passing the flue gas through the boiler within a set of parallel smaller diameter tubes.
2. Water tube boilers in which the water is heated and boiled with the water flowing inside tubes that are exposed to the fire and combustion products on the outside.

Fire tube boilers are used for smaller installations, particularly lower-pressure installations and for portable uses such as railway locomotives. Their applications are limited by the need to build a high-pressure vessel large enough to contain the fire tubes.

Figure 8.2 shows a cross section of an early water tube boiler, which was built in 1877. Natural convection caused water to circulate to the bottom of the bank of inclined tubes and the boiling mixture of water and steam to rise. Steam separated from the liquid water in the upper drum. Blowdown water was removed from the boiler in order to limit the concentration of dissolved solids.

Modern, large, high-pressure boilers are of the water tube type. The boiler shown in Figure 8.3 shows the type of construction employed in a large, modern water tube boiler. This boiler generates steam within vertical tubes that form the

Figure 8.1 Haycock Boiler (1720) (from Babcock and Wilcox 1972)

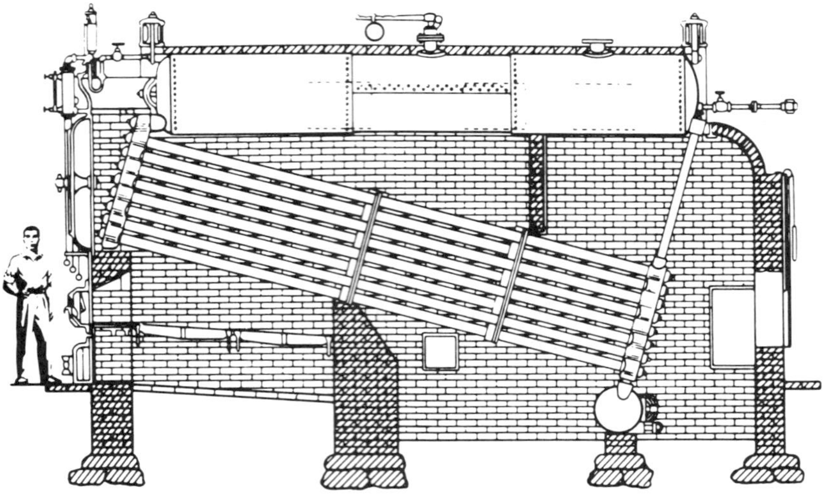

Figure 8.2 Coal-Fired, Babcock and Wilcox Water Tube Boiler (1877) (from Babcock and Wilcox 1972)

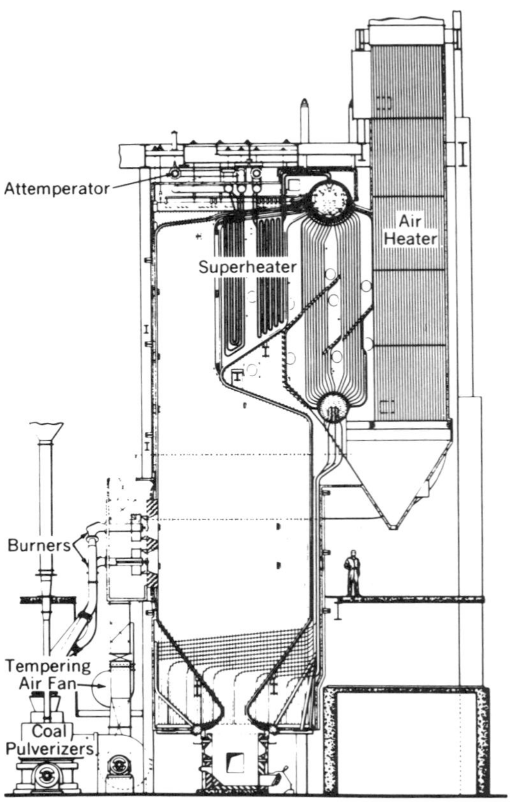

Figure 8.3 Stirling Boiler for 925 psi and 900°F Steam Temperature. This Boiler Is Fired By Pulverized Coal. The Walls of the Combustion Chamber Are Lined With Water Tubes Which Absorb Radiation from the Flames. Sensible Heat from the Combustion Gas Product Is Absorbed In Convection Tube Banks (from Babcock and Wilcox 1972)

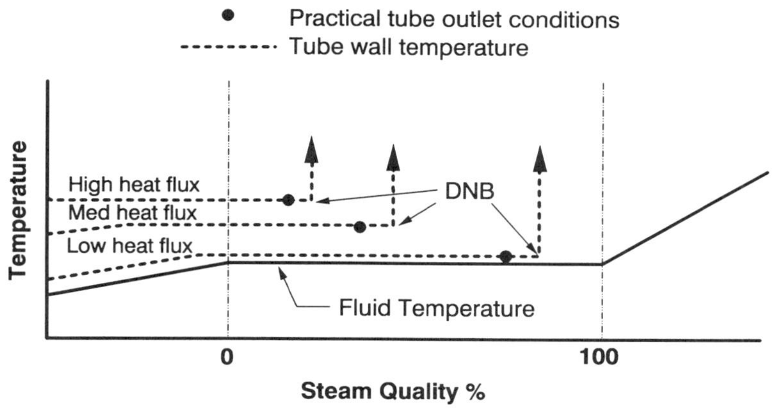

Figure 8.4 Boiler Tube Fluid and Tube Wall Temperatures (after Babcock and Wilcox 1972)

wall of the combustion chamber and also in an independent bank of tubes. The steam is superheated in a separate section of tubes.

Heat is transferred through the surface of the boiler tubes both by direct radiation from the fire and by convection from the hot gases. Flow through the tubes may be caused by natural circulation, as in the design shown here, or it may be created by pumping.

Water tube boilers are fired using any of the common fuels—gas, oil, or coal; and they can also be adapted for special fuels such as refinery coke.

The diagram in Figure 8.4 shows the temperature along a tube that is heated externally by a furnace; water is flowing inside the tube and boiling. The horizontal scale represents the cumulative heat transfer represented by the quality of the steam. The ordinate depicts the temperature of the tube wall and also of the fluid. The three upper broken curves show the metal temperature of the tube for three different levels of heat flux.

Near the entrance to the tube, assuming that the water enters below its boiling point, the temperature of the tube rises along with that of the liquid water that is being heated. The temperature of the tube wall is higher than that of the fluid, and a larger temperature difference between the wall and the fluid is required for a larger heat flux. At the point where the water begins to boil,[1] the temperature difference between the wall and the fluid decreases because of the increased heat-transfer coefficient that results from the agitation due to the boiling.

As boiling continues, there comes a point along the tube where the boiling mechanism changes from that of nucleate boiling, where separate bubbles of vapor form at the hot surface, to that of film boiling, where the surface becomes covered by a layer of vapor through which the heat must be transferred. At this point a

[1]The temperature gradient at the heating surface causes the liquid to boil at the surface before the liquid is generally heated to the boiling point. At moderate heat fluxes, the bubbles of vapor that are formed collapse as they mix with the bulk of the liquid. At extremely high heat fluxes, it would be possible to have a complete film of vapor covering the inner wall of the tube even though the bulk of the liquid has not reached the boiling point; this is not a desirable condition.

much larger temperature difference is required to maintain the heat flux; as a result, the temperature of the boiler tube tends to rise rapidly. The film of vapor forms an insulating blanket through which the heat must be transferred. The point where the boiling mechanism changes is known as the *departure from nucleate boiling* (DNB).

Boilers are normally designed to maintain nucleate boiling within the tubes in which evaporation is occurring. For large heat fluxes, this limits the evaporation per pass to low values. In once-through oil field steam generators, the evaporation (in one pass) is higher than in typical drum-type boilers, but the heat flux is much lower. The expense of the additional heat-transfer surface that is required is offset by the mechanical and operating simplicity. Typical heat fluxes are given in Table 8.1.

TABLE 8.1 Typical Average Heat Flux in kBTU/ft^2 h[(1)] in Radiant Section

High capacity power boilers	90–180[(2)]
Oil field generators	15– 18[(3)]

[(1)]1 kBtu/ft^2 h = 3.1546 kW/m^2.
[(2)]Based on projected area of tube wall. Lower value is for pulverized coal firing and upper value is for fuel oil (Blokh 1988); see also Delibert (1987).
[(3)]Based on tube area $3\frac{1}{2}$-in. OD on 6-in. spacing (Kerby, Kense, and Peachey 1984).

The lower flux in oil field steam generators makes them much more tolerant to the overheating caused by the deposition of scale on the heating tubes. Scale deposits due to such causes as water hardness provide a heat-transfer resistance between the wall and the water. The heat flux flowing through this resistance produces an increased temperature drop that is proportional to the resistance and to the flux. Although this effect is less in oil field generators, it is still very important to soften the feedwater to minimize scale formation. There is, however, considerable tolerance to other dissolved solids such as sodium chloride.

The feedwater flow rate to a steam generator must be kept within a certain range. Low velocities, for a given firing rate, result in excessive vaporization, DNB, and tube failure. On the other side, high feed rates result in low-quality steam and tube erosion.

EFFECT OF WATER IMPURITIES

The quality of the feedwater employed for conventional steam boilers is frequently critical to their operation if corrosion and scale formation are to be avoided. Table 8.2 gives specifications for feedwater and for the water within the boiler that have been recommended by the ASME Committee on Water in Thermal Power Systems. These recommendations are for typical water tube boilers; oil field steam generators are much more tolerant because of the lower heat flux.

TABLE 8.2 Recommended Feedwater Guidelines for Modern Industrial Water Tube Boilers for Reliable Continuous Operations (from Kirk-Othmer 1978–84)

	Boiler feedwater			Boiler water		
Drum pressure MPa[1]	Iron, ppm Fe	Copper, ppm Cu	Total hardness, ppm $CaCO_3$	Silica ppm SiO_2	Total alkalinity[2], ppm $CaCO_3$	Specific conductance, $\mu S/cm$
0.1–2.2	0.100	0.050	0.300	150	700[3]	7000
2.2–3.2	0.050	0.025	0.300	90	600[3]	6000
3.2–4.2	0.030	0.020	0.200	40	500[3]	5000
4.2–5.3	0.025	0.020	0.200	30	400[3]	4000
5.3–6.3	0.020	0.015	0.100	20	300[3]	3000
6.3–7.0	0.020	0.015	0.050	8	200[3]	2000
7.0–10.4	0.010	0.010	0.000	2	0[4]	150
10.4–13.9	0.010	0.010	0.000	1	0[4]	100

[1]To convert MPa to psi, multiply by 145.
[2]Minimum level of OH alkalinity in boilers <6.9 MPa (1000 psi) must be individually specified with regard to silica solubility and other components of internal treatment.
[3]Alkalinity not to exceed 10% of specific conductance.
[4]Zero in these cases refers to free sodium or potassium hydroxide alkalinity. Some small variable amount of total alkalinity is present and measurable with the assumed congruent control or volatile treatment used at these high pressures.

The control of dissolved oxygen is necessary in order to avoid corrosion.[2] When oxygen is present at appreciable concentrations, it reacts with steel to form ferric oxide; this does not protect the steel; and corrosion is excessive. In the absence of oxygen, a protective layer of magnetite is formed.

$$3\ Fe + 4\ H_2O = \underset{\text{Magnetite}}{Fe_3O_4} + 4\ H_2$$

It is recommended that the oxygen concentration should be kept below 0.04 ppm up to 3 MPa pressure and below 0.007 ppm for pressures up to 14 MPa.

Appreciable hardness (dissolved salts of calcium and magnesium) is disastrous in boiler feedwaters because these compounds deposit scale as a result of their inverse solubility curves. The solubilities of calcium and magnesium hardness compounds decrease with increasing temperature and these compounds, notably $CaCO_3$ and $Mg(OH)_2$, tend to precipitate where the temperature is highest—i.e., at the heating surface.

Silica is undesirable in boiler feedwater because of its tendency to form scale deposits consisting of inorganic silicates. However, with oil field steam generators it

[2]A common potential problem is the boiler corrosion that can occur when steam generators are shut down and idle. Precautions must be taken to ensure that water containing air does not come in contact with the boiler internals during such periods. Steam generator manufacturers will supply recommended guidelines for boiler layup.

is possible to tolerate considerable dissolved silica provided that the water is soft[3] (so that the formation of calcium and magnesium silicate compounds is avoided), the pH is alkaline (so that the silica can remain in solution as silicate ions), and the iron content is very low (so that iron silicates are not formed). It is necessary to maintain the feedwater alkaline and oxygen free so that iron solution is minimized. Iron should be considered as a hardness ion and treated as such.

In conventional boilers dissolved solids become concentrated in the water within the boiler and they are controlled by the blowdown of water. Maximum concentrations of dissolved and total solids recommended by the ABMA (data from Delibert 1987) are given in the following table:

TABLE 8.3

Steam pressure psig	Total solids ppm	Suspended solids ppm
0–300	3500	300
301–450	3000	250
451–600	2500	150
601–750	2000	100
751–900	1500	60
901–1000	1250	40
1001–1500	1000	20
1501–2000	750	10
2001+	500	5

In once-through oil field steam generators, where the solids dissolved in the feedwater leave concentrated in the liquid part of the wet-steam product, the concentrations of dissolved solids are typically much higher than those given in the preceding table.

DEAERATION AND OXYGEN CONTROL

The oxygen concentration in all boiler feedwater must be minimized to reduce corrosion. It is usual to do this by first heating and stripping the water with live steam in a *deaerator* to remove dissolved gases.

Figure 8.5 shows a schematic diagram of a spray-tray deaerator. It consists of a countercurrent scrubbing system in which low-pressure steam contacts the feedwater in two stages. Much, but not all, of the steam is condensed in the top spray chamber, where the incoming feedwater is raised to its boiling point.[4] Most of the

[3]The hardness should be less than 1 ppm and preferably less than 0.5 ppm measured as $CaCO_3$.

[4]In many boilerhouses, the steam used in deaeration comes from the exhaust from steam turbines driving the feed pumps and other boilerhouse auxiliaries. Using steam-driven pumps can make the boilerhouse a dependable source of steam even in the event of a failure of the electrical power supply.

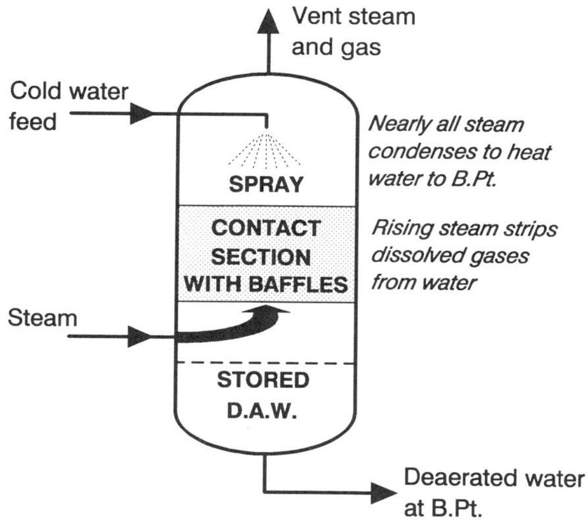

Figure 8.5 Diagram of a Spray-Tray Boiler Feedwater Deaerator

remaining dissolved gas is stripped by contact with the incoming steam in the lower stage. This section often contains trays or baffles.

The residual traces of oxygen in the water coming from the deaerator are controlled by the addition of a chemical that will react with the oxygen to convert it into a harmless product. With low-pressure boilers (below about 1000 psi, or 7 MPa) it is common to employ sodium sulphite as the reactive chemical. The oxygen is consumed by the oxidation of this compound to sodium sulphate.

$$\underset{\text{Sodium sulphite}}{2\ Na_2SO_3} + O_2 = \underset{\text{Sodium sulphate}}{2\ Na_2SO_4}$$

At higher pressures (and, hence, higher steam temperatures) the use of sodium sulphite can present a problem because it decomposes to form hydrogen sulfide and sulphur dioxide. In conventional drum-type boilers, it is common to maintain a residual level of sodium sulphite in the water to ensure the removal of oxygen; typically a level of about 40 ppm is maintained. Above 1000 psi (546°F), this practice is unsuitable because of the decomposition of the sodium sulphite. In these cases, it is the practice to use the chemical hydrazine as the oxygen scavenger. Hydrazine reacts with oxygen to form nitrogen and water.

$$\underset{\text{Hydrazine}}{N_2H_4} + O_2 = N_2 + 2\ H_2O$$

This approach can be and also has been used with oil field steam generators. However, there is a problem because of the carcinogenic nature of hydrazine and the hazards that this involves. One solution is to use chemicals that are safe to handle but that liberate hydrazine within the boiler.

Another approach is to use sodium sulphite by adding it well upstream of the generator, which also protects the water-treating equipment from corrosion, and to adjust the quantity so that there is only a very small residual amount within the generator (less than 5 ppm). With this technique, the sulphite is converted to sulphate (which is thermally stable) before it enters the steam generator, and thus the sulphite is not exposed to high temperatures.

OIL FIELD STEAM GENERATORS

Figure 8.6 is a diagram of a typical horizontal oil field steam generator. Feedwater under pressure is usually preheated in a heat exchanger to a temperature high enough to prevent condensation of water from the flue gas on the exterior of the tubes in the convection section. With hot recycle water feed, preheating is unnecessary.

Condensation is avoided to prevent corrosion of the tubes. In the convection section, the flue gases exchange heat with the water. The partially heated feed passes back through the exchanger and then to the heating tube that surrounds the combustion zone. If the feedwater is preheated by steam deaeration or otherwise, then the preheat exchanger is usually omitted.

In the radiant section heat transferred to the water boils about 60 to 80% of the water to give the steam product. The horizontal cylindrical furnace has, typically, a length-to-diameter ratio of 3:1 to 4:1.

The figure also shows a typical heat and material balance for an oil field steam generator. A generator having a duty of 50 million Btu/h, such as that shown, is the largest that has been employed in oil field practice until recently, when units having about three times this capacity were placed in service by Esso Resources Canada in their field at Cold Lake (Peachey 1984).

The unit shown in Figure 8.6 has a single 3-in.-diameter boiler tube; the larger units recently built by CE-Natco and Struthers-TIW for Esso have three 3-in.

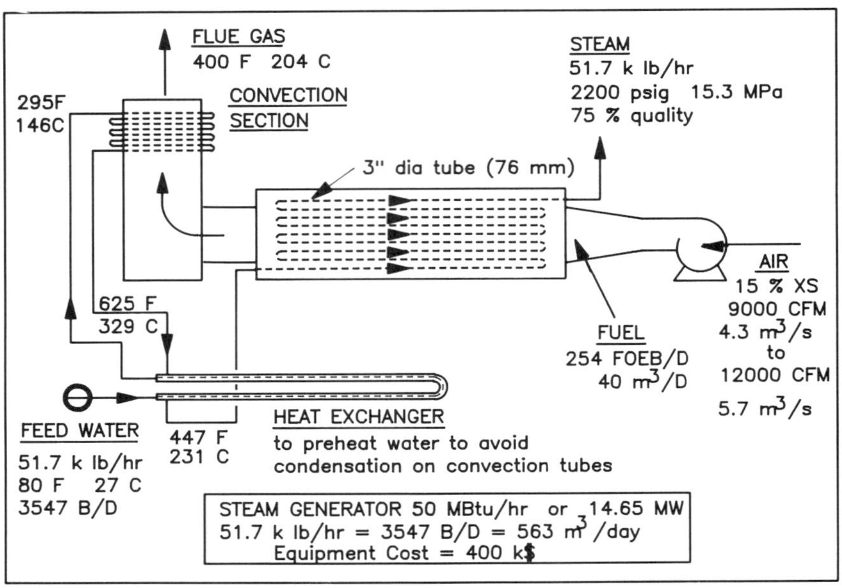

Figure 8.6 Steam Generator 50×10^6 Btu/h (14.675 MW)

tubes in parallel. Larger units are possible, provided that it is not required to transport the fabricated equipment along normal highways; it is this transportation limitation that has determined the size of the Esso units.

The generator shown in Figure 8.6 produces 3547 B/d of 75% quality steam. For an oil-steam ratio of 0.3 (a fairly high value), this would be sufficient to produce somewhat over 1000 B/d of oil; the generator requires the equivalent of about 235 B/d of heavy oil as fuel.

Generators of this general type burn either gas or oil as fuel. In California, when oil (usually the produced crude) is employed as fuel, it is usual to scrub the flue gases to remove sulphur dioxide. Typical installations use caustic soda or sodium carbonate solutions for scrubbing; however, this is costly and troublesome. There is a growing tendency to use natural gas as fuel as this reduces the environmental protection problems. The removal of SO_2 from stack gases by scrubbing with produced water (containing added alkali if required) is described by Snavely and Bertness (1975). Wendt (1978) has reviewed the operation of a scrubber using sodium hydroxide solutions.

Horizontal oil field generators are the most common types used. A general cutaway view of a typical horizontal generator is given in Figure 8.7.

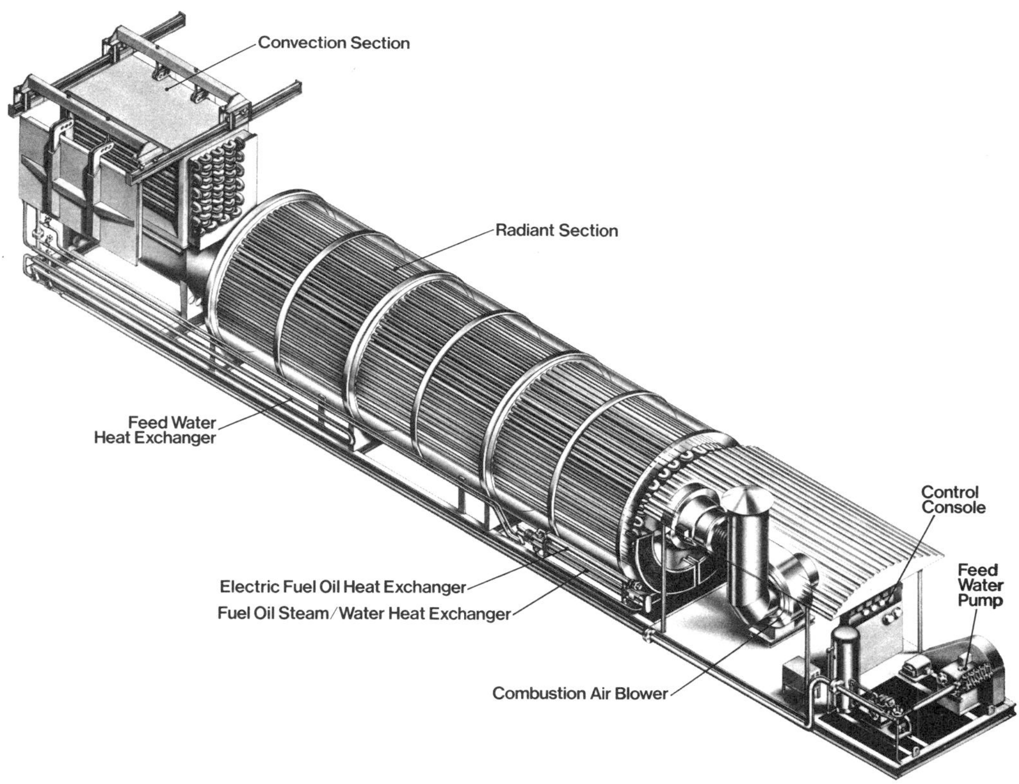

Figure 8.7 Cutaway Diagram of Horizontal Steam Generator (courtesy Natco)

Figure 8.8 Photograph of Part of a Natco Horizontal Steam Generator with a Capacity of 160,000 lb/h Being Transported to BP's Thermal Recovery Site in the Cold Lake Oil Field (Courtesy of Totran Services Limited, Calgary and Edmonton)

Figure 8.8 is a photograph of the main section of a 160,000 lb/h steam generator being transported to a pilot site in the Cold Lake oil field.

The technology of once-through steam generators has been reviewed by Fanaritis and Kimmel (1965). Kirby, Kense, and Peachey (1984) reviewed the means for improving and maintaining the efficiency of oil field steam generators in a very practical paper.

The need to prevent condensation on the convection section tubes in order to avoid corrosion was mentioned earlier. The dew point of the combustion gases depends upon the sulphur content of the fuel—larger sulphur contents result in higher dew points.

The main reason for the effect of fuel sulphur content on the dew point is the presence of SO_3 in the flue gas. This is extremely soluble in water and forms sulphuric acid. In contrast, SO_2, which is the main sulphur oxide in flue gas, is much less soluble and has little effect on the dew point. The effect of SO_3 on the dew point can be predicted by a correlation given by Verhoff and Banchero (1974):

$$(1/T_{DP}) = 0.002276 - 0.00002943 \ln(p_{H_2O}) - 0.0000858 \ln(p_{H_2SO_4}) + 0.00000620 \ln(p_{H_2O}) \ln(p_{H_2SO_4})$$

where T_{DP} is the dew point in K and the partial pressures, p, are in mm Hg.

The problem in using this correlation is that the distribution of fuel sulphur between SO_3 and SO_2 is variable.

The following recommended minimum wall temperatures to avoid condensation in economizers burning fuel oil have been read from a chart given by Babcock and Wilcox (1972):

Wt% Sulphur in fuel oil:	0.1	2	3	4	6
Minimum temperature °F:	240	240	249	253	260

Steam Quality

One of the measurements which should be made in the control of a steam generator is that of the quality of the steam being produced. A common method of doing this is to measure the chloride content of the feedwater by chemical analysis and then to analyze a sample of the condensate in the steam (see Figure 8.9). Sodium can be used as the tracer element instead of chlorine using atomic absorption techniques for analysis. The steam quality is calculated by assuming that the chloride or the sodium is nonvolatile.

In taking a sample of the liquid associated with the steam, the valve should be cracked open enough to maintain a flow at a rate that will allow only condensed water to leave the system. It is necessary to cool the collected sample to avoid evaporation. These sample systems work best if they are left to drip continuously between the collection of samples.

Another way to measure the quality of the steam leaving a generator is to employ an orifice flowmeter (Palm, Anderson, and Kirkpatrick 1968; Wilson 1975). For a given mass flow the reading from an orifice meter is strongly dependent upon the quality of the steam. The quality of the steam may be determined from this reading and the water feed rate to the generator.

A common method of adjusting the quality of the produced steam involves the control of the ratio of fuel flow to feedwater flow. A practical arrangement involves a ratio controller, which measures the feedwater flow and controls the fuel flow proportionally.

Convection Section

Figure 8.10a shows a schematic diagram of a convection section, or economizer, for a typical steam generator. The tubes, except for the lower couple of rows, are usually finned in order to improve the heat-transfer coefficient. The lower tubes are usually not finned because with the very high gas temperatures found in this section, overheating of the fins would otherwise occur. Also because of the high temperature at the bottom of the convection section, the gas velocity tends to be high, and the elimination of the fins can be desirable for pressure-drop reasons.

The convection section, or economizer, plays a very important role in determining the thermal efficiency of the steam generator. It is the unit that recovers heat from the flue gas on its way to the stack. It is essential to keep the finned tubes clean. Many early designs of steam generators had undersized economizers

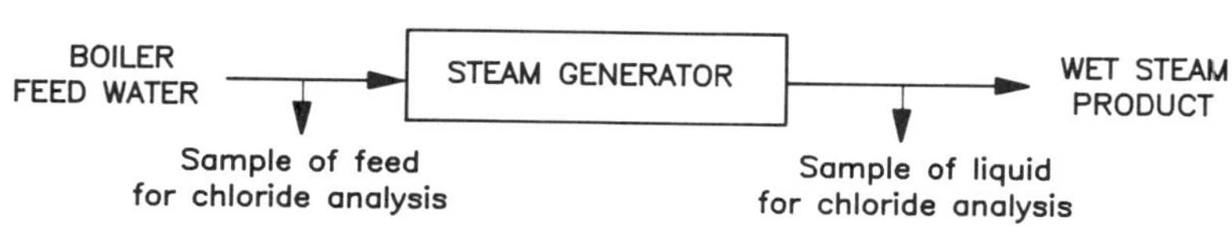

$$\% \text{ Quality} = \left(1 - \frac{\text{ppm Cl in feed}}{\text{ppm Cl in condensate}}\right) \times 100\%$$

Figure 8.9 Steam Quality from Water Analyses

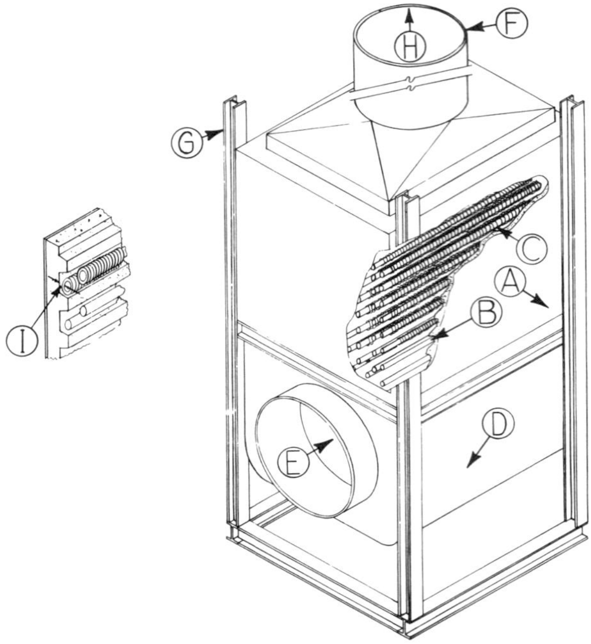

Figure 8.10 Steam Generator Construction
a) Field Steam Generator—Convection Section
A. Convection box
B. Bare shield tubes
C. Finned tubes
D. Hog trough
E. Flue gas inlet from radiant section
F. Stack
G. Girder
H. Flue gas exit
I. One-half tube diameter between tube and side wall.
(Courtesy of Foster Wheeler Fired Heaters Ltd., Calgary)

and wasted heat in the flue gas. A well-designed and well-operated economizer should be able to cool the flue gas to about 340°F. Some early units, after years of operation, gave stack gas temperatures of 650°F to 800°F even though they were producing steam satisfactorily. A decrease in stack gas temperature of 100°F is

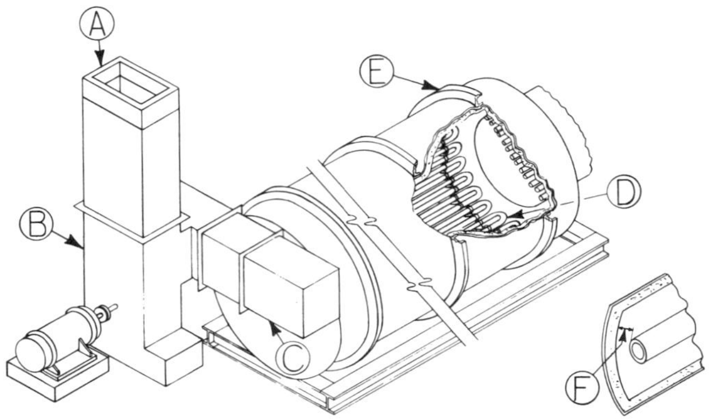

Figure 8.10 b) Field Steam Generator—Radiant Section
A. Air inlet duct
B. Forced draft fan
C. Burner
D. Radiant coil
E. Casing stiffener
F. One tube diameter spacing between tube and wall insulation.
(Courtesy Foster Wheeler Fired Heaters Ltd., Calgary)

equivalent to a fuel savings of about 3% (Kerby, Kense, and Peachey 1984). These authors give an excellent discussion of means for improving the efficiency of steam generation.

Radiant Section

Figure 8.10b shows a section through the radiant section. The tubes run longitudinally backward and forward around the perimeter of the cylindrical combustion chamber with a spacing between centers of about 6 in. They are connected sequentially in a single pass (except in the new large Natco and Struthers-TIW units, where there are three or four parallel passes).

The tubes are supported by hangers away from the insulated wall of the combustion chamber. With this construction, the backs of the tubes away from the fire receive considerable energy by reflection from the hot insulation.

The burner for the generator is mounted axially and produces a long flame, which passes down the center of the chamber. It is important to keep the burner clean and in adjustment so that direct impingement of the flame on the tubes, with its accompanying high local heat flux, will not occur.

Vertical Steam Generators

Another style of generator that has found popularity, particularly with Shell, is the vertical Mitchell Engineering (ME) generator manufactured by Babcock and Wilcox. This equipment was developed for Shell's operations in Venezuela and is also used at their thermal recovery pilot at Peace River, Alberta.

As well as conventional fluid-fueled units, Babcock and Wilcox also offer a coal-fired unit.

The main distinguishing features of these generators is that the firing is vertically upwards with a single burner. The boiler tube is in the form of two concentric, closely-spaced helical coils surrounding the flame. The inner coil absorbs direct radiation from the flame. The combustion gases from the top of the combustion chamber are directed downward between the two coils and then upward, past the outside of the outer coil; the outer coil forms the convection section.

The symmetrical design of the vertical steam generator is an attractive feature that contributes to even heating and stressing of the boiler tubes. Also, the symmetrical location of the burner pointing upward gives easier alignment and reduced flame impingement.

STEAM DISTRIBUTION SYSTEM

It is normal practice in larger projects to use a number of oil field steam generators grouped together to provide sufficient capacity. In California, one of the reasons for doing this is to allow the flue gas from several generators to be collected so that it can be treated in a single scrubber.

A factor of general significance is that the operation of the system can be centralized and combined with the operation of central water-treating facilities. Less worker time is required to operate a centralized facility.

It is normal to provide separate oil and steam lines and to run these to individual wells or groups of wells from the central plant(s).

A feature of steam distribution systems is the need to allow for the thermal expansion of the pipeline. Figure 8.11 shows the expansion loops that were used in Esso's Leming pilot and a more recent design.

The measurement of the flow of wet steam using orifice or flow nozzle meters is discussed by Miller (1983).

A problem associated with the distribution of steam in thermal projects is that of the division of liquid and vapor flow at pipe junctions. There is always a tendency for uneven division; as a result, some wells will tend to get steam of a different quality from others (Hong 1978; Saba and Lahey 1984).

This problem has been widely recognized, but there are, as yet, no generally accepted solutions. Gulf Canada developed a dividing system that has been reported to give improved results, but it is complicated and has not yet found general use.

Another system for splitting wet steam flows has been described by Konak (1985 and 1986). The principle is shown in Figure 8.12.

This method depends on the observation that, under the usual steam-flow conditions, the water tends to flow as an annulus just within the perimeter of the pipe. If the pipe is horizontal, this water ring tends to be thicker at the bottom. As a result, a line connected into the bottom of the main pipe tends to produce wetter steam than the average, and a top connection, drier steam. In the Konak device the two streams are blended to give the desired quality. This can be automated by using a steam-quality meter downstream.

A very promising development is the neutron-scattering steam-quality meter that was described by Lim of Esso Resources (Lim 1985). An americium/berylium source is placed beside the steam pipe; this emits fast neutrons, which penetrate the pipe. Neutrons are scattered by hydrogen atoms much more than by other elements. The scattered neutrons that leave the pipe at right angles are measured, and their flux is an indication of the quality of the steam. A larger flux of scattered neutrons indicates a higher proton density, i.e., a lower steam quality. Meters operating on this principle have been constructed, and Lim describes interesting and promising results. An instrument that depends upon the same principle has been described by Woiceshyn et al. (1985).

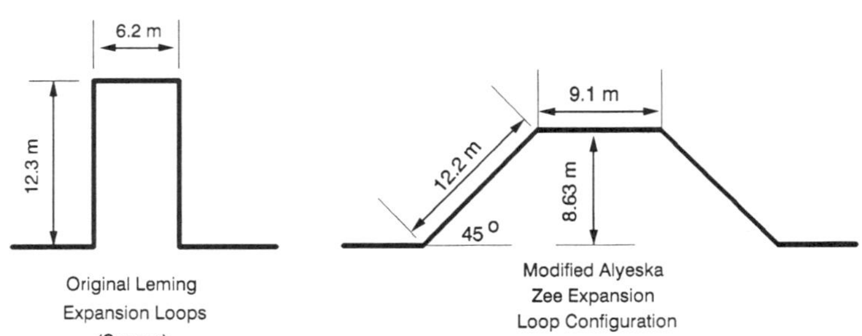

Figure 8.11 Cold Lake Steam Line Expansion Loops (after Peachey and Nodwell 1981)

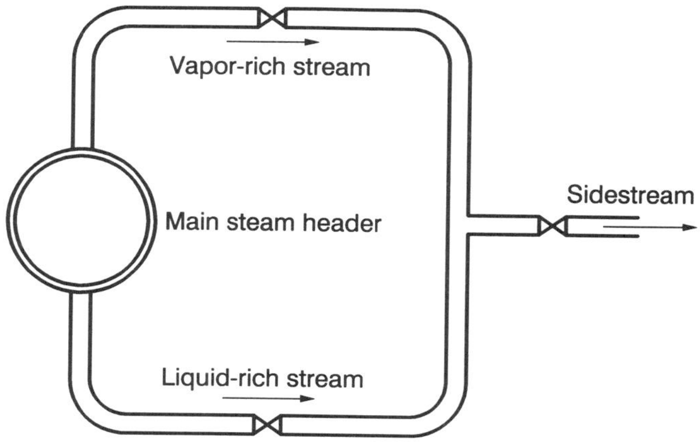

Figure 8.12 Esso's Wet Steam Splitter (after Konak 1985)

CLUSTERED DEVIATED WELLS

A growing trend in steam projects is to drill deviated rather than vertical wells and to locate the well heads in groups in close proximity to each other. This practice was developed by Esso Resources in their Leming pilot at Cold Lake. In this pilot the well completions formed hexagonal patterns, with a distance between adjacent wells of 500 to 600 ft.

The well layout for the original part of the Leming pilot is shown in Figure 8.13. The diagram shows the location of the completions of the wells in the reservoir. The heads of the wells and the pumps are located in groups of 7 at the centers of the hexagonal cells (except for the J pattern, which involved 13 wells arranged in two concentric hexagons).

Subsequent expansions of the Leming pilot have involved the same principle of clustered, deviated wells. Figure 8.14 is a perspective view of such a pattern design.

The use of clustered wells greatly reduces the length and complexity of the above-ground facilities—roads, steam and production lines, and power lines. This also reduces the area that must be cleared and thus makes the project more acceptable environmentally. Also, the manifold facilities can be located more flexibly. This is important in areas where much of the ground is muskeg.

Figure 8.15 shows the effect of varying the number of wells per satellite on the drilling cost and on the overall cost of the well facilities. The costs were found to be relatively insensitive to the number of wells per satellite, at least for satellites having 12 wells or more.

THERMAL WELL-COMPLETIONS

A typical completion for Cold Lake wells is shown in Figure 8.16. The 7-in. casing is cemented to the surface with 50% silica flour thermal cement. The producing zone is perforated with six shots per meter; an insert pump and $3\frac{1}{2}$-in. tubing are

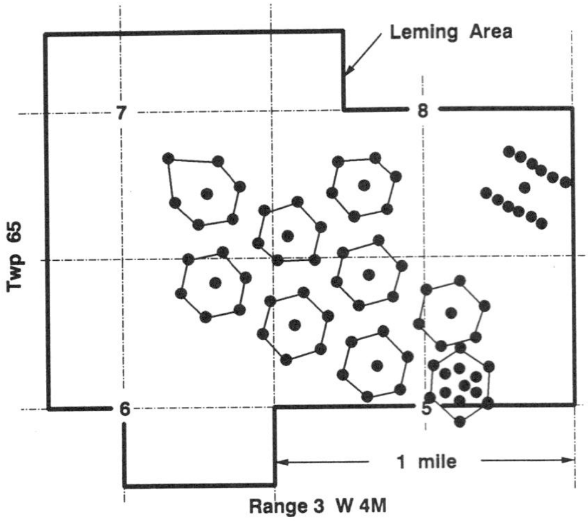

Figure 8.13 Bottom Hole Location for Esso's Leming Pilot (after Buckles 1979)

employed. A 5-in. slotted liner is placed inside the perforations to keep sand from the pump.

During injection, steam flows down the annulus between the casing and tubing. It would be preferable to inject down the tubing to keep the casing cooler, but this is not done by Esso Canada because of the difficulty presented by the high-temperature packer that would be required to isolate the casing. One of the problems in using thermal packers in steam stimulation is the need to bypass the packer

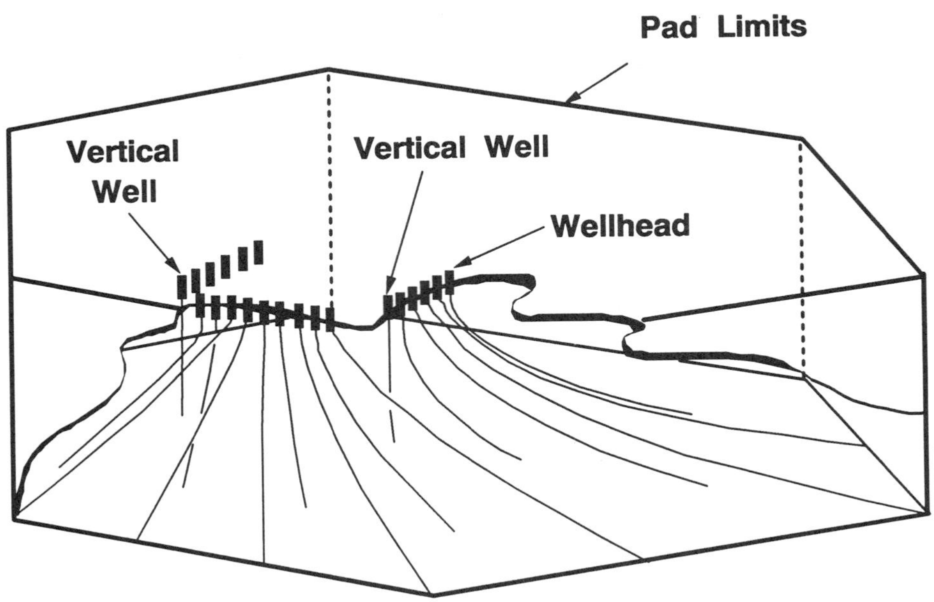

Figure 8.14 Esso Design for a Cold Lake Steam Stimulation Well Pad (after Peachey and Nodwell 1981)

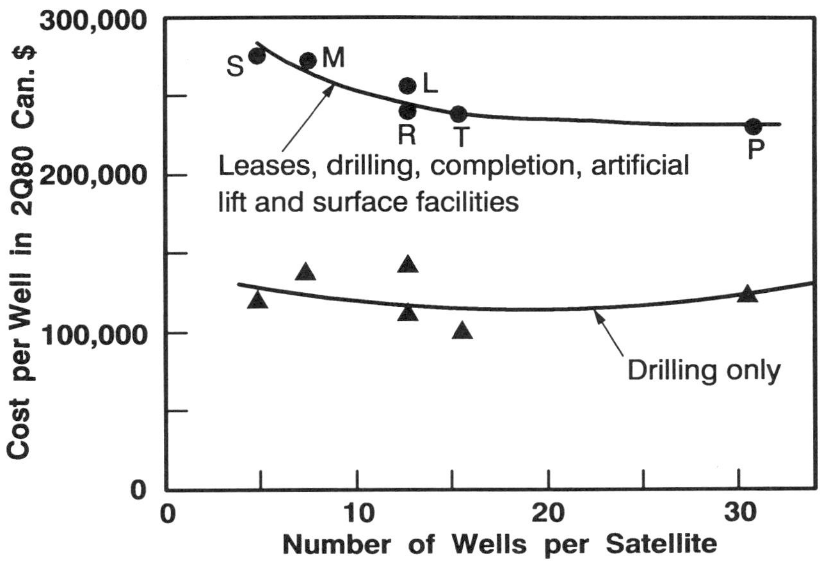

Figure 8.15 Unit Well Cost at Cold Lake. Letters Identify Particular Satellites (after Peachey and Nodwell 1981)

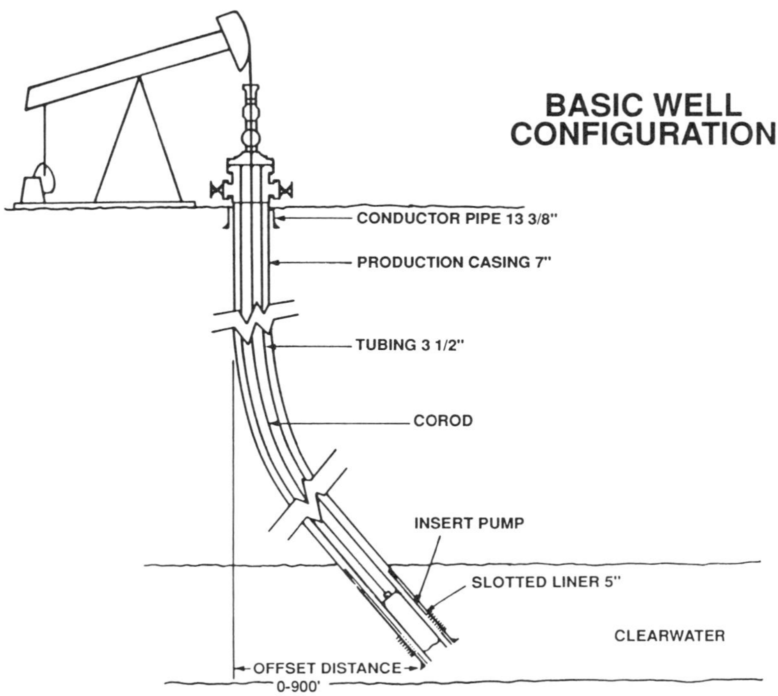

Figure 8.16 Cold Lake Steam Stimulation Wells (after Peachey and Nodwell 1981)

so that gas can be vented to the annulus during the pumping period (Gates and Holmes 1967).

The coefficient of thermal expansion of steel is about 7×10^{-6} per degree Fahrenheit of temperature change. Thus, for example, if the temperature of a casing that is 1500 ft (457 m) in length is raised by 600°F (333°C), then the increase in length would be 6.3 ft (1.92 m) if it were unconstrained. There are three possible solutions to this problem:

1. Use high-strength casing and restrain the movement by cementing the well throughout its length. In this example, the compressive stress would be

$$S = \left(\frac{\Delta L}{L}\right)E = \left(\frac{6.3}{1500}\right)(29 \times 10^6) = 122{,}000 \text{ psi}$$

   where E is Young's modulus. In cases such as this the yield point of the steel will be exceeded and the steel will yield under the compressive stress. As a result, if the well is cooled later, it will be under tension.
2. Prestress the casing of the well during cementing by applying an initial tension to offset the compressive stresses that will be caused later by heating. To do this, a quick-setting cement mixture is placed next to the bottom 10% of the well, with a slower-setting mixture above. Tension is applied to the well by pulling with the rig after the lower cement has set but before the upper material hardens.
3. Allow the casing to expand and the wellhead to rise. One way of doing this involves covering all the pipe except for the lowest 10% of its length with bitumen. The wellhead is allowed to rise above the surface of the ground by using swivel connections ("Chicksans") similar to those used in loading racks. This technique is not used anymore.

Peachey and Nodwell no longer favor the use of the extreme high-strength steels that would be necessary to avoid inelastic yield, since these materials are susceptible to catastrophic failure due to sulphide stress-corrosion cracking. Instead they believe that L-80 or MN-80 casing with special collars is most suitable. The inelastic yield of the casing material that occurs on heating seems to be acceptable. It is important that the well cement should provide essentially complete radial support to the casing to reduce the tendency of buckling under the compressive stress.

A survey of thermal well completion practices (Farouq Ali and Meldau 1979) is reproduced in Table 8.4. A general review of thermal well completions is given by Gates and Holmes (Gates and Holmes 1967).

The design of steam-injection wells is discussed in a paper by Willwhite and Dietrich (1967).

TEMPERATURE LOGGING

Techniques for logging the temperature profile in thermal recovery wells were discussed by Leschyshyn and Seyer (1989). They found that consistent measurements

TABLE 8.4 Thermal Well Completion Practices

Test	Field	Operator	Depth (ft)	Maximum Steam-Injection Pressure (psig)	Casing Size (in.)	Casing Weight (lbm/ft)	Casing Grade	Casing Type Coupling	Prestress
1	Kern River	Getty	900	350	7	23	K-55	STC	No
2	Midway-Sunset	Chevron	1,200	450	8⅝	36	K-55	LTC	No
3	Cold Lake	Esso	1,500	2,000	7	23	SOO-95[(1)]	BUTT	Yes
4	Tia Juana	Maraven	1,600	1,500	9⅝	43.5	N-80	BUTT	No
5	Mount Poso	Shell	1,800	500	9⅝	36	K-55	STC	No
6	San Ardo	Texaco	2,350	800	8⅝	32	K-55	STC	No
7	Guadalupe	Union	3,000	1,800	7	26	N-80	BUTT	No
8	Cat Canyon	Conoco	3,500	2,500	7	23	SOO-95	BUTT	Yes
9	Morichal	Lagoven	3,500	1,500	7	23	N-80	BUTT	Yes

Test	Field	Operator	Annulus During Cyclic	Liner Length (ft)	Liner Size (in.)	Liner Weight (lbm)	Liner Grade	Liner Type Coupling
1	Kern River	Getty	Steam	None	—	—	—	—
2	Midway-Sunset	Chevron	Steam	400	6⅝	28	K-55	BUTT
3	Cold Lake	Esso	Steam	200	5½	15½	H-40	FJ
4	Tia Juana	Maraven	Vent	100	7	29	N-80	BUTT
5	Mount Poso	Shell	—	100	7	23	K-55	STC
6	San Ardo	Texaco	Vent	150	6⅝	28	K-55	SF
7	Guadalupe	Union	Gas pack	100	5½	17	N-80	BUTT
8	Cat Canyon	Conoco	Steam	400	5½	17	N-80	X-Line
9	Morichal	Lagoven	Crude	250	3½	13	N-80	FJ

[(1)]L-80 or MN-80 has been used more recently (Peachey and Nodwell 1981).

(Farouq Ali and Meldau 1979)

can be made if the temperature measuring tool is at the bottom end of the logging string. The design of the logging tool is discussed in their paper.

When they employed a conventional, "stackable" temperature-logging tool in which the sensing element was contained within an open section in the center of the tool, it was found that the logs obtained were variable and depended on the direction of logging (up or down) and on the rate of movement of the tool. This problem was caused by the thermal capacity of the fluid that was carried along by the tool.

CONTROL OF HEAT LOSS IN STEAM-INJECTION WELLS

The effect of using the tubing for injection of steam with insulation between it and the casing was discussed in Chapter 2. This is desirable from the point of view of reducing heat losses, providing higher-quality steam at the sandface and reducing the mechanical stresses within the casing. The latter is a particularly important factor if it is planned to employ existing wells that are not designed for thermal operation in a thermal project. The current practices in reducing well bore heat losses have been reviewed by Meldau (1988).

In many cases, because of the complication of using a thermal packer and because of the need to vent gases up the annulus between the tubing and the casing during cyclic production, operation with no insulation has been used; the casing must be designed to operate at the steam temperature. This approach is common in the Cold Lake area.

Some operators inject steam down the tubing and inject a small amount of gas, either natural gas or nitrogen, down the annulus in order to prevent heat transfer by refluxing in the annulus—i.e., by liquid water boiling on the hot tubing and condensing on the casing. This has been done both with bare tubing and, more recently, with insulated tubing (Meldau 1988; Cormier 1987).

Insulation by isolating the annulus with a packer and venting the annulus is often used as a simple form of insulation in steamfloods; this is common practice in California. If the annulus is vented, any steam leaking at the packer or couplings passes through the annulus without condensation.

The use of insulated tubing is quite practical as a means of saving heat and reducing casing temperature. However, it is expensive; costs are of the order of $25 to $35 U.S. per foot (Meldau 1988).

The general arrangement for the use of insulated tubing is shown in Figure 8.17.

As is shown in Figure 8.18, insulated tubing is constructed with a hollow wall with the annulus filled with layers of foil (to reduce radiation losses) and ceramic fiber. The insulation space is evacuated. A major problem with earlier versions of insulated tubing was the heat loss that occurred at the couplings between sections of the insulated tube. The problem is aggravated by *refluxing*; this involves the boiling of water in contact with the hot coupling followed by its condensation on the casing. The heat loss has been greatly reduced by lining the couplings with insulation. Typical construction is shown in Figure 8.18.

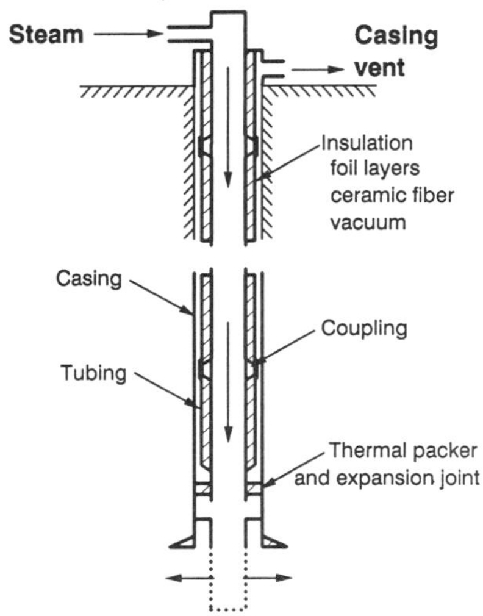

Figure 8.17 Use of Insulated Tubing for Steam Injection (after Meldau 1988)

SELECTIVE STEAM INJECTION

Several means have been developed to allow the selective injection of steam into specific horizontal layers. Borregales (1977) and Burkill (1977) describe methods that have been used in Venezuela. These allow the partial plugging of the gravel pack with a cement material that is forced through a special *port collar* (see Figure 8.19). This equipment allows ports in a blank section of the liner to be opened so the plugging agent can be squeezed into the gravel.

The use of specially sized and placed perforations to allow selective injection of steam during a steamflood is described by Gates and Brewer (1975).

ARTIFICIAL LIFT

Artificial lift is necessary in both steamflooding and cyclic steam projects. In cyclic steam production, reservoir pressure drives the fluids up the well initially, but pumping is required when the reservoir pressure falls.

Conventional pump jacks with tubing insert pumps[5] are generally employed for lifting oil from thermal wells (Peachey and Nodwell 1981). The pumps have a metal-to-metal piston seal that will withstand high temperatures in place of the conventional nipple seals. Longer and slower strokes are employed than in normal practice. The construction of a typical pump is shown in the diagram in Figure 8.20.

The wear of the couplings used with conventional sucker rods has presented a significant problem, particularly when pumping deviated wells. The effect of this wear can be reduced by rotating the rods during operation. This spreads the wear around the perimeter of the couplings rather than allowing it to concentrate at one

[5]An insert pump can be run into the hole as a complete assembly on the end of the sucker rod string. It is seated within a seal that is previously installed in the tubing.

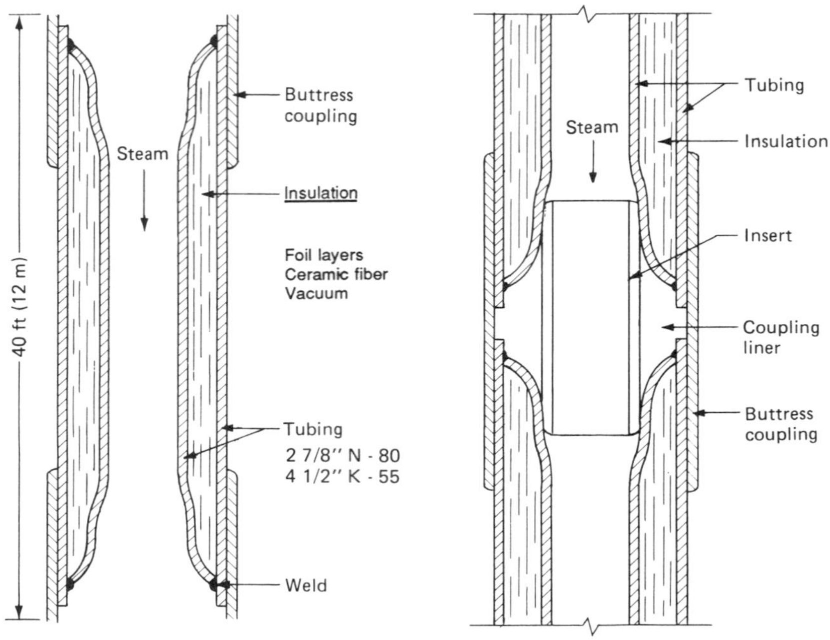

Figure 8.18 Typical Insulated Tubing and Coupling (after Meldau 1988)

location. The rotation is achieved by a mechanical device on the pump jack that imparts a slight rotation during each stroke.

Another approach to reducing coupling wear is to use a type of coupling containing small wheels that roll on the surface of the well tubing. These are available with plastic wheels for operation below 250°F and with steel wheels and journals for high-temperature operation (DDS Calgary).

Another means for alleviating the problem of sucker rod wear involves the use of a continuous sucker rod without couplings (*Corod*).[6] This continuous sucker rod is supplied in long lengths in large-diameter (18-ft) coils, and these are welded into still longer lengths at the site as the rod is installed into the well. Figure 8.21 illustrates the procedure for feeding the rod from the transportation reel into the well.

Another feature of Corod that reduces wear is that it can be made with a flattened rather than a circular cross section. If the oval section is used, it is chosen so that the flattened faces have approximately the same radius of curvature as the tubing in which it is installed; this increases the area of contact and reduces the wear.

Pump maintenance is a major source of expense in many heavy oil projects. Elgert, Chambers, and Suzuki (1989) report that at the Esso cyclic steam project in Cold Lake, average pump life was only 200 d, as compared to 1 to 2 y in conven-

[6]Corod® is a registered trademark of Corod Manufacturing Ltd., Nisku, Alberta.

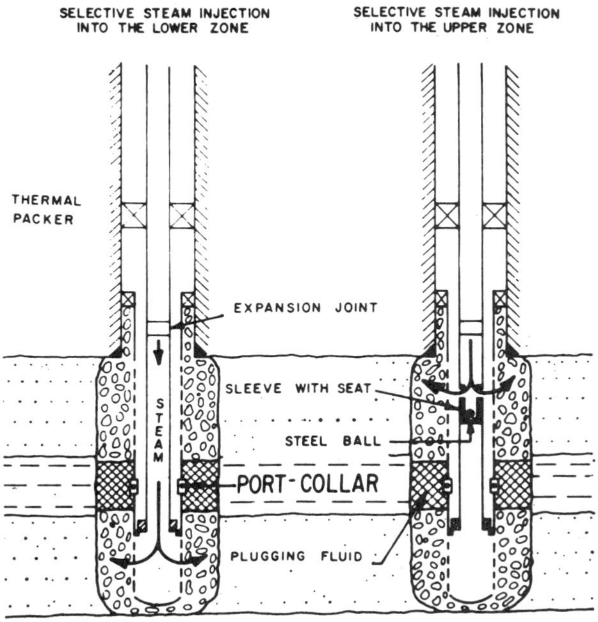

Figure 8.19 Selective Injection of Steam (after Borregales 1977)

tional oil fields. Pump repairs and related service work cost $2 million per year for the 1500 operating wells in the project.

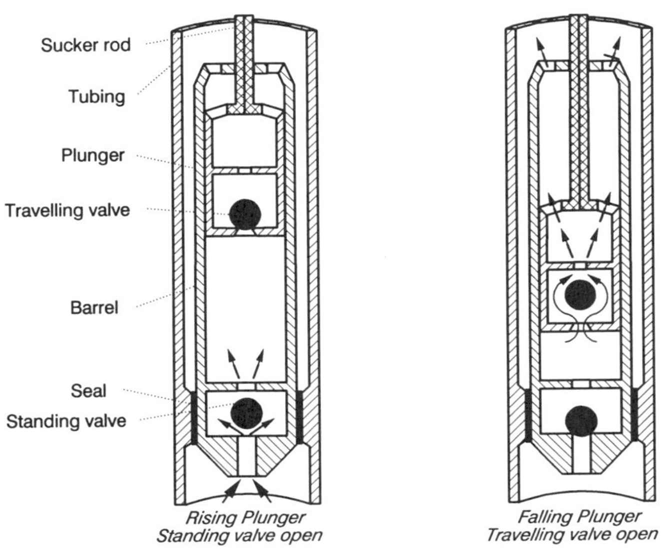

Figure 8.20 Diagram of Tubing Insert Pump

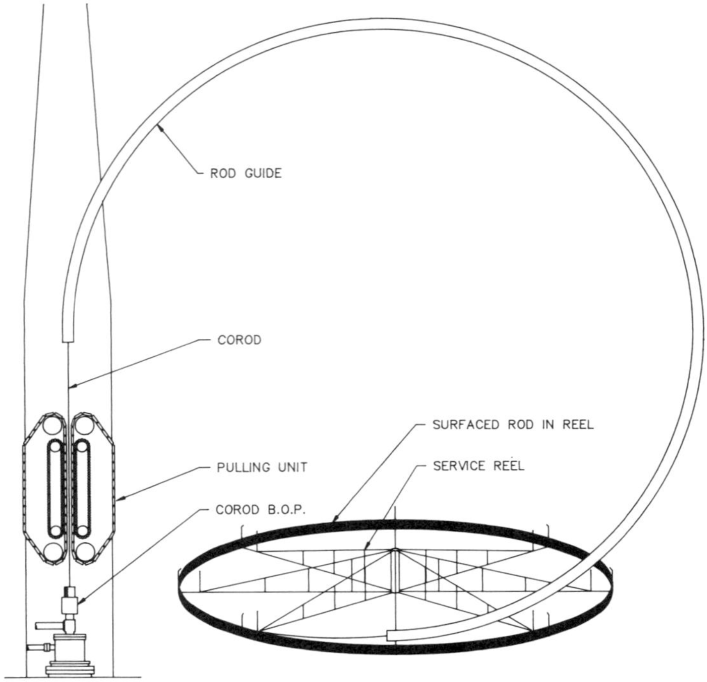

THE COROD SYSTEM

Figure 8.21 The Installation of Continuous Sucker Rod (Courtesy of Corod Manufacturing Ltd.)

Their analysis of the problem indicated that the primary causes of pump failure were related to the following:

1. Sand flowing into the wells, particularly during flowback and particularly during the first cycle.
2. Scales containing $CaCO_3$ and SiO_2 adhering to pump barrels.

These problems have been alleviated by these measures:

1. Throttling the production during initial blowback to reduce sand production. A choke-operating guideline has been developed for well operation in which the well choke is progressively opened as the ΔP across the choke decreases.
2. Using chromium-plated pump barrels. These are resistant to corrosion, and scale does not adhere to the smooth surface. It is not practical to use chromium-plated plungers inside chromium-plated barrels because the two hard materials gall. However, Esso has found that plungers with a sprayed nickel coating work effectively with chromium-plated barrels.

In many thermal recovery projects the production of sand with the oil is a very significant problem. Some reservoirs are particularly prone to this problem.

T. R. Vonde (1979) describes the production of 6° API Cat Canyon crude in California where the average production contains 27 wt% of sand. Some production in this field has contained as much as 70%. To handle this, Husky has developed the special pump shown in Figure 8.22.

This pump has the following features:

1. There are two tubing strings. One carries diluent to the pump and the other carries the diluted production to the surface.
2. The sucker rod is contained within the diluent tubing and does not come into contact with the production stream. This avoids the slow fall that is experienced when the rod must move through the viscous product, and it also protects the rod and plunger from the abrasive action of the sand.
3. The pump delivers fluid to the surface on the downstroke (most pumps pull the fluid upward; this one pushes it). In order to minimize compressive loading in the sucker rod, a counterweight consisting of 2900 to 3850 kg of 2-in.-diameter steel bars is fastened to the end of the sucker rod.
4. The diluent flow is controlled by adjusting the addition rate at the surface. In Husky's application at Cat Canyon, the rate is controlled to give a 12° API product.

The pump is considered successful and, although its cost is more than for a conventional pump, the extra is said to be paid for by the reduced maintenance and the smaller power requirement. It is reported that, in one lease, installing five of the new pumps increased production from 200 to 700 B/d of oil.

The sand problem just described is an extreme case. At the other end of the scale is the experience of Esso at Cold Lake, where sand production is usually very small and is handled by the small section of a slotted liner shown in Figure 8.19. Small amounts of fine solids, which are carried with the crude, tend to accumulate slowly in the separators and tanks, and these require occasional cleaning.

The chamber-lift principle is another means that has been used for pumping heavy oil wells (Elfarr 1976). In this technique, a downhole chamber connected to the production tubing is allowed to fill with produced fluid through a nonreturn valve. After the chamber has filled, the fluid contents are displaced upward through another nonreturn valve, through the production tubing, and to the surface by forcing a gas into the top of the chamber. The gas thus replaces the conventional plunger piston. After the discharge cycle, the pressure is reduced in the chamber by shutting off the gas supply and allowing excess gas to blow to the surface with the rising liquid.

In the chamber-lift system two tubing strings are employed: one for the production and the other to transfer the lifting gas down to the pump chamber periodically. The lift is assisted by the buoyant effect of the exhausted gas from the chamber. The pump has been used successfully in the Slocum field in Texas.

Another lifting technique which has been tried is the use of pumps operated by downhole hydraulic motors. The Canterra-Tenneco in situ pilot in Athabasca has

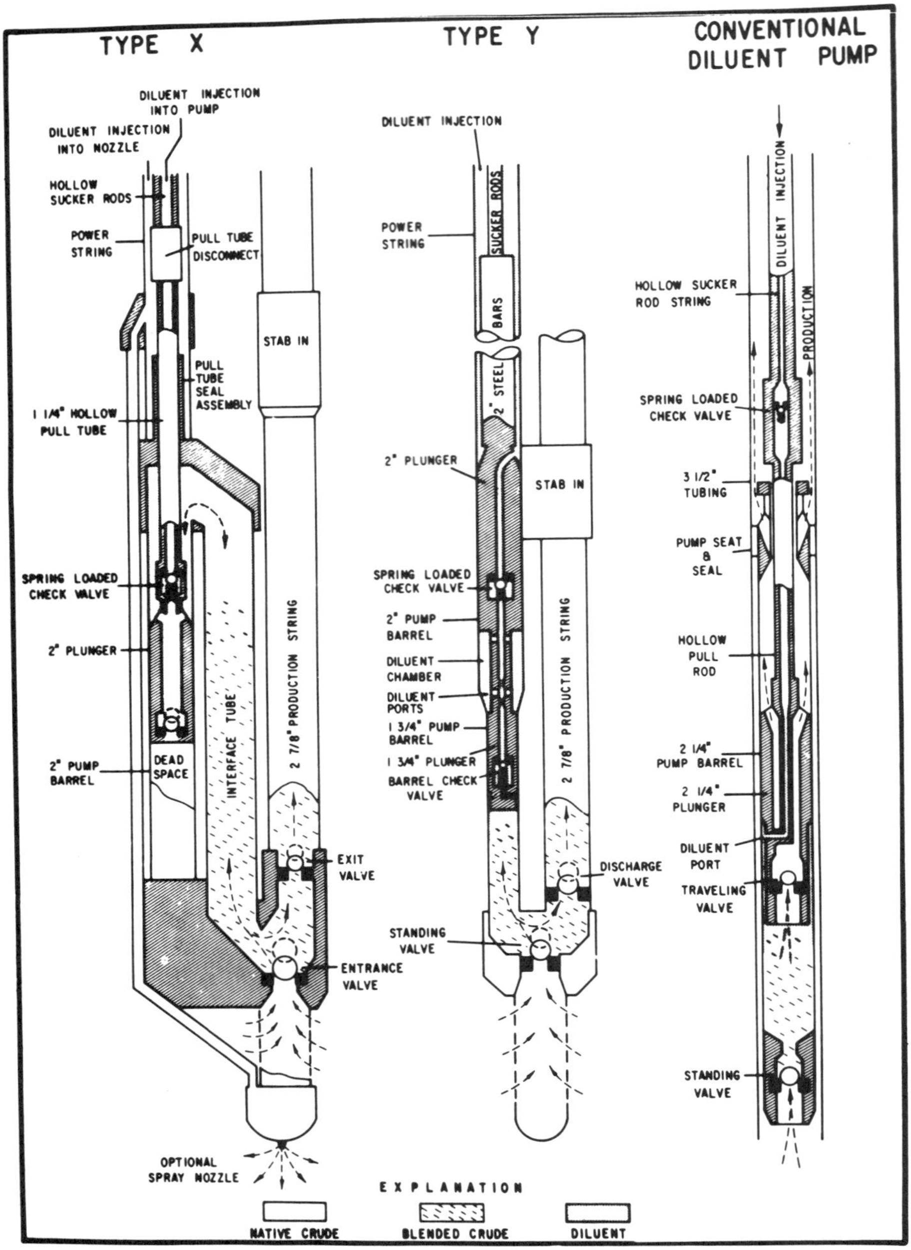

Figure 8.22 Husky Diluent Pump (from Vonde 1977)

used Kobe downhole hydraulic pumps; these have also been used at the Suncor World Wide Energy pilot at Fort Kent.

Another development of interest is the use of drivers, driven by hydraulic cylinders, in place of conventional crank-driven pump jacks. One of these is the

HEP unit, which is manufactured in Calgary by Foremost Engineering; another is the Hydrabeam unit; a third is the Curtis Hoover Hydraulic Pump Jack.

An important potential advantage of this type of driver is its ability to adjust the down- and upstroke timing independently. This is important with heavy oil production where the speed of the downstroke, due to the viscous nature of the oil and the need to avoid compressive buckling of the rod, can be a limiting factor. In some cases there have been problems with hydraulically operated pumping equipment in very cold weather.

IMPROVING WELL PERFORMANCE

In cyclic steam projects, each well is subjected to a series of changing conditions.

Keeling (1985) has discussed means for improving well performance. The temperature and pressure of a typical well during the production cycle in the Esso cyclic steam project at Cold Lake are shown in Figure 8.23.

As production continues, the wellhead temperature and the bottom hole pressure fall as indicated. Initially, the entire production from the well flows, driven by reservoir pressure, through the tubing and a choke to the production flowline. During this period, there is considerable steam in the product. At some point the annulus of the well is connected to the flowline, and the gas flows separately. Shortly after this the production pump is seated and pumping starts. At this point the gas is flowing by itself to the production line, and the oil is being lifted by the pump. The gas is separated from the produced liquids at the bottom of the well and flows separately up the annulus. This separation of the gas at the bottom of the well makes the pumping much more efficient and allows a lower bottom hole pressure to be maintained.

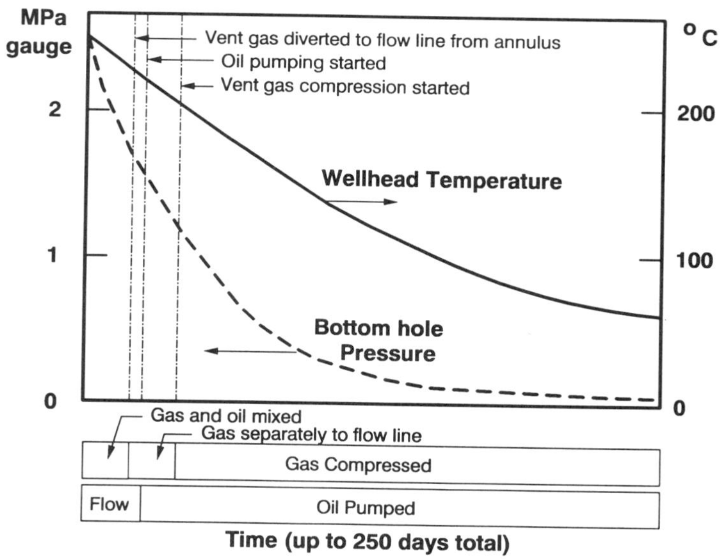

Figure 8.23 Well Temperature and Pressure During Production Cycle for Cold Lake Reservoir (from Keeling 1985)

When the wellhead pressure falls to the point where gas flow into the production line is no longer possible, the vent gas-compression system is started and the gas is compressed separately.

A flow plan showing the compression system employed by Esso at Cold Lake is shown in Figure 8.24; it is the result of considerable development.

The casing head product from the wells flows to the vent gas separator. The liquid that separates is pumped to the product line. The gas from the separator is cooled, and some of the liquid is condensed in an exchanger. The cooled stream passes to another separator. The liquid from this separator is pumped to join the plant product stream. The gas from the second separator passes to a liquid ring compressor; the compressed gas, after passing through a knockout drum to remove the compressor liquid, joins the remainder of the product stream.

One of the problems in designing a well-gas-compression facility such as this is the variability of the amount of gas in the well feed stream from the wells. Figure 8.25 shows the flow of gas to be treated in such a plant as a function of time.

In order to handle the variability of flow, Esso's design calls for two compressors. Both compressors are used in the first two cycles; after that, the second compressor becomes an on-line spare.

TREATING PRODUCED FLUIDS

The production from steamed heavy oil wells is usually a mixture of hot oil and water with some gas. In many cases it contains substantial portions of solids such as clay and sand. The production is nearly always emulsified; some of the water is emulsified in the oil and some of the oil is emulsified in the water.

Figure 8.26 shows a typical scheme of the treatment of the production from thermal recovery operations.

A demulsifier chemical is added before the production stream from the wells reaches the treatment plant; this promotes the separation of droplets of water from

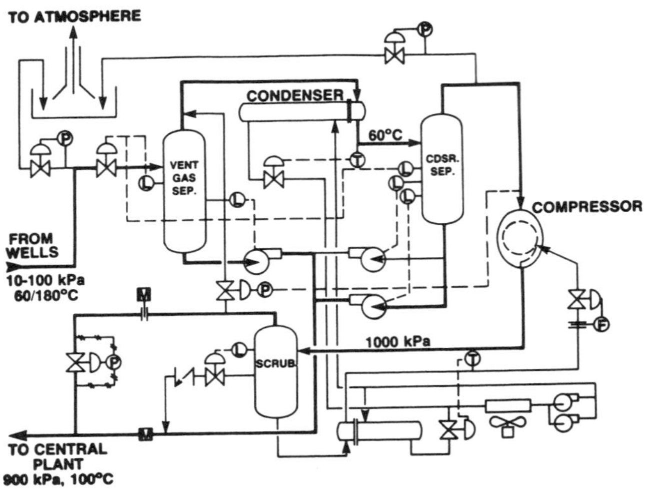

Figure 8.24 Esso's Cold Lake Casing Gas-Compression Scheme (from Keeling 1985)

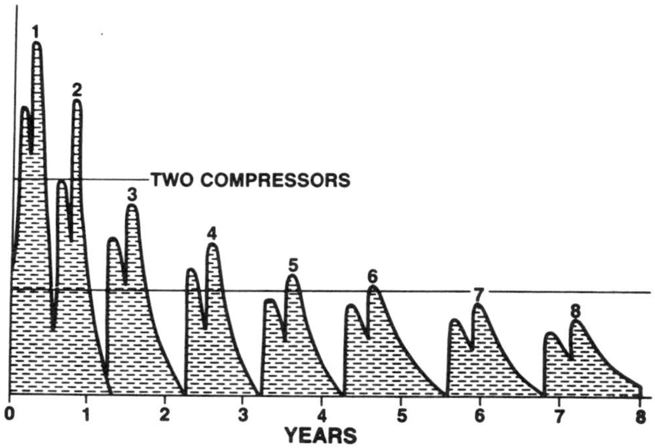

Figure 8.25 Expected Flow of Vent Vapor in Esso's Cold Lake Commercial Project (from Keeling 1985).
The Diagram Shows the Vapor Flow from a Gas Compression Facility Serving a Pad of 20 Wells. The Wells are Steamed 10 at a Time. The Numbers above the Peaks Refer to the Production Cycle. There Are Two Peaks for Each Cycle, Corresponding to the Two Batches of 10 Wells Going on Production

the oil. A variety of demulsifier chemicals is available from the companies that supply oil field additives. There is an art in selecting the best one, and it is found that what works well in one location is unsuitable in another.

Water-in-oil emulsions are more viscous than the oil itself,[7] and the addition of demulsifier at the wells can reduce the pressure drop in the gathering lines as

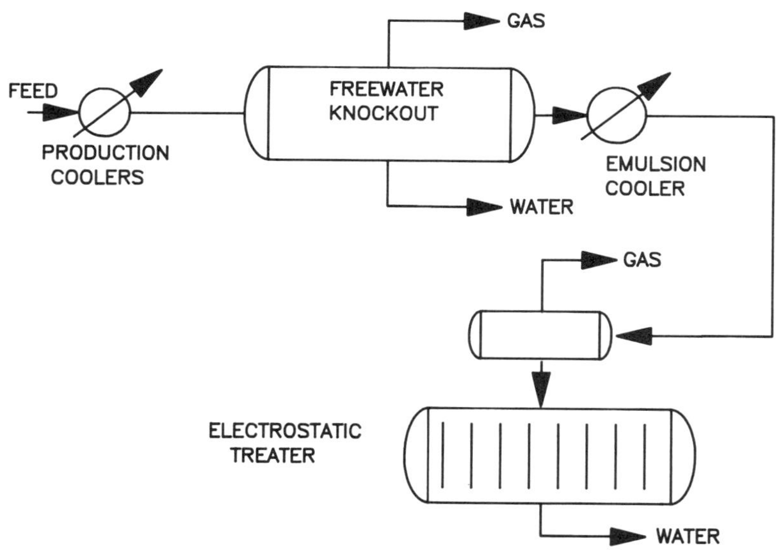

Figure 8.26 Production Treatment Used by Esso at Cold Lake (after Peachey and Nodwell 1981)

[7]Emulsions are dispersions of one liquid in another. There is a large difference between water-in-oil emulsions, where the oil is the continuous phase and which have higher viscosities than the oil, and oil-in-water emulsions, which are less viscous than the oil although more viscous than the water. Attempts have been made to create oil-in-water emulsions within the reservoir to promote flow by the addition of chemicals such as caustic soda to the steam (Doscher et al. 1963). As was discussed in Chapter 1, there is currently interest in making concentrated oil-in-water emulsions to allow pipeline transportation of very heavy oils.

well as allowing premixing of the demulsifier. Premixing also gives the demulsifier more time to act.

Figure 8.27 shows the measured viscosities of Cold Lake crude containing various fractions of emulsified water as functions of temperature.

At the plant, the stream is cooled and introduced into a baffled horizontal separator vessel. The oil, which still contains emulsified water (about 3%), is cooled further and treated in an electrostatic treater to produce the final bitumen product (0.5% BS & W) and more water, which can be sent to the recycle plant or disposed of otherwise. Figure 8.28 is a diagram of a modern electrostatic treater. Electrostatic treaters are often combined with a fired heater (it is then a "heater treater") to raise the temperature of the oil. This is required for conventional production which is cold. In thermal projects, it is more common for the oil to be cooled.

Electrostatic treaters utilize an electrical field between immersed electrode grids to promote coalescence of the suspended water droplets. The role of the coalescer should be confined to removing small residual amounts of water from the oil rather than large quantities. The separation depends upon the effect of the electrical field in causing a motion of the water droplets. Although both AC and DC fields have been used, AC is more popular because it is simpler, although perhaps not quite so effective. More recent technology uses both AC and DC fields in the same unit—the so-called dual polarity treater such as that shown in Figure 8.28. The electrodes in this equipment are connected to the electrical power in the manner shown in Figure 8.28.

Electrostatic fields promote droplet coalescing because

1. The water droplets become polarized and tend to align themselves in the electrical field, with one side of the droplet positively charged and the other, negatively charged. There is thus a tendency for droplets to attract each other; this promotes coalescence.

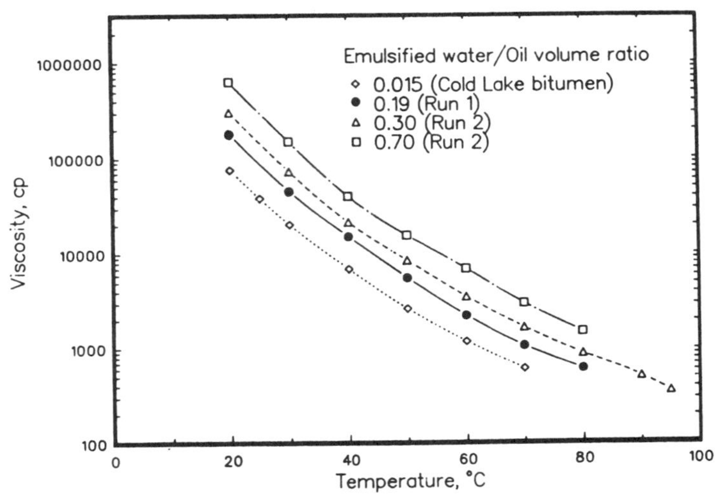

Figure 8.27 Effect of Emulsified Water on the Viscosity of Cold Lake Bitumen (from Chung and Butler 1988)

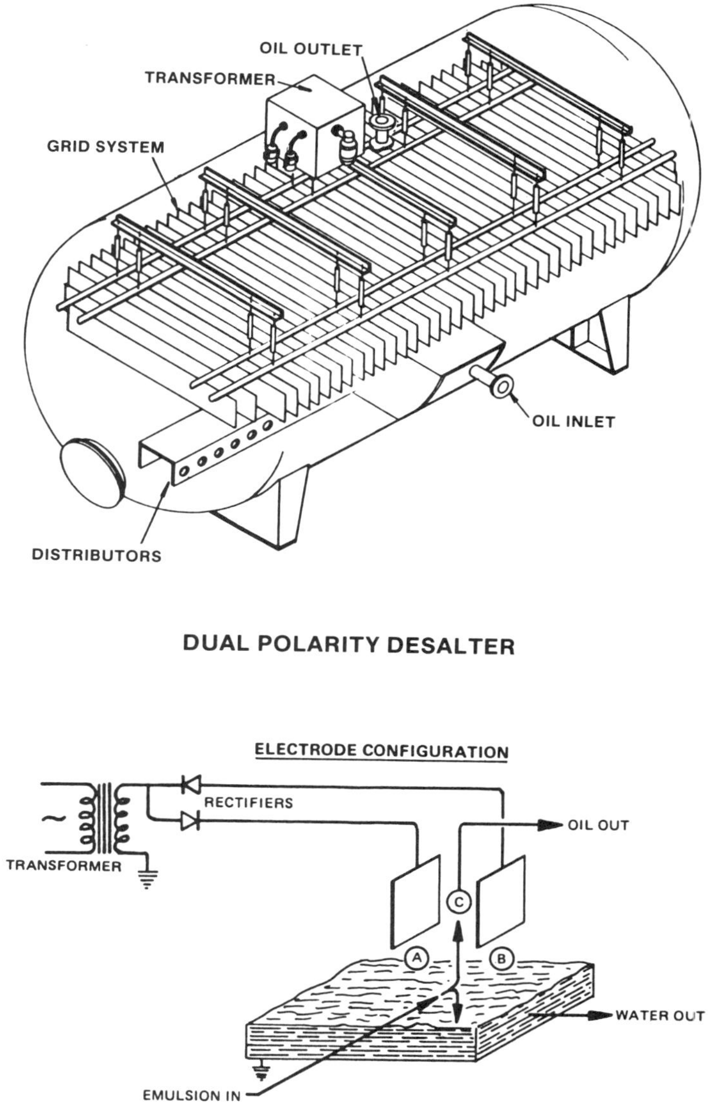

Figure 8.28 Diagram of a Dual-Polarity Electrostatic Treater. Water in Oil Emulsion Feed Is Introduced Beneath the Inverted Distributor Trough and It Flows Upwards Between the Vertical Electrodes. These are Charged Alternately + and − by the Electrical System Shown in the Lower Right. At the Same Time an Alternating Voltage is Applied to the Whole Electrode System with Respect to Ground (Courtesy Natco)

2. The droplets tend to have electrical charges initially and to migrate in the electrical field. However, there is a difference in the velocity with which large droplets and smaller ones migrate, and there is an increased tendency for col-

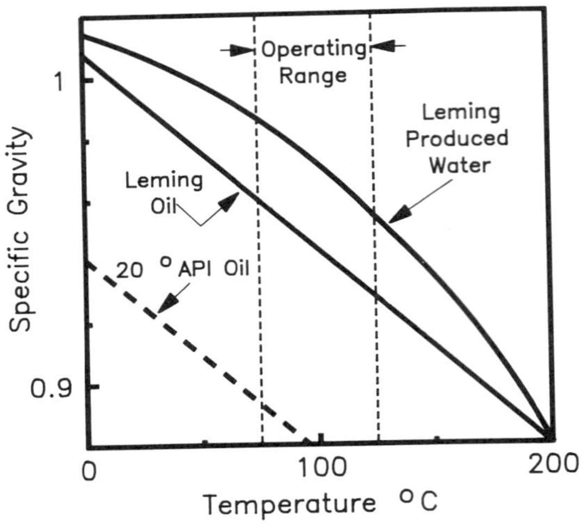

Figure 8.29 Densities of Cold Lake Oil and Water (after Peachey and Nodwell 1981)

lision. Also, particularly in a DC field, droplets having opposite electrical charges move in opposite directions and tend to collide with each other.

3. The electrical field may weaken the film of emulsifier on the surface of the droplets. This promotes coalescence upon collision.

If there is too much water suspended in the oil, then the droplets may form chains between the electrodes and produce a short circuit. Adjustments that can be made to the operation of an electrostatic treater include the spacing of the grids and the applied voltage.

The temperature of operation of the separating facilities is important as it controls the density difference between the oil and the water (Figure 8.29) and also the oil viscosity (Figure 8.30). Both of these affect the rate of settling.

The rate at which water droplets settle from oil is determined by Stokes law. This may be written for a single spherical water droplet as:

$$V = \frac{2}{9}\frac{R^2(\rho_w - \rho_o)g}{\mu_o} \tag{8.1}$$

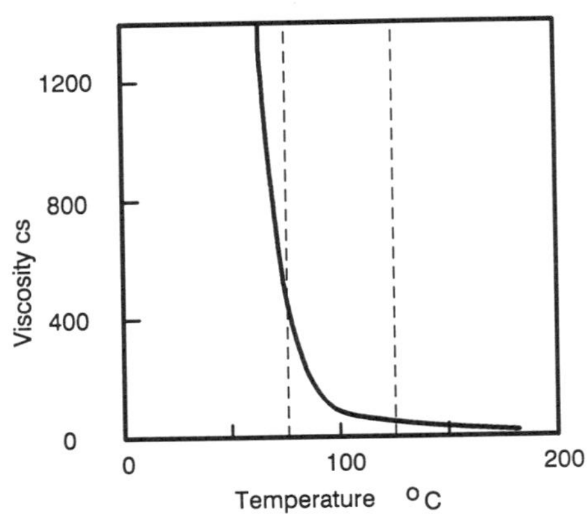

Figure 8.30 Viscosity of Cold Lake Oil (after Peachey and Nodwell 1981)

where V is falling viscosity of water droplets (m/s)
ρ_w is density of water (kg/m^3)
ρ_o is density of oil (kg/m^3)
g is acceleration due to gravity (9.81 m/s^2)
μ_o is oil viscosity (Pa · s)

Other sets of dimensionally consistent units can be used in place of the SI units shown.

For the case of an actual emulsion, the velocity is lower because the falling droplets interfere with each other. The effect may be calculated by multiplying 8.1 by a factor F_S (Steinour 1944).

$$V = \frac{2}{9}\frac{R^2(\rho_w - \rho_o)g}{\mu_o}F_S \tag{8.2}$$

where $F_S = \epsilon^2(10^{-1.82(1-\epsilon)})$ and ϵ = volume fraction of oil in the emulsion.

As the temperature is raised, the viscosity of the oil decreases, and this tends to increase the settling velocity. However, the density difference passes through a maximum, and there is thus an optimum temperature that provides the maximum settling rate.

Both factors just mentioned can be improved by adding diluent to the system. This is particularly valuable if the densities of the bitumen and water are very close. If the addition of diluent is practicable, as it is, for example, when the bitumen is to be diluted to transport it eventually through a pipeline, then it is very desirable to add diluent to the mixture before separating the water. This will increase the performance (capacity and/or product quality) of the separation equipment. Most pipeline companies require BS & W to be less than 1% by volume and usually less than 0.5% in order to prevent corrosion and to reduce the frequency of pigging in the pipelines.

PRODUCTION TREATMENT WITH HIGH SAND PRODUCTION

Figure 8.31 shows the facilities used by Husky in the treatment of their production at Cat Canyon. The average production from this project contains 27 wt% sand. Somewhat over half of the sand is separated in the direct-fired desander vessel, which is fitted with a conical bottom and contains hydraulic jets to flush the sand out periodically. The remainder of the sand is removed from the bottom of the emulsion treater. The produced oil contains less than 0.2 wt% sand and less than 3% BS & W.

MAKEUP WATER SUPPLY

Even if the fraction of the produced water that is recycled is very high (greater than 90%), there will still be a need for makeup water supply. In the reservoir, the oil that has been removed is normally replaced by injected water of an equivalent volume. There will also be a need for additional water in a recovery project for service and cooling purposes.

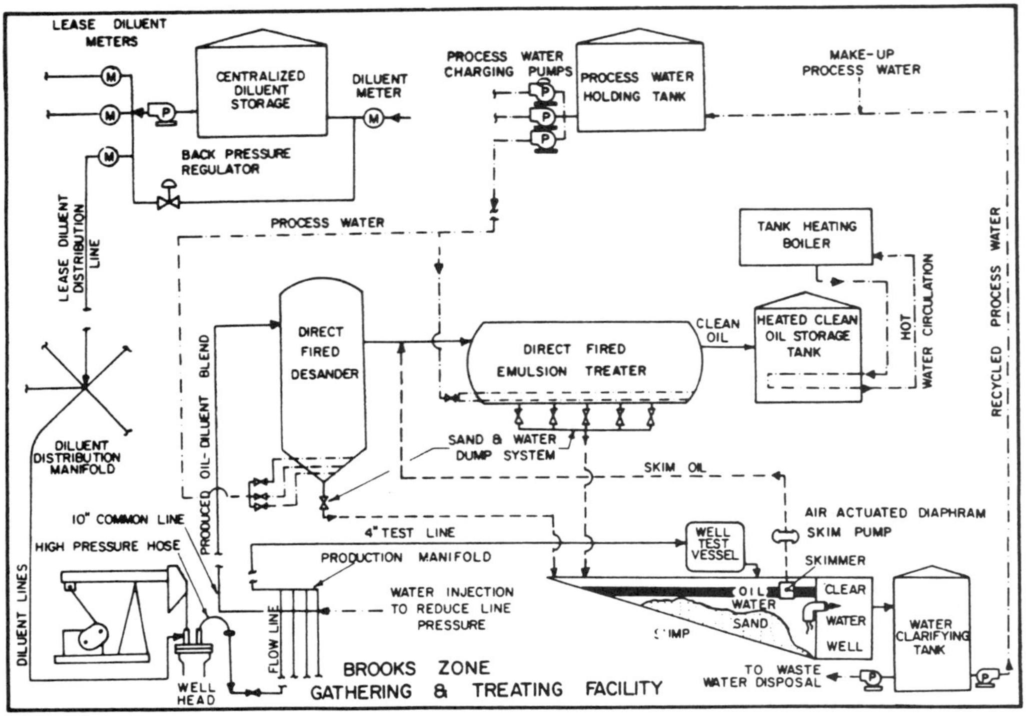

Figure 8.31 Process for Treating High-Sand-Content Heavy Crude (from Vonde 1979)

Conventional sources of water supply such as from underground reservoirs, rivers, and lakes are normally used. However, other water supplies may be considered if there is a shortage of water in the area. For example Suncor, in their thermal recovery project at Bonneville, Alberta, uses wastewater effluent from a nearby municipality.

Fresh water is normally treated by chlorine oxidation, lime/soda softening, filtration, ion exchange, and deaeration (Kloepfer, Card, and Kus 1983).

RECYCLING PRODUCED WATER

In the previous section it was shown how the production from steam recovery projects can be separated into a marketable oil product and a produced water stream. In many cases it is desirable to treat this water so that it may be recycled to the steam generators.

Recycling water reduces the impact of the thermal recovery project on the environment; it not only provides an acceptable means for disposing of the tainted produced-water stream, but it also greatly reduces the need for fresh water. In areas where water is scarce, as in California, this reduction in the need for fresh water is a very significant factor. Another advantage of water recycling is that the heat in the recycled water reduces the heat requirement for steam generation somewhat.

Produced Water Analyses

Analyses of a typical produced water sample from the Esso pilot at Cold Lake and a corresponding treated produced water sample that is suitable for feeding to a once-through, oil field steam generator are given in Table 8.5.

TABLE 8.5 Composition of Produced and Treated Recycle Water[1] at The Leming Cold Lake Pilot[2]

	UNTREATED	ppm	TREATED
Oil and Grease	5,000–10,000		0–100[3]
TDS	4,000–10,000		4,000–10,000
SS	10–120		0–5
Ca	40–60		0–1
Mg	4–8		0–1
Na, K	1,000–4,000		1,000–4,000
Cl	2,000–6,000		2,000–6,000
SO_4	40–200		40–200
HCO_3	100–400		0–10
Sulphide	10–40		5–10
SiO_2	150–300		15–30
Temperature (°C)	80–90		80–90
pH	7–8		8–9

[1]From Electrostatic Treaters and Water Recycle Pilot

[2]Konak and Grisard (1979).

[3]Suspended oil (usually near 0). There are also about 250 ppm dissolved organic materials measured as *total organic carbon* (TOC).

The produced water contains relatively large quantities of dispersed oil. The dissolved solids are largely sodium chloride, arising from the reservoir water, but the water also contains appreciable amounts of hardness salts and silica, arising from the leaching of the reservoir rocks. There is also substantial dissolved organic material.

In the treatment of this material for feed for steam generation, it is necessary to remove the suspended oil and the hardness. It is also desirable to reduce the silica, although considerable silica can be tolerated if the calcium, magnesium, and iron contents are kept low and if the water is alkaline when it is fed to the generator. These conditions will tend to keep the silica in solution as sodium silicate and prevent the formation of insoluble iron, magnesium, and calcium silicate scales.

Table 8.6 shows an analysis of produced water from the Texaco pilot in Athabasca. The analysis is similar to that from Cold Lake but the concentrations of the impurities are lower.

The lower silica content may reflect the lower solubility of silica at the temperatures of the steamed reservoir; the Texaco pilot is at a shallower depth than the Esso pilot at Cold Lake, and one would expect the injection pressures to be lower. It seems likely that the lower level of dissolved solids (salt) arises because of a lower concentration in the reservoir connate water—possibly the original seawater has become more diluted by surface water.

TABLE 8.6 Produced Water Analysis from Texaco Steamflooding Pilot Near McMurray Athabasca

Quantitative Analyses of Dissolved Components of In Situ Production Water[1]

Sodium	529
Potassium	11
Calcium	56
Magnesium	20
Chloride	820
Sulfate	80
Aluminium	<1
Iron	<2
Silicon	47

[1]Milligrams of solute per liter of solution at 23°C.
(Burchfield and Hepler, 1979)

The water analyses shown from the Getty plant at Kern River, California, in Table 8.7 show even less dissolved salt but are otherwise rather similar.

TABLE 8.7 Water Analyses from Getty's Kern River Operation

CONSTITUENTS	JUNCTION WATER PLANT (mg/kg) PHYSICAL PLANT INLET	PHYSICAL PLANT OUTLET
Bicarbonates, HCO_3	292	278
Chlorides, Cl	185	200
Sulfates, SO_4	82	65
Sulfides, S	0.0	0.0
Nitrates, NO_4	0.0	0.0
Silica, SiO_2	125	111
Boron, B	1.3	1.2
Sodium, Na	190	260
Iron, Fe	3.3	0.1
Hardness as $CaCO_3$	113	0.3
Total dissolved solids	622	793
pH value at 25°C	7.2	7.4
Conductivity: micromhos/cm at 25°C	950	1150
Resistivity: ohm-meters at 25°C	10.8	8.9
OIL AND GREASES		
Freon extractables (soluble and insoluble)	50	20
Millipore (insoluble)	20	0

(Carrell 1979)

Treating Recycled Water

Various schemes for the treatment of recycled water were investigated during the development of the design for the large commercial plant at Cold Lake that was proposed by Esso. These were discussed by Whalley and Wilson at the Unitar Conference in Edmonton in 1979 (Whalley and Wilson 1979). Much of this early thinking was directed toward the development of schemes that would allow the use of large utility-type boilers because of the ability to build these with very large capacities and very high thermal efficiencies.

The process finally adopted (Figure 8.32) for use at Cold Lake involves the treatment of the recycled water to remove oil and hardness and the generation of the steam in multiple, once-through, oil field steam generators. As has been discussed previously, these operate with low heat flux and are able to tolerate considerable dissolved solids, provided that the water is soft. The process adopted by Esso also removes much of the silica from the feedwater by adsorption and/or coprecipitation with magnesium hydroxide in the hot lime treater.

A simplified flowplan of the treatment system used by Esso in their Leming plant expansion is given in Figure 8.33, and operating conditions for the plant are summarized in Table 8.8. As has been mentioned previously, the commercial production facilities employed by Esso employ once-through steam generators containing three parallel tube passes, which generate about 160,000 lb/h of steam each.

TABLE 8.8 Operating Conditions for Esso's Leming Water Recycle Plant

CURRENT REUSE PLANT OPERATION	
IGF inlet rate	170 m^3/h
IGF inlet suspended oil	200 ppm
IGF outlet suspended oil	21 ppm
Hot lime treater inlet suspended oil	9 ppm
Ion exchanger effluent:	
Rate	100 m^3/h
Total hardness	0.2 ppm
Silica	26 ppm
pH	9.1
Chloride	1600–2200 ppm
Sulphides	20 ppm
Oxygen	5 ppb

(Peachey and Nodwell 1981)

The water from the primary separation of the oil is treated in an *induced gas flotation cell*.[8] An apparatus of this type is shown in Figure 8.34. Its operation depends upon the dispersion of small bubbles of gas by means of centrifugal impellers located near the bottom of the contact stages. The bubbles of gas rise to the surface and carry the oil with them as adhering droplets. At the surface they are removed as a foam. The equipment shown contains four separation stages, in which the water is treated successively.

[8]This process is commonly referred to as *induced air flotation* in other applications. In the treatment of recycle water, it is desirable to use a natural gas rather than air to reduce corrosion. The buildup of H_2S in the gas within the process is a problem that needs consideration in the design of the facilities.

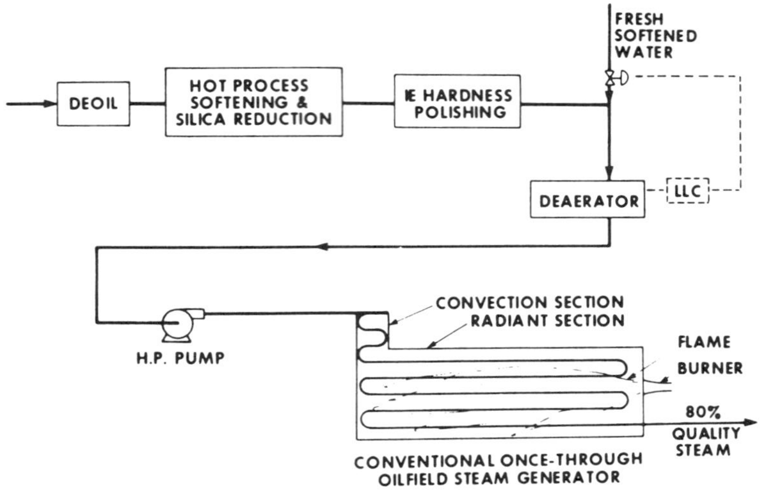

Figure 8.32 Conventional Oilfield Steam Generators (from Whalley and Wilson 1979)

The water leaving the induced gas flotation cell is filtered and treated in a hot-lime treater. In the hot-lime treatment process, the water is softened and silica is reduced by adsorption and coprecipitation with magnesium hydroxide. A diagram of a hot-lime treater (of the type used by Esso resources at Cold Lake) is shown in Figure 8.35. The equipment shown softens the water, removes much of the silica, and also removes dissolved gases; a separate deaerator is not required.

Calcium is removed from solution as calcium carbonate, and magnesium is precipitated as magnesium hydroxide. The chemical reactions are as follows:

$$Ca(HCO_3)_2 + Ca(OH)_2 = 2\ CaCO_3 \downarrow + 2\ H_2O$$

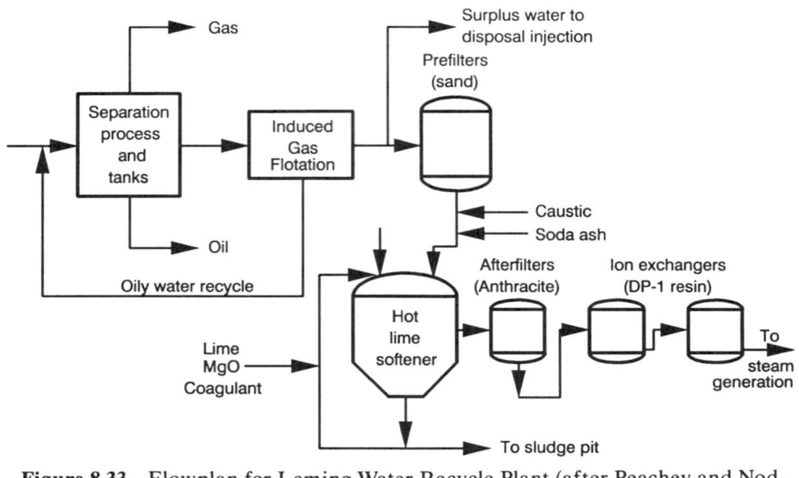

Figure 8.33 Flowplan for Leming Water Recycle Plant (after Peachey and Nodwell 1981)

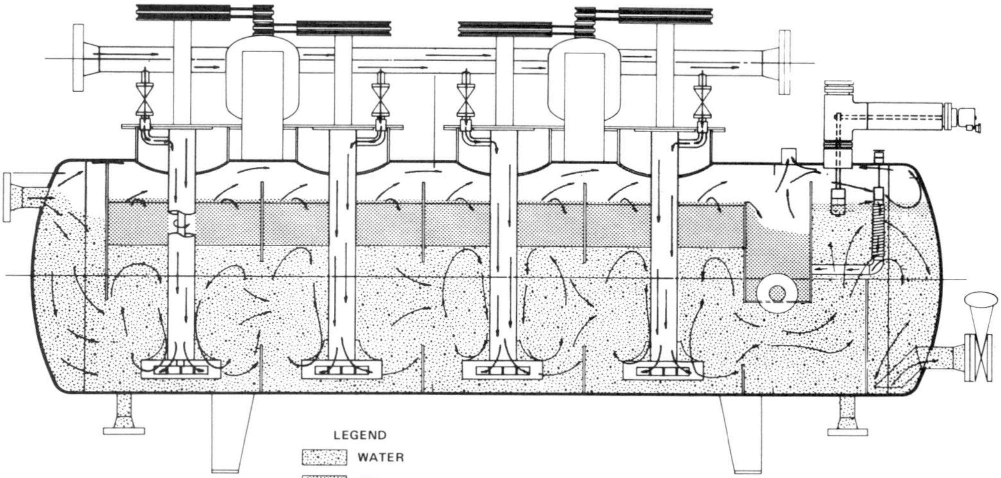

Figure 8.34 Four-Stage Induced Gas Flotation Cell (Courtesy Natco)

$$Mg(HCO_3)_2 + Ca(OH)_2 = MgCO_3 + CaCO_3 \downarrow + 2\ H_2O$$

$$MgCO_3 + Ca(OH)_2 = Mg(OH)_2 \downarrow + CaCO_3 \downarrow$$

Depending upon the concentrations of the ionic species in the feed, it is sometimes necessary to add other reactants besides lime to the hot-lime treater. Insufficient carbonate or bicarbonate ion in the feed is augmented by the addition of carbonate as soda ash (sodium carbonate). Similarly, if there is insufficient magnesium in the feed, additional magnesium oxide may be added to adsorb silica.

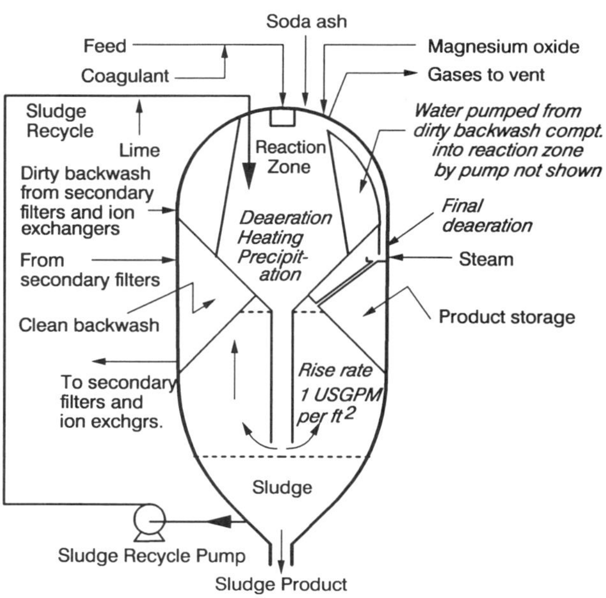

Figure 8.35 Sludge-Contact Hot-Process Softener (after Bridle 1986)

The successful operation of a hot-lime treater requires the maintenance of a sludge inventory of a proper consistency. Chemicals may be added to promote coagulation.

It is also very important to control the amount of oil in the feed to very low levels (less than 20 ppm). Failure to do this results in a putty-like sludge, which is ineffective as a contacting medium.

Following the hot-lime treatment, the water is filtered to remove sludge that has been carried over and is then given a "polish," or softening treatment, in two stages of ion exchangers.

Conventional ion exchange treatment of water for boiler feed involves contact with sodium "zeolite" resin. Although this resin is commonly called zeolite, it is no longer made from zeolite minerals but is usually a sulphonated cross-linked polystyrene resin. Calcium and magnesium ions in the water exchange with the sodium ions in the resin. As a result, the calcium and magnesium in the water are replaced by sodium. When the resin is spent, it is regenerated by contact with a sodium chloride solution. This is the same process used in domestic water softening.
With low dissolved solids the following reactions occur:

Softening: $$2\,R^-Na^+ + Ca^{++} = R_2^-Ca^{++} + 2\,Na^+$$

$$2\,R^-Na^+ + Mg^{++} = R_2^-Mg^{++} + 2\,Na^+$$

Regeneration: $$R_2^-Ca^{++} + 2\,Na^+ = 2\,R^-Na^+ + Ca^{++}$$

In these equations R^- represents the sulphonate anions in the polymeric resin material.

The zeolite softening process loses its effect when the water that is to be softened contains a high concentration of sodium ions. In this circumstance the equilibrium of the softening reaction is suppressed by the sodium ions already in solution, and the capacity of the sodium form of the resin to remove calcium and magnesium ions is greatly reduced. This problem prevents zeolite softening from being effective at Cold Lake.

The softening of high-brine-content water can be carried out using a weak carboxylic acid resin in which the acidic groups on the resin are carboxylic rather than sulphonic acid groups. The disadvantage of this resin is that it is much more difficult and costly to regenerate. A two-stage regeneration with acid and then caustic is required rather than the simple reaction with brine. The process and the regeneration are shown by the following chemical equations:

Softening $$2\,R^-Na^+ + Ca^{++} = R_2^-Ca^{++} + 2\,Na^+$$

$$2\,R^-Na^+ + Mg^{++} = R_2^-Mg^{++} + 2\,Na^+$$

Regeneration:

1. With acid $R_2^-Ca^{++} + 2\,H^+ = 2\,R^-H^+ + Ca^{++}$
2. With caustic $2\,R^-H^+ + 2\,NaOH = 2\,R^-Na^+ + 2\,H_2O$

Getty, at Kern River, has a very large water-recycle operation using the process sequence shown in Figure 8.36.

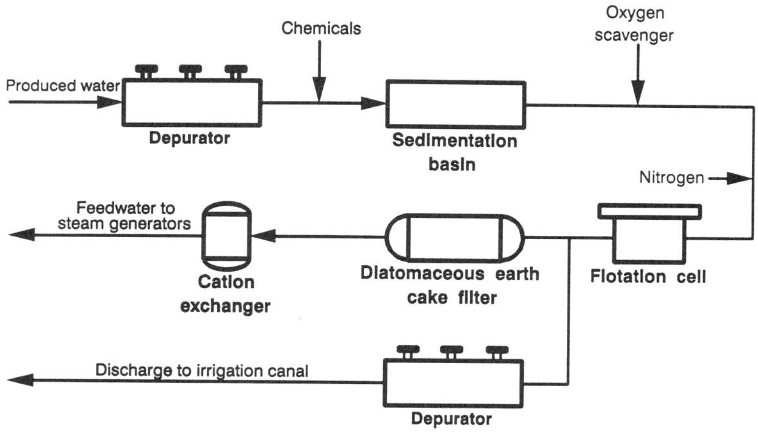

Figure 8.36 Getty's Water-Recycle Process at Kern River (after Carrell 1979)

A recycle rate of 80,000 tonnes per day is used and the plant supplies water to 230 individual steam generators. The process involves deoiling with a depurator (equipment similar to an IGF), further flotation, and filtration through diatomaceous earth. Conventional zeolite softening is employed because the salt concentration of the water is relatively low. Impurity concentrations reported by Carrell (1979) for Getty's operation are shown in Figure 8.37.

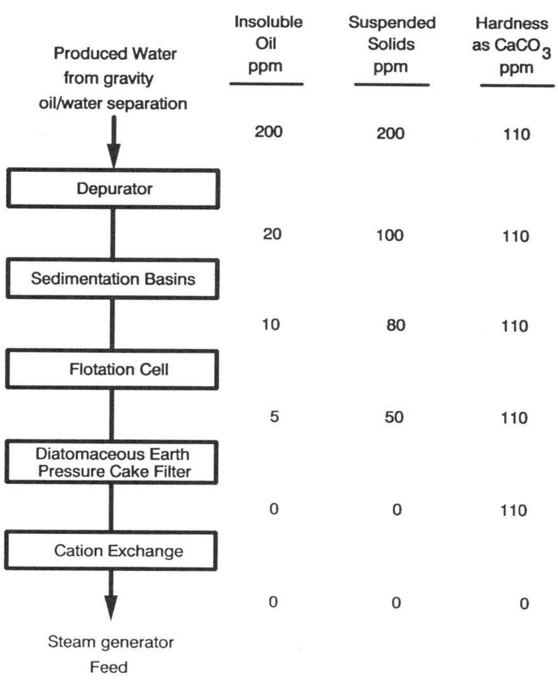

Figure 8.37 Composition of Streams in Getty's Water-Recycle Process (after Carrell 1979)

Wastewater Management

In a large steam recovery project, there is a considerable complexity in the handling of the various water streams. The situation for the Esso project at Cold Lake has been described by L. A. Courtnage (1987). Figure 8.38 gives an idea of the factors that have to be considered.

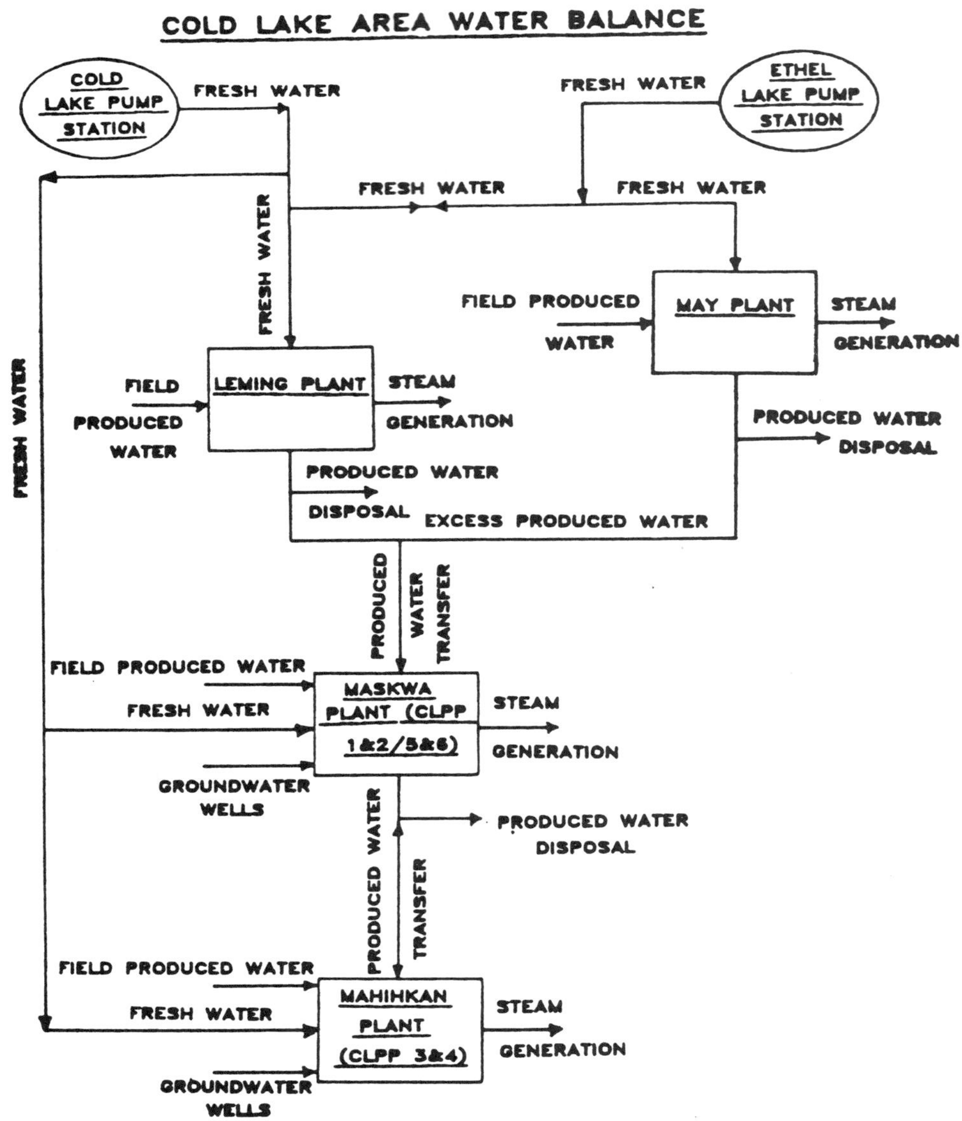

Figure 8.38 Water Flows for Esso's Cold Lake Facilities. Leming and May are Esso's Pilot Facilities; Maskwa and Mahihkan are Commercial Facilities (after Courtnage 1987)

In general, water is supplied from recycle and from various freshwater sources. The system must be managed to minimize freshwater usage and minimize disposal of produced water (in deep disposal wells). There is also a considerable scheduling problem involved because of the necessity at times to shut down major equipment for repairs. There can be both planned and unplanned shutdowns.

Esso's Thermal Softening Process

Esso has described a process (Lim and Konak 1985) for the softening of water that involves heating the water to a high temperature by direct mixing with steam, followed by the removal of the precipitated solid by filtration. The principle is shown in Figure 8.39.

Heating the feedwater by using direct steam avoids the scaling of heat-transfer surfaces that would occur otherwise. Removal of some silica also occurs, as in the hot-lime process by coprecipitation with magnesium hydroxide. The inventors find that it is desirable to add caustic soda or sodium carbonate to the feed to bring the pH of the effluent into the range 9 to 10. Some experimental results reported by Lim and Konak are given next.

Experimental Data Reported for Thermal Softening Process

Average Data from Seven Tests
Reactor Temperature 211°C; Effluent pH 9.7

Feed ppm			Product ppm		
Ca	Mg	SiO_2	Ca	Mg	SiO_2
121	31	306	0.49	0.07	204

REDUCING TOTAL DISSOLVED SOLIDS

The water-recycle process that has been described removes oil, dissolved oxygen, and hardness but does not reduce the total dissolved solids (TDS). From the early days of water recycling, there has been much concern over the level of total dissolved solids that can be tolerated in steam generator feed. Sodium chloride makes up the bulk of these solids. This has a high solubility and will not crystallize under normal conditions.

In its commercial operations at Cold Lake, Esso is successfully recycling water containing about 8000 ppm of TDS. Thielen et al. (1988) demonstrated that water containing 25,000 ppm of total dissolved solids can be used to generate steam, without problems, in a test rig that simulates an oil field steam generator.

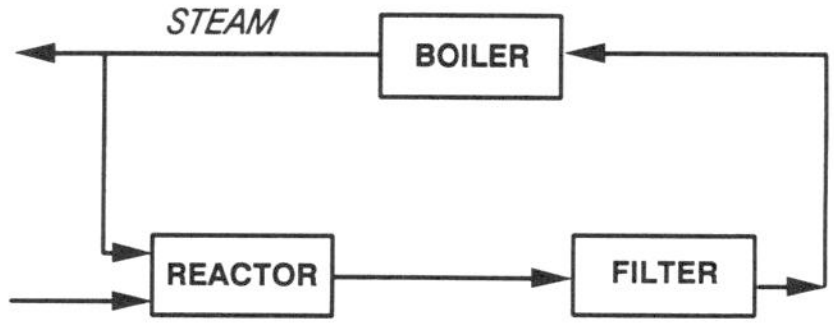

Figure 8.39 Esso's Thermal Softening Process (from Lim and Konak 1985)

The removal of most of the total dissolved solids from water of even higher salinity can be done by methods such as evaporation, reverse osmosis, electrodialysis, and freeze desalination. This field has been reviewed by Zaidi, Kok, and Schmidt (1988); they have arrived at the following cost estimates for the production of 3000 m^3/d of once-through oil field steam generator feed.

Cost of Reducing TDS in Produced Water

	Canadian Dollars per Cubic Meter	
	With Capital Depreciation	Without Capital Depreciation
Vapor compression evaporation	2.07	1.01
Electrodialysis	1.66	1.05
Freeze desalination	1.33	0.80

(Zaidi, Kok, and Schmidt 1988)

For a water-to-oil ratio of 3 to 5, the costs of desalination just given would be $4 to $10 per cubic meter of oil product.

ALTERNATE STEAM GENERATORS

Coal-Fired Steam Generators

In areas where coal is abundant it may be a more economic fuel for the generation of steam than gas or oil. There has been considerable interest in this possibility in Alberta and a number of ongoing joint industry government studies have been described (Alberta Energy 1989).

Two approaches are being followed. The first approach is the design of a new coal-burning, oil field steam generator which will have a capacity of 180 million BTU/h and which will use feedwater containing a high concentration of dissolved solids. Reactive solids will be injected into the furnace to absorb about 40% of the sulphur in the coal fuel. The second approach involves the design of a coal burner which can be added to existing oil field steam generators as a retrofit. This too will involve a reduction of sulphur emissions by employing a new type of slagging burner in which part of the sulphur is removed in the slag.

Both approaches involve the use of pulverized coal. The first was proposed by Combustion Engineering Canada and would involve the coal being pulverized so that 70% would pass through a 200 mesh screen. The second approach was proposed by Struthers-TIW; it would use micronized coal (i.e. coal ground so that 90% of the coal would pass through a 325 mesh sieve). A new Low NO_x/SO_x (LNSB) burner that has been developed by Rockwell International will be used. A diagram showing how this burner would be connected to a conventional horizontal oil-field steam generator is shown in Figure 8.40. A project to demonstrate the operability of the concept has been initiated at the Esso operations at Cold Lake by the TransAlta Resources Investment Corporation. The project is expected to be com-

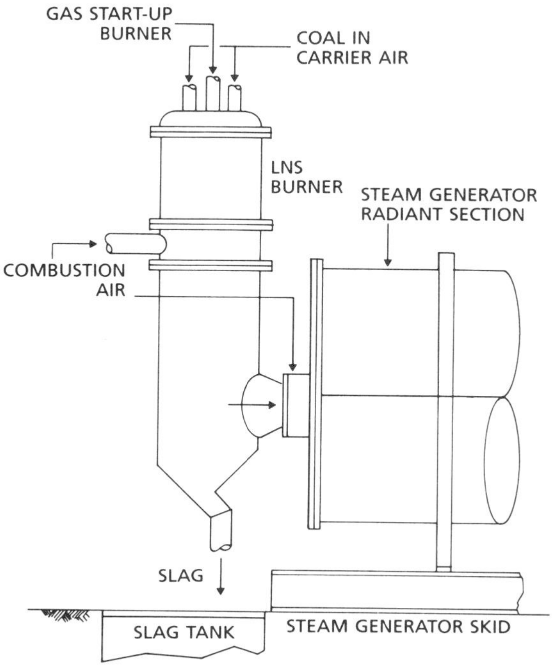

(Source: LNS Burner on EOR Steam Generator Feasibility Study, TransAlta Resources Investment Corporation, 1988)

Figure 8.40 Low NO_x/So_x Burner Shown Retrofitted to a Horizontal Steam Generator (from Alberta Energy)

plete in 1991/92. The emission goals for the program for coal containing 0.36% sulphur, are 0.3 lb of SO_2 and 0.2 lb of NO_x per million BTU; these compare to the existing emission regulations of 0.6 SO_2 and 0.6 NO_x (0.2 NO_x for gas fuel).

Downhole Steam Generation

There has been a significant effort to develop downhole steam generators for oil field steam generation. One of the main advantages seen for this approach is the reduction of well bore heat losses and, because of this, improved economics for production in very deep deposits.

There are two basic approaches:

1. Low-pressure combustion, in which the downhole combustion is carried out at relatively low pressure and in which the flue gas products are vented up the injection well. This approach requires a heat exchanger down the well to isolate the low-pressure combustion zone from the high-pressure steam.
2. High-pressure combustion, in which the products of combustion are mixed directly with the steam and pass into the reservoir to be collected at the production wells.

An important possible variation of the second approach involves the use of oxygen rather than air for the combustion. This also has the potential advantage that the resulting high concentration of carbon dioxide may improve the effect of the steam in recovering oil. Considerations similar to this also apply to the use of oxygen for in situ combustion and to the wet oxidation steam generators, such as the Zimpro equipment that is discussed later.

A major advantage seen for the use of downhole steam generators with the direct injection of the flue gas into the reservoir is that the sulphur and nitrogen oxides will be absorbed in the reservoir, either as anions in the water or by the rocks directly; flue gas scrubbing is avoided.

Figure 8.41 shows a cross-sectional diagram of the high-pressure downhole steam generator developed by Sandia National Laboratories in the DOE "Deep Steam" project described by B.W. Marshall (1982).

The walls of the generator are cooled by the feedwater that flows from the annulus around the "combustion can" onto the wall at the lower part of the combustion chamber. Here it evaporates, and the steam mingles with the flue gases. Versions for using either air or oxygen have been studied. Diesel fuel was used as fuel in the prototype model.

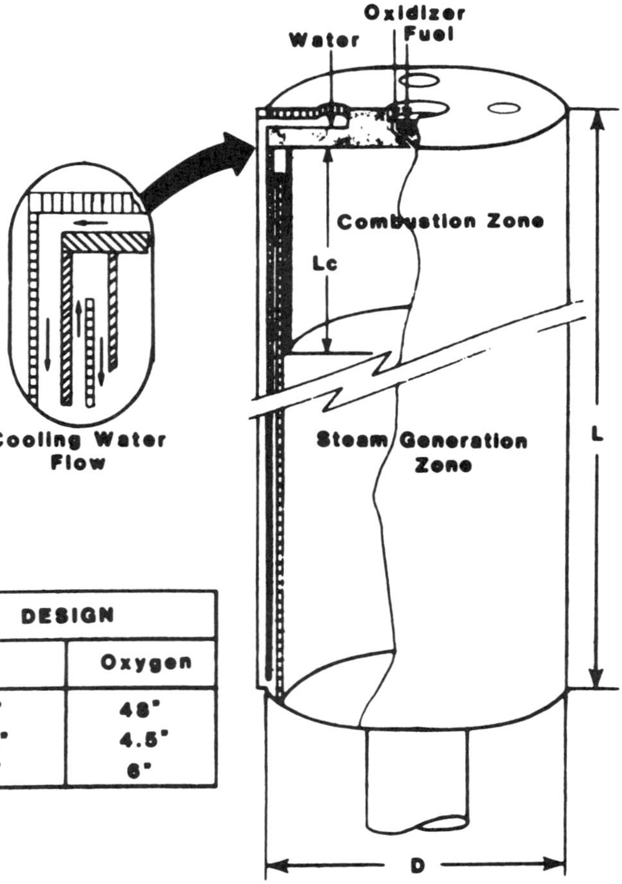

DIMENSION	DESIGN	
	Air	Oxygen
L	48"	48"
D	4.5"	4.5"
Lc	18"	6"

Figure 8.41 Sandia's High Pressure Downhole Steam Generator (from Marshall 1982)

Severe corrosion resulting from the action of the sulphur dioxides was experienced. This was controlled by the direct addition of caustic soda to the product below the combustion zone. The caustic was conveyed to the generator in a separate small-diameter tube. The need to install several separate tubing strings to the surface is a disadvantage of the downhole generator approach.

A downhole generator of this type has operated successfully in California for a period of several months; performance is shown in Figure 8.42. Ignition of the flame has been obtained both by means of an electric glow plug or, alternatively, by adding a hypergolic (self-igniting) liquid, triethyl borane, to the diesel fuel; both methods worked. Another high-pressure downhole generator developed by the Chemical Oil Recovery Co. (Eson 1982) is shown in Figure 8.43.

Up until now, downhole steam generation has not advanced to the point where it is accepted as a commercial alternative. The equipment that has evolved is complicated and not competitive. Although the use of downhole steam generators may become practical for steamflooding, it is unlikely to be so for steam stimulation, where the periodic requirement for steam would make it more expensive.

Fluidized Bed Combustion Boilers

The use of fluidized bed combustion boilers is another approach to the generation of steam that may be of value in the recovery of heavy oil. In these boilers, solid fuel is added to a bed of solid particles, which are maintained in a fluid state by means of the combustion air stream.

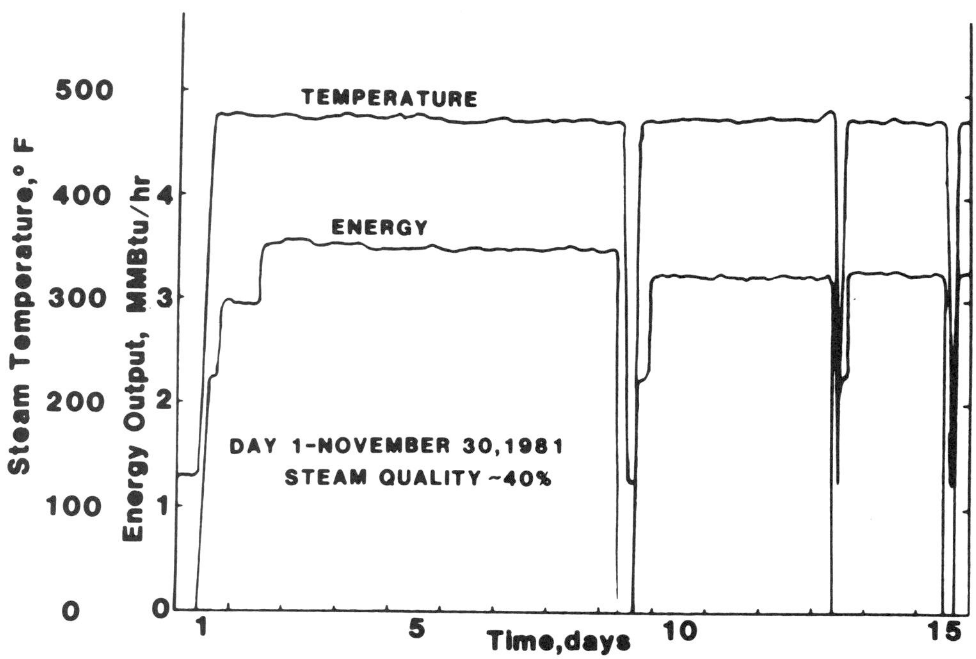

Figure 8.42 Performance of Sandia's Downhole Steam Generator (from Marshall 1982)

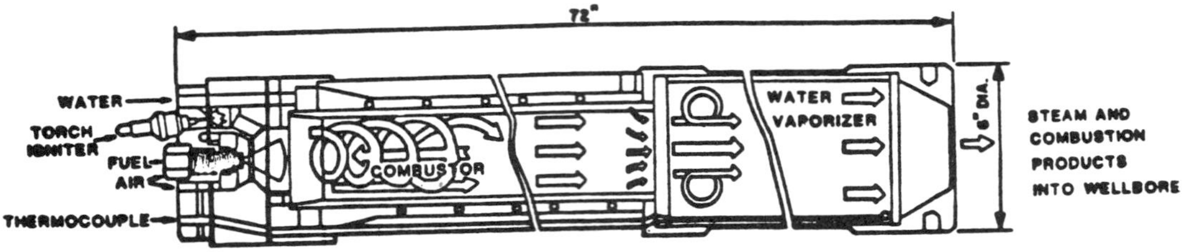

Figure 8.43 Chemical Oil Recovery Company's Downhole Generator (from Eson 1982)

Limestone is also added to the combustion zone bed; it reacts with the sulphur oxides that result from the combustion of the fuel to yield calcium sulphate, and this is removed as a byproduct. It is necessary to add limestone in excess of the stoichiometric quantity required to react with the sulphur.

Three types of fluid bed combustors are being developed:

1. Atmospheric-pressure combustors, in which the fluidized bed operates at a pressure slightly above atmospheric.
2. High-pressure combustors, in which the gases leaving the combustion furnace are used to drive a gas turbine. Although this approach is attractive for the generation of increased power because of its higher potential efficiency, there has been trouble in obtaining reliable mechanical turbine operation because of the entrained solids in the gas stream.
3. A multi-solid fluidized bed generator, which is being developed by Battelle; this equipment is being promoted by Struthers Wells for the generation of steam for thermal recovery (Berry 1979).

A schematic flowplan for a multi-solid fluid bed combustion boiler is shown in Figure 8.44. Fluidized silica particles are used to convey heat from the combustion zone to the external boiler. These silica particles circulate from the combustion zone to the external boiler and then back to the construction zone when they become reheated. Separation from the fly ash is obtained because of the considerably higher density of the silica as compared to that of the fly ash. The surplus limestone and calcium sulphate are removed as a separate stream from the fly-ash collector.

Although more lime is used in this process than would be required for a wet slurry flue gas treatment system, it is simpler to dispose of the dry solid product. Possible uses for the waste include agricultural liming, a neutralizing agent for municipal wastewater treatment, and aggregate for construction and road building.

Other companies involved in fluidized bed combustion (mostly with a view to power generation) are Babcock Wilcox, BP, Riley Stoker Corp. (Worcester, Mass.), Foster Wheeler, Fluidyne Engineering, and Johnston Boiler Co. (Ferrysburg, Mich.).

Vapor Therm Steam Generators

The Vapor Therm steam generation process (Sperry 1981) has been developed by Carmel Energy Inc. The principle is shown in Figure 8.45.

An air and fuel mixture under a pressure slightly higher than the required steam pressure is burned in an external combustion chamber, and the gases are di-

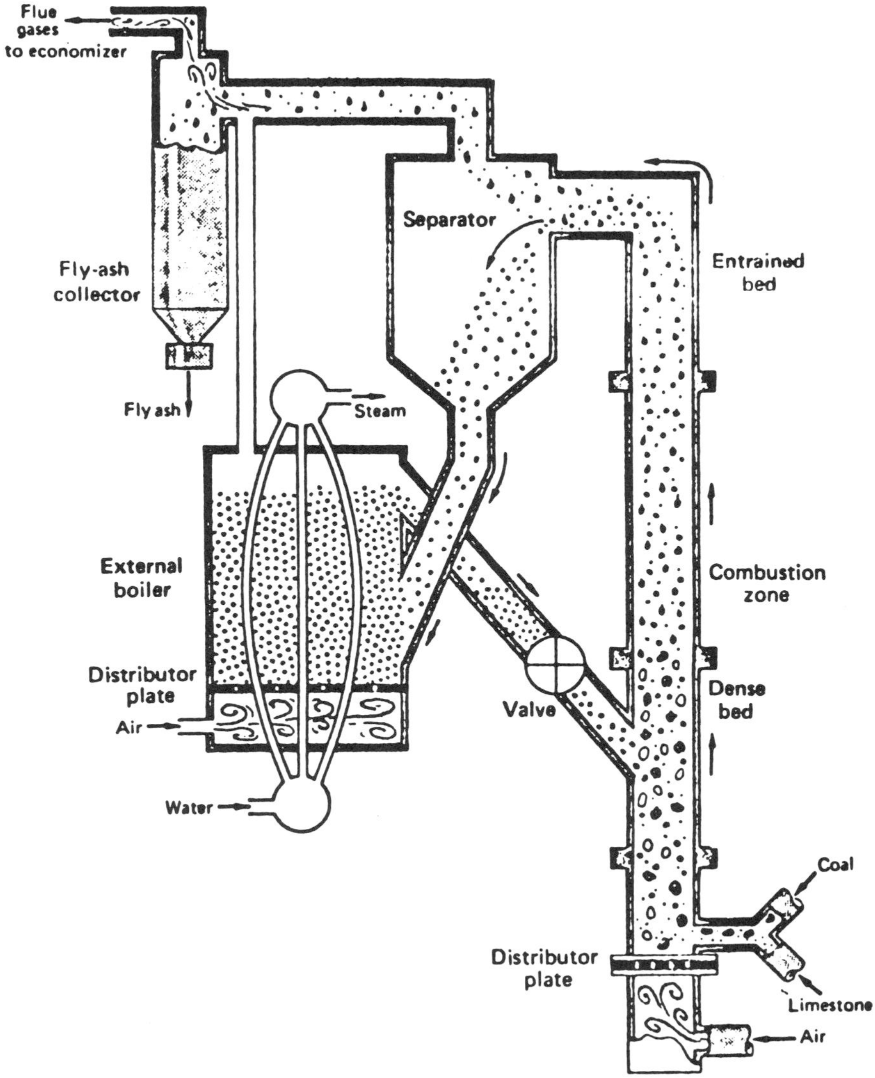

Figure 8.44 Battelle's Fluid Bed Combustion Boiler. Fluidized Solids Which Are Retained within the Plant Circulate and Carry Heat from the Combustion Zone to the External Boiler (from Berry 1979)

rected into the pool of boiling water. The mixture of steam and flue gas is conveyed directly to the wells that are being steamed. The fuel that has been used is a mixture of crude oil and diesel fuel. There have been several successful tests of the generator on a relatively small scale, and it is available commercially. It is recommended by Sperry for use where the high-pressure inert gas in the steam can provide useful drive in pressure-depleted reservoirs. Sperry also suggests that in some cases the output from the generator can be combined with steam from a conventional generator to give a cheaper mixture, which may, for some applications, contain a more optimum amount of inert gas.

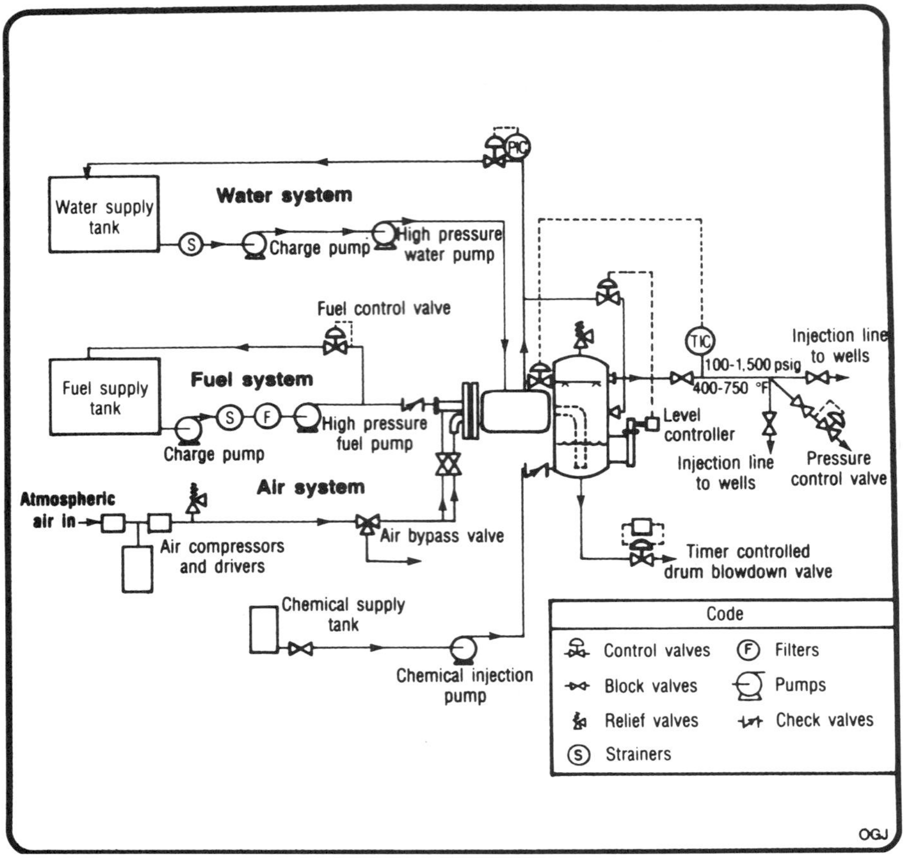

Figure 8.45 Carmel Energy's Vapor Therm Process (from Sperry 1981)

For the Vapor Therm process to be widely utilized, it will be necessary to demonstrate that the added cost of air compression and the usually more expensive fuel are justified by improved recovery performance. It is an interesting idea.

The Zimpro-AEC Steam Generator

The Zimpro-AEC steam generation is another process in which steam mixed with flue gas is produced for injection into the reservoir. However, unlike the Vapor Therm approach, this can utilize solid and other poor-quality fuels.

Poor-quality fuels that come to mind for Canadian bitumen projects include coal, petroleum coke, coker cycle oil, residual crude, and high-melting-point asphalt. To be able to dispose of such materials in an integrated in situ project should allow significant reductions in the cost of upgrading.

The principle of the process is shown in Figure 8.46. It is based upon technology that is well known in the disposal of aqueous effluents such as sewage and pulp mill effluents. Many commercial disposal plants are in operation. Fuel, water, and oxygen (or air) are introduced continuously into a reactor maintained at the steam pressure.

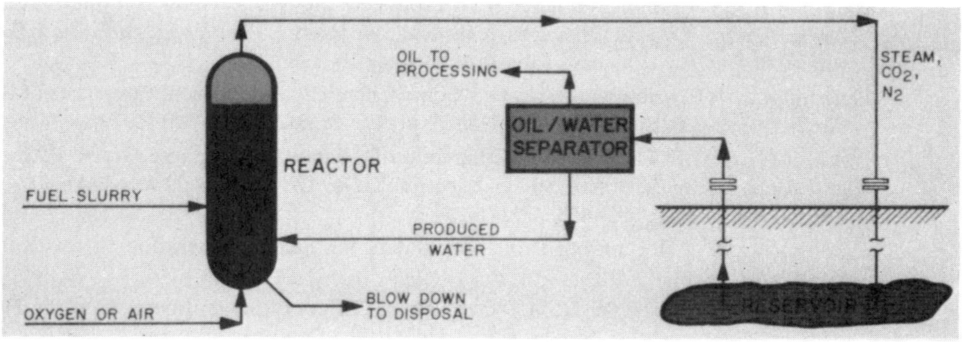

Figure 8.46 Zimpro's Wet-Air Oxidation Process (from AEC/Zimpro)

The conditions within the reactor are such that it is largely full of liquid water, which boils as a result of the oxidation of the fuel. The heat is liberated within the liquid phase. Unreactive solids and some water are blown down periodically to prevent accumulation.

The process has the advantage of being able to consume a wide range of fuel types and of avoiding pollution by sulphur oxides; the sulphur leaves as dissolved sulphates and sulphites within the water. The process can use untreated water, produces 100% quality steam, and is potentially very energy efficient.

Disadvantages of the process are the need to compress all the combustion air or oxygen to the steam pressure and the corrosivity of the steam–carbon dioxide product. Also, the reactor has to be built from special corrosion-resistant alloys, and it has to have thick walls to withstand the high pressure; these factors make it expensive.

A critical factor in determining the success of this process will be whether it can be shown that the steam–carbon dioxide mixture is more effective for recovery than simple steam. Also, it will have to compete with wet, in situ combustion using oxygen, which also generates a mixture of steam and carbon dioxide. Other processes whose success depends largely on the effectiveness of steam-gas mixtures include high-pressure downhole steam generation and Carmel's Vapor Therm generation.

BIBLIOGRAPHY

ALBERTA ENERGY, "Development of a Coal-Fired Boiler for Steam Injection in Heavy Oil Recovery," Alberta Office of Coal Research and Technology, Pub No I/293, Edmonton, (1989).

BABCOCK and WILCOX, *Steam — Its Generation and Use,* New York: Babcox and Wilcox Co., (1972).

BERRY, R. I., "Fluid-Bed Gets the Nod," *Chem. Eng.,* 60–62 (Oct. 8, 1979). © 1979 SPE.

BETZ *Handbook of Industrial Water Conditioning,* 8th Ed., Trevose, Penn.: (1980).

BLOKH, A. G., *Heat Transfer in Steam Boiler Furnaces,* Hemisphere Pub. Co. (1988).

BORREGALES, C. J., "Steam Soak on the Bolivar Coast," in *The Oil Sands of Canada-Venezuela,* CIM Special Volume 17, 561–583, (1977).

BRIDLE, M. K., "Esso's Experience with Produced and Waste Water Recycle Systems," 3rd Annual Heavy Oil and Oil Sands Technical Symposium, *Challenges and Innovations,* University of Calgary, Calgary, Alberta (February 19, 1986).

BURCHFIELD, T. E. and HEPLER, L. G., "Some Chemical and Physical Properties of Produced Water from an In Situ Oil Sands Plant," *In Situ* 3, no. 4: 383–390 (1979).

BURKILL, G. C. C., "Thermal Well Completion Design with Openhole Gravel Packed Liners and Methods for Selective Steam Injection," *The Oil Sands of Canada-Venezuela,* CIM Special Volume 17, 595–608, (1977).

CARRELL, N. A., "Reclaiming Produced Waters for Steam Generation in the Kern River Field," *SPE* 8411 (1979).

CHUNG, K. H. and BUTLER, R. M., "Geometrical Effect of Steam Injection on the Formation of Emulsions in the Steam-Assisted Gravity Drainage Process," *JCPT,* 36 (January/February 1988).

CORMIER, K.W., "Insulated Tubing at Shell's Peace River Project," 4th Annual Heavy Oil and Oil Sands Symposium, University of Calgary, Calgary, Alberta (February 18, 1987).

COURTNAGE, Lisa A., "Water Management Challenges at Esso's Cold Lake Operation," 4th Annual Heavy Oil and Oil Sands Symposium, University of Calgary, Calgary, Alberta (February 18, 1987).

DELIBERT, J. C., "Steam Boilers," in *Marks Standard Handbook for Mechanical Engineers, 9th Edition,* McGraw Hill (1987).

DOSCHER, T. M., LABELLE, R.W., SAWATSKY, L. H., and ZWICKY, R.W., "Steam-Drive—A Process for In-Situ Recovery of Oil from the Athabasca Oil Sands," *K. A. Clark Volume,* 123–141, Edmonton, Alberta: Research Council of Alberta, Information Series No. 45 (1963).

ELFARR, J., "The Chamber Pump and its Application to Hot Wells," ASME 76-Pet-91, presented at Joint Petroleum Mechanical Engineering and Pressure Vessels and Piping Conference, Mexico City (September 19–24, 1976).

ELGERT, B. D., CHAMBERS, N. A., and SUZUKI, F., "Cold Lake Artificial Lift Optimization," 6th Annual Heavy Oil and Oil Sands Symposium, University of Calgary, Calgary, Alberta (March 8, 1989).

ESON, Rod L., "Downhole Steam Generator—Field Tests," *SPE* 10745 (March 1982). © 1982 SPE.

FANARITIS, J. P. and KIMMEL, J. D., "Review of Once-Through Generators," *JPT,* 409–416 (April 1965).

FAROUQ ALI, S. M. and MELDAU, R. F. "Current Steamflood Technology," *JPT,* 1332–1342 (October 1979). © 1979 SPE.

GATES, C. F. and BREWER, S.W., "Steam Injection into the D and E Zone, Tulare Formation, South Belridge Field, Kern County, California," *JPT,* 343–348, (March 1975).

GATES, C. F. and HOLMES, B. G., "Thermal Well Completions and Operation," Paper PD11, 7th World Petroleum Congress, Mexico City (1967).

HONG, K. C., "Two-Phase Flow Splitting at a Pipe Tee," *JPT,* 290–296, (Feb. 1978).

IMPERIAL OIL LIMITED, "The Cold Lake Project," Appl. 770866 to Alberta Energy Resources Conservation Board, May 1978.

KEELING, L., "Venting Well Casings at Cold Lake," Heavy Oil and Oil Sands Technical Symposium, University of Calgary, Calgary, Alberta (February 20, 1985).

KERBY, S. L., KENSE, K. and PEACHEY, B. R., "Upgrading of Oilfield Steam Generator Performance," Heavy Oil and Oil Sands Technical Symposium, University of Calgary, Calgary, Alberta (February, 1984).

Kirk-Othmer Encyclopedia of Chemical Technology 3rd Ed., Volume 24, 369 (1984), Wiley (1978–84).

KLOEPFER, J. G., CARD, R., and KUS, J., "Process Design for the Re-Use of Wastewater for Oil Production," Presented at AOSTRA's 4th Annual Conferences on Advances in Petroleum Technology, Calgary, Alberta (May, 1983).

KONAK, A. R., "Method and Apparatus for Splitting Two-Phase Flow at Pipe Tees," *US,* 4, 522, 218 (1985).

KONAK, A. R., "Method and Apparatus for Splitting Two-Phase Flow at Pipe Tees," *US,* 4, 574, 827 (1986).

KONAK, A. R. and GRISARD, P. J., "Deoiling of Produced Water by Acidification," Paper No. 80-31-30, Petroleum Society of CIM, Annual Technical Meeting, Calgary, Alberta (1980).

LESCHYSHYN, T. and SEYER, W., "In Situ Oilsands Temperature Logging," 6th Annual Heavy Oil and Oil Sands Technical Symposium, Calgary, Alberta (March 8, 1989).

LIM, G. B., "Steam Quality Survey at Esso's Cold Lake Pilots Using a Non-Intrusive Neutron Scattering Device," Heavy Oil and Oil Sands Symposium, University of Calgary, Calgary, Alberta (February 20, 1985).

LIM, G. B. and KONAK, A. R., "Thermal Softening Process," *US* 4, 518, 505 (1985). © 1985 SPE.

MARSHALL, BILLY W., "Operational Experiences of a Downhole Steam Generator," *SPE* 10744 (March 1982).

MELDAU, R. F., "Reducing Well Bore Heat Loss," Reprints of papers in the Thermal Well Completions Seminar held at the 4th International Conference on Heavy Crudes and Tar Sands (UNITAR), Edmonton, Alberta (August 7–12, 1988). Presented and published by the Canadian Heavy Oil Association.

MILLER, R.W., *Flow Measurement Engineering Handbook,* McGraw Hill (1983).

PALM, J.W., ANDERSON, W. H. and KIRKPATRICK, J.W., "Determination of Steam Quality Using an Orifice Meter," *JPT,* 587 (June 1968).

PEACHEY, B. R., "Design Conditions for Very Large Oilfield Steam Generators," Paper No. 84-35-81, Petroleum Society of CIM, Annual Technical Meeting, Calgary (June 10–13, 1984).

PEACHEY, B. R. and NODWELL, J. A., "Design Considerations for Heavy Oil In Situ Pilots," Seminar on Advances in Petroleum Recovery and Upgrading, Calgary, Alberta (March 24–26, 1981). Organized by AOSTRA and CANPET.

PETROLEUM EXTENSION SERVICE, University of Texas and API Division of Production, *Treating Oil Field Emulsions,* 3d Ed. (1974).

SABA, N. and LAHEY JR., R.T., "The Analysis of Phase Separation Phenomena in Branching Conduits," *Int. J. Multiphase Flow,* 10, No. 1 p 1–20, (1984).

SNAVELY JR., E. S. and BERTNESS, T. A., "Removal of Sulfur Dioxide Emissions by Scrubbing with Oilfield-Produced Water," *JPT,* 227–232 (Feb. 1975).

SPERRY, J. S., "Heavy-Oil Recovery System Completes Three Field Tests in Mid-Continent Region," *Oil Gas J.,* 225–237 (July 27, 1981). © 1981 SPE.

STEINOUR, H. H., "The Rate of Sedimentation," *Ind. Eng. Chem.,* 36, 618–624, 840–847, 901–907 (1944).

THIELEN, W., WALDMANN, H., MORICEL, M., PADRON, and CAMACHO, F., "Steam Generation Using High TDS Water and Heavy Fuels," Paper 183, 4th UNITAR International Conference on Heavy Crudes and Tar Sands, Edmonton, Alberta (August 7–12, 1988).

VERHOFF, F. H. and BANCHERO, J.T., "Predicting Dew Points of Flue Gases," *Chem. Eng. Progress,* 70, No. 8, 71–72, (Aug. 1974).

VONDE, T. R., "Specialized Pumping Techniques Applied to a Very Low Gravity Sand-Laden Crude, Cat Canyon Field, California," 1st UNITAR Conference, Edmonton, Alberta (June 4–12, 1979), reported in *The Future of Heavy Crudes and Tar Sands,* New York: McGraw-Hill (1981), 574–585.

WENDT, R. E., "Review of Stack Gas Scrubber Operating Experience in an Oil-Fired Steam Generator," *SPE,* 7125 (1978).

WHALLEY, M. J. and WILSON, T. M., "Water Conservation in a Steam Stimulation Project," 1st UNITAR Conference, Edmonton, Alberta (June 4–12, 1979), reported in *The Future of Heavy Crudes and Tar Sands,* New York: McGraw-Hill (1981), 734–738.

WILLWHITE, G. P. and DIETRICH, W. K., "Design Criteria for Completion of Steam Injection Wells," *JPT,* 15–22 (January 1967).

WILSON, T. M., "Steam Quality and Metering," Preprint 26th Ann. Tech. Mtg. of the Pet. Soc. of CIM, Banff, Alberta, June 11–13, (1975).

WOICESHYN, G., MARTIN, W., MONTIN, J., YUEN, P., and MANZANO, J., "Steam Quality Measurements by Neutron Transmission," 3rd UNITAR/UNDP International Conference on Heavy Crude and Tar Sands, Long Beach, Calif, July 22–31, 1985. pub. by AOSTRA, Edmonton (July 1988).

WOICESHYN, G. E., YUEN, P. S., JOHN, H., and MANZANO-RUIS, J. J., "Measurement of Steam Quality, Mass Flow Rate and Enthalpy Using Combined Densitometer and Nozzle," *SPE/DOE* 14907 (1986).

ZAIDI, A., KOK, S. and SCHMIDT, J.W., "Process Options for Recycle of High TDS Produced Water during In Situ Recovery of Heavy Oil," in R. F. Meyers and E. J. Wiggins (Editors), The Fourth UNITAR/UNDP International Conference on Heavy Crude and Tar Sands, Vol 3: Mining, Drilling, AOSTRA, Edmonton, (1989), pp 449–465.

9

In Situ Combustion

INTRODUCTION

In steamflooding, a considerable amount of the energy generated by the fuel is lost without providing useful heat in the reservoir. There are losses in the steam generator, in the steam transmission lines, from the well bore, and to the overburden. A significant factor is that in steamflooding, it is necessary to keep the reservoir and overburden heated behind the condensation front even after the oil has been depleted.

Estimates of the ranges of the various heat losses are shown in Table 9.1. The heat losses from the steam-generation facilities, from the steam distribution lines, and from the well bore within the overburden can amount to over half of the energy in the fuel and even in favorable circumstances, to about one-quarter of the energy.

TABLE 9.1 Steamflooding—Thermal Efficiency Basis Heating Value of Fuel = 100

	% Efficiency of Step	Approximate Range Cumulative
Steam Generator	75–85	75–85
Transmission to well[1]	75–95	56–81
Flow down well to reservoir	80–95	45–77
Flow in reservoir to condensation front	25–75	11–58

[1]Loss ≈ 270 Btu/h ft of insulated 6-in. pipe.

Within the reservoir, it is necessary for the steam-swept region to remain at steam temperature so that the steam can condense at the advancing condensation front. Because of this, the heat losses from the steam-swept region continue after the front has passed.

With in situ combustion the situation is different in that there are no heat losses until the oxygen in the injected gas reacts with the fuel at the fire front. Heat behind the front preheats the advancing air, and there is thus some conservation of heat. This effect is enhanced in wet combustion by the addition of water to the injected air. The water cools the swept reservoir behind the front, boils, and then supplies steam that passes through the combustion front and condenses farther along, where it preheats the reservoir ahead of the front.

In situ combustion thus has features that give it the potential for being more efficient and economical. The fuel for in situ combustion comes from the residual material in the reservoir and there is no need to supply combustion fuel. However, it is necessary to provide energy to compress the air and if oxygen is used, to separate the oxygen from the air. This energy for compression or, in the case of oxygen, for separation and compression is much less than that required for steam generation.

In Table 9.2, the fuel requirements (excluding the in situ combustion fuel) for in situ combustion with air or with pure oxygen are compared to that for thermal recovery using steam. The basis for the comparison is the supply of a constant amount of heat, 1 million Btu, to the reservoir. The quantity of steam generation fuel is compared to that required to compress the air or to separate and compress the oxygen. It is assumed that the compressed air or oxygen must be supplied at 1000 psig.

The column headed efficiency in the table relates the amount of heat supplied to the reservoir to the amount of energy, assumed to be supplied as natural gas fuel

TABLE 9.2 Fuel Requirements to Supply 1 Million Btu to Sand Face at 1000 psig

	EFFICIENCY %	FUEL REQUIRED Millions of Btu	ENERGY COST \$/Million Btu in Res.(4)
Steam(1)	45–77	1.3–2.2	2.6–4.4
Air in situ combustion(2)	190	0.5	1.0
O_2 in situ combustion(3)	315	0.3	0.6

(1)Based on Table 9.1.

(2)Assuming compressor driven by gas engine:
- Engine efficiency 34% based on LHV of 908 Btu/SCF for CH_4;
- Compressor drive requirement 6.35 HP h/1000 SCF;
- Heat of combustion 100 Btu/SCF of air (476 Btu/SCF of O_2).

(3)The mechanical energy required to separate 1000 SCF of air is approximately 2.9 HP-h.
O_2 required to generate 1 million Btu is 2101 SCF.
This can be produced by separating 2101/0.21 = 10,004 SCF air.

	Work HP h	Equivalent Fuel Btu
Air separation	29.0	217,160
O_2 compression	13.3	99,904
		317,064

(4)Fuel cost assumed to be \$2/Million Btu. Capital costs are not included.

to fire the boiler, to fuel an engine to compress the air or to separate and compress the oxygen. On this basis, the in situ combustion options compare very favorably to the steam option. In this comparison air in situ combustion requires only about 23% to 39% of the fuel needed for steam.

For oxygen injection, the potential fuel savings are even greater than those for air injection. The reason for this is that only about one-fifth as much gas needs to be compressed to 1000 psig. Although it is also necessary to separate the air by liquification and distillation, the energy requirements for doing this are relatively small compared to those for high-pressure compression.

In a modern tonnage air separation plant, the air feed needs to be compressed to about 100 psia, and this work supplies essentially all of the energy for separation (Newton 1979). A comparison of the energy required for the compression of the O_2 to that for separation is shown in the footnotes to the table.

The cost of the energy for supplying 1 million Btu to the reservoir is only \$0.60 to \$1 for in situ combustion as compared to \$2.60 to \$4.40 for steam. These numbers do not take into account either the capital costs or the nonfuel operating costs.

In spite of the potential advantages for in situ combustion, it is much less widely used than steam. Figure 9.1 shows the results of a survey of thermal EOR production in the USA.

In 1988 there were nine commercial U.S. ISC projects, having a total capacity of 6525 B/d. The largest U.S. steam projects were those of Shell, Belridge (101,000 B/d) and Texaco Kern River (87,600 B/d), both in Kern County, California.

In situ combustion has found less success than steam because of the difficulty in controlling the process. Fire fronts tend to advance much more erratically than steam fronts, and it is much harder to obtain an even sweep of the reservoir. Nevertheless there are successful in situ combustion projects.

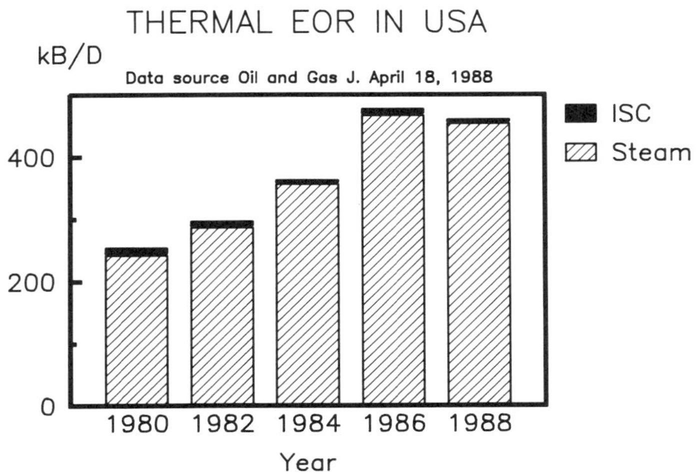

Figure 9.1 Oil Production Using Thermal EOR in USA

DRY COMBUSTION

Description of Phenomena

Figure 9.2 shows typical temperature and saturation profiles for a dry combustion process carried out in a laboratory combustion tube. The term *dry* means that water is not introduced intentionally into the reservoir. The gas saturation at a distance ahead of the front is high enough to allow the combustion gases to flow to the production end without undue pressure drop. This is frequently already the case in laboratory experiments where the tube has been filled by packing. Some distance beyond the front, water is condensing, and the oil is swept forward by the gas and steam. If the initial gas saturation is high, the oil displaces some of the gas and forms an oil bank. If there is a low gas saturation, the initial gas flow tends to displace oil and build gas saturation. Under these circumstances, injectivity for air may be very low.

Immediately beyond the combustion zone, residual oil loses volatile material and cracks thermally, leaving behind a carbonaceous, nonvolatile deposit, coke. It is this coke that provides the fuel to react with the injected air. At the fire front, the temperature is at maximum. Upstream of the front, the temperature falls because of the cooling effect of the injected air. Downstream beyond the front, the temperature falls as the heat carried forward by the flowing gas is consumed heating the reservoir and residual oil and by supplying the heats of cracking and evaporation.

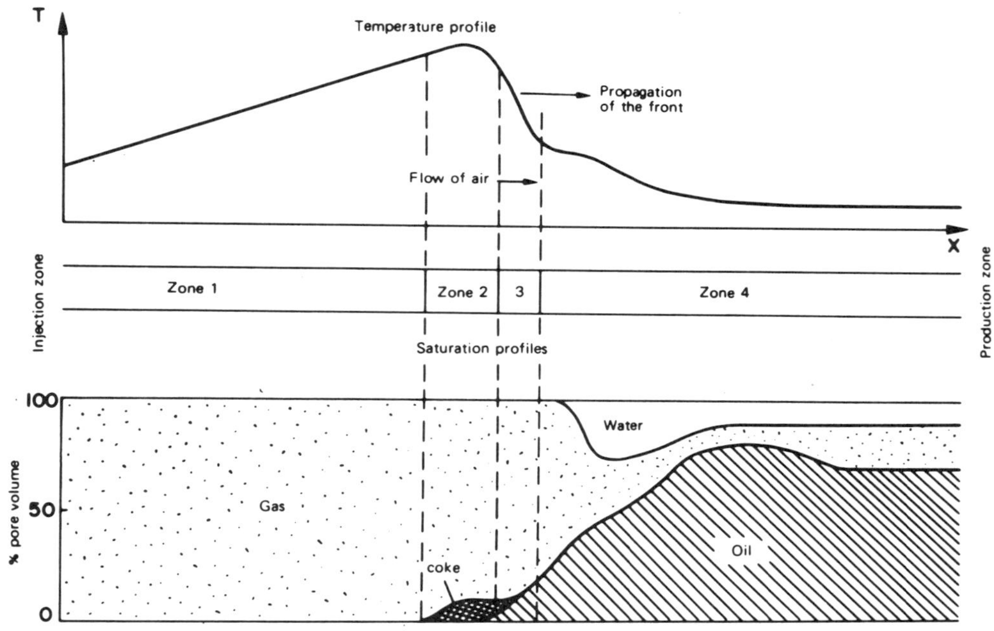

Figure 9.2 Temperature and Saturation Profiles in Dry Forward Combustion (from Latil 1980)

In the condensation region beyond the fire front, there is frequently a temperature plateau that corresponds to the condensation of the steam. This is much more pronounced when considerable water is present; in dry combustion it may be almost absent.

The process shown in Figure 9.2 is known as *forward dry combustion.* Another possibility called *reverse combustion* (Trantham and Marx (1966; Berry and Parrish (1960)) has received attention and has been investigated experimentally (see, for example, the laboratory studies described in Wilson et al. 1963).

In reverse combustion, as shown in Figure 9.3, the process is started in the same manner as for forward combustion, and then the direction of air injection is reversed, so that the original injector becomes the producer and vice versa. This can be compared to smoking a cigarette by lighting it in the normal manner and then carrying on the combustion by blowing into the cigarette rather than sucking. The combustion zone moves up the cigarette but against the flow of air.

Such a process was thought to be particularly suited for the production of very viscous bitumens and tar because the produced fluid remains much hotter and it does not have to be forced through the unheated reservoir as in Figure 9.2.

There is an analogy between the reverse combustion described here and the SAGD process (see Chapter 7), in which the injector is close to the producer. In the SAGD process, the produced oil remains hot as it flows to the production well, whereas in conventional steamflooding, oil that is displaced from the steam chamber has to flow through a cooler reservoir to reach the production well. Similarly, in reverse combustion, the displaced oil flows through the hot burned zone to reach the producer. In forward combustion, the displaced oil must be forced through a cold reservoir to reach the production well. This is very difficult in cold tar sands.

Unfortunately, practical attempts to use the reverse combustion process have usually proven to be unsuccessful because of the formation of new flame fronts near to the injector. These are ignited by spontaneous combustion in a manner that is described later. Reverse combustion is an ingenious idea with attractive incentives for oil sands, but it has turned out to be unsuccessful (Dietz and Weijdema 1968).

Combustion Tubes

There have been many studies of in situ combustion in the laboratory using combustion tubes. A recent design is shown in Figure 9.4 (Moore et al. 1987).

In order to minimize the effect of the heat capacity and thermal conductivity of the combustion tube, it is constructed with a very thin wall. The tube is filled with reservoir material, and air or oxygen is passed through it. Combustion is

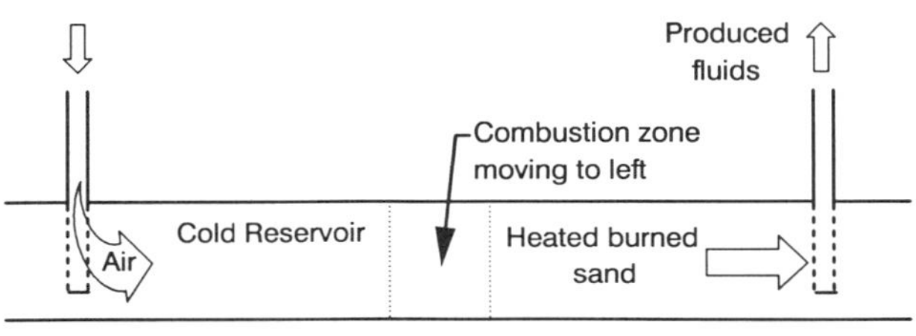

Figure 9.3 The Reverse Combustion Concept

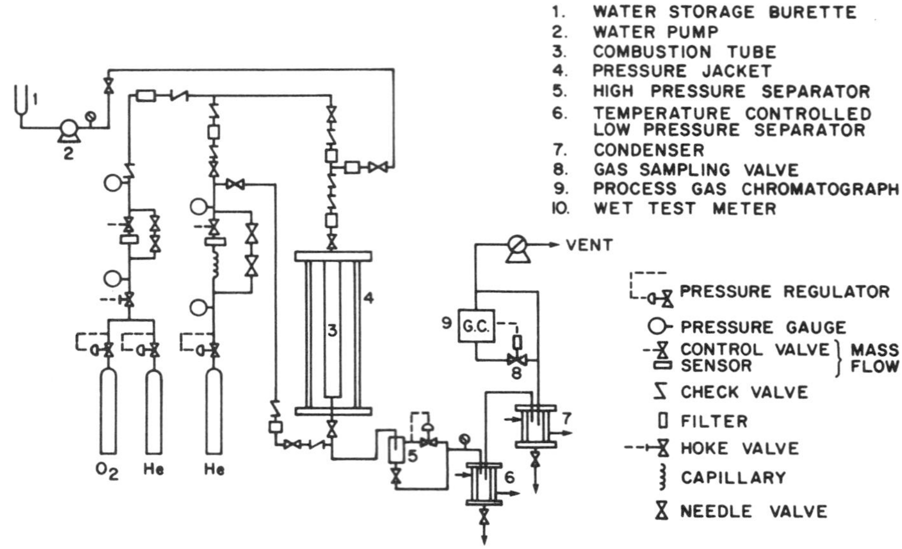

Figure 9.4 University of Calgary in Situ Combustion Apparatus (from Moore et al. 1987)

started from one end, and the progress of the combustion is followed by means of thermocouples and produced gas analyses. Ignition is usually accomplished by means of an electric resistance heater.

The thin-walled combustion tube is contained within a strong outer vessel that can withstand the desired operating pressure. The annular space between the tube and the pressure vessel is filled with insulation and with a gas under a pressure high enough to support the weak inner tube.

A major consideration is the minimization of heat loss from the tube so that the adiabatic conditions within a large reservoir can be simulated. In some tubes this heat loss is minimized by insulation alone; in others, short compensating electrical heaters are placed outside the inner tube and controlled in order to minimize the temperature differences. A problem with this sort of system is caused by the possibility of leading, or "helping," the combustion if the external heaters get too hot. Since the axial temperature gradients can be quite steep, it is desirable, if external heaters are to be employed, to use numerous small heaters so that the internal gradient can be duplicated.

The design of the adiabatic compensating heaters has been studied by Leaute and Collyer (1984). They have shown that it is important to use a high grade of insulation between the heaters and the tube and to employ narrow heaters. The large effects of these variables are shown in Figure 9.5, which is taken from their paper.

It is usually found desirable to employ higher front velocities in the laboratory than would be found in the field so that the heat loss problem in the laboratory apparatus can be minimized.

It is customary to operate combustion tubes in a vertical position so that gravity will not cause overide. Another approach that has been used is to make the tube horizontal but to rotate it so that the effects of gravity are canceled out (Latil 1980).

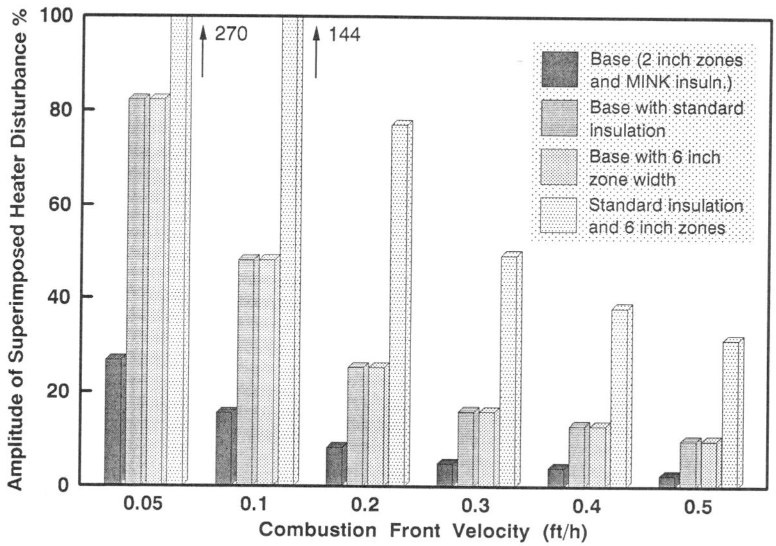

Figure 9.5 Calculated Effect of Heater Width and Insulation on the Performance of an Electrically Heated 3-in. Diameter Adiabatic Combustion Tube. Using a Mathematical Model the Authors Investigated the Effect of Varying the Length of the Heated Zones and of Using a Superior Insulation (Johns-Manville MINK) Which Has One-Third of the Thermal Conductivity of the Standard Material (after Leaute and Collyer 1984).

Figures 9.6 through 9.10 from Penberthy and Ramey (1966) show results obtained in a classic combustion tube experiment using a 5-darcy sandpack and a 21° API oil.

Figure 9.6 shows temperature measurements from a typical experiment. The very sharp gradient ahead of the advancing combustion front is a notable characteristic. A temperature plateau ahead of the front, which corresponds to the condensation of water, can also be seen. This is caused by the connate water in the sample; the steam plateau becomes very much more pronounced if water is added intentionally to the air.

As may be seen from Figure 9.7, the burning front for a constant air injection rate moves at a steady and essentially constant rate down the tube; it moves more slowly than the steam front.

The produced gas rate and the injection pressure are shown in Figure 9.8. The main point of interest here is the buildup in pressure gradient caused by the banking of the oil ahead of the combustion front. In this example, the gas saturation ahead of the combustion zone decreases as oil is forced down the tube. The increased oil saturation represents a "bank." If the initial gas saturation in the reservoir were nearly zero, then the pressure gradient would be high initially, and the gas saturation would increase to accommodate the flow of flue gas to the production end. In this case, oil would flow almost immediately, reflecting the effect of the gas flood beyond the combustion front. This is analogous to the waterflooding beyond a steam-condensation front that was discussed in Chapter 5.

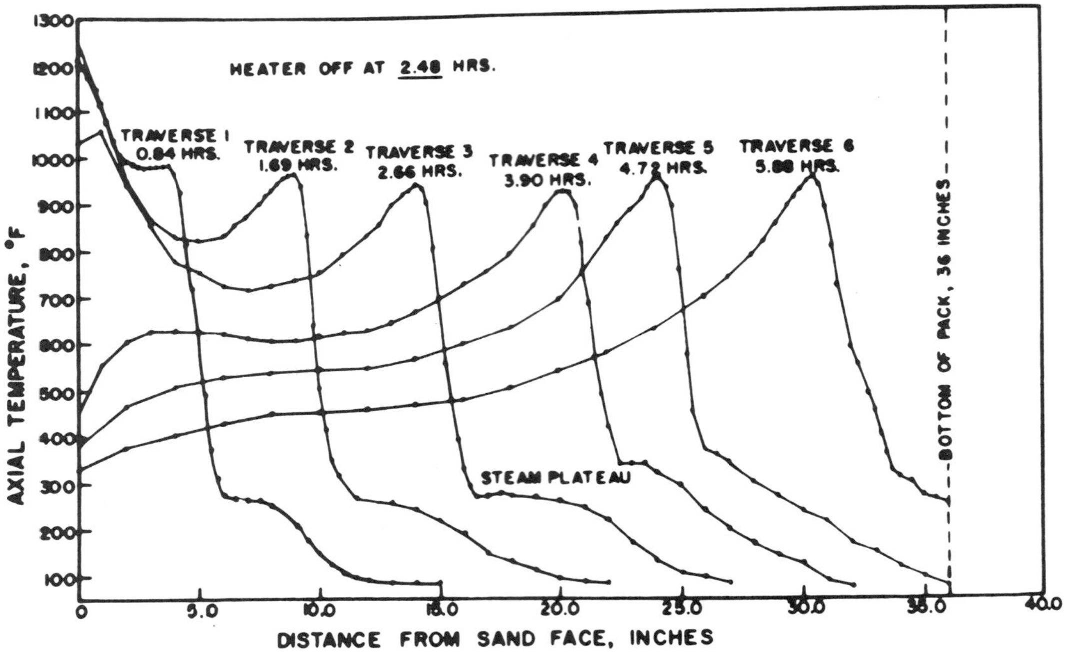

Figure 9.6 Temperatures along Axis of Combustion Tube at Various Times (from Penberthy and Ramey 1966)

Figure 9.9 shows the produced gas analyses for the same experiments. Once combustion is underway, the composition does not change very much. Only a small amount of unreacted oxygen is present in the produced gases. This is a common re-

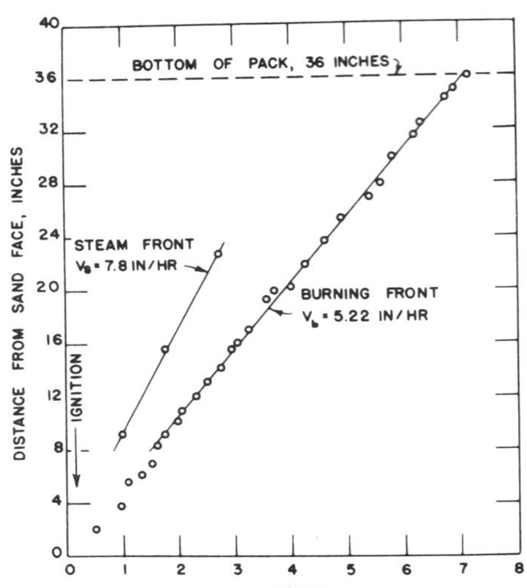

Figure 9.7 Position of Burning and Steam Fronts (from Penberthy and Ramey 1966)

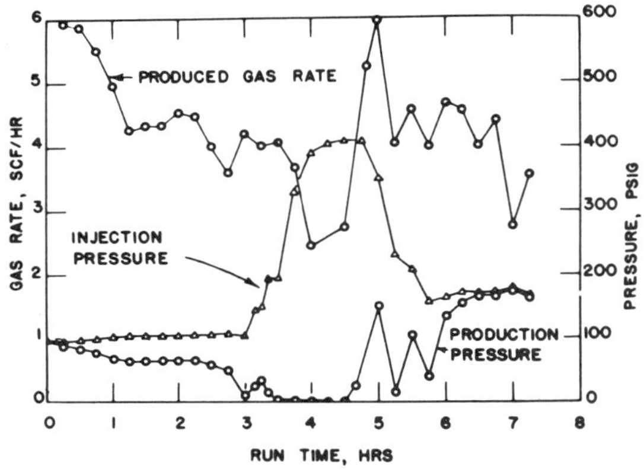

Figure 9.8 Gas Production Rate and Injection and Production Pressures versus Time (from Penberthy and Ramey 1966)

sult for one-dimensional combustion tubes, but in the field there can be a number of problems caused by the bypassing of unreacted air to the production wells.

The cumulative productions of oil and water are shown in Figure 9.10. There is a considerable delay before oil appears at the outlet. This is caused by the need to build a bank before the oil can flow to the production well. The delay is dependent mostly upon the amount of gas saturation present in the original reservoir, since this has to be filled before appreciable flow to the producer can be achieved. This delay in production can also occur in field projects and should be anticipated; it is discussed later in this chapter.

Alexander's Fireflood Pot

Another type of apparatus that has proved useful in in situ combustion research is the *fireflood pot* shown in Figure 9.11.

In this apparatus, air flows radially inward through a core sample to a tube inserted in a hole drilled in its center. The whole apparatus is heated at a controlled rate by an electrical resistance element wound around the external pressure con-

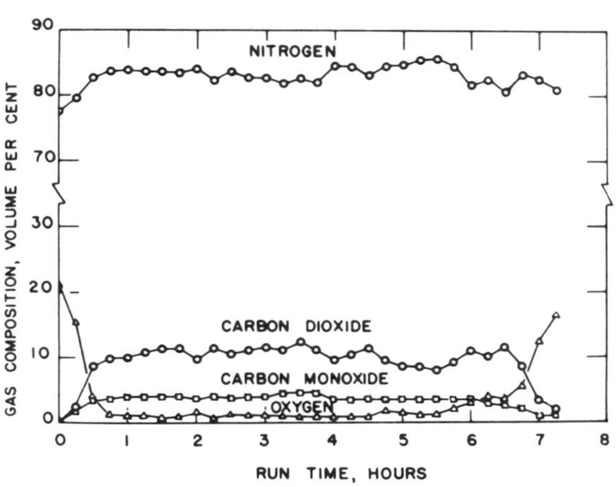

Figure 9.9 Produced Gas Composition (from Penberthy and Ramey 1966)

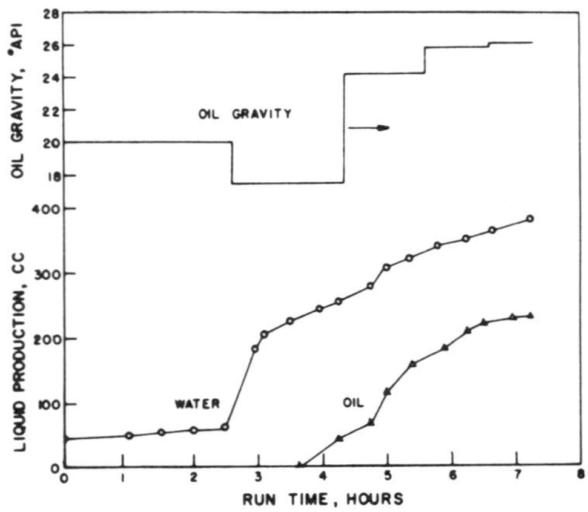

Figure 9.10 Liquid Production and Oil Gravity (from Penberthy and Ramey 1966)

tainment vessel. A typical experiment involves raising this temperature at 200°F/h up to a maximum combustion temperature of 800°F. Resulting gas analyses for an experiment are plotted in Figure 9.12.

Calculation of H/C Ratio for Fuel

Stoichiometric calculations can provide the H/C atomic ratio for the fuel that is being consumed in combustion tubes.[1] To do this we proceed as follows:

Suppose that the balanced chemical equation for the combustion process can be represented by:

$$\underset{\text{Fuel}}{CH_n} + \underset{\text{Air}}{S(O_2 + 3.76\ N_2)} = a\ CO_2 + b\ CO + c\ H_2O + d\ O_2 + e\ N_2$$

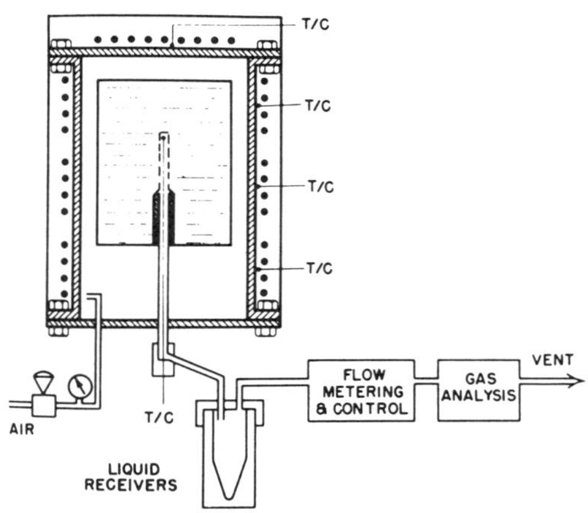

Figure 9.11 Fireflood Pot (from Alexander, Martin, and Dew 1962)

[1]This technique can be applied usefully to other combustion areas—for example, for the analyses of the operations of steam generators or process furnaces. Such studies are frequently directed toward the estimation of the excess combustion air. Large excesses of air introduced into a furnace result in low thermal efficiency because of the heat carried from the furnace by the heated excess air.

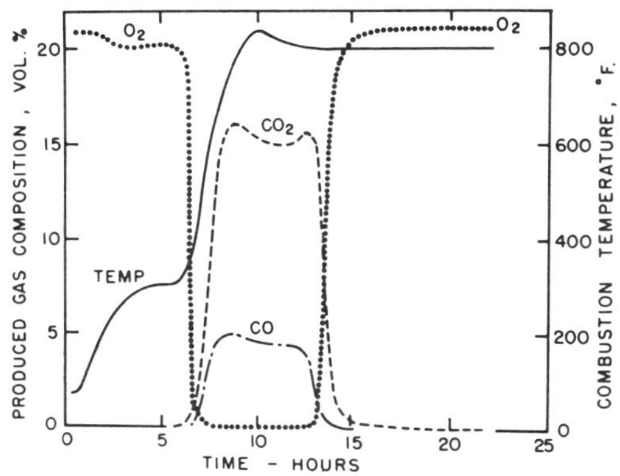

Figure 9.12 Temperature and Gas Analysis for an Experiment Using a Fireflood Pot (from Alexander, Martin, and Dew 1962)

There are seven unknowns in the preceding equation: n, S, a, b, c, d, and e. However, they are not all unrelated. Four algebraic equations involving the variables can be written from the material balances for each of the four elements. In addition, an analysis of the flue gas for CO_2, CO, and O_2 is normally available. This is usually on a dry basis—i.e., the concentration of water is excluded. These three analyses provide three more algebraic equations,which can be solved together with the equations from the material balances to give all seven of the unknown variables. This is simple to do in practical cases; an example is given next.

Example of Stoichiometric Calculation for Combustion Process

The flue gas analysis[2] for a combustion process is

Carbon dioxide	14.0%
Carbon monoxide	3.0%
Oxygen	0.2%
Nitrogen (by diff)	82.8%

Find the apparent H/C ratio n and the number of standard cubic meters of air fed per kilogram of fuel burned. Write an overall stoichiometric equation for the reaction.

1. Obtain S from ratio of C:N in feed and products by setting these equal:

$$\frac{1}{3.76S} = \frac{0.14 + 0.03}{0.828}$$

$$S = 1.2954$$

2. Calculate e from nitrogen balance:

$$e = 3.76 \times 1.2954$$

$$= 4.8706$$

[2]Gas analyses are usually reported on a volume basis. This is the same as a mole basis. Since water vapor normally condenses, it is usually not included in the analysis. The analysis is said to be on a dry basis.

3. Calculate a, b, and d from ratios of gas components to N_2:

$$a = \frac{0.14}{0.828} \times 4.8706 = 0.8235$$

$$b = \frac{0.03}{0.828} \times 4.8706 = 0.1765$$

$$d = \frac{0.002}{0.828} \times 4.8706 = 0.0118$$

4. Write a material balance for oxygen and solve for c:

$$1.2954 = \frac{0.8235 + 0.1765}{2} + \frac{c}{2} + 0.0118$$

$$c = 2(1.2954 - 0.8235 - 0.0118) - 0.1765$$

$$= 0.7437$$

5. Obtain n from the hydrogen balance and complete the problem:

$$n = 2c = 1.4874 = 1.49 \qquad \text{(for example)}$$

$$\text{kg moles air/kg fuel} = 1.2954 \times \frac{4.76}{12 + 1.49}$$

$$= 0.457$$

$$\text{Volume of air/kg fuel} = 0.457 \times 22.4 \text{ m}^3 \quad \text{at NTP}$$

$$= 10.24$$

The overall stoichiometric equation may be written:

$$CH_{1.49} + 1.2954(O_2 + 3.76\ N_2) = 0.8235\ CO_2 + 0.1765\ CO + 0.7437\ H_2O + 0.0118\ O_2 + 4.8706\ N_2$$

The Alberta ERCB defines standard conditions as 1 atmosphere and 15°C. Volume of air at standard conditions per kilogram of fuel are

$$\frac{10.24 \times 288}{273} = 10.80 \text{ m}^3$$

In the former British system, standard temperature was 60°F; 15°C is equivalent to 59°F.

Fuel Deposition

The most important measurement that comes from a combustion tube or fireflood pot run is that of the fuel production. This is frequently expressed as the weight of fuel per unit volume of reservoir or, alternatively, as the weight of fuel per unit weight of reservoir rock. It is calculated from a material balance.

The available experimental data can be largely explained by considering two factors that influence the fuel availability:

1. The propensity of the particular crude oil to deposit coke as it is heated and cracked. Light crudes contain larger proportions of volatile materials that distil without forming coke. As a result, these crude oils give relatively less fuel with in situ combustion than do heavy crudes. Also, the residues from paraffinic crudes tend to give higher yields of cracked products and less coke.
2. The conditions that affect the saturation of the crude oil within the reservoir pores as the combustion front advances. These include the initial oil saturation and the effect of steam moving oil ahead of the coke-forming zone. Oil that is moved ahead by the flood does not contribute to the fuel (coke) formation in the cracking zone. The movement of oil ahead of the combustion front is also dependent upon the nature of the reservoir matrix, so this too affects the fuel production.

In summary, the formation of fuel is determined by the nature of the crude and also by the conditions under which the fuel is formed.

Figures 9.13 through 9.16 show the effect of some of the significant variables upon fuel availability as measured by Alexander, Martin, and Dew (1962).

Figure 9.13 shows the effect of initial oil saturation for a number of different reservoir materials (both cores and crushed cores) using a 21.8° API crude. Increasing the initial oil saturation tends, in general, to raise the fuel availability. The relatively clean Ottawa sand gives much less fuel at the higher saturations because of the improved displacement of the crude ahead of the combustion front. This may also be seen from Figure 9.14, where data for the same series of experiments are plotted with an abscissa equal to the residual oil saturation that was left behind by the steam plateau ahead of the combustion front.

Figure 9.15 shows the effect of the crude oil type. Low °API crudes give much more fuel because of their greater tendency to form coke.

This is also shown in Figure 9.16, where the fuel availability has been plotted against the Conradson carbon for the different crude oils. Conradson carbon is a

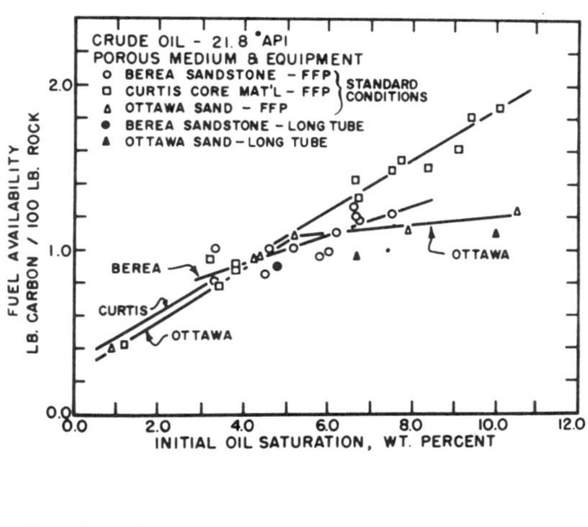

Figure 9.13 Effect of Initial Oil Saturation on Fuel Availability (from Alexander, Martin, and Dew 1962)

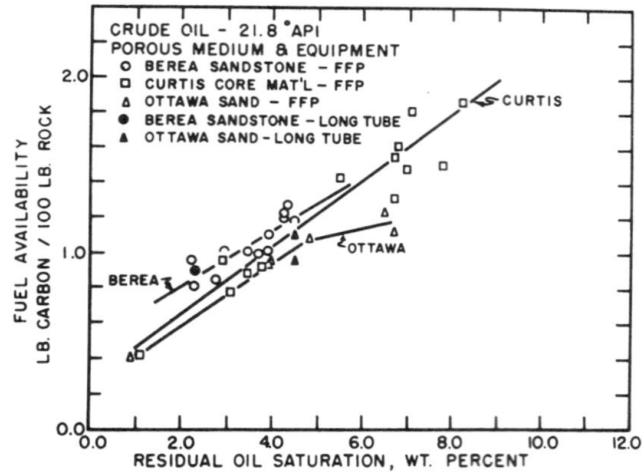

Figure 9.14 Effect of Residual Oil Saturation on Fuel Availability (from Alexander, Martin, and Dew 1962)

direct measure of the percentage of coke that is laid down by the crude oil in a standard high-temperature coking test (ASTM 1985). This test is used extensively in the design of petroleum refining equipment such as cokers. High Conradson carbon crudes give high fuel availabilities.

Low-Temperature Oxidation

When crude oils are contacted at high temperatures with oxygen, the products are mostly carbon oxides and water. However, at low temperatures, reactions occur in which the oxygen tends to combine and remain with the crude and also to cause it to polymerize by oxidative dehydrogenation type reactions. A particularly undesirable result of these reactions is that they increase the viscosity of the crude oil.

Figure 9.17 shows the effect of preoxidizing the crude at a lower temperature before measuring the fuel availability by combustion at 800°F. There is a very strong effect with a maximum at about 425°F. Above this, the fuel availability (measured in the subsequent oxidation at 800°F) drops off because the fuel is already partially converted to gaseous products in the preoxidation. At 425°F the polymerization caused by the preoxidation greatly increases the fuel availability.

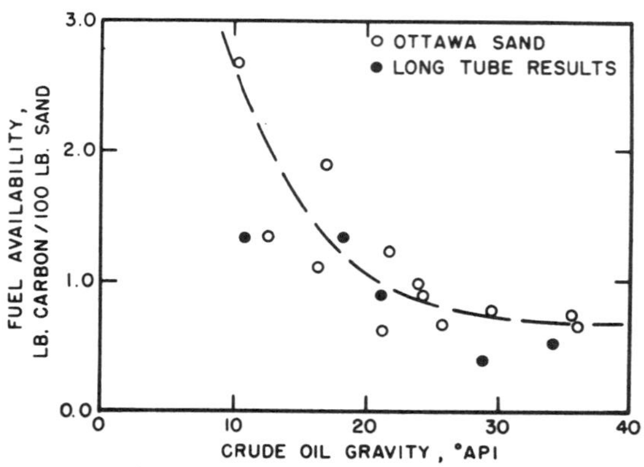

Figure 9.15 Fuel Availability versus °API (from Alexander, Martin, and Dew 1962)

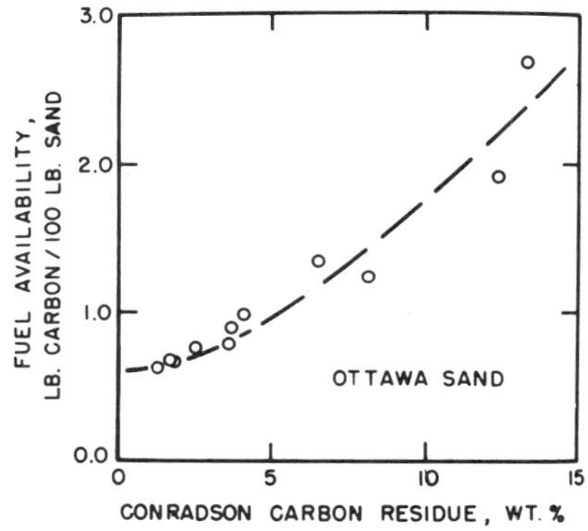

Figure 9.16 Fuel Availability versus CCR (from Alexander, Martin and Dew 1962)

The overall effect is shown in Figure 9.18, where the gross fuel availability (total from pre and final oxidations) is plotted against the preoxidation temperature. At the peak condition, the availabiilty of fuel is increased about three times.

Previously it was shown how the H/C ratio for the fuel could be determined from the analysis of the gaseous product. In this calculation, it was assumed that all of the oxygen not in the product as CO_2 and CO had reacted to form water. This is not true, particularly for low-temperature oxidations, because part of the oxygen can react and remain with the crude and coke.

Alexander, Martin, and Dew measured the "apparent" H/C ratio for a series of experiments carried out with a range of maximum reaction temperatures; the results are plotted in Figure 9.19. For reaction temperatures below 650°F, the apparent H/C ratio for the fuel is even larger than the ratio for the whole original crude.

High apparent H/C ratios (i.e., low CO_2 in flue gas) are an indication of a low combustion temperature and may be used as an indicator of what is happening within the reservoir.

With combustion temperatures of the order of 800°F, the apparent H/C ratio is a real indication of the type of fuel that is being consumed. Figure 9.20 shows a correlation of the total air requirements against fuel availability for a wide range of normal high-temperature combustions. The average is 189 SCF/lb of carbon burned.

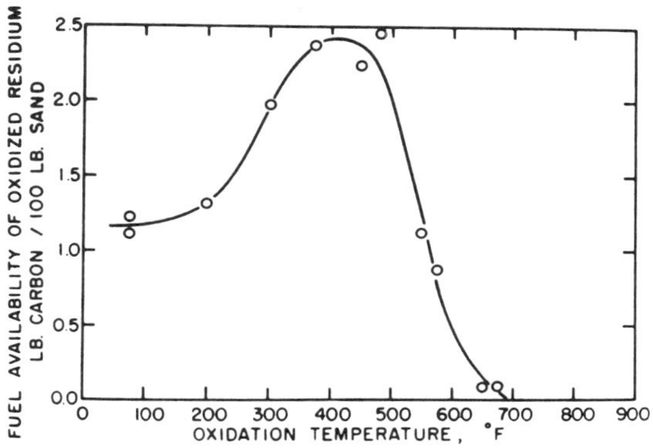

Figure 9.17 Effect of Low Temperature Oxidation on Fuel Availability (from Alexander, Martin, and Dew 1962)

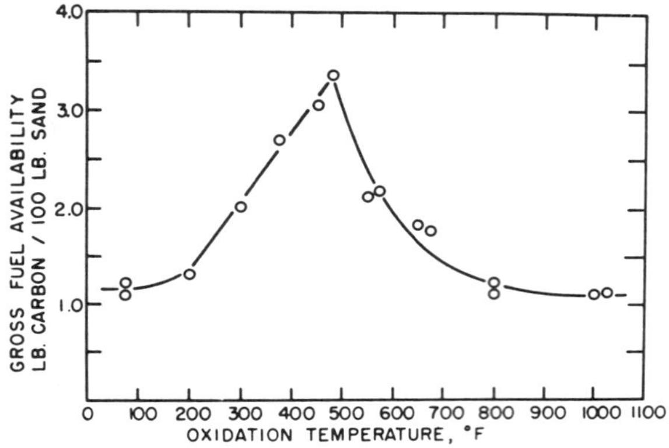

Figure 9.18 Effect of Preoxidation on Gross Fuel Availability (from Alexander, Martin, and Dew 1962)

Figure 9.21 is typical of a number of results that have been published by workers at IFP. These results were obtained in an experiment in which a sample of the reservoir material was heated at a 100°C/h in a laboratory apparatus with air passing through it. The marked difference between the oxygen consumed and that produced as CO_2 and CO indicates the prevalence of low-temperature oxidation around 250°C.

A significant difference between laboratory and field results is that the concentration of CO found in the field is much less than that in the laboratory. This has the effect of increasing, somewhat, the air requirements needed in the field.

In Situ Combustion Experiments Using Oil Sands

Leaute and Collyer (1984) have studied the effect of initial reservoir temperature in a series of combustion tube experiments using Cold Lake bitumen and dry conditions. From this work there are two very interesting findings:

1. With low reservoir temperatures where the unheated bitumen is essentially immobile, there is little mingling of the mobilized bitumen with the bitumen downstream of the combustion front; instead it moves through the cold reser-

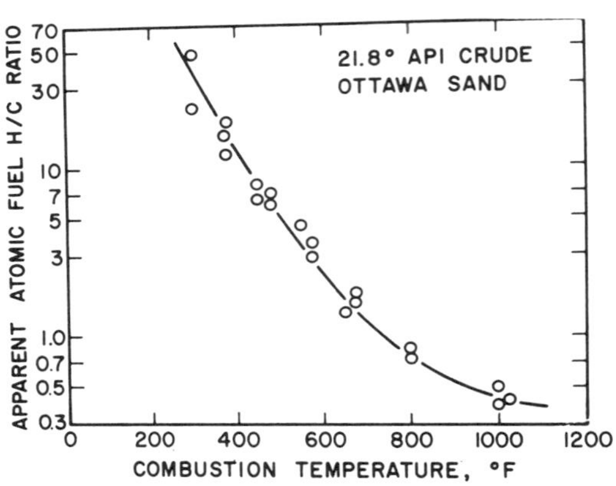

Figure 9.19 Apparent H/C Ratio of Fuel (from Alexander, Martin, and Dew 1962)

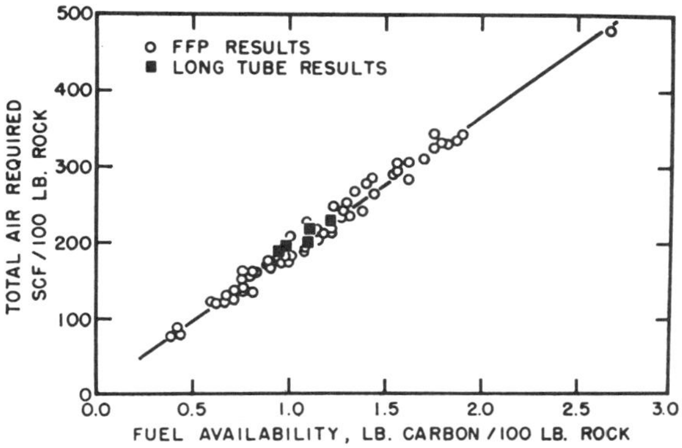

Figure 9.20 Effect of Fuel Availability on Air Requirements. Laboratory Tests above 650 °F (from Alexander, Martin, and Dew 1962)

voir in fingers. This was found from measurements of the viscosity of the product with time. The data are shown in Figure 9.22.

2. The fuel load is very dependent upon the initial reservoir temperature; higher temperatures result in less fuel laydown. This is shown in Figure 9.23.

These results indicate that better results should be obtained from oil sands that have been preheated. The workers from Esso conclude that in situ combustion

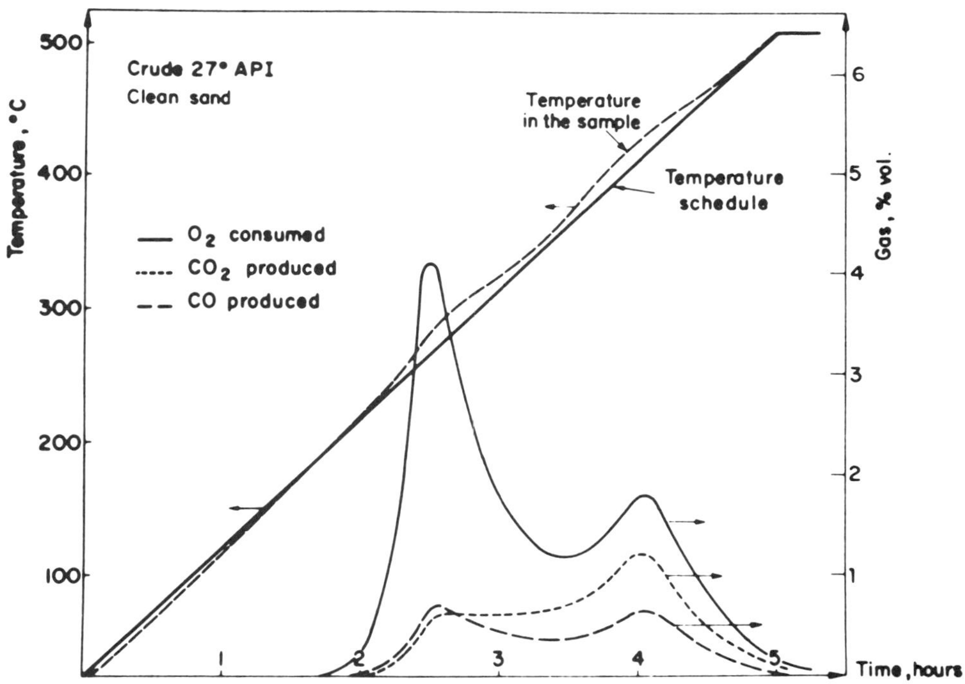

Figure 9.21 Comparison of O_2 Consumed and CO_2 and CO Produced During Oxidation of 0.89 SG Crude Oil (from Burger and Sahuquet 1972)

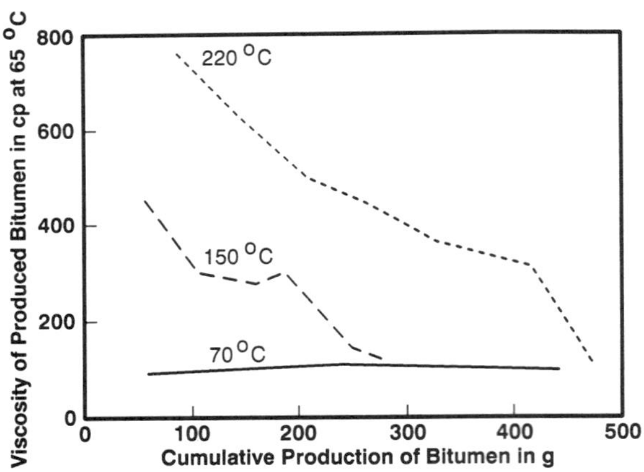

Figure 9.22 Effect of Preheat Temperature on Viscosity of Produced Bitumen (after Leaute and Collyer 1984). Parameter is the Preheat Temperature.

should be employed in oil sand reservoirs that have been first produced and preheated by steaming. Later in this chapter field projects are discussed in which in situ combustion is used to produce bitumen from reservoirs that have been produced initially by cyclic steam stimulation.

Ignition

The combustion process is frequently initiated by heating the reservoir around the injection well by means of an external source of heat. Both electrical and gas heaters have been employed, and each method is successful. Information on ignition methods is given in a paper by J.T. Moss, who is one of the pioneers in this area (Moss 1965). Also see White and Moss (1983).

In electrical ignition, a heating element with a power capacity of 10 to 40 kW is lowered into the well on the end of an armored power supply cable through a pressure seal. Means should be provided for measuring the downhole temperature in the vicinity of the heater. Air injection is commenced at a low rate, and power is supplied to the heater until ignition is obtained. Moss discusses various means that can be used to determine whether ignition has occurred:

- Rise in temperature of the air above that expected for the given input of electrical power.

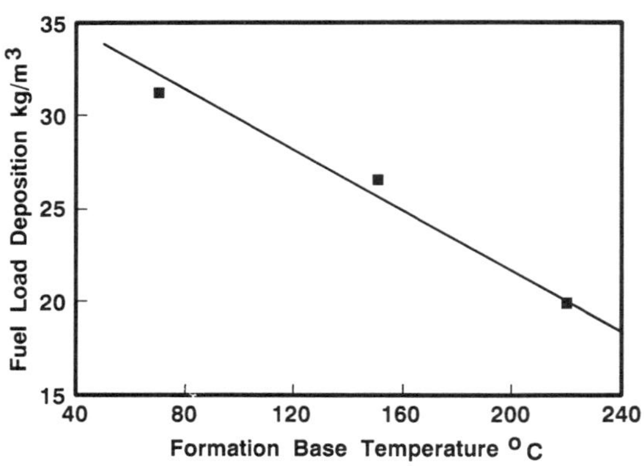

Figure 9.23 Effect of Formation Temperature on Fuel Load for Cold Lake Crude (after Leaute and Collyer 1984)

- Decreased air injectivity.
- Rise in temperature detected in observation well(s) near to the injector.
- Decrease in oxygen content of gas produced from producers. There is often a delay in the corresponding rise in carbon dioxide content because of reservoir absorption effects.

Moss also describes gas burners used for ignition. The burners are contained within a cyclindrical heat shield to protect the well from direct radiation. The burners themselves may be ignited using a chemical which flames spontaneously. The large Rumanian in situ combustion projects employ electrical heaters for ignition (Aldea, Turta, and Zamfir 1988).

Another means of achieving ignition is to rely upon the autoignition of the oil. The oxidation rate of an oil is often significant at the original reservoir temperature, and the heat produced by this low-temperature oxidation is sufficient to heat the reservoir gradually. As the temperature rises the oxidation reaction rate increases and eventually the temperature runs away to give high temperature combustion.

This mechanism caused the accidental ignition of oil reservoirs during the 1920s when air was being injected for pressure maintenance. It is these accidents that caused the discovery of the in situ combustion process (Ramey 1971).

The autooxidation process is much more rapid if the initial temperature of the reservoir is elevated. One way of doing this is to steam the reservoir first to heat it and then to inject air. This is a reliable means for achieving ignition (see, for example, Counihan 1977). As was mentioned earlier, attempts to use reverse combustion in the field fail because of the spontaneous ignition that occurs; this results in secondary combustion fronts near to the injection well.

Although ignition can generally be achieved in a straightforward manner, there are examples where it has been very difficult to achieve. In the Golden Lake project in Saskatchewan, Husky Oil had little difficulty in igniting their initial pilots using a gas heater. However, they subsequently failed to ignite two additional patterns that bordered on the original ones. They ascribe the difficulty to the partial depletion of the patterns before ignition was attempted (Miller and Jacques 1987).

The kinetics of the spontaneous ignition process have been studied by Tadema and Weijdema of Shell (Tadema and Weijdema 1970). Figure 9.24 shows their measurements of the oxidation rate of several crude oils as a function of temperature.

They represent the straight lines that are shown by the Arrhenius equation 9.1.

$$K = A_0 P_x^n e^{-B/T} \tag{9.1}$$

where K is mass O_2 consumed per unit time per mass of oil
A_0, B, n are constants that depend on oil and sand
P_x is O_2 partial pressure

By relating the rate of heat release to the heat needed to raise the temperature of the reservoir, they were able to derive the approximate equation 9.2. This predicts the time required for the spontaneous ignition of the reservoir as a function of the properties of the oil and reservoir and of the initial reservoir temperature.

$$t_i = \frac{\rho_1 C_1 T_0(1 + 2T_0/B)e^{B/T_0}}{\phi S_o \rho_o H A_0 P_x^n \, B/T_0} \tag{9.2}$$

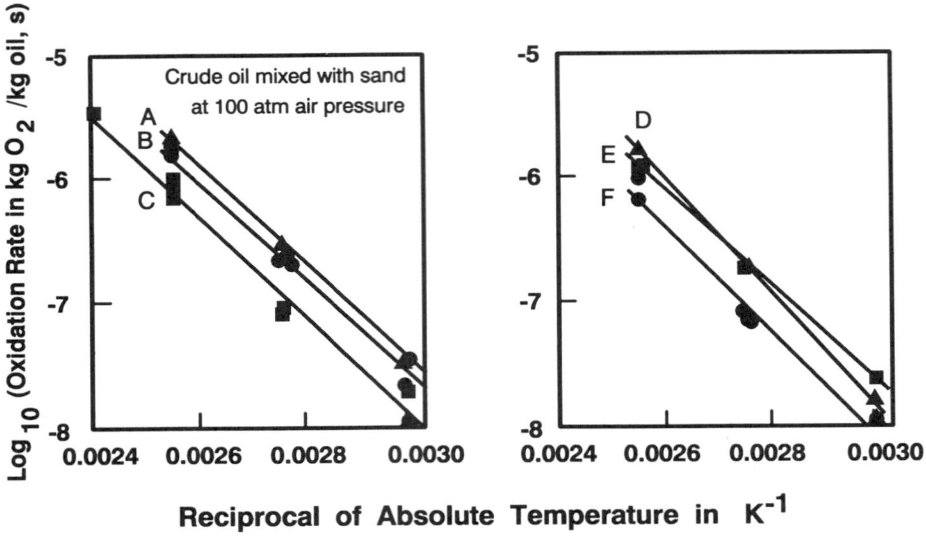

Figure 9.24 Oxidation Rate of Various Oils as a Function of Temperature (after Tadema and Weijdema 1970)

where t_i is ignition time in seconds
$\rho_1 C_1$ is heat capacity/unit volume formation
T_0 is absolute initial formation temperature
A_0, B, n are as in (9.1)
H is heat of reaction per unit mass of O_2
ϕ is porosity
S_o is oil saturation
ρ_o is oil density

Typical values of the constants A_0, B, and n for different crude oil-sand mixtures are given in Table 9.3.

A comparison of the ignition time that was measured in two field tests with those predicted from equation 9.2 is given in Table 9.4.

TABLE 9.3 Values of A_0, B, and n

CRUDE	A_0	B	n
A	3,080	8,860	0.46
B	925	8,640	0.57
C	498	8,880	0.79
D	84,800	10,270	0.48
E	1,210	8,680	0.45
F	7,380	9,480	0.31

TABLE 9.4 Spontaneous Ignition Times for Two Oilfields

	SOUTH BELRIDGE	VENEZUELA
Formation temperature T_0, °K	303.8	312.2
Formation pressure, atm	15.3	29.6
O_2 partial pressure, atm	3.20	6.18
A_0	3080	1210
B	8860	8680
n	0.46	0.45
H, kCal/kg O_2	2940	2940
Porosity, ϕ	0.37	0.34
Oil saturation, S_o	0.60	0.66
Water saturation, S_w	0.37	0.34
ρ_o, kg/m^3	970	980
$\rho_1 C_1$, kCal/m^3 °C	553	527
Ignition times in days		
Calculated	99	49
Observed	106	35

The agreement is very satisfactory. Using the data for the Venezuelan case from Table 9.4, the ignition time has been calculated assuming different reservoir temperatures, with the following results.

Effect of Reservoir Temperature on Autoignition Time

	(Same properties as for Venezuela case)					
Temperature, °C	10	20	30	40	50	60
Ignition time, days	706	266	107	46	21	10

It is apparent that a small amount of reservoir heating will greatly reduce the time required to obtain autoignition.

Temperature at the Combustion Front

One of the important variables in an in situ combustion process is the air flux at the combustion front. Air arriving at the front reacts with the fuel, and heat is generated. This heat raises the rocks in the vicinity of the combustion front to the combustion temperature and also supplies the heat that is conducted away from the front both ahead and also vertically.

The following discussion is based upon a paper by Ramey (1959). If the air flux is very large, then the heat lost by conduction will be a relatively small fraction of the total, and the combustion front temperature will approach the adiabatic temperature. In this circumstance the heat of reaction per unit volume of reservoir will just be equal to the sensible heat gained by the rocks. This adiabatic temperature rise is given by equation 9.3.

$$T_{max} - T_i = \frac{\Delta H W}{\rho C} \tag{9.3}$$

where T_{max} is adiabatic maximum temperature
T_i is initial temperature
ΔH is heat of combustion of fuel
ρC is reservoir heat capacity per unit of volume
W is fuel concentration per unit volume

It is important to note that, other things being equal, the temperature rise is proportional to the fuel concentration. This implies that if the fuel concentration is too low, it may not be possible to raise the rock to the required combustion temperature even if all of the other conditions are adequate. This can be a problem in attempts to use in situ combustion to produce oil from reservoirs having low oil saturation. Calculated values of $T_{max} - T_i$ are given in Table 9.5.

TABLE 9.5 Typical Values of $T_{max} - T_i$

Assume ΔH = 18000 Btu/lb and ρC = 35 Btu/ft^3 °F

Fuel Load W lb/ft^3	$T_{max} - T_i$ °F
0	0
0.5	257
1.0	514
1.17	600
1.5	771
2.0	1029

Effect of Conductive Cooling upon the Combustion Temperature

In general, the temperature rise will be lower than calculated from equation 9.3 because of the effect of the conductive heat losses. If the rate of heat generation is low because of a low air flux, then the relative effect of the heat loss will be greater, and the temperature of the combustion front will be lower. If the flux is low enough, the temperature of the front will fall to the point where vigorous combustion can no longer be sustained and only low temperature oxidation will occur. The "fire" will be extinguished. This effect can be analyzed mathematically by writing a differential equation that equates the heat conduction from the front plus the heat used to raise the front temperature to the heat evolved by combustion.

Ramey considered two cases. In both, the combustion front is considered to be growing radially from a vertical injector. In the first case, the flame front is considered to be an infinite vertical cylinder (i.e., vertical conduction losses are ignored); in the second, the vertical losses are included.

Figure 9.25 shows calculated results for the infinite vertical case. The temperature rise at the front (as a fraction of the adiabatic rise calculated from equa-

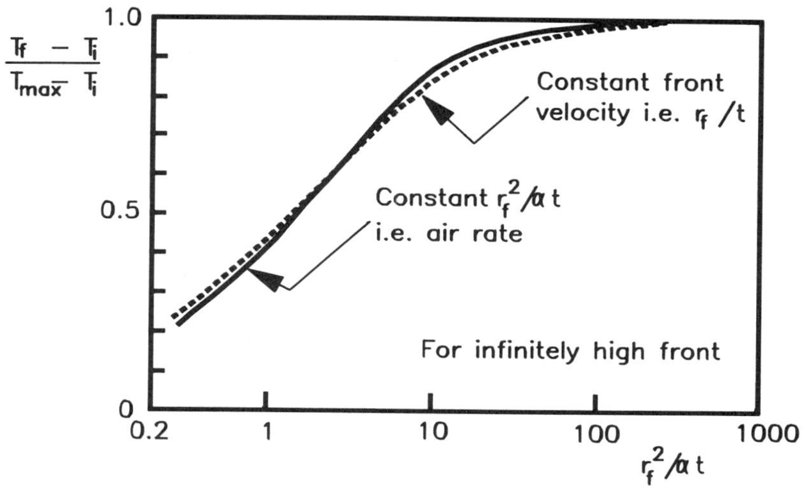

Figure 9.25 Temperature at Fire-Front (after Ramey 1959)

tion 9.3) is plotted against a dimensionless number, $r_f^2/\alpha t$, where r_f is the distance of the front from the starting location, α is the thermal diffusivity of the reservoir, and t is the time from the start.

Two curves are shown:

1. The broken curve is for the case where the velocity of the front is constant; i.e., $V = r_f/t$ is constant. In this case the abscissa is Vr_f/α. As the front advances, the temperature approaches T_{max}. In order to achieve the constant front velocity, which is assumed in this case, it would be necessary to raise the air-injection rate in proportion to t.
2. The solid curve in Figure 9.25 is for a constant value of r_f^2/t. This is approximately equivalent to the case where the air-injection rate is maintained constant. In this case the flame-front temperature remains constant corresponding to the value of the ordinate which is determined by $r_f^2/\alpha t$.

Figures 9.26–9.29 show the results of Ramey's calculations for the case where the effects of vertical conduction are included. The figures give the calculated temperature rise for four different horizontal planes in the reservoir. The largest effect of vertical heat loss is for the case corresponding to the top (and also the bottom) of the reservoir; the results for this case are shown in Figure 9.29. The calculated temperature rise is much smaller for this case than for the planes nearer to the center. Also, the curves for the other planes are very similar to each other (note the different vertical scales).

This result introduces the important possibility that as the flame front moves outward from the injection well, the vertical heat losses at the top and bottom of the reservoir may cause the combustion to be extinguished there before it is extinguished in the center.

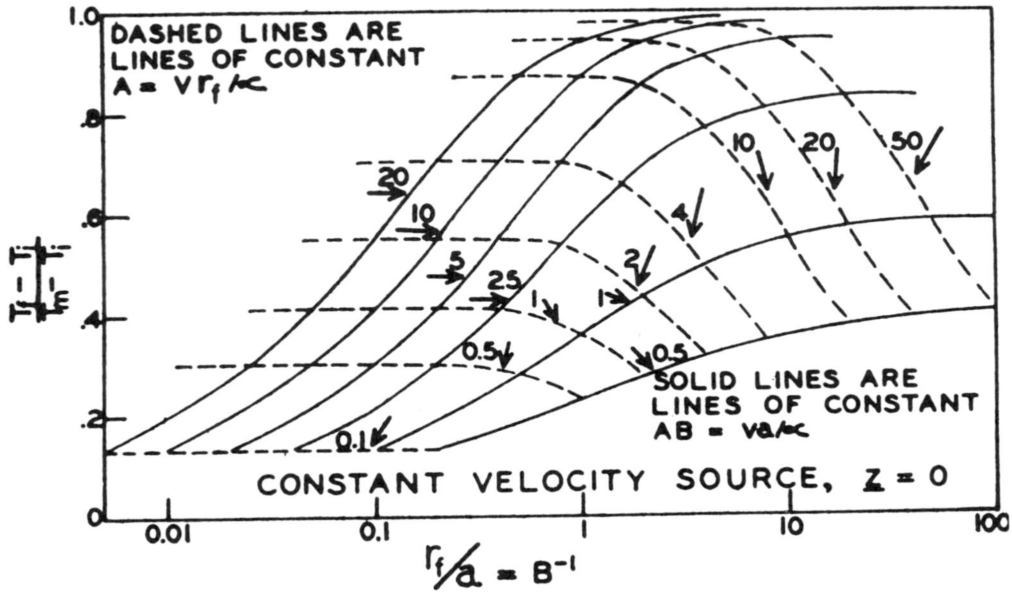

Figure 9.26 Dimensionless Temperature Rise for a Heat Source of Finite Height Moving at Constant Velocity—Temperature at Center Plane Edge ($z = 0$). (from Ramey 1959)

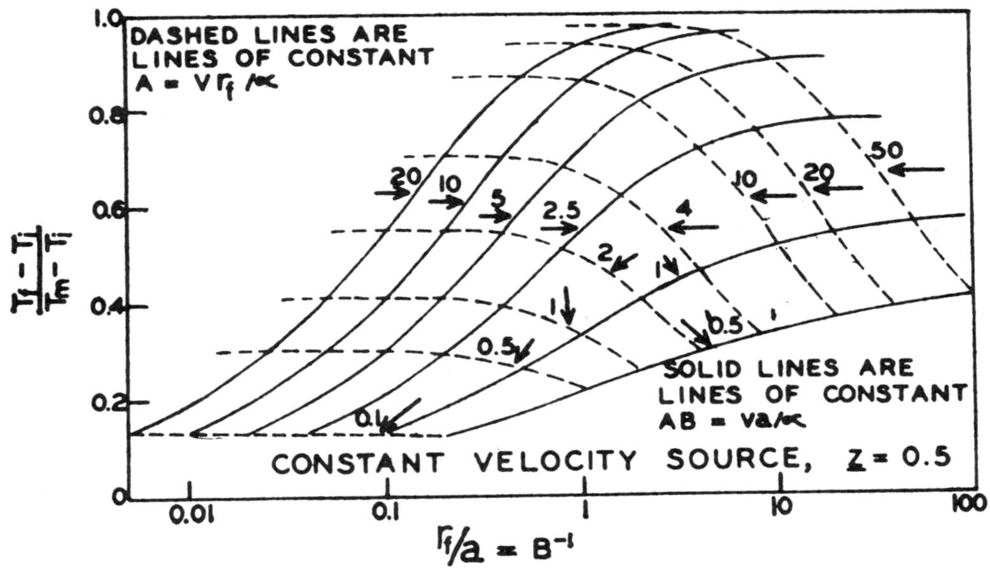

Figure 9.27 Dimensionless Temperature Rise for a Heat Source of Finite Height Moving at Constant Velocity—Temperature Halfway Between Center Plane and Edge ($z = 0.5$) (from Ramey 1959)

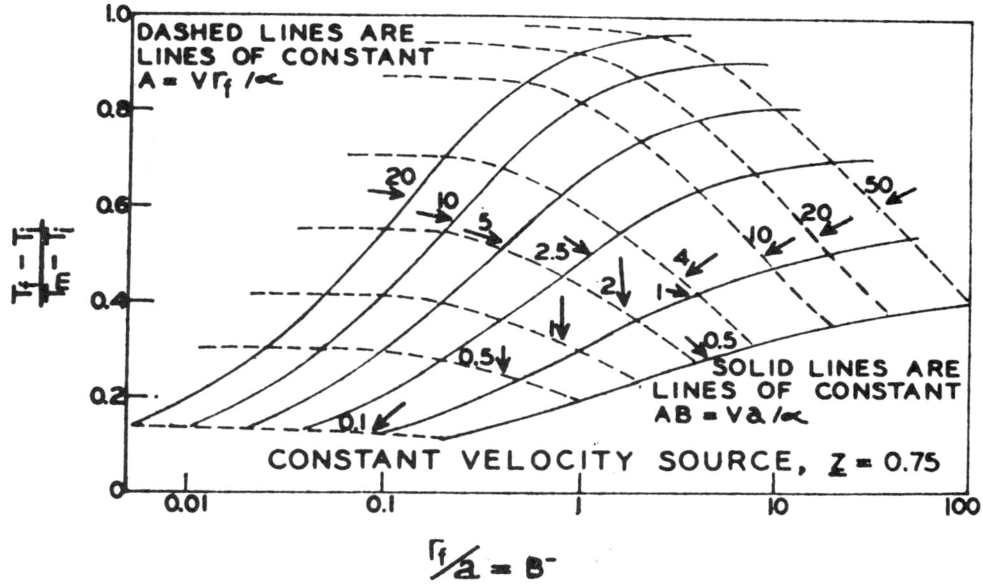

Figure 9.28 Dimensionless Temperature Rise for a Heat Source of Finite Height Moving at Constant Velocity—Temperature Three-Quarters of the Distance from the Center Plane to the Edge ($z = 0.75$) (from Ramey 1959)

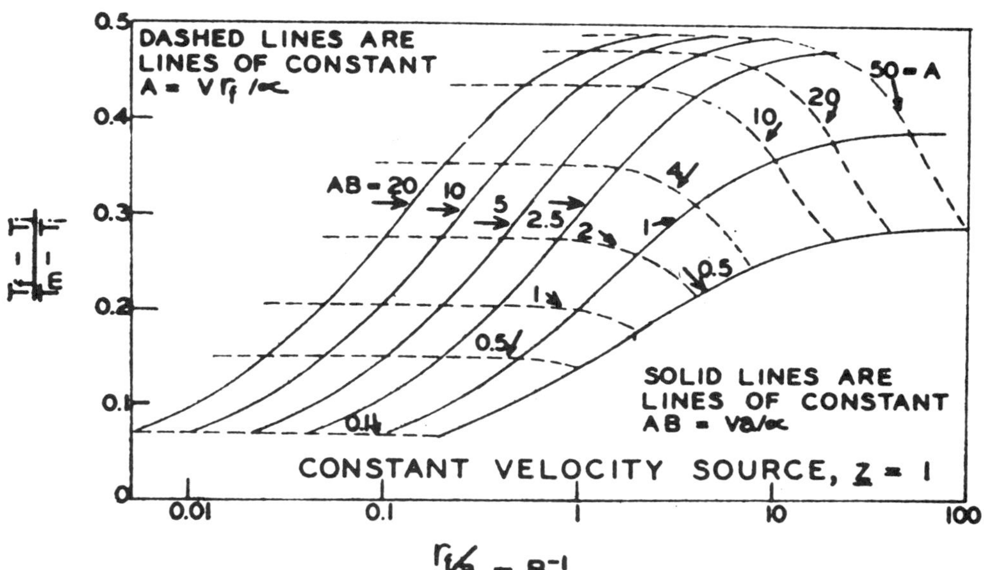

Figure 9.29 Dimensionless Temperature Rise for a Heat Source of Finite Height Moving at Constant Velocity—Temperature at the Edge ($z = 1$) (from Ramey 1959)

Examples of the Use of Ramey's Solutions

Figure 9.30 shows the combustion-front temperatures calculated by Ramey for the advance, at a constant rate, of an infinitely high, cylindrical combustion front for the conditions shown. As the front advances the temperature increases because of the decreasing heat flux loss in front of the cylinder; the cylinder curvature is decreasing, and an increasing temperature is required to conduct the heat away. In each case the front approaches T_{max} as an asymptote.

Curves are drawn for several fuel loads. The higher the fuel load, the higher is the temperature of the front. For the front to advance at all, it is necessary for it to reach the ignition temperature.

For example, if it is assumed that the temperature has to be greater than 400°F for ignition to occur, then, as shown by the dotted line on the diagram, the front would have to have a radius of greater than 0.9 ft in order for it to be able to advance if the fuel load is 1.5 lb/ft^3. External heat would have to be applied to "kindle the fire" in order to move the 400°F isotherm 0.9 ft from the well.

The curves in Figure 9.30 do not allow for the effect of convection on the temperature of the combustion front. The reader is referred to Ramey's paper for an analysis of this. The effect is to reduce the amount of fuel required to attain a particular front temperature. For the case shown in Figure 9.30, Ramey calculates that the fuel concentration could be reduced by 18%.

The curves just discussed are for an infinite well with a constant rate of advance of the combustion front. In order to sustain the constant radial velocity, the air-injection rate would have to be increased continuously. If the air-injection rate is

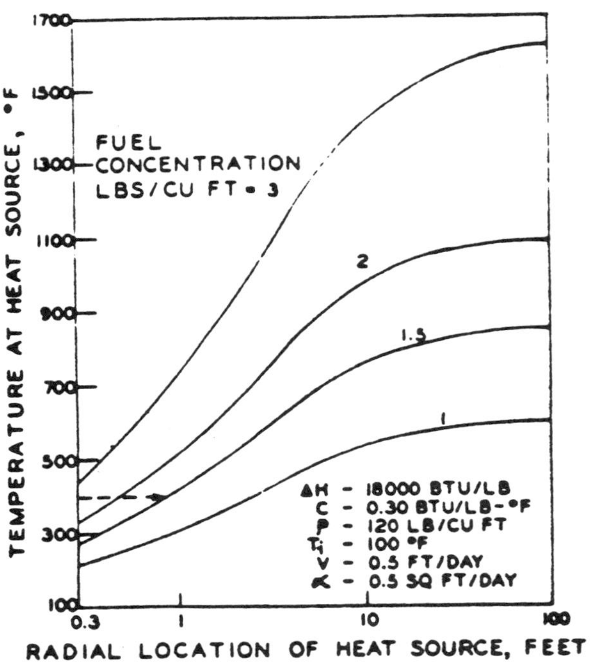

Figure 9.30 Effect of Fuel Concentration on the Temperature of a Combustion Front of Infinite Height Moving Radially at a Constant Velocity (from Ramey 1959)

maintained constant, then the front temperature for an infinite front remains constant as the front advances.

For a finite front, the temperature falls with a constant air-injection rate, and it is lowest at the top and bottom of the reservoir.

Figure 9.31 shows the calculated temperature of the front (expressed as a fraction of the adiabatic temperature rise) for such a case. This diagram includes the effect of vertical heat losses, and it may be seen that, at any particular radius (the abscissa is the ratio of the radius from the well to the half-reservoir height a), the front temperature is lowest at the very top and bottom (i.e., where $Z = 1$) and highest at the center plane. In this figure the parameter U is a constant that determines r_f^2/t; it is defined by equation 9.4.

$$r_f = (2Ut)^{1/2} \tag{9.4}$$

The horizontal dotted line in Figure 9.31 indicates, for a particular example (it would be at a different vertical location if the fuel concentration were different), the limit at which high-temperature combustion could continue. In the case shown, the combustion process would decline at the very top and bottom of the reservoir at a radius of about eight times the reservoir half-height, and at larger radii, it would decline nearer to the center of the reservoir. In order to maintain high temperature combustion beyond this, it would be necessary to increase the air-supply rate.

In practical cases there may be difficulty in increasing the air-injection rate because of the excessive gas velocity that is created at the production wells; this can cause sandblast-type erosion damage. Another solution, and one that is increasingly being studied, is to use enriched air or oxygen. This allows higher heat-generation

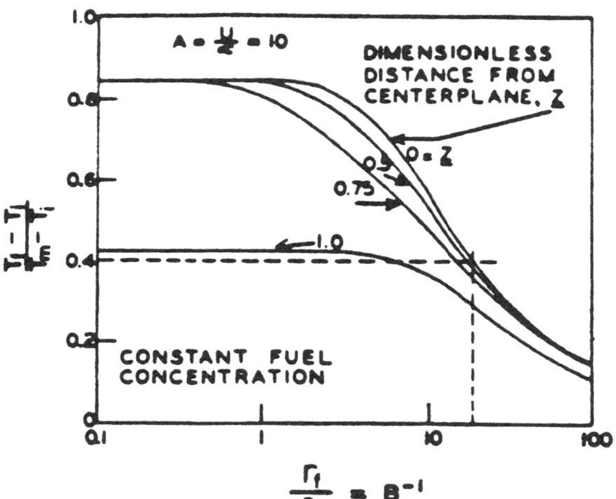

Temperature rise ratio at a moving heat source of finite height and velocity inversely proportional to radial source location and U/α of 10.

Figure 9.31 Effect of Vertical Height on the Temperature of a Combustion Front of Finite Height Moving with a Velocity Inversely Proportional to the Radius (from Ramey 1959)

rates at the combustion interface and, as a result, allows larger spacing between wells without excessive gas production rates.

Figure 9.32 is similar to Figure 9.30, except that it includes the effect of vertical heat conduction and is drawn for a constant fuel load of 1.5 lb/ft^3; the front velocity is assumed constant at 0.5 ft/d. In this case it would be necessary to, in some manner or other, raise the upper and lower boundaries of the reservoir to above the ignition temperature (400°F) at a radius of 7 ft in order to have the front extend vertically through the reservoir.

Properties of Produced Oil

One of the features of in situ combustion is that there is some improvement in the properties of the oil. The material that is burned is essentially coke, whereas the oil produced consists of unchanged oil that has been displaced from the reservoir diluted with distilled and cracked materials. This mixture usually has a significantly lower density (higher API gravity) and a lower viscosity than the original oil; it contains a smaller proportion of high boiling materials.

Inspections on produced oils from firefloods are compared with those for the original oil in Table 9.6.

WET COMBUSTION

In the previous discussion, it has been shown how in situ combustion can be used to produce a combustion front, which, as it advances through the reservoir, drives the oil ahead of it and consumes the residual coke that is laid down as fuel.

One of the main parameters in this process is the fuel concentration, which is determined by the nature of the crude oil and by the conditions. Oil that has been driven forward or converted to cracked volatile material cannot remain as coke to be burned.

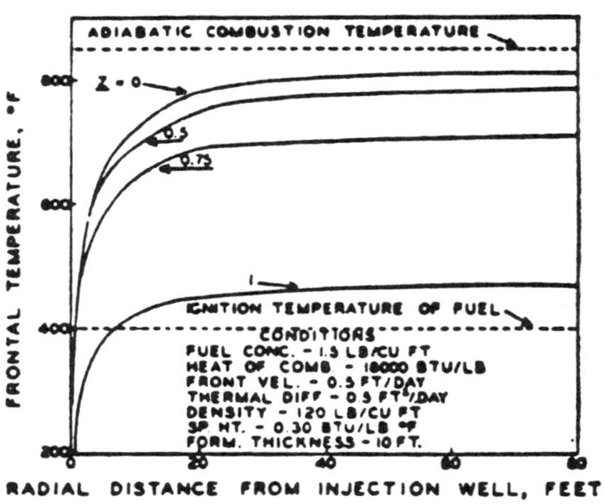

Temperatures at the combustion front, sample problem.

Figure 9.32 Temperatures for Combustion Front within a Finite Reservoir Which Is Advancing with a Constant Velocity (from Ramey 1959)

TABLE 9.6 Changes in Oil Properties after Firefloods

FIELD, LOCATION	°API		TEMPERATURE	VISCOSITY cp	
Operator	Before	After	°F	Before	After
South Belridge, California	12.9	14.2	87	2,700	800
General Petroleum			120	540	200
			160	120	54
West Newport, California	15.2	20.0	60	4,585	269
General Crude			100	777	71
			210	32	10
East Venezuela	9.5	12.2			
Mene Grande		Then 10.5			
Kyrock, Kentucky	10.4	14.5	60	90,000	2,000
Gulf Oil			210	120	27
South Oklahoma	15.4	20.4	66	5,000	800
Magnolia				After 1 month	
					5,000
Asphalt Ridge, Utah	14.2	20.3			
U.S. DOE(1)					

(1)Changes in other properties:

	Before	After
Pour point, °F	140	25
1000 + °F wt%	62	35

(Chu 1982)

In the dry combustion process, the air requirement is not determined by the need for heat but rather by the availability of fuel. With heavy crudes, there tends to be more fuel than is required simply to heat the reservoir. Also, much of the heat generated remains behind the combustion front in the rocks that have been completely depleted.

The two diagrams in the upper left part of Figure 9.33 show idealized profiles for dry combustion. The temperature is high behind the combustion front, and much heat is being left behind. No water vaporizes behind the combustion front. There is a small temperature plateau ahead of the front, where some water products from combustion, together with connate water vaporized by the advancing combustion front, condense.

In wet combustion, water is added to the air. This addition can be either continuous or, with essentially the same result, intermittent. In some cases, separate wells are used for water injection.

Some of the added water remains in the burned zone as water saturation. Without water addition, the burned rocks would be dry, since the connate water is evaporated in front of the combustion zone. If sufficient water is added to the air, liquid water flows toward the combustion front, and as the water approaches the combustion front, it evaporates; this cools the rocks. Heat is transported through the combustion front by the steam, largely as latent heat. Heat is released ahead of the front when the steam condenses.

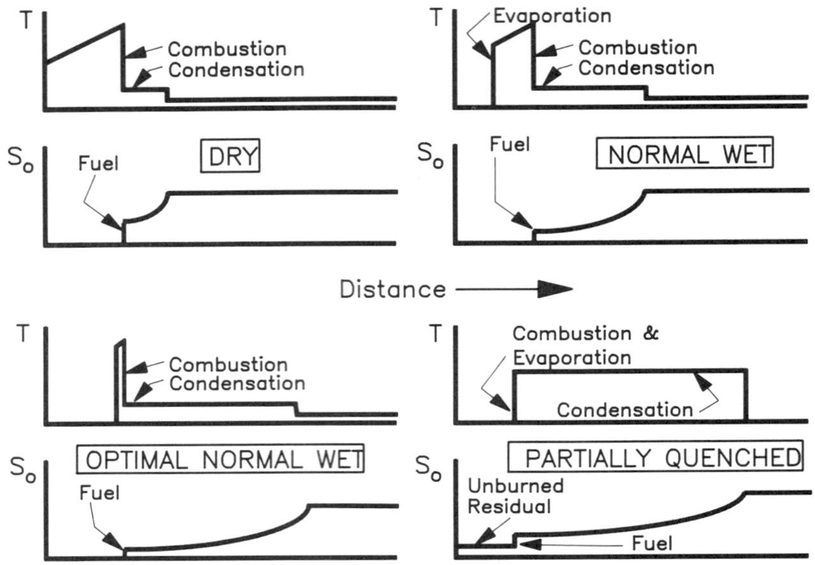

Figure 9.33 Effect of Water Addition

Normal wet combustion is shown by the two diagrams in the upper right part of the figure. Steam passes through the front and later condenses. The steam moves oil ahead of the combustion front as well as moving heat. There are thus two beneficial effects:

- The rocks are preheated before the combustion front reaches them, and this tends to increase the temperature of the combustion front (T_i is higher in equation 9.3).
- The steam reduces the concentration of residual oil remaining in the path of the advancing combustion front. This results in a substantial reduction in the fuel concentration and reduces the air required to burn through a given volume of reservoir. This effect tends to lower the temperature of the combustion front (W is lower in equation 9.3).

Also, because of the displacement caused by the steam ahead of the front, it may not be necessary to burn all the way through the reservoir in order to achieve effective recovery. In the extreme, wet combustion can be looked upon as a means of generating steam within the reservoir rather than in a surface steam generator. When sufficient fuel has been burned to generate the steam required for the recovery, then the process can be terminated, leaving the remaining residual oil uncoked and unburned.

The lower left diagrams in Figure 9.33 show the condition obtained if the amount of water is increased to the point where the evaporation front just trails the combustion front. In this condition the maximum amount of steam is generated without liquid water reaching the combustion zone. The combustion zone is not itself being cooled by the direct evaporation of water. The steam condensation zone extends far beyond the combustion front, and the fuel laydown is decreased because of the effect of the steam in reducing the residual. Essentially all the oxygen is consumed in the high-temperature combustion process at the front.

If the water-to-air ratio is increased still further, the situation shown in the diagram in the lower right-hand corner is produced. Now liquid water can enter the combustion zone, and this zone is cooled below the levels found with less water. Some unburned fuel is left behind. In this condition there can be excessive low-temperature combustion and production of viscous tars.

When some liquid water enters the combustion zone and vaporizes completely, the process is described as *partially quenched combustion*. At higher water-to-air ratios, liquid water passes right through the oxidation zone, and the operation is described as *quenched combustion*, or *superwet combustion*.

Figure 9.34 shows typical temperature and saturation profiles for a normal wet combustion process. It may be compared with the similar diagram given earlier for dry combustion (Figure 9.1).

Laboratory Results

Figure 9.35 shows temperatures measured in a combustion tube run in which water was injected into the air feed partway through the experiment. The temperatures were measured by a series of thermocouples placed along the path of the combus-

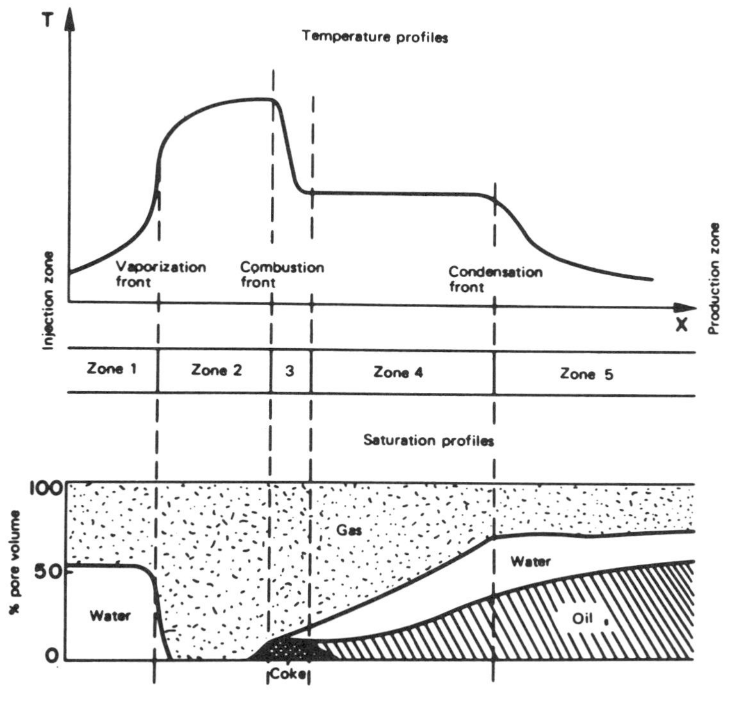

Figure 9.34 Temperature and Saturation Profiles for Wet Combustion (from Latil 1980)

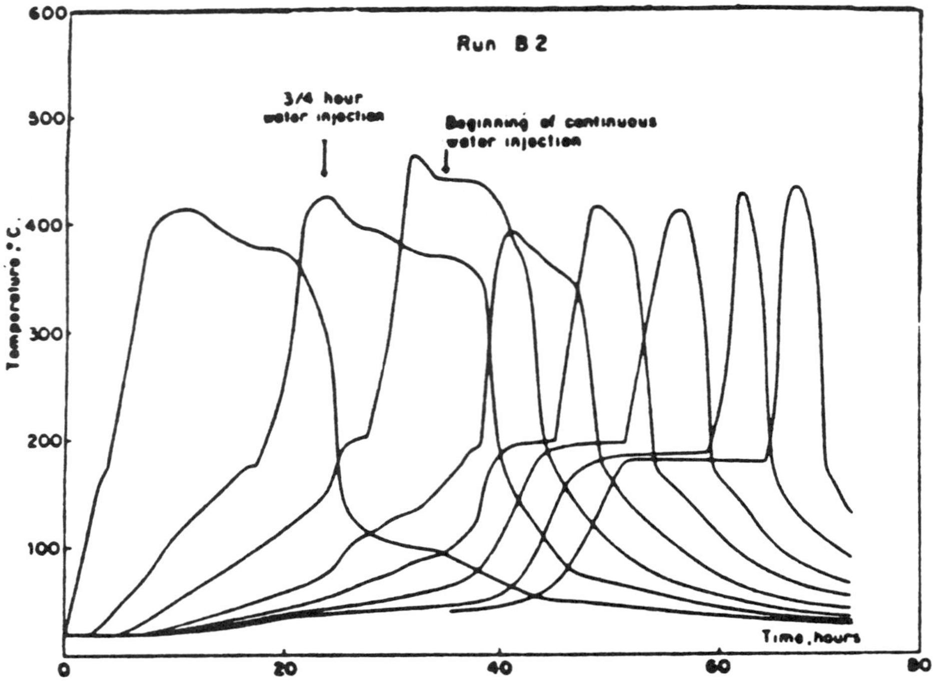

Figure 9.35 Temperatures as a Function of Time for Various Locations along Combustion Tube (from Burger and Sahuquet 1973)

tion. As the combustion front approached each measurement point, the temperature increased, reached a maximum, and then fell as the front passed along.

Several features should be noticed:

- The width of the high-temperature zone decreased markedly when the water was added.
- Water addition did not greatly affect the peak temperature. Presumably, the two effects of water on the combustion front temperature that were discussed previously almost compensated each other.
- A distinct steam plateau formed when the water was added.

The same temperature measurements are plotted as instantaneous temperature profiles in Figure 9.36, and the same observations can be made from an examination of this diagram.

Figure 9.37 shows analytical data for the produced gas from the same experiment. The gas composition showed no change as a result of the water addition. This indicates that the fuel was of the same composition with water addition as it was without it.

Although the composition of the fuel did not change with the addition of water, there was a large change in the quantity of fuel. This is shown in Figure 9.38, where the positions of the combustion front, the condensation front, and the vaporization front are plotted against time. At the point where the water was added, the

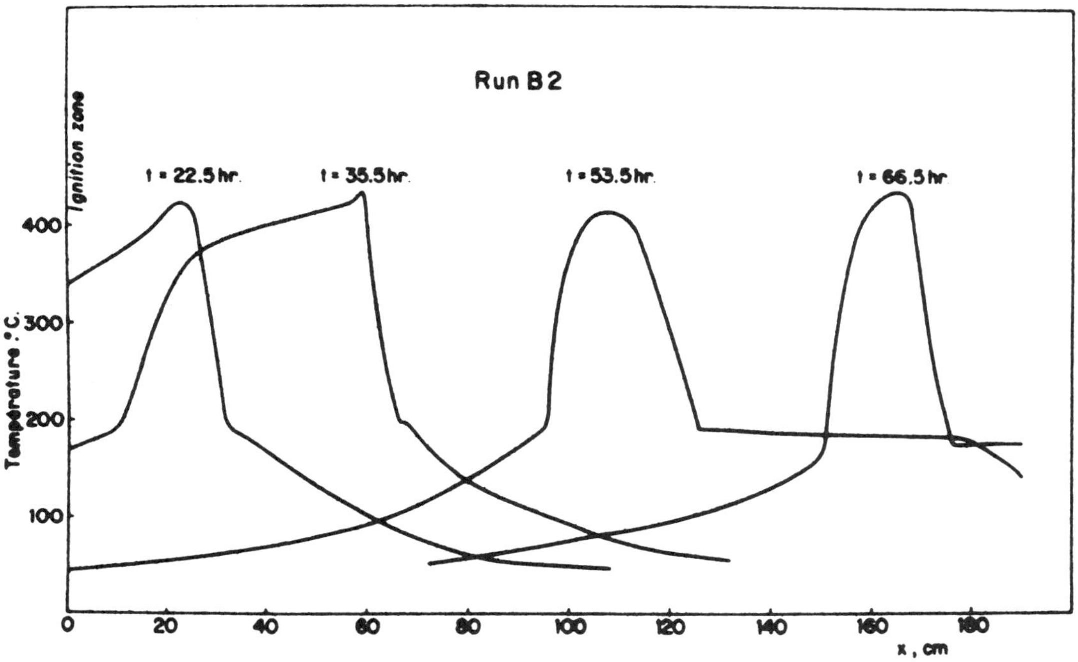

Figure 9.36 Temperatures as a Function of the Distance along the Combustion Tube (from Burger and Sahuquet 1973)

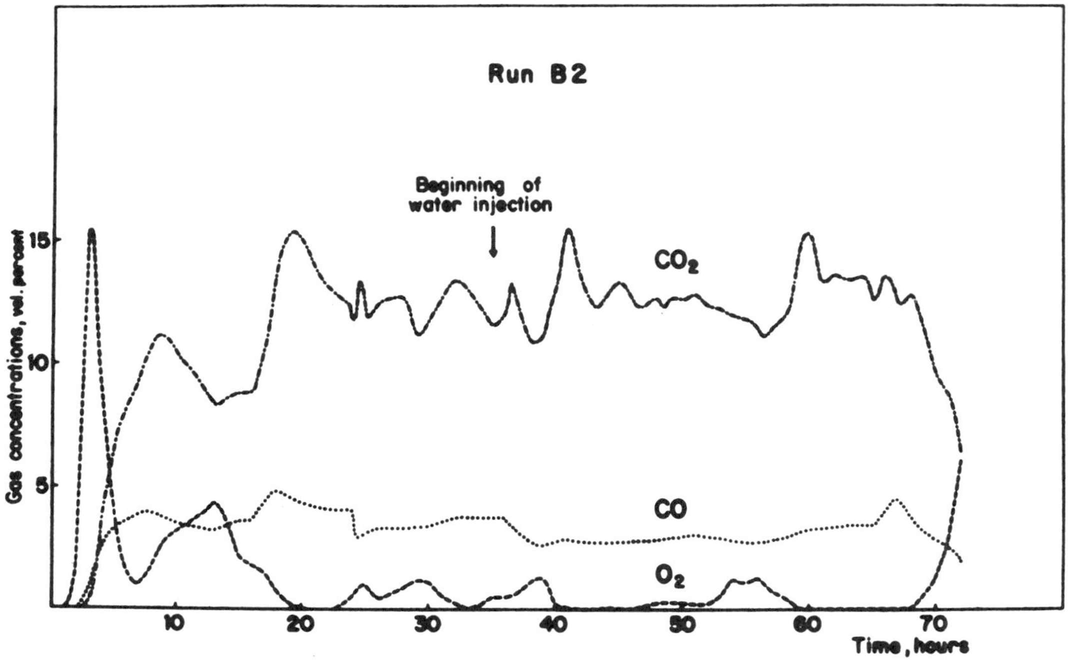

Figure 9.37 Composition of Produced Gas from Combustion Tube (from Burger and Sahuquet 1973)

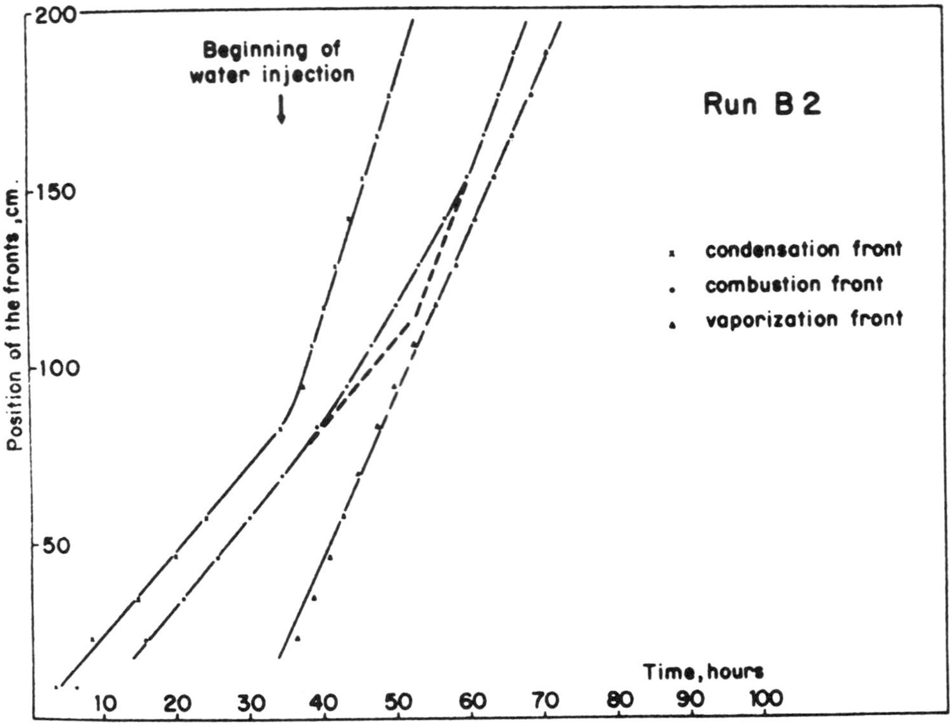

Figure 9.38 Effect of Water Addition upon the Velocities of the Condensation, Combustion, and Vaporization Fronts (from Burger and Sahuquet 1972)

condensation front accelerated as more heat was transported ahead of the front. The additional steam had the effect of reducing the residual oil left in the path of the combustion front, and, as this decreased the available fuel, the burn could move faster even though the air rate was kept constant.

Similar results have been reported by Joseph and Pusch (1980) for a field pilot study in the Bellevue field in Louisiana. The test, carried out by Cities Service involved two side-by-side five-spot patterns. One of these patterns was operated dry and the other, wet. Some results from this paper are shown in Figure 9.39, where the heated reservoir volumes are compared as a function of the volume of air injected. It was concluded that with wet combustion, higher recoveries of oil could be expected because of better volumetric sweep, that the air volume required to process a given reservoir volume was reduced by 63%, and that less time would be required because of the lower air requirement.

Water-to-Air Ratio

The ratio of water to air that should be used in wet combustion depends upon such factors as the fuel concentration, the water content already present in the reservoir, and the possibility of water intruding into the combustion region from outside the pattern. In principle, the amount of water to be added can be calculated from heat and material balances using the objective of removing most of the excess heat from the rocks left behind the combustion front. In many cases, the amount of water added is about 200 to 250 B per million SCF of air (0.92 to 1.15 lb water per pound of air).

The choice of the water-to-air ratio to be employed has been discussed by Chiu (1988). If low water-addition rates are employed, the added water partially fills the pores behind the combustion front and does not remove heat from the combustion zone by forming steam to condense beyond the combustion front. The minimum water-to-air ratio required for water to be available for evaporation can be estimated from the following equation; this assumes that, at the limit, water occupies 80% of the pore volume behind the front.

$$R_{wa} = \frac{0.8\phi\rho_w B_g}{B_g R_{ar} + 0.2\phi} \quad \text{(Chiu 1988)} \tag{9.5}$$

where R_{wa} is the mass of water per unit volume of air
R_{ar} is the volume of air required to burn a unit volume of reservoir
ϕ is the porosity
B_g is the formation volume factor of the air behind the front
ρ_w is the density of water.

Chiu also calculates the heat carried forward from the combustion front into the steam zone. This is equal to the heat in the dry gas at the combustion temperature plus the heat in the water vapor—both the vapor from the vaporization of the injected water and that from the water formed by combustion. He expresses this as a percentage of the heat of combustion. As the quantity of injected water increases, so does this percentage. At the limiting condition for normal wet combustion, this percentage becomes 100%. This is an absolute upper limit. Chiu recommends that

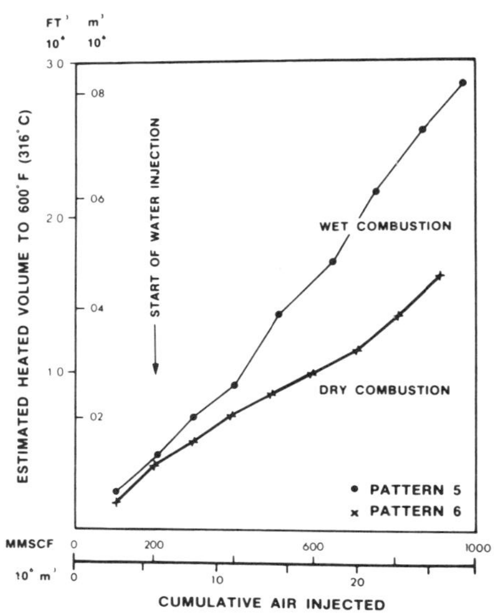

Figure 9.39 Comparison of Heated Volumes for Side-by-Side Wet and Dry Five-Spot Patterns (from Joseph and Pusch 1980)

the percentage of the heat of combustion carried forward to the steam zone should not exceed 85% and that combustion tube experiments should be conducted to show that this is workable for a particular reservoir situation.

Figures 9.40 and 9.41 show the results of Chiu's calculations of the water-to-air ratio required to carry various percentages of the heat of combustion forward for typical reservoir conditions. Figure 9.40 shows the water-oil ratio as a function of the air requirement for air ISC. Figure 9.41 shows the effect of using enriched air for a particular reservoir.

IN SITU COMBUSTION IN TAR SANDS

A process for utilizing in situ combustion within tar sand deposits must overcome two fundamental obstacles:

- There is little or no initial injectivity in tar sand deposits.
- The low volatility of bitumen, together with its asphaltic nature, makes the fuel deposition load very high. The deposition can amount to 2 to 3 lb/ft^3 of reservoir, whereas 1 lb/ft^3 would be sufficient to raise the reservoir to 500°F.

There have been some attempts to produce bitumen from tar sands by in situ combustion without prior heating. AMOCO operated several pilots at Gregoire Lake in Athabasca. Their process is described in a paper by Jenkins and Kirkpatrick (1979). It involved the injection of air into wells located within inverted five-spots. The formation was ignited by first injecting steam and then air. The process was carried out in three phases.

In the first phase of the process, the objective was to heat the reservoir by combustion. Injection was continued until the combustion front approached the production wells. Relatively little bitumen was produced during this phase.

In the second phase, air injection was stopped and the reservoir pressure was lowered by allowing production at the production wells. It was thought that this production was assisted by the flashing of connate water to steam within the heated reservoir.

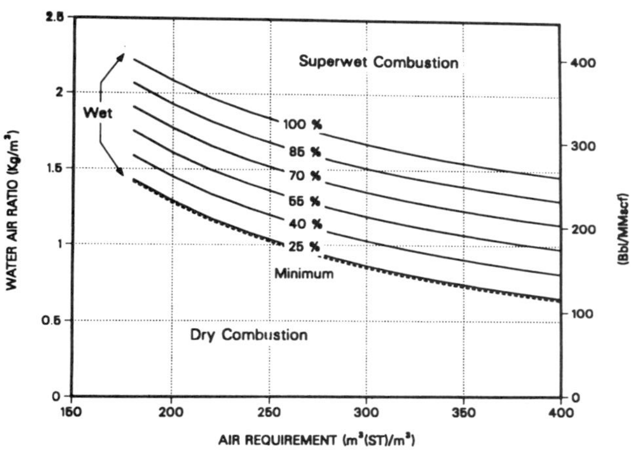

Figure 9.40 Effect of Air Requirement upon the Water-Air Ratio Required to Achieve Various Percentages of Carry Forward of Heat of Combustion. The Parameter Is the Percent of Heat of Combustion That Is Carried Forward to the Steam Zone (from Chiu 1988).

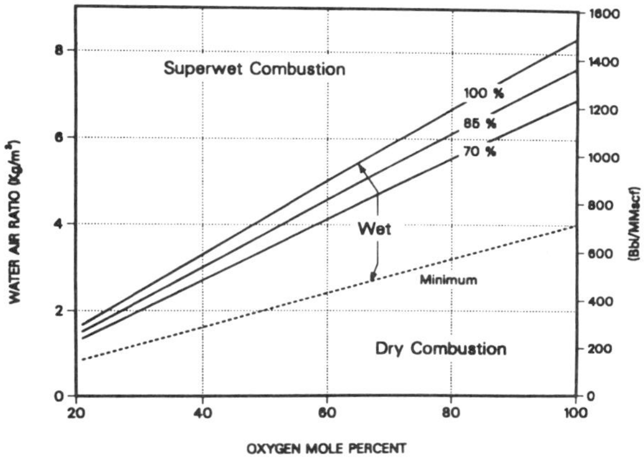

Figure 9.41 Effect of Oxygen Upon the Water-Air Ratio Required to Carry Forward Various Percentages of the Heat of Combustion. The Parameter Is the Percent of Heat of Combustion That Is Carried Forward to the Steam Zone (from Chiu 1988)

In the final stage of the process, air and water were injected simultaneously to produce bitumen by wet combustion.

Apart from the paper just referred to, there appears to be little published information on the AMOCO project. An expanded pilot was built jointly with AOSTRA, but it was apparently unsuccessful.

When in situ combustion occurs in an initially cold, bitumen-containing reservoir, the gas presumably flows along relatively thin permeable paths. These may already exist in the reservoir, or they may be formed by fracturing. With ignition, combustion can occur along the surfaces of these paths, with a combustion front advancing at right angles to the direction of flow. However, as the combustion front advances, the supply of air to it will tend to decline, and the front temperature may fall. Under these circumstances, low-temperature oxidation may take over from the normal combustion process with little mobilization of the oil. Also, just as in the case of steamflooding along a fracture, there may be little pressure differential available to drive the oil toward the producer.

This problem may be overcome by preheating the reservoir with steam before using in situ combustion. It appears to be practical to use in situ combustion as a means of producing additional oil from a reservoir that has been produced using cyclic steam stimulation. Cyclic steaming can thus serve as a means for the economic production of perhaps 15% of the oil in place as well as a means for preheating the reservoir to allow the production of further oil by combustion.

This combination of processes was pioneered by the Chanslor-Western Oil Company in the Midway Sunset field in California. This reservoir contains an 11.5° API crude oil, which can be produced by cyclic steaming. In the Chanslor-Western pilot, about 15% of the oil in place was produced by cyclic steaming, and it was predicted that the ultimate recovery without combustion would have been less than 30%. The project was converted to dry in situ combustion using the engineering design procedures described by Nelson and McNeil (1961). The project was successful, and it was concluded that oil was produced more economically than was possible by continuing the cyclic steam recovery. The production rate was at least as high as that which could be obtained by steam drive in the same pattern, and it was concluded that the in situ combustion process was more economic in this instance. An expanded project involving 10 injection and 40 production wells is operated by Sante Fe Energy on the same lease. The project is described as promising and is producing 1000 B/d of oil, of which 800 is enhanced recovery production (Aalund 1988).

BP Resources Canada (Donnelly, Hallam, and Duckett 1985; Nzekwu, Hallam, and Williams 1988) is following a similar approach at its combustion pilot at Marguerite Lake in the Cold Lake oil sands deposit in Alberta. In its process, BP first produced oil from the tar sands by steam stimulation; this required the initial steam injection to be at above fracturing pressure. After a number of cycles of steam stimulation, the operation was converted to in situ combustion. Initially, BP used air in situ combustion, as in its original plans. Subsequently they converted from the use of air to oxygen and have now had considerable experience with pure oxygen injection.

BP found that the combustion zone overrides the reservoir and moves rapidly along the fracture paths opened during cyclic steaming. To avoid production well damage, they developed cyclic techniques in which the injection of oxygen is interrupted. Water injection, either as slugs or intermittently, is important in controlling the temperatures and in distributing the heat (Nzekwu, Hallam, and Williams 1988).

The cyclic steaming phase of the process requires 5 to 7 y to produce 15 to 20% of the original bitumen in place. It is anticipated that with combustion, the recovery will be doubled to about 30 to 50%. BP started pilot operations at Marguerite Lake in 1977 and carried out cyclic steam experiments, wet-air combustion, and, since March 1983, enriched air and high-purity oxygen combustion.

A larger semicommercial operation was built in 1983 at nearby Wolf Lake; this started operations in early 1985. This project has 192 start-up wells that were directionally drilled from 10 satellites; it is designed to produce 1100 m^3/d of bitumen. After the cyclic steam phase, it is planned to incorporate combustion (probably using oxygen) in the early 1990s.

Wet combustion has considerable potential and many advantages when used with tar sands. It provides a means for reducing the quantity of injected air that would be needed to consume the very high fuel load that arises because of the composition of the bitumen. Wet combustion produces a steamflood ahead of the combustion front that drives oil ahead, and this reduces the fuel load. In addition, it is probably practical to look upon the wet in situ combustion mechanism as an in situ steam generator. This leads to the concept of not burning completely through the reservoir, but stopping after adequate steam for recovery has been generated. Thus, in practical bitumen, in situ combustion projects, it is probably economic to leave behind considerable volumes of unburned reservoir.

USE OF OXYGEN OR ENRICHED AIR

One of the most interesting current areas of development in in situ combustion is the use of enriched air or oxygen in place of air. This possibility was suggested by H. Ramey (1954).

For the use of oxygen to become an economic, commercial reality, it is necessary for its advantages to outweigh its disadvantages. As it is seen at present, the following are the major advantages and disadvantages.

Potential Advantages for the Use of Oxygen

1. Elimination of nitrogen reduces the gas velocity at the producing wells for a given oxygen-injection rate. This in turn can allow a higher rate of oxygen injection and can give much more rapid heat production and shorten production time. As has been mentioned previously, it is also possible to use larger well spacing. With air, the rate of production of heat within the reservoir from an injection well is frequently less than can be achieved by using the same well for steam injection. With oxygen injection, the rate of heat production becomes approximately equal to that which is achievable with steam.[3]
2. The concentration of carbon dioxide in the gas within the reservoir is much higher, and it has been suggested that this will improve recovery by increasing the amount of carbon dioxide dissolving in the crude; this will tend to lower the viscosity and cause swelling. Figures 9.42 and 9.43 compare the flue gas analyses found in combustion tube experiments using air and oxygen, respectively (Moss and Cady 1982).

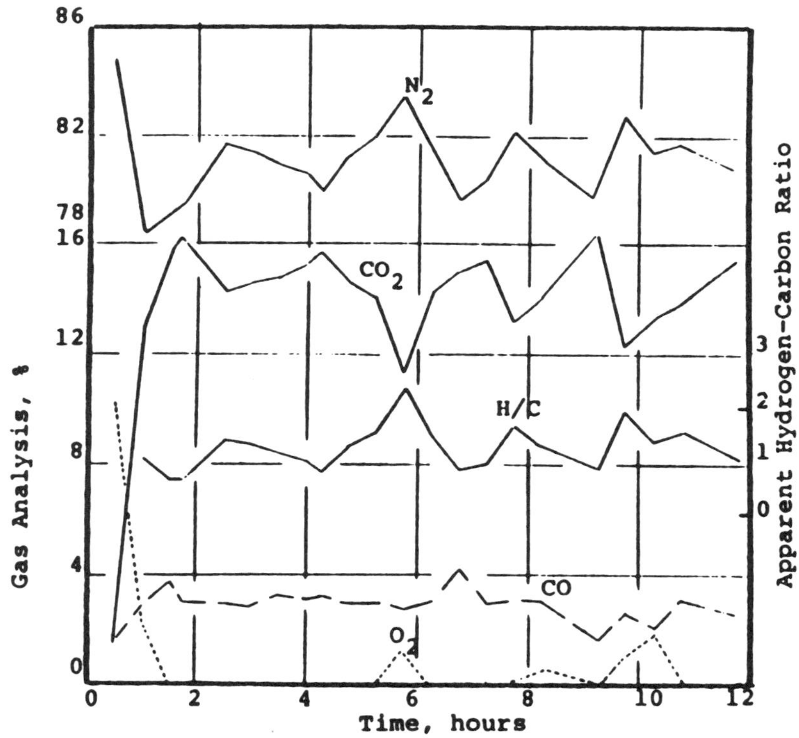

Figure 9.42 Combustion Tube Run with Air 12° Lindbergh Crude (from Moss and Cady 1982)

[3]One SCF of air generates about 100 Btu when the oxygen within it reacts with the fuel in the reservoir. Steam injection gives about 1000 Btu per pound of steam. Thus, to be equivalent to the injection of 1000 B/d of steam (350×10^6 Btu/d), one would have to inject 3.5×10^6 SCF/d of air, or 0.7×10^6 SCF/d of O_2. In practice, the higher efficiency of utilization of the heat from combustion may reduce these requirements.

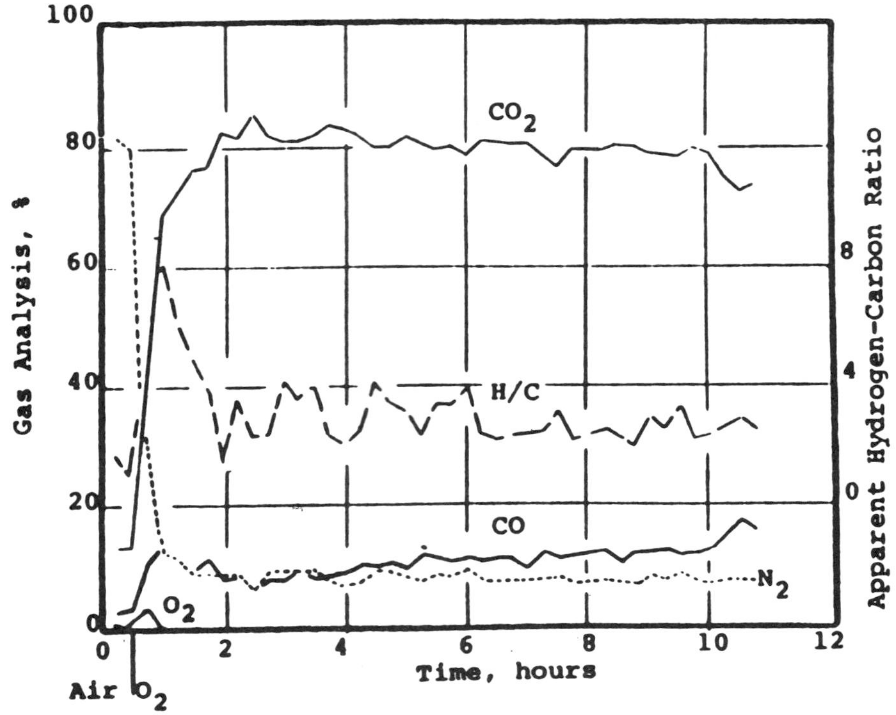

Figure 9.43 Combustion Tube Run with O_2 12° Lindbergh Crude (from Moss and Cady 1982)

3. The heating value of the produced gas will be much higher because it is not diluted with nitrogen. Also, the high concentration of carbon dioxide may make the gas useful as a source of carbon dioxide. It is possible, then, that the produced gas may have a positive rather than a negative economic value. It could, for example, be possible to separate the CO_2 for sale and utilize the resultant high-heating-value tail gas as a fuel. The combustion of the tail gas from conventional in situ combustion is difficult and requires special equipment.
4. There may be less override than with air for the same oxygen-injection rate because of the reduced volume of gas and perhaps because of its higher density.

Possible Disadvantages of the Use of Oxygen

1. There is a hazard in using high-pressure, high-concentration oxygen in circumstances where it can be mixed with hydrocarbons. A particular concern is to eliminate mixing of oxygen with oil in the injection well. This can occur if the pressure in the injection well is allowed to fall during operation and thus allow oil to back up into the well. A likely cause for such an event would be the failure of the oxygen supply. Precautions that can be used to prevent this include the provision of backup oxygen and means for injecting water into the

well in the event of failure of the oxygen supply. The mechanical problems with oxygen in situ combustion have been largely solved in pilot operations. For example, BP in their Cold Lake pilot has injected oxygen for a considerable period and are contemplating its use on a large commercial scale (Donnelly, Hallam, and Duckett 1985).

2. Special precautions are also necessary to handle oxygen. For example, precautions must be taken to avoid sources of ignition that could cause the combustion of ordinary steel pipe carrying high-pressure oxygen. Although these precautions are well known in the conventional handling of oxygen, they introduce a new degree of complexity into oil field operations. It is, for example, necessary to minimize dust particles and to stay within maximum line velocities to avoid ignition from static electricity (Henningson and Duckett 1984; Hvizdos, Howard, and Roberts 1983). The choice of materials that will minimize corrosion problems and be compatible with oxygen in situ combustion is discussed by Zawierucha et al. (1988).
3. The investment cost for oxygen-separation plants is higher than that for simple air compressors. This is partially—and in some cases completely—offset by the reduced power requirement to compress the smaller volume of oxygen.
4. As is discussed later, low-temperature oxidation seems to occur more readily with oxygen than with air—particularly at higher pressures.

The Cost of Oxygen

There is already a vast experience in the manufacture of oxygen on a large scale. The most economic method for manufacture, and the only developed one that is practical for consideration for large-scale operations, involves the liquefaction of air and fractional distillation to separate the oxygen. The process is very efficient. Cooling is achieved by countercurrent heat exchange with the products and adiabatic expansion of the compressed feed. The free energy to operate the process comes from the compression of the feed. An input air pressure of about 75 to 90 psig is sufficient to drive the process and produce pure oxygen at atmospheric pressure (see Table 9.2). Newton (1979) gives a concise description of the technology for air liquefaction and fractionation.

Oxygen for small combustion pilots is usually transported to the site as liquid in insulated trailers. Production pilots of this type have been operated in Canada by British Petroleum, by Husky Oil, and by Dome Petroleum. There have been tests by Arco and by the Greenwich Oil Company in Texas (Hvizdos, Howard, and Roberts 1983).

The liquid oxygen is pumped under pressure through a vaporizer. It is usual to provide a similar facility for the vaporization of liquid nitrogen so that a blend of oxygen and nitrogen can be injected. A blend corresponding to air is used for startup, and the oxygen concentration is increased gradually as the burn progresses. Systems for the supply of oxygen to fire floods are discussed by Henningson and Duckett (1984). Figure 9.44 shows a system suitable for a field pilot.

The compression of the oxygen from a separation plant requires only about a fifth of the energy that would be required for the compression of the same amount

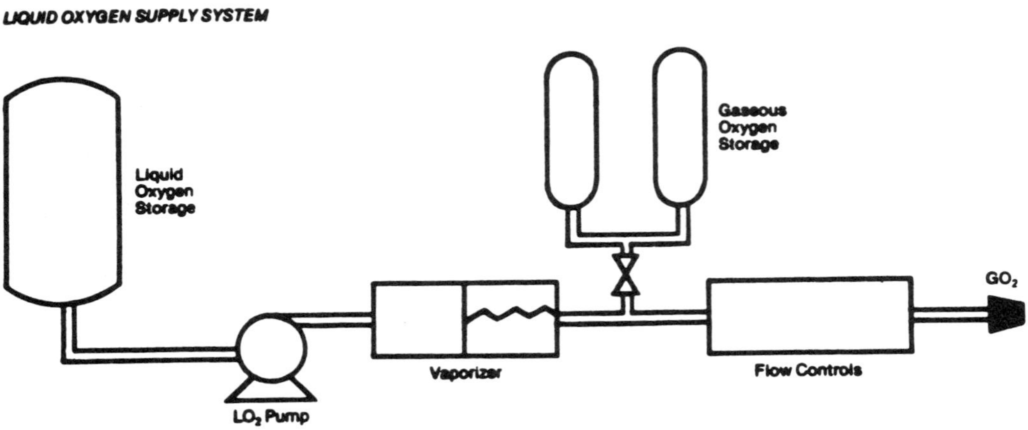

Figure 9.44 Liquid Oxygen Vaporization System for In Situ Pilot Operations (from Henningson and Duckett 1984)

of oxygen as air. This saving in compression work can offset the work needed to separate the oxygen from air.

Whether the overall energy is lower or higher depends upon the final delivery pressure. Figure 9.45 (Hvizdos, Howard, and Roberts 1983) shows a comparison of the power requirements to produce 4 million SCF/d of oxygen as air and as pure oxygen. At pressures above about 175 psia, the production of oxygen requires less power.

The costs from the same reference are compared in Figure 9.46. Which source of oxygen is cheaper depends not only upon the pressure required but also on the volume. This is because of the very substantial economy in scale in the manufacture of oxygen; large plants are more economical than small ones. This is a significant problem when it is desired to experiment on a small pilot scale with oxygen to develop the method. The discontinuities in the curves of Figure 9.46 reflect the economies that can be made in the cost of air compression by switching from reciprocating to turbocompressors if the requirement is large enough.

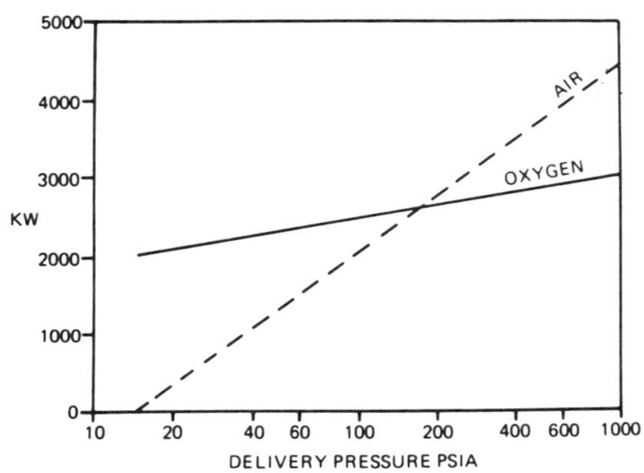

Figure 9.45 Electrical Energy to Produce 4MSCF/d of O_2 or Air Containing 4MSCF/d of O_2 (from Hvizdos, Howard and Roberts 1983)

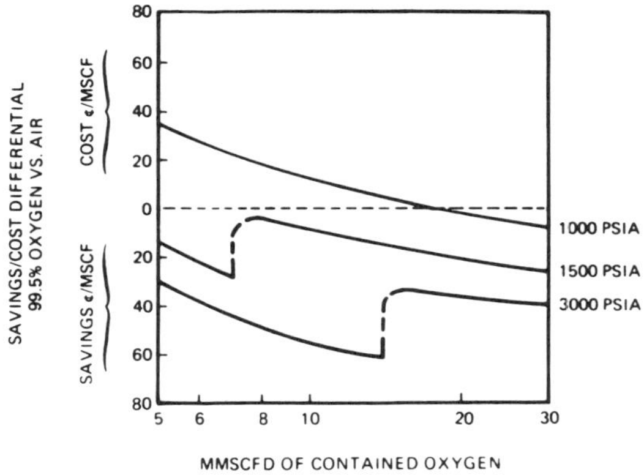

Figure 9.46 Differential Cost for Oxygen Compared to Air (from Hvizdos, Howard, and Roberts 1983)

Rather than produce pure oxygen there is some economy, in oxygen cost at least, in using enriched air. This is shown in Figure 9.47. Although there are minima in these curves, the differences are not very great, and it seems likely that other factors such as reservoir performance will determine the best oxygen concentration. It seems reasonable to expect that this will likely turn out to be essentially pure oxygen.

An interesting discussion of the potential for in situ combustion using oxygen is given by Fairfield and White (1982). A state-of-the-art review of oxygen in situ combustion is presented by Garon, Kumar, and Cala (1986). In their review, they compare the physical characteristics of nine different field projects that use oxygen or enriched air.

The Effect of Pressure on Combustion Performance with Oxygen

The preceding discussion of oxygen costs presumes that burn performance is unaffected by degree of oxygen enrichment. Combustion tube experiments performed by Moore and Bennion at the University of Calgary suggest that this may not be the case (Moore et al. 1987). Using a 4-in.-diameter, 6-ft-long adiabatic combustion tube, they ran a series of dry combustion tube tests using 95% oxygen-enriched air

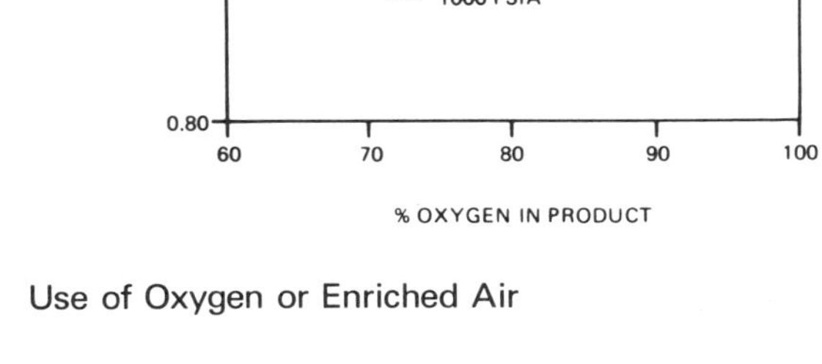

Figure 9.47 Effect of Oxygen Purity on Cost (from Hvizdos, Howard, and Roberts 1983)

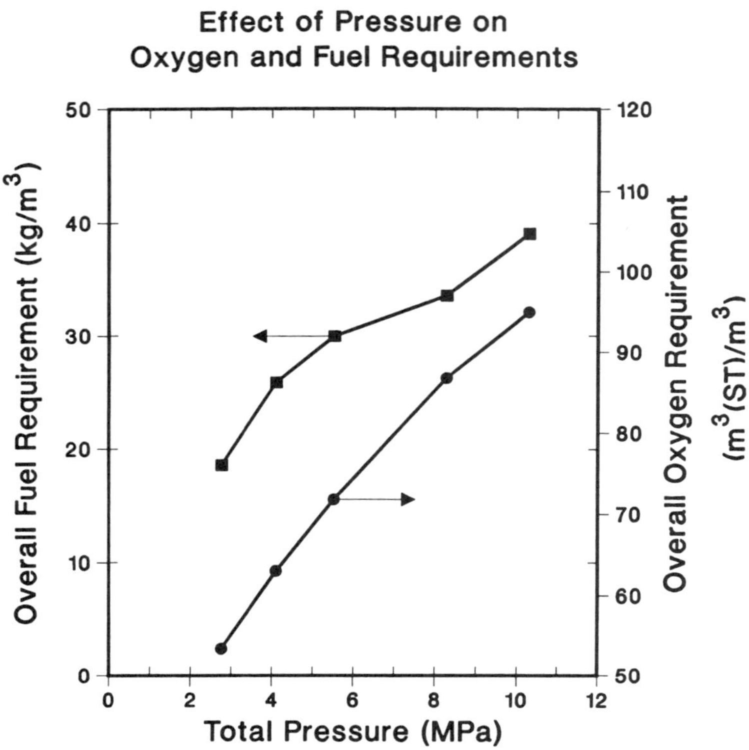

Figure 9.48 Effect of Pressure on Oxygen and Fuel Requirements—Combustion and Tube Experiments with Athabasca Sand Core and 95% O_2 (from Moore et al. 1987)

and Athabasca oil sand core. Figure 9.48 shows that the overall oxygen and fuel requirements appeared to increase linearly with operating pressure, nearly doubling over the range of 2700 to 10,300 kPa (400–1500 psi). They attributed this increase to the preoxidizing effect caused by the high oxygen partial pressures. This is consistent with the observations of Alexander, Martin, and Dew (1962) already presented (see Figure 9.17). Both Moore and Alexander noted that oxygen partial pressure had a much smaller effect on normal air (21% oxygen) combustion parameters. Further observations on the relative performance of oxygen and air in situ combination in a large number of combustion tube tests are summarized by Moore, Bennion, and Ursenbach (1988).

DESIGN OF IN SITU COMBUSTION PROJECTS

The practical design and sizing of facilities for in situ combustion projects have been discussed by Nelson and McNeil (1961) and by Gates and Ramey (1980). These are two excellent papers that will be of considerable assistance to an engineer faced with the planning and design of a new project. Chiu (1988) discusses a related analytical model that extends the theory.

Nelson and McNeil describe means for making simple estimates of the following:

- Total air requirements.
- Rate and pressure at which the air must be supplied.
- Total amount of oil that will be produced.
- Rate at which the oil will be produced.
- Operating expense.

Total Fuel Load

The fuel load measured in a combustion tube test is adjusted to allow for the different porosity in the reservoir as compared to the laboratory sand pack.

$$\text{lb fuel/acre-ft burned} = 43{,}560\, WF \tag{9.6}$$

where W is lb fuel/ft^3 lab test
F is $(1 - \phi_R)/(1 - \phi_p)$
ϕ_R is reservoir porosity
ϕ_p is sand-pack porosity

The acre-foot measure of reservoir volume is often used when employing conventional petroleum-measuring units.

$$1 \text{ acre-ft} = 43{,}560 \text{ ft}^3$$

Air Requirement

The air consumption in the laboratory tube is expressed as SCF per pound of fuel, and this is prorated as shown in equation 9.6 to give the air requirement per unit volume of reservoir.

$$A = \frac{V_a}{W_F} WF \tag{9.7}$$

where A is SCF air/CF reservoir
V_a is SCF air in lab test
W_F is total lb of fuel burned in lab test

The total air required for the project is estimated by multiplying the requirement per unit volume by the estimated volume of the reservoir that will be burned. This volume is equal to the volume of the pattern multiplied by an estimated volumetric sweep efficiency. For a five-spot, Nelson and McNeil suggest that a sweep efficiency of 62.6% should be employed; this leads to equation 9.8.

$$\begin{aligned}\text{Air per acre-ft burned} &= 43{,}560A/10^6 \text{ MSCF/acre-ft burned}\\ \text{Air per acre-ft pattern} &= 0.626 \times 43{,}560A/10^6 \\ &= 27{,}269A/10^6 \text{ MSCF/acre-ft pattern}\end{aligned} \tag{9.8}$$

Air Rate and Pressure

The rate at which air is introduced controls the burning rate.

$$U = A\nu \tag{9.9}$$

where U is air flux SCF/ft^2 d
A is SCF air/CF reservoir
ν is burning front advance, ft/d

As has been discussed previously, a very low burning rate is insufficient to maintain combustion, and a very high rate causes erosion problems at the production wells. The practical maximum gas rate at a production well seems to be about 500,000 to 600,000 SCF/day.

It is common to choose an injection rate that will maintain, initially, a constant burning front velocity. A velocity of 0.5 ft/d is recommended by Nelson and McNeil. As the front advances, the air-injection flow is increased up to the capacity of the compressor. Following this, the injection is maintained at the maximum rate. During this main period, the rate of advance decreases because of the increasing area of the front.

Figure 9.49 shows how the air-injection rate and cumulative injection vary with time. The gradual decrease in the rate at the end of the production is required to minimize oxygen bypassing. In a large multiple-pattern development, the air capacity that is made free during this period can be utilized to start up a new pattern. Nelson and McNeil discuss the scheduling of air for production from a number of staggered pattern operations.

Nelson and McNeil consider the lowest burning velocity at which satisfactory combustion can be obtained is about 0.125 ft/d. The methods of Ramey described earlier can also be used to study this.

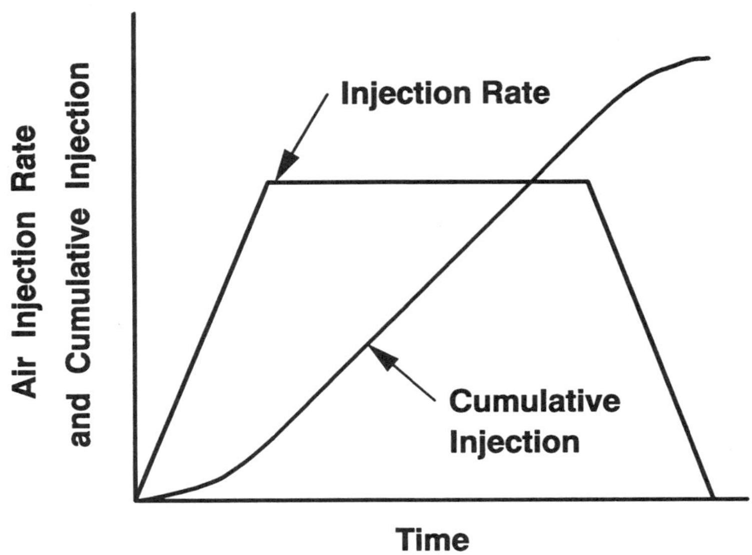

Figure 9.49 Air Injection Programm (after Nelson and McNeil 1961)

Gates and Ramey (1980) consider that the air capacity should be great enough to give a minimum burning rate of 0.15 ft/d or an air flux of 2.15 SCF/h ft^2, but this flux should be calculated as if the air were passing through only $\frac{1}{3}$ to $\frac{1}{4}$ of the reservoir thickness.[4]

The existence of a limiting gas-production rate per well and the need to maintain a minimum burning advance rate tend to make the use of oxygen attractive, since it will allow larger rates and wider patterns. The injection pressure required is best obtained from actual field injection test data.

As a means of making a prior estimate, Nelson and McNeil recommended equation 9.10. This requires an estimate of the permeability of the cold formation to the flow of gas, i.e., the permeability of the formation multiplied by the relative permeability. If there is no specific information, they suggest that a relative permeability of 5% can be used.

Equation 9.10 gives the pressure at the time the air-injection rate first reaches the maximum value. This is the point where the pressure reaches its maximum value.

$$P_{iw}^2 = P_w^2 + \left(\frac{i_a \mu_a T_f}{0.703 k_g h}\right)\left[\ln\left(\frac{a^2}{r_w v_1 t_1}\right) - 1.238\right] \qquad (9.10)$$

where P_{iw} is injection well bottom hole pressure, psia
P_w is production well bottom hole pressure, psia
i_a is maximum air rate, SCF/d
μ_a is air viscosity at T_f, cp
T_f is absolute formation temperature, R
a is well spacing for the five spot pattern, ft
t_1 is time to reach maximum rate, d
k_g is effective permeability to air, mD
h is formation thickness, ft
r_w is production well radius, ft
v_1 is initial rate of advance of burning front, ft/d

Oil Displaced

Assuming that there is no oil left in the burned-out zone, then the amount of oil displaced must be equal to the original oil in place minus that consumed as fuel. This is given by equation 9.11.

$$\text{Oil displaced} = \text{oil at start} - \text{fuel}$$

$$= 43{,}560\left(\frac{S_o \phi_R}{5.61} - \frac{WF}{350}\right) \text{ B/acre-ft} \qquad (9.11)$$

where S_o is fractional oil saturation
ϕ_R is reservoir porosity, fraction
43,560 ft^2/acre
5.61 ft^3/B
350 lb fuel/B (assumed)

[4]The temperature at the firefront for a given rate of advance depends upon the fuel load as well as upon the air rate. Thus a very heavy oil that gives a high fuel load will give a higher front temperature for the same rate of advance. Thus, for example, in the Midway Sunset field, Counihan (1977) reports excellent combustion results with a design rate of only 1 in./d.

Gates and Ramey pointed out that, at an intermediate time, the cumulative amount of oil produced can be greater or less than the amount of oil displaced from the burned zone because of two opposing factors:

1. Oil ahead of the front may be displaced to the production well by other mechanisms "including hot water drive, steam drive, hot gas drive, vaporization, miscible phase displacement, expansion and gravity drainage." These mechanisms are all made more effective by the elevated temperatures.
2. Oil may have to form a bank ahead of the combustion zone in order to fill some of the gas saturation that is present initially in the reservoir.

Based on this concept and using experimental data from the long-standing and successful South Belridge pilot, Gates and Ramey developed the chart in Figure 9.50.

This correlation predicts the cumulative oil production as a function of the fraction of the reservoir that has been burned and the initial gas saturation. It is based upon considerable data from cored wells at the pilot and also upon laboratory experiments that demonstrated the effect of initial gas saturation.

Figure 9.50 predicts that there is a delay period before oil is produced that corresponds to the formation of the oil bank and that the initial production, when it eventually occurs, is at a much higher rate than the average. A delay in the initial production of oil is shown by the data shown previously in Figure 9.10.

In using the correlation it is necessary to know the fuel concentration. This may be obtained from combustion tube experiments or, if these are not available, from the correlation in Figure 9.51. Figure 9.52 may be used to estimate the air requirement to burn the fuel.

Using the correlation in Figure 9.50, we can measure the slope of the appropriate curve and calculate the instantaneous oil rate as a function of the fraction burned. Using this information and the fuel requirement, we can calculate the instantaneous air-oil ratio as a function of recovery. The results of such a calculation for the South Belridge field are shown in Figure 9.53.

The initial current air-oil ratio is much lower than the average rate shown by the dotted horizontal line. The cumulative ratio is also lower than the average except at the end, where it converges.

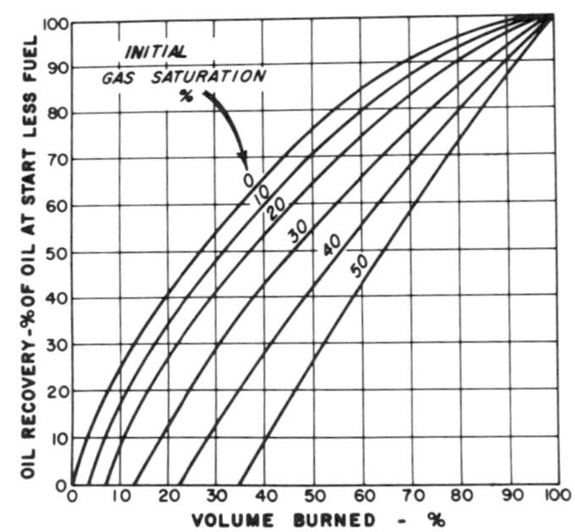

Figure 9.50 Chart for Estimating Oil Recovery as a Function of the Percent of the Reservoir Burned (from Gates and Ramey 1980)

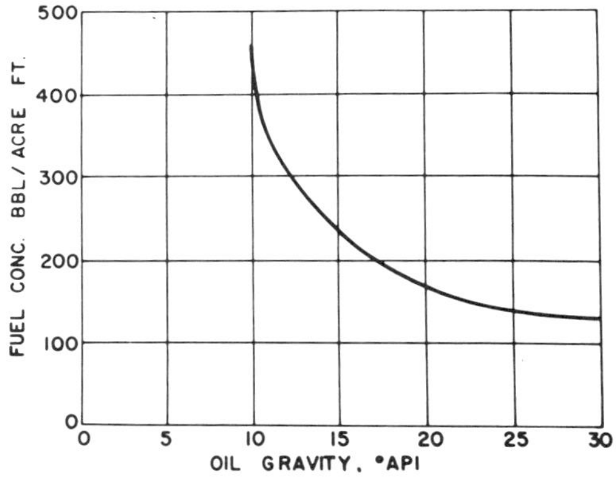

Figure 9.51 Chart for Predicting Fuel Concentration (from Gates and Ramey 1980)

In practice it is likely that the process would not be taken to the very end but that the production would be terminated when the current air-oil ratio reached an uneconomic level. In this case the final cumulative ratio would also be lower than the dotted line.

The calculation so far has assumed that the oxygen is completely consumed in the combustion process. In practice there is some unreacted oxygen in the produced gas, and allowance should be made for this in estimating the size of the burned zone. The excess air requirements may be estimated from Figure 9.54; however, Gates and Ramey consider that this figure may overestimate the amount of excess air and that it is conservative.

Figure 9.55 shows the experimental air-oil ratios from South Belridge compared to predictions made using the preceding method. Gates and Ramey point out that the spikes in the experimental data in this figure correspond largely to periods when the production wells at the pilot were partly shut in for mechanical reasons.

The reader is referred to the original paper for further details, including a very interesting description of several parallel methods for estimating the fuel concentration from the field data.

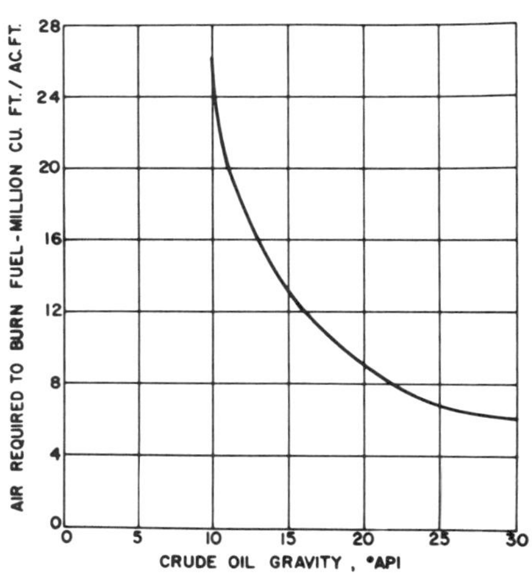

Figure 9.52 Air Requirements per Acre-Foot of Reservoir (from Gates and Ramey 1980)

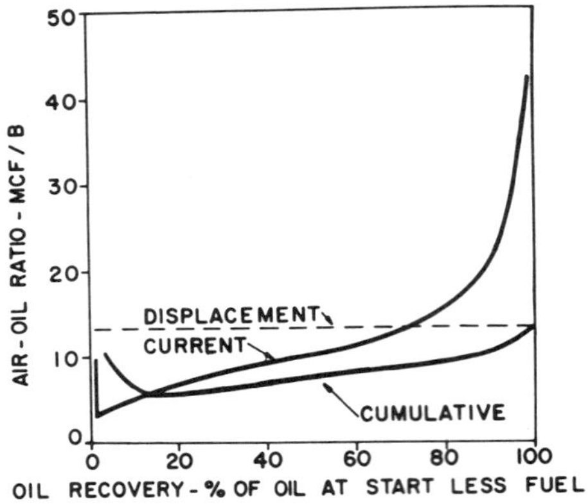

Figure 9.53 Computed Air-Oil Ratio for South Belridge Field (from Gates and Ramey 1980)

Effect of Water-Air Ratio on Oil Recovery per Volume Burned

The correlation of Gates and Ramey (Figure 9.50) for predicting the oil recovery as a function of the burned reservoir fraction and the initial gas saturation is for dry combustion in a particular reservoir. Different correlations may be expected for different situations. For example, in wet combustion, oil is moved ahead by the steam, and oil production occurs more rapidly. This phenomenon has been studied by Moore et al. (1988) in a series of combustion tube tests with varying ratios of water to air. Some of their data are shown in Figures 9.56–9.59.

Figure 9.56 shows the effect of adding 2 kg/m^3 (ST) of water to the combustion air. Oil production starts when a smaller percentage of the core has been burned, and for a given fraction burned, the oil recovery is larger. Also, because the fuel load is less for wet combustion, less air is required to achieve the same fraction burned. As a result, the curves of percent oil recovery versus quantity of air injected (Figure 9.57) show an even larger difference.

Results from similar experiments using cores of Athabasca reservoir sand and bitumen that were reconstituted by flooding are shown in Figures 9.58 and 9.59. In this series of experiments, runs with much higher water-injection ratios are compared to a dry run.

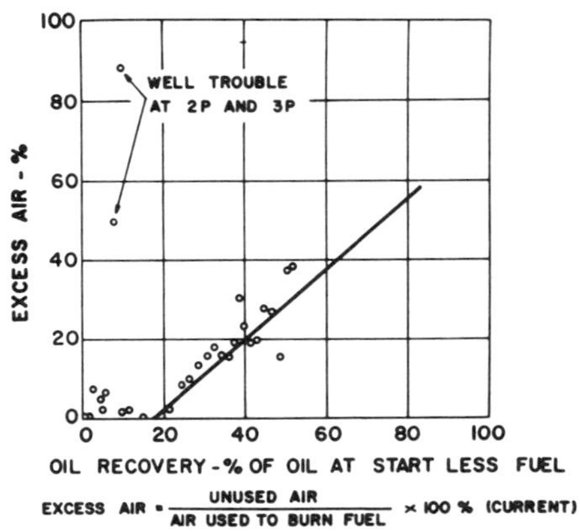

Figure 9.54 Excess Air from South Belridge Pilot (from Gates and Ramey 1980)

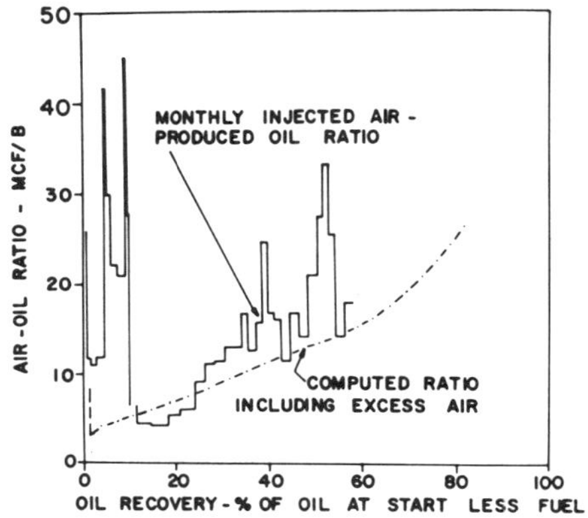

Figure 9.55 South Belridge Current Air/Oil Ratio (from Gates and Ramey 1980)

The "optimal" water-air ratio for this series of runs was about 4.6. In this experiment, the average peak combustion front temperature was 608°C. This was slightly higher than that for the dry run (569°C). On the other hand, the average peak temperature for the run with a ratio of 6.9 was only 225°C. In it, liquid water was cooling the combustion zone, and unburned fuel was left behind. This is the reason for the relatively low final recovery.

It can be seen from Figure 9.58 that, as in Figure 9.56, the addition of water caused oil to be recovered sooner. However, in the superwet run, the early advantage of more production was not maintained because of the material left behind as unburned fuel. The difference is even more pronounced when the recovery is plotted against the cumulative air injection in Figure 9.59.

Further insight into the factors that affect the oil recovery as a function of the quantity of air injected can be obtained from the mathematical analysis of the problem described by Chiu (1988). In his paper, Chiu shows how the results of combustion tube tests such as those shown in Figures 9.56–9.59 can be developed from a mathematical model involving steamflooding ahead of the combustion zone, with the quantity of steam being calculated from a heat balance. The heat balance involves the calculation of the vertical heat losses from the steam zone using a modification of the Marx-Langenheim method described in Chapter 4.

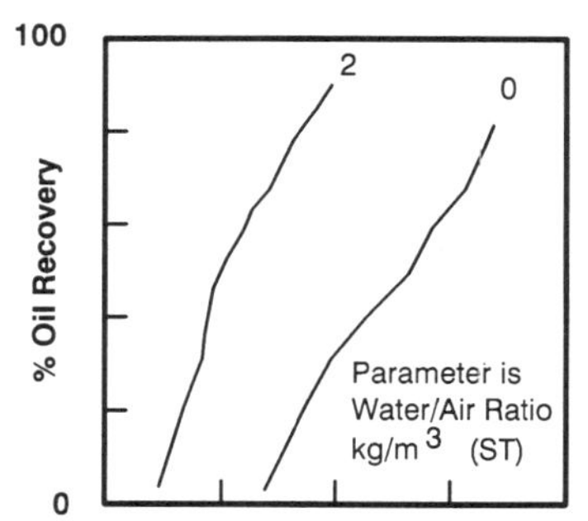

Figure 9.56 Effect of Water-Air Ratio on Oil Recovery; Combustion Tube Test with Pack of Athabasca Bitumen and Silica Sand (after Moore et al.)

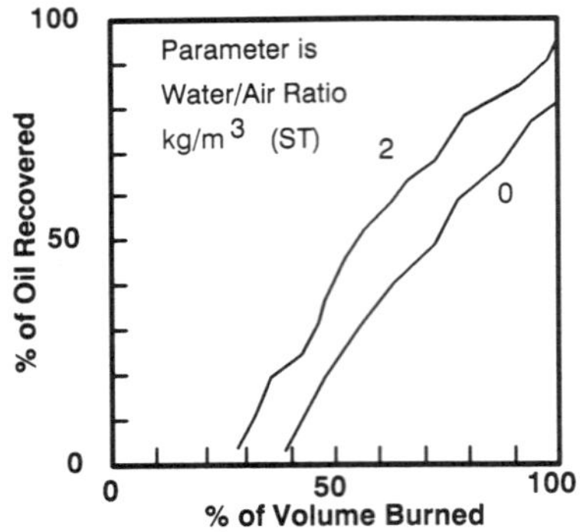

Figure 9.57 Effect of Water Addition upon Oil Recovery as a Function of Cumulative Air; Injection Tests with Athabasca Bitumen–Silica and Sand Pack (after Moore et al.)

FIELD PROJECT RESULTS

Lloydminster, Golden Lake

As Fairfield and White (1982) have discussed, the Lloydminster area contains many reservoirs of heavy oil contained in thin, Lower Cretaceous sands. The oils are very viscous, and the oil saturation is usually high; the sands are very permeable. The reservoirs are estimated to contain a total of 50 to 70 billion barrels of oil.

The Lloydminster oils are less viscous than those in the Cold Lake field to the north, and primary production is possible; well productivities of the order of 20 B/d are obtained, but the primary recovery is only 3 to 8%. Because of the high viscosity and the fingering that occurs, waterflooding is not very effective; Fairfield and White indicate that the incremental recovery from waterflooding is of the order of only 2%.

Figure 9.60 shows that about 90% of this oil occurs in sands thinner than 20 ft, and 50% occurs in sands less than 10 ft. Because the sands are so thin, steam drive is generally not applicable. For thermal efficiency reasons, a thickness of about 25 ft or more is necessary for steam drive to be practical.

In situ combustion is suitable for sands down to about 10 ft in thickness and has promise for allowing efficient recovery of much of the Lloydminster crude. It has the advantages of using cheaper energy than steam recovery, of not requiring all of the reservoir to be maintained at the thermal recovery temperature throughout the project, and of removing some of the most refractory material as fuel.

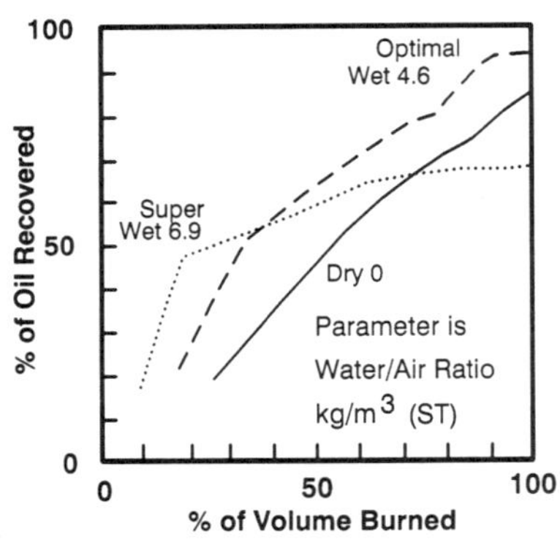

Figure 9.58 Effect of Water-Air Ratio on Oil Recovery Tests with Reconstituted Athabasca Tar Sand (after Moore et al.)

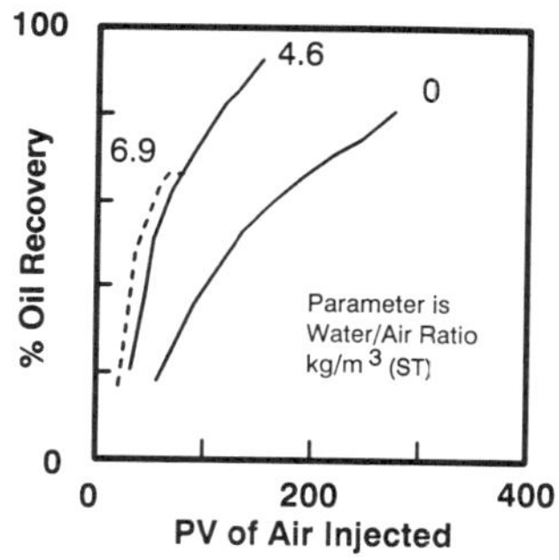

Figure 9.59 Effect of Water Addition upon Oil Recovery as a Function of Cumulative Air Injection Tests with Reconstituted Athabasca Tar Sand (after Moore et al.)

TABLE 9.7 Properties of Golden Lake Reservoir

Sand	Sparky
Depth, ft	1600
Net sand, ft	
Original pilot	23
Expansion #1	20.8
Porosity, %	35
Permeability, md	
Core data	1200
Calculated from production data	8000
Saturation, %	
Oil	82
Water	18
Original reservoir pressure, psig	510
Reservoir temperature, °F	70
Reservoir fluid properties	
Oil gravity, °API	12–13
Oil formation volume factor	1.01
Solution gas oil ratio, SCF/B	45
Dead oil viscosity at 70°F., cp	6300
Live oil viscosity at 500 psig, cp	3500

(Fairfield and White 1982)

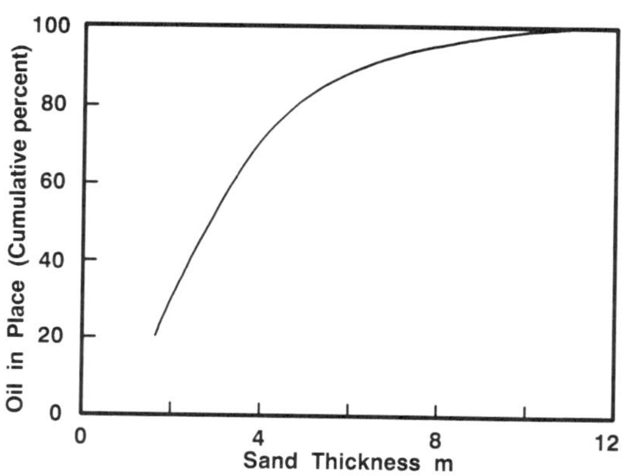

Figure 9.60 Oil in Place in the Lloydminster fields as a Function of Sand Thickness (after Fairfield and White 1982)

TABLE 9.8 Combustion Characteristics of Golden Lake Reservoir Material

Molecular weight (unit)	13.4
Atomic hydrogen—carbon ratio	1.4
Fuel, lb/ft^3	1.93
Fuel, B/acre-ft	250
Unit air requirement at (100% efficiency)	
SCF/lb carbon	200
SCF/lb fuel	179
SCF/ft^3	345
Water formed by combustion, B/acre-ft	224
Oil displaced by combustion front, B/acre-ft	1957
Theoretical AOR, dry combustion, kSCF/B	7.68
Water-air ratio, B/MSCF	205
Residual oil saturation, steam zone:	
% pore space (avg.)	22.6
B/acre-ft (avg.)	614
Displacement in steam zone, B/acre-ft	1593

(Fairfield and White 1982)

Properties of the Golden Lake Reservoir and the combustion characteristics are shown in Tables 9.7 and 9.8. The oil has a viscosity of 3500 cp at reservoir conditions. There is a large difference between the permeabilities measured on core samples and those calculated from production data. This is thought to be due to "worm holes" in the reservoir, possibly formed as a result of sand production.

The well layout for the Golden Lake project is shown in Figure 9.61. The original five-spot pattern around well B15-11 was ignited in July 1969 and expansion 1, consisting of two seven-spot patterns, was ignited in 1974. Water injection was started at a design rate of 205 B per million SCF in 1972 in the original pattern and in July 1976 in the expansion pattern.

Production results are shown in Figure 9.62 and are summarized in Table 9.9. An analysis of the produced gas is given in Table 9.10.

TABLE 9.9 Golden Lake Injection and Production (81-09-30)

Pattern	Cumulative Air Million SCF	Cumulative Water kB	Burned Volume Percent[(1)]	Oil Recovered kB	Recovery Percent OOIP
Original Pattern	1240	165	17.1	567	39.8
Expansion 1					
D7 Pattern	507	107	4.4	438	18.3
B9 Pattern	562	116	5.7	378	19.7
Total Expansion 1	1069	223	5.0	816	19.0

[(1)]It was anticipated that it would be economic to continue the burn until the burned volume was about 20%. There are thus quite a few years of production ahead of even the original pattern (Fairfield and White 1982).

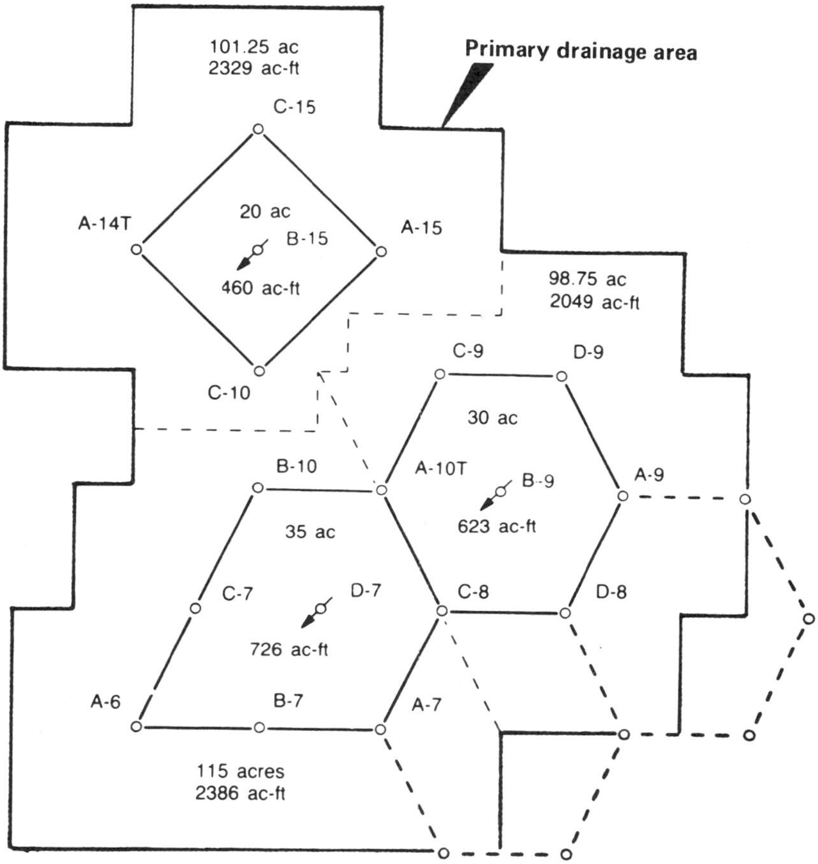

Figure 9.61 Well Arrangement for Husky's Pilot Projects at Golden Lake (from Fairfield and White 1982)

TABLE 9.10 Golden Lake Produced Gas Analysis

Component	Volume Percent
Carbon dioxide	16.0
Carbon monoxide	0.4
Methane	1.2
Nitrogen	81.4
Oxygen	0.0
Argon	1.0
	100.0

(Fairfield and White 1982)

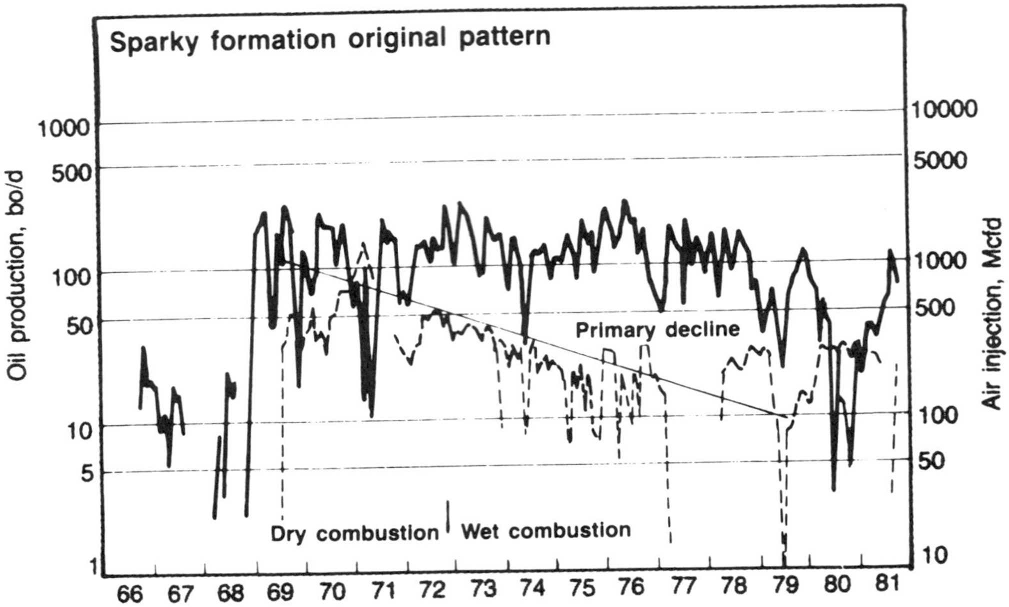

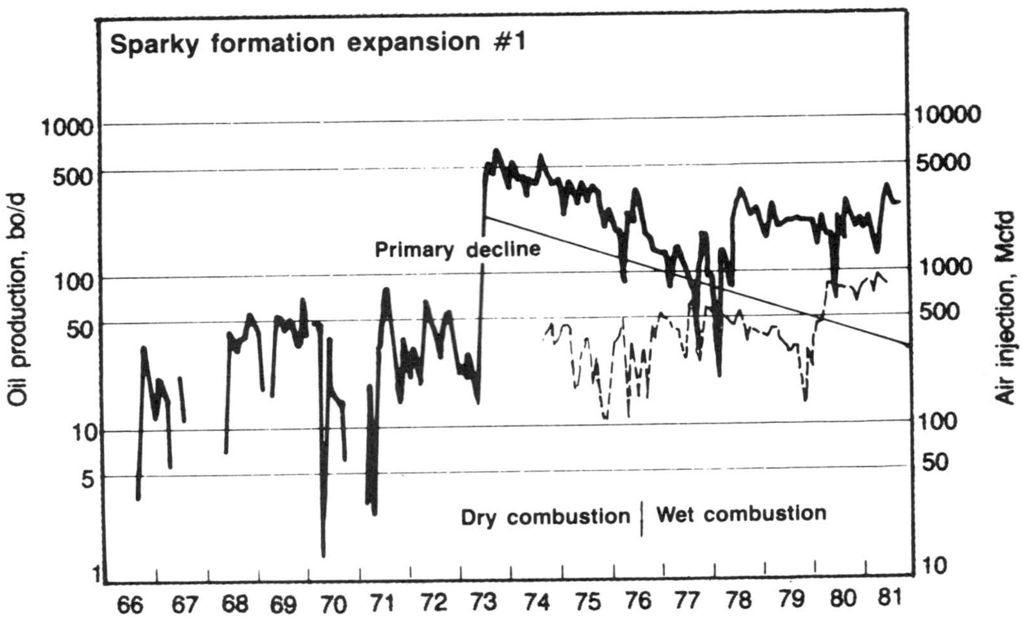

Figure 9.62 Production from Golden Lake In Situ Combustion Pilots. The Air Injection Is In kSCF/d. (from Fairfield and White 1982)

Trouble was found with air short-circuiting from the injector through a high permeability thief zone in the original pattern, and this eventually required the drilling of a new injection well. Several well workovers and squeeze jobs to shut off water layers have also been required. Nevertheless, the operation has been found to be very promising, and a second expansion has been completed.

Sand control is a continuing problem, and Husky has found it better to allow the oil to be produced rather than to try to restrict it from the well bore. They use heavy rods and large (3.5-in.) tubing.

Very high permeability channels form when high injection pressures are employed, and they limit the pressure in order to prevent these.

The possible advantages of using oxygen rather than air are emphasized by Fairfield and White:

- Lower gas velocities
- Enhanced oil mobility (see Figure 9.63)
- A set of generalized correlations for predicting the solubility of carbon dioxide and its effect on swelling and viscosity is given by Simon and Grave (1965). A series of papers by Mehrotra and Svrcek (Svrcek and Mehrotra 1982; Mehrotra and Svrcek 1984, 1985a, 1985b, 1985c) contain measurements for bitumen–carbon dioxide systems.
- Rapid contact with carbon dioxide and swelling
- Valuable produced gas
- Less override

It appears that the in situ combustion approach to the production of Lloydminster crude is successful and that the use of oxygen may be very promising. The ideal type of reservoir for this process is one in which oxygen could be injected at the top of a dipping reservoir, and the authors consider that recoveries of the order of 30 to 50% of the oil in place should be practicable.

More recent experience at the Golden Lake fireflood project is described by Miller and Jacques (1987). The pilot was expanded from the three patterns described by Fairfield and White by adding two further inverted seven-spot patterns, which were ignited in 1982, and three adjacent patterns in 1985. There was difficulty in igniting the last three patterns, and this was not solved by the time the pilot was shut down in 1986 because of the low price of oil. There were also problems encountered because of the encroachment of water into the pilot area.

Getty's Bellevue Field in Louisiana

This project is the largest in situ combustion project in the United States. Since the original pilot began operations in 1963, it has been expanded numerous times, as may be seen from Figure 9.64. In 1978 about 2900 B/d of oil was produced from about 350 wells (most of it from 223 fireflooded wells).

The reservoir is shallow (depth is 300–420 ft) and relatively thin (20–90 ft). It is broken up by numerous faults and by a 4-ft layer of noncontinuous limestone. There is vertical communication between the upper and lower zones.

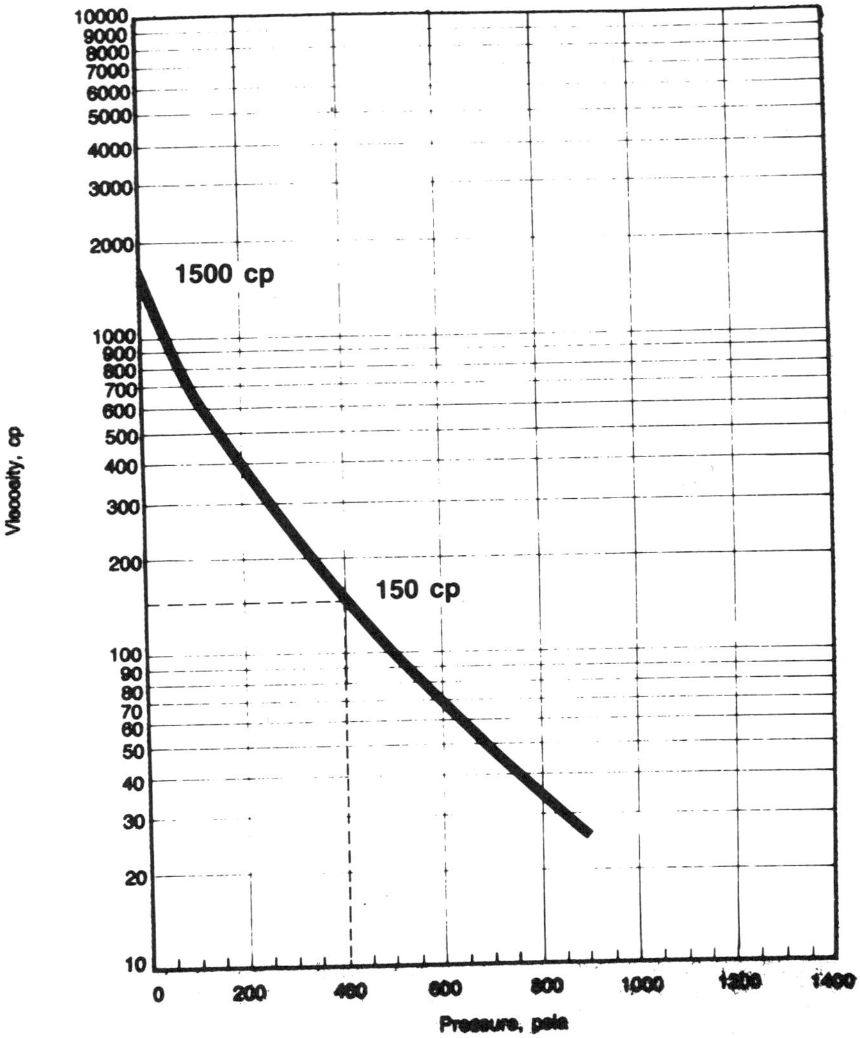

Figure 9.63 Effect of CO_2 on Viscosity of Lloydminster Crude (from Fairfield and White 1982)

Although the project is very successful, it requires a very high air-to-oil ratio—about 19,000 SCF per barrel. The high air-to-oil ratio is believed to be due to the crude depositing an abnormal amount of fuel, to the relatively low oil saturation (about 52%), and to the difficulty in moving all the heated oil to the production wells because of the reservoir heterogeneities.

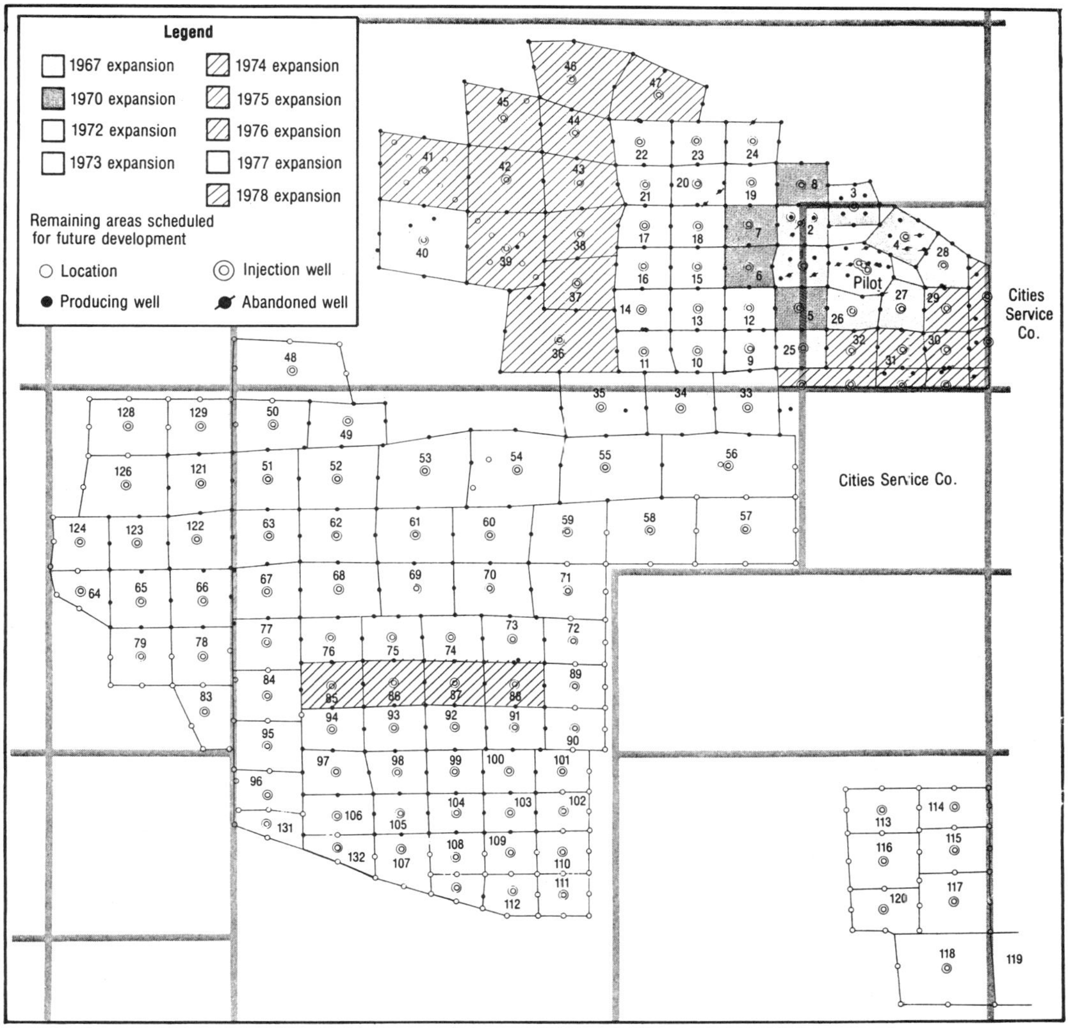

Figure 9.64 Getty's In Situ Combustion Project in Bellevue Field, Bossier Parish, Louisiana, U.S.A. (from Bleakely 1978). Reprinted by permission of Edgell Communications. Copyright © November, 1978, PETROLEUM ENGINEER INTERNATIONAL.

Field Expansions at Bellevue

1963:	Pilot				
1967:	1–4	1973:	25–28	1976:	32
1970:	5–8	1974:	29–31	1977:	60–73 & 121–123, except 64
1972:	9–24	1975:	36–47	1978:	85–88
				1979:	52–55

A major factor in the economic success of the project is the shallow depth of the reservoir and the low reservoir pressure. This makes wells cheap to drill and reduces compression costs; the discharge pressure of the compressors is only about 100 psi. Although the air consumption is large, the compressed air is cheap.

The wells are drilled on inverted nine-spot patterns. Originally, the pattern sizes were chosen to give reservoir volumes of 185 acre-ft per pattern; i.e., the wells were spaced farther apart where the reservoir was thinner. Some patterns were as small as 2.2 acres and others, as large as 8.5 acres. Slow response was found in the large patterns, and these were infilled with additional producers. This gave much better results.

Originally dry injection was used; then, after about 50% of the theoretical burn was achieved, air injection was stopped and waterflooding was used to scavenge and utilize the residual heat. This procedure has now been modified to include a period of simultaneous water and air injection after the dry operation and before the waterflood.

Getty expects to recover 60% of the original oil in place in the Bellevue field using in situ combustion. To do this will require the combustion to 6 of 15% of the oil in place as fuel.

Cities Service Company has a wet combustion project similar to that developed by Getty on their lease in the Bellevue field; this lease lies directly to the east of the Getty lease (Figure 9.64). Joseph and Pusch (1982) report that Cities Service expects to recover nearly 40% of the oil in place. Well productivity is about 20 to 30 B/d. In their paper Joseph and Pusch also provide an interesting breakdown of the operating costs for the field, which, exclusive of taxes, were $17.27/B in 1980–1.

In Situ Combustion Projects in Rumania

There are some very interesting and large in situ combustion projects being carried out in Rumania (Gadelle et al. 1981; Turta and Pantazi 1982; Carcoana 1982; Aldea, Turta, and Zamfir 1988). These four papers also contain many related references.

TABLE 9.11 Suplacu de Barcau Reservoir Properties

Sand character	Unconsolidated
Depth	35–220 m
Effective thickness	4–24 m
Dip	5°–8°
Initial reservoir pressure	0.4–2.2 MPa
Reservoir temperature	18°C (at a depth of 80 m)
Porosity	0.32
Absolute permeability	1.7 μm^2
Initial oil saturation	0.85
Oil specific gravity	0.96 (16° API)
Oil viscosity at 18°C (heavy asphalt-base crude oil)	1.8–2 Pa · s

(Turta and Zamfir 1988).

The largest project is in the Suplacu de Barcau field, which is in the northwest of the country. The reservoir consists of a high-quality, shallow sand saturated with a 2000-cp oil; the reservoir dips at 5 to 8° toward the south. Reservoir properties are shown Table 9.11. Figure 9.65 shows a map of the field.

A pilot test was initiated in 1964, and semi-industrial operation was started in 1967. This has since been expanded to the point where "given the amounts of air injected and oil produced, this project may be considered the most important combustion in the world" (Gadelle et al. 1981).

Oil production due to combustion has increased from 340 t/d (2200 B/d) in 1974 to 800 t/d in 1977 and to 1000 t/d in 1979. Production rates are plotted in Figure 9.66.

The increase in production was achieved by the addition of air-compression capacity and by the extension of the length of the combustion front; at the end of 1979 this was more than 4 km long.

The number of injection wells was increased from 11 in 1974 to 38 in 1979. During this period the air-compression capacity rose from 0.55 million standard m^3/d (19.4 million SCF/d) to 1.8 million. The air-to-oil ratio achieved in long-term operation is between 1500 and 2000 std m^3/t (8.4 and 11.2 kSCF/B) of oil.

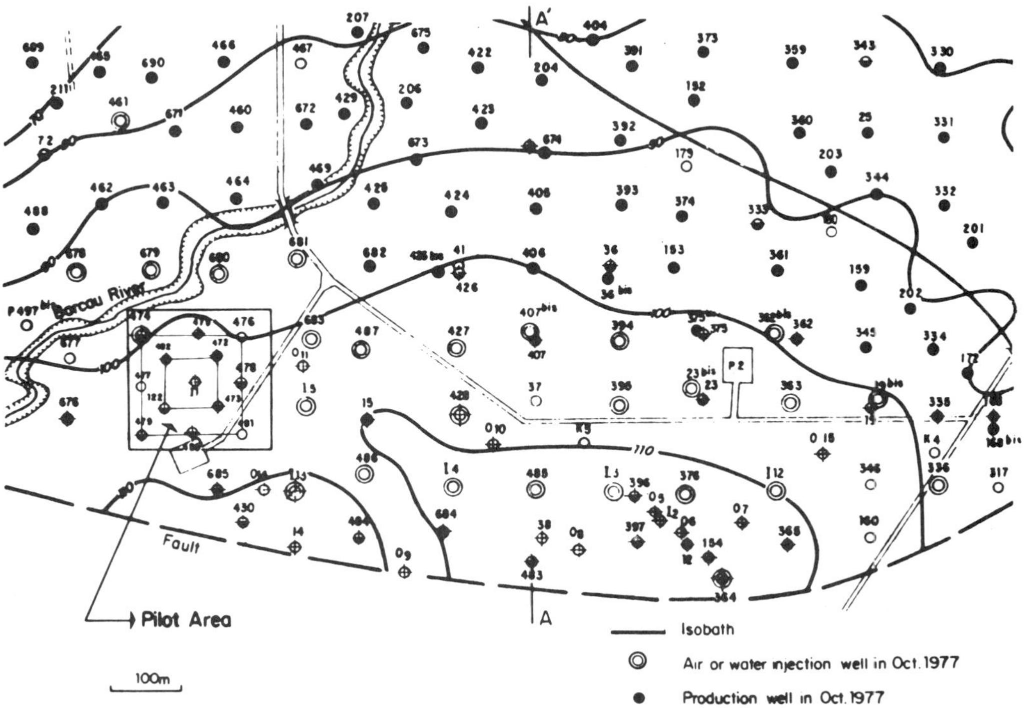

Figure 9.65 Suplacu de Barcau Field (from Gadelle et al. 1981)

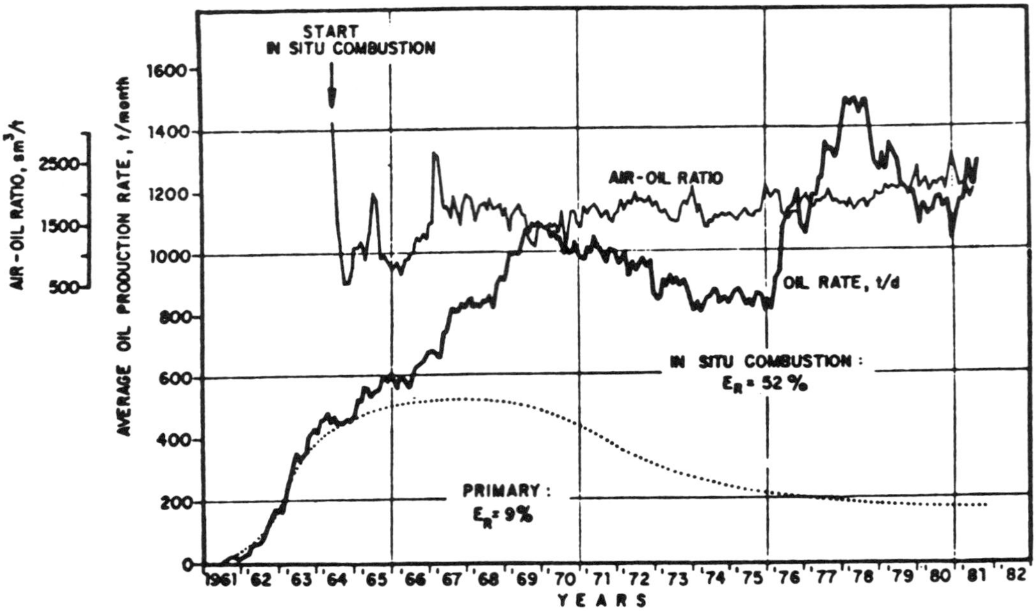

Figure 9.66 Suplacu de Barcau Field Performance (from Carcoana 1982)

Production history up to 1981 is shown in Figure 9.66. In a more recent paper (Aldea, Turta, and Zamfir 1988), results up to 1985 are described. At that time, the oil production was 1400 t/d, with an air-to-oil ratio of about 2250 $m^3(ST)/m^3$ (12,645 SCF/B). In 1988, air injection was 3.5 million $m^3(ST)/d$ into 120 wells, and production was from 600 wells. The combustion front was 7.5 km long, and recovery in the combustion front area was greater than 50%. The project has many interesting features:

- The operation is a line drive with air injected at the upper edge of the sloping reservoir; this is bounded by a fault, which forms the trap.
- Water is injected into wells behind the combustion front: 6–10 water injectors with about 50 t/d per well. The paper contains some interesting combustion tube data, which show that the air requirement is reduced by about one-third for wet combustion as compared to dry.
- Stable emulsion formation was associated with water injection, but this has been overcome by a thermal-chemical treatment with a final stripping distillation (Aldea, Turta, and Zamfir 1988).
- The position of the combustion front in the field is determined by plotting isothermal contour maps of the well head production temperatures. A second method involves the analysis of such data as oil and gas rates, O_2 content of produced gas, and downhole temperatures from north-south rows of production wells, i.e., for rows of wells approximately normal to the combustion front.
- Various methods are used to control the development of the front.

These include

1. Decreasing the gas flow by throttling production
2. Stimulating specific wells using steam
3. Creating secondary combustion fronts perpendicular to the main front
4. Combining air and water injection.

- An oil-based, weighted killing fluid is used for well workover.
- Production wells are cooled as the combustion front approaches by injecting 3 to 4 t/d of water.
- Acid treatment and prolonged water injection have been used to restore the air injectivity of partially plugged injection wells.
- Production gases containing carbon monoxide are vented through high stacks.

Other large Rumanian combustion operations, in various stages of development, are described in the papers cited.

Of particular interest is the enormous in situ combustion project being developed for the Videle field in Rumania (Turta and Pantazi 1982). This is a line-drive scheme that will burn downdip and that will have a combustion front more than 18 km in length; an air-compression capacity of about 7 million std. m^3/d will be required (247,203,000 SCF/d). It is forecast that, after about 45 to 50 years of operation, the ultimate oil recovery will be about 38 to 40%.

The Rumanian Suplacu de Barcau project is very large compared to EOR in situ combustion projects elsewhere.

	1988 Capacity, B/d
Surplacu de Barcau	9200 (Aldea, Turta, and Zamfir, 1988)
U.S. total (9 projects)	6525 (*Oil G. J.,* April 18, 1988)
Canada total (8 projects)	8133

There are also nine other commercial in situ combustion operations in Rumania.

BIBLIOGRAPHY

AALUND, L. R., "Annual Production Report," *Oil & Gas J.,* 33–73 (April 18, 1988).

ADEGBESAN, K. O., DONNELLY, J. K., MOORE, R. G., and BENNION, D. W., "Low-Temperature-Oxidation Kinetic Parameters for In-Situ Combustion: Numerical Simulation," *SPE Res. Eng.,* 573–582 (November 1987).

ALDEA, G. H., TURTA, A. L., and ZAMFIR, M., "The In-Situ Combustion Industrial Exploitation of Suplacu de Barcau Panonian Field, The Socialist Republic of Rumania," in R. F. Meyers and E. J. Wiggins (Editors), The Fourth UNITAR/UNDP International Conference on Heavy Crude and Tar Sands, Vol 4: In Situ Recovery, AOSTRA, Edmonton, (1989), pp. 841–848.

ALEXANDER, J. D., MARTIN, W. L., and DEW, J. N., "Factors Affecting Fuel Availability and Composition during In-Situ Combustion," *JPT,* 1154–1162 (October 1962). © 1962 SPE.

ASTM, "Conradson Carbon Residue of Petroleum Products. D189–81" in Annual Book of ASTM Standards 05.01, p. 131–137 (1985) or in other Annual Books.

BARTA, L. and HEPLER, L. G., "Kinetics and Energetics of Oxidation of Athabasca Bitumen," *J. Energy and Fuels,* 2: 309–316 (1988).

BERRY, V. J., JR. and PARRISH, D. R., "A Theoretical Analysis of Heat Flow in Reverse Combustion," *Trans. AIME,* 219: 124–131 (1960).

BLEAKELY, W. B. (Ed.), "Getty's Bellevue Field Still Going and Growing," *Pet. Eng. International,* 54–68 (November 1978).

BURGER, J. G. and SAHUQUET, B. C., "Chemical Aspects of In-Situ Combustion—Heat of Combustion and Kinetics," *SPEJ,* 410–422, (October 1972). © 1972 SPE.

BURGER, J. G. and SAHUQUET, B. C., "Laboratory Research on Wet Combustion," *JPT,* 1137–1146, Oct 1973 and also SPE 4144 (1972).

CARCOANA, A. N., "Enhanced Oil Recovery in Rumania," *SPE* 10699 (April 1982).

CHIU, K.W., "Analytical Analysis of Air/Oxygen Wet Combustion by Energy Balance," in R. F. Meyers and E. J. Wiggins (Editors). The Fourth UNITAR/UNDP International Conference on Heavy Crude and Tar Sands. Vol 4: In Situ Recovery, AOSTRA, Edmonton, (1989), pp. 807–825.

CHU, C., "State-of-the-Art Review of Fireflood Field Projects," *JPT,* 19–36 (January 1982). © 1982 SPE.

COUNIHAN, T. M., "A Successful In-Situ Combustion Pilot in the Midway-Sunset Field," *SPE* 6525 (1977).

DIETZ, D. N., "Wet Underground Combustion, State of the Art," *JPT,* 605–617 (May 1970).

DIETZ, D. N. and WEIJDEMA, J., "Reverse Combustion Seldom Feasible," *Producers Monthly,* 32, No. 5: 10 (1968).

DONNELLY, J. K., HALLAM, R. J. and DUCKETT, J. A., "An Oil Sands Oxygen In-Situ Combustion Project," 3rd Unitar Conference on Heavy Crude and Tar Sands, Long Beach, California (1985).

FAIRFIELD, W. H. and WHITE, P. D., "Lloydminster Fireflood Performance, Modifications Promise Good Recoveries," *Oil Gas J.,* 101–112 (February 8, 1982).

FARQUHARSON, R. G. and THORNTON, R. W., "Lessons from Eyehill," *JCPT,* 47–53 (March-April 1986).

GADELLE, C. P., BURGER, J. G., BARDON, C. P., MACHEDON, V., CARCOANA, A., and PETCOVICI, V., "Heavy-Oil Recovery by In-Situ Combustion–Two Field Cases in Rumania," *JPT,* 2057–2066 (November 1981). © 1981 SPE.

GARON, A. M., KUMAR, M., and CALA, G. C., "The State of the Art of Oxygen Fireflooding," *In-Situ,* 10, no. 1: 1–26 (1986).

GATES, C. F. and RAMEY, H. J., "A Method for Engineering In-Situ Combustion Oil-Recovery Projects," *JPT,* 285–294 (February 1980). © 1980 SPE.

HALLAM, R. J., MOORE, R. G., KROUSE, H. R. and BENNION, D.W., "Carbon Isotope Analysis: A New Tool for Monitoring and Interpreting the In-Situ Combustion Process," *SPE* 17418 (1988).

HENNINGSON, C. J. and DUCKETT, J. A., "Oxygen Fireflooding for In-Situ Production of Heavy Oils and Tar Sands," 5th Annual AOSTRA Conference on Advances in Petroleum Recovery and Upgrading Technology, Calgary, Alberta (June 14–15, 1984).

Hvizdos, L. J., Howard, J. V., and Roberts, G.W., "Enhanced Oil Recovery via Oxygen-Enriched In-Situ Combustion: Test Results from the Forest Hill Field in Texas," *JPT,* 1061–1070 (June 1983). © 1983 SPE.

Jenkins, G. R. and Kirkpatrick, J. W., "Twenty Years' Operation of an In-Situ Combustion Project," *J. Can. Pet. Tech.,* 60–66 (January-March 1979).

Joseph, C. and Pusch, W., "A Field Comparison of Wet and Dry Combustion," *JPT,* 1523–1528 (September 1980). © 1980 SPE.

Joseph, C. and Pusch, W., "Engineering and Economics of the Bellevue In-Situ Combustion Project," 2nd Unitar International Conference on Heavy Crude and Tar Sands, Caracas, Venezuela (1982); New York: McGraw-Hill (1984), 844–850.

Kaye, S. E., Ting. V. C., and Fair, J. C., "Development of a System to Utilize Flue Gas from Enhanced Oil Recovery Combustion Projects," *JPT,* 181–188 (January 1982).

Latil, M., *Enhanced Oil Recovery,* Paris, Editions Technip (1980).

Leaute, R. P. and Collyer, C. J., "Laboratory Studies of In-Situ Combustion with Cold Lake Crude," 5th Annual AOSTRA Conference on Advances in Petroleum Recovery and Upgrading Technology, Calgary, Alberta (June 14–15, 1984).

Mehrotra, A. K. and Svrcek, W.Y., *AOSTRA J. Res.* 1: 51–62 (September 1984); *AOSTRA J. Res.,* 1: 215–230 (June 1985); *AOSTRA J. Res.* 1: 269–280 (June 1985); *AOSTRA J. Res.* 2: 83–93 (December 1985).

Miller, K. A. and Jacques, D. D., "Recent Observations at the Golden Lake Sparky Fire-flood Pilot," Paper 87-38-61, presented at the 38th Ann. Tech. Mtg. of the Pet. Soc. of CIM, Calgary (1987).

Moore, R. G., Bennion, D.W., Belgrave, J. D. M., Gie, D. N., and Ursenbach, M. G., "New Insights into Enriched Air In-Situ Combustion," *SPE* 16740 (1987).

Moore, R. G., Bennion, D.W., and Ursenbach, M. G., "A Review of In-Situ Combustion Mechanisms," in R. F. Meyers and E. J. Wiggins (Editors), The Fourth UNITAR/UNDP International Conference on Heavy Crude and Tar Sands, Vol 4: In Situ Recovery, AOSTRA, Edmonton, (1989), pp 775–784.

Moore, R. G., Bennion, D.W., Ursenbach, M., Gie, D., and Millour, J. P., "Experimental basis for extending the oil recovery/volume burned method to wet and superwet combustion," *JCPT,* 27, no. 6: 24–32 (1988).

Moss, J.T., "Practical Notes on Ignition in Fireflooding," *Pet. Eng.,* 72 (May 1965).

Moss, J.T. and Cady, G.V., "Laboratory Investigation of the Oxygen Combustion Process for Heavy Oil Recovery," *SPE* 10706 (1982).

Nelson, T.W. and McNeil, J. S., "How to Engineer an In-Situ Combustion Project," *Oil Gas J.,* 58–65 (June 5, 1961).

Newton, C. L., "Cryogenics," in *Kirk-Othmer Encyclopedia of Chemical Technology,* 3rd ed., 7: 227–242, New York: John Wiley (1979).

Nzekwu, B. I., Hallam, R. J., and Williams, G. J. J., "Interpretation of Temperature Observations from a Cyclic Steam/In-Situ Combustion Project," *SPE* 17389 (1988).

Penberthy, W. L. and Ramey, H. J., "Design and Operation of Laboratory Combustion Tubes," *SPEJ,* 183–198 (June 1966). © 1966 SPE.

Ramey, H. J., "Discussion on the Development of an Underground Heat Wave for Oil Recovery," *Trans. AIME,* 201: 108 (1954).

Ramey, H. J., "In Situ Combustion," Proc. 8th World Pet. Cong., London, 3: 253–262 (1971).

RAMEY, H. J., "Transient Heat Conduction During Radial Movement of a Cylindrical Heat Source—Applications to the Thermal Recovery Process," *Pet. Trans. AIME,* 216: 115–122 (1959). © 1959 SPE.

SIMON, R. and GRAVE, D. J., "Generalized Correlations for Predicting Solubility, Swelling and Viscosity Behaviour of CO_2-Crude Oil Systems," *JPT,* 102–106 (January 1965).

SVRCEK, W.Y. and MEHROTRA, A. K., "Gas Solubility, Viscosity and Density Measurements for Athabasca Bitumen," *JCPT,* 31–38 (July-August 1982).

TADEMA, H. J. and WEIJDEMA, J., "Spontaneous Ignition of Oil Sands," *Oil Gas J.,* 77–80 (December 14, 1970).

TRANTHAM, J. C. and MARX, J.W., "Bellamy Field Tests: Oil from Tar by Counterflow Underground Burning," *JPT,* 109–115 (January 1966).

TURTA, A.T. and PANTAZI, G., "Development of the In-Situ Combustion Process on an Industrial Scale at Videle Field, Rumania," *SPE* 10709 (April 1982).

WHITE, P. D. and MOSS, J.T., *Thermal Recovery Methods,* Tulsa, Okla.: Pennwell Books (1983).

WILSON, L. A., REED, R. L., REED, D.W., CLAY, R. R., and HARRISON, N. H., "Some Effects of Pressure on Forward and Reverse Combustion," *SPEJ,* 127–137 (June 1963).

ZAWIERUCHA, R., DRNEVICH, R. F., MCILROY, K., and KNECHT, P., "Material Compatibility and Systems Considerations in Thermal EOR Environments Containing High-Pressure Oxygen," *JPT,* 1477–1483 (November 1988).

Appendix 1

Symbols

Except where indicated otherwise, the equations in this book are dimensionally consistent and they may be used by substituting values for the variables in any desired set of consistent units. The table on page 21 lists some possible unit systems.

In the following list, the dimensions of the quantities which are represented by the symbols are indicated using the following letters:

L	length
M	mass
T	time
H	heat
θ	temperature

Definitions of some variables that occur only locally are not included here if they are defined in the text adjacent to the equation where they appear. In some cases, such symbols may have special meanings and require particular units.

LOWER CASE

a, b, c	Arbitrary constants
c	Compressibility Pressure^{-1} = $\mathsf{LT^2M^{-1}}$
f	Dimensionless function of time
f_{hv}	Ratio of injected latent heat to injected total heat
f_L	Fractional flow at L

f_s	Steam quality
f_w	Fractional flow of water (dimensionless)
f_{wf}	Fractional flow of water just upstream of front
g	Acceleration due to gravity LT^{-2} (standard value is 9.80665 ms^{-2})
g_G	Geothermal temperature gradient $\theta\mathsf{L}^{-1}$
h	Reservoir height L
$\bar{h}_{st}$	Average thickness of steam chamber L
h_R, h_C	Heat transfer coefficients for radiation and convection, respectively $\mathsf{HL}^{-2}\mathsf{T}^{-1}\theta^{-1}$
k	Permeability L^2
$k_o k_w, k_g$	Effective permeabilities L^2
k_{ro}, k_{rw}, k_{rg}	Relative permeabilities (dimensionless)
k'_{ro}, k'_{rw}	End point relative permeabilities (dimensionless)
m	Parameter which depends on oil viscosity temperature curve T_R and T_S
n	Exponent (dimensionless)
p	Temporary variable, also used as arbitrary constant
q	Flow rate $\mathsf{L}^3\mathsf{T}^{-1}$
q_c	Cumulative volume of displaced oil L^3
q_o, q_w, q_g	Oil, water and gas flow rates $\mathsf{L}^3\mathsf{T}^{-1}$
q_t	Total flow rate $\mathsf{L}^3\mathsf{T}^{-1}$
r	Radius L
r	Thermal resistance $\theta\mathsf{T/H}$
s	Interwell distance L
t	Time T
t_D	Dimensionless time
t', t'', and t^*	Dimensionless times used in gravity drainage theory
$\vec{u}$	Heat flux vector $\mathsf{HL}^{-2}\mathsf{T}^{-1}$
x, y, z	Distance L
x, y, z	also used as temporary variables
x_f	Position of front L
z	Dimensionless distance defined by equation 2.8

UPPER CASE

A	Area L^2
$\dot{A}$	Rate of increase of area $\mathsf{L}^2\mathsf{T}^{-1}$
A_R	van Lookeren's dimensionless steam injection number for radial flow

B	Dimensionless heat factor (Equation 2.52)
B_i	Dimensionless constants $B_1, B_2, \ldots$, etc.
C	Heat capacity $HM^{-1}\theta^{-1}$
C_P	Heat capacity at constant pressure $HM^{-1}\theta^{-1}$
C, C_i	Arbitrary constants
C_h	Chuoke's constant (dimensionless)
CCR	Conradson Carbon Residue (ASTM Standard test D 189)
E	Young's modulus $ML^{-1}T^{-2}$
E_h	Fraction of injected heat which remains in the reservoir
E_{hs}	Fraction of injected heat which remains within the steam-saturated zone
F	Function
H_o	Total heat injection rate HT^{-1}
H_λ	Latent heat injection rate HT^{-1}
K	Thermal Conductivity $HT^{-1}L^{-1}\theta^{-1}$
L	Distance between well centres in equations 4.15 through 4.32
L	Length L
N	Number
N_{Ra}	Rayleigh number
OSR	Oil steam ratio
ΔP_S	Pressure drop across skin $ML^{-1}T^{-2}$
P	Pressure force/area $ML^{-1}T^{-2}$
P_c	Capillary pressure $ML^{-1}T^{-2}$
P_o, P_{oi}, P_{op}	Constant pressures in equations 4.10 et seq.
P_{wi}, P_{wp}	Pressures at injector and producer $ML^{-1}T^{-2}$
Q	Heat flow rate H/T
Q_c	Cumulative heat flow H
Q_{ci}	Cumulative injected heat H
Q^*	Dimensionless oil drainage rate in SAGD defined in Figure 7.13
Q_{LC}	Cumulative heat loss H
Q_{SC}	Cumulative heat in steam chamber H
Q_{TC}	Cumulative total heat injected H
R	Ratio of volumetric water flow to steam flow, equation 5.63 et seq.
R	Radius L
R_i, R_p	Distance from injector and producer L
R_{wi}, R_{wp}	Radii of injector and producer well bores L
S	Skin factor (dimensionless)

S_o	Fractional oil saturation (dimensionless)
S_{or}	Residual oil saturation (dimensionless)
SOR	Steam oil ratio
ΔS_o	$S_o - S_{or}$
S_w	Fractional water saturation (dimensionless)
S_{wi}	Irreducible water saturation (dimensionless)
S_{wf}	Water saturation just upstream of water flood front
$\overline{S}_w$	Average water saturation behind front
$\overline{S}_{wf}$	Average water saturation behind front at breakthrough
S_g	Fractional gas saturation (dimensionless)
S_s	Fractional steam saturation
$\overline{S}_w, \overline{S}_o$	Average water and oil saturations
T	Temperature θ
T_S	Steam temperature θ
T_R	Undisturbed reservoir temperature θ
T^*	Dimensionless temperature
U	Advance velocity of moving front LT^{-1}
U	Overall heat transfer coefficient $HT^{-1}L^{-2}\theta^{-1}$
V	Velocity LT^{-1}
V	Volume L^3
V_C	Volume of steam chamber
W	Mass flow per unit area $MT^{-1}L^{-2}$
X	Dimensionless variable used by Marx and Langenheim

Enthalpies

H	Enthalpy per unit mass HM^{-1}
H, h	Enthalpy of steam and of water HM^{-1}
U	Internal energy per unit mass HM^{-1}
V	Volume per unit mass L^3M^{-1}
H_s	Enthalpy of steam per unit volume (of steam) HL^{-3}
H_w	Enthalpy of water per unit volume HL^{-3}
H_o	Enthalpy of oil per unit volume HL^{-3}
H_r	Enthalpy of rock per unit volume HL^{-3}

Greek

α	Thermal diffusivity L^2T^{-1}
$\alpha_1, \alpha_2, \ldots$	constants in some expressions
β	Gravity term
β	Coefficient of thermal expansion θ^{-1}

ρ	Density ML^{-3}
ψ	Dimensionless function of exponent m
ϕ	Fractional porosity
ρ	Density ML^{-3}
σ	Interfacial tension MT^{-2}
σ	Stefan-Boltzmann constant $HT^{-1}L^{-2}\theta^{-4}$
σ	Stefan-Boltzmann constant $HT^{-1}L^{-2}\theta^{-4}$
σ^*	Effective interfacial tension MT^{-2}
μ	Dynamic viscosity $ML^{-1}T^{-1}$
ϵ	Emissivity (dimensionless)
λ	Latent heat of evaporation HM^{-1}
ν	Kinematic viscosity L^2T^{-1}
ν_S	Kinematic viscosity of oil at steam temperature L^2T^{-1}
ν_R	Kinematic viscosity of oil at initial reservoir temperature L^2T^{-1}
θ	Angle between interface and horizontal radians
θ	Ratio of volumetric heat capacities
θ	Reservoir dip in Table 4.2 and in equation 5.1 et seq.
ξ	Distance from front L
ξ^*	Dimensionless distance from front
δ	Gap L
ψ	Dimensionless function defined in equation 2.28

Subscripts

abs	absolute (temperature)
BT	breakthrough
c	cold
c	critical
conv	convective
cum	cumulative
D	Dimensionless
e	perimeter
eff	effective
f	front
g	gas
h	hot
H,V	horizontal and vertical
i	inside
inj	injected
L	liquid

L	loss
o	oil and outside
r	rock
R	reservoir
s	skin
s	stored
S	steam
t	total
w	water
w	well

Units and Conversion Factors

atm	atmosphere (760 mm = 14.696 psia = 101325 Pa)
B	barrel (42 USG = 34.97 IG = 0.15899 m^3)
Btu	British thermal unit (1.055 kJ = 778 ft. lb.)
cp	centipoise (0.01 poise = 0.01 g cm^{-1} s^{-1} = 1 mPa.s = 2.419 lb ft^{-1} h^{-1} = 58.06 lb f^{-1} d^{-1})
cs	centistoke (0.01 stoke = 0.01 cm^2 s^{-1} = 1×10^{-6} m^2 s^{-1})
°C	degree Celsius or Centigrade
°F	degree Fahrenheit
d	day (86,400 s)
D	darcy (0.9869 μm^2 = 0.9869×10^{-12} m^2)
ft	feet (0.3048 m)
g	gram (0.00220462 lb)
h	hour (3600 s)
ha	hectare (10,000 m^2 = 2.471 acres)
HP	horsepower (550 ft lb/s = 0.7457 kW)
K	degree Kelvin (°C + 273.15)
kg	kilogram (2.20462 lb)
lb	pound (mass) (0.45359237 kg)
lb	pound (force) (4.448222 N or kg ms^{-2})
m	meter (3.2808 ft)
md	millidarcy (0.001 D)
p	poise (g cm^{-1} s^{-1})
Pa	pascal (Nm^{-2} = 1.45×10^{-4} psi)
psi	pounds per square inch (6.8948 kPa)
°R	degree Rankin
st	stoke (cm^2 s^{-1})
t	tonne (1000 kg = 2204.6 lb.)
W	watts
y	year

Appendix 2

Densities of Oil Reservoir Materials

Water at Boiling Point

$$\rho_w = 1001.7 - 0.1616T - 0.00262T^2$$

where ρ_w is water density in kg/m^3 at saturation pressure
T is temperature in °C (10 to 290°C).

Saturated Steam

The specific volume of saturated steam can be estimated with satisfactory engineering accuracy from the following correlations. The maximum absolute error is slightly greater than 1%.

$$50 < P < 500 \qquad V_s = 363.9\,P^{-0.9588}$$

$$500 < P < 1500 \qquad V_s = (490.386/P) - 0.04703$$

$$1500 < P < 2500 \qquad V_s = (551.74/P) - 0.0887$$

In these equations P is the steam saturation pressure in psia and V_s is the steam volume in ft^3/lb.

The first equation is from (Farouq Ali 1974) and the others are from (Ejiogu and Fiori 1987).

Brine Solutions

The specific gravity of aqueous sodium chloride solutions can be estimated from $S.G._{20/20} = 1 + 7.5 \times 10^{-7}\, W_{ppm}$ $0 < W_{ppm} < 260{,}000$ where W_{ppm} is the salt content in ppm by weight (or mg/L).

Reservoir Oil

$$(\rho_o)_{15^\circ C} = \left(\frac{141.5}{131.5 + {}^\circ API}\right)(999) \text{ kg/m}^3$$

The effect of temperature on the density of petroleum fractions and crude oils can be predicted from the nomograph shown in Figure A.2.1.

A prediction of the effect of temperature up to temperatures of 260°C can be made from the tables for petroleum measurement published by ASTM (1968).

For oils in the range of 0 to 15 °API, these tables give the following factors:

Temperature °F	60	100	200	300	400	500
°C	15.6	37.8	93.3	148.9	204.4	260
$\rho/\rho_{15.6}$ ASTM Tables:	1	0.9861	0.9520	0.9187	0.8864	0.8549
Figure A.2.1 $K = 11.3$ 10 °API	1	0.982	0.951	0.920	0.885	0.850

In the preceding table, the values are compared with values read from Figure A.2.1; the agreement is excellent.

These values can be represented accurately by the equation,

$$\rho = \rho_{15}\left[1 - 0.06285\left(\frac{T - 15}{100}\right) + 0.001426\left(\frac{T - 15}{100}\right)^2\right]$$

or by the simpler but slightly less accurate formula,

$$\rho = \rho_{15}\left[1 - 0.0603\left(\frac{T - 15}{100}\right)\right]$$

where T is in degrees Celsius.

The variation of the densities of four samples of Athabasca bitumen were studied by Bulkowski and Prill (1978) over the temperature range 0 to 150°C. They found that the results were in reasonable agreement with the ASTM tables and developed the following correlation for their data.

$$\rho = \rho_o - 0.62T$$

where ρ is the density in kg/m^3 at T°C
ρ_o is the density at 0°C.

Gewers (1965) reported the densities of samples of Cold Lake and Athabasca crudes over the temperature range 0 to 150°C. His data can be represented by the equations

$$\rho = 1024 - 0.645T \qquad 8.1\ {}^\circ\text{API Athabasca crude}$$

$$\rho = 1009 - 0.634T \qquad 10.4\ {}^\circ\text{API Cold Lake crude}$$

Again, these results are in reasonable agreement with the correlations described previously.

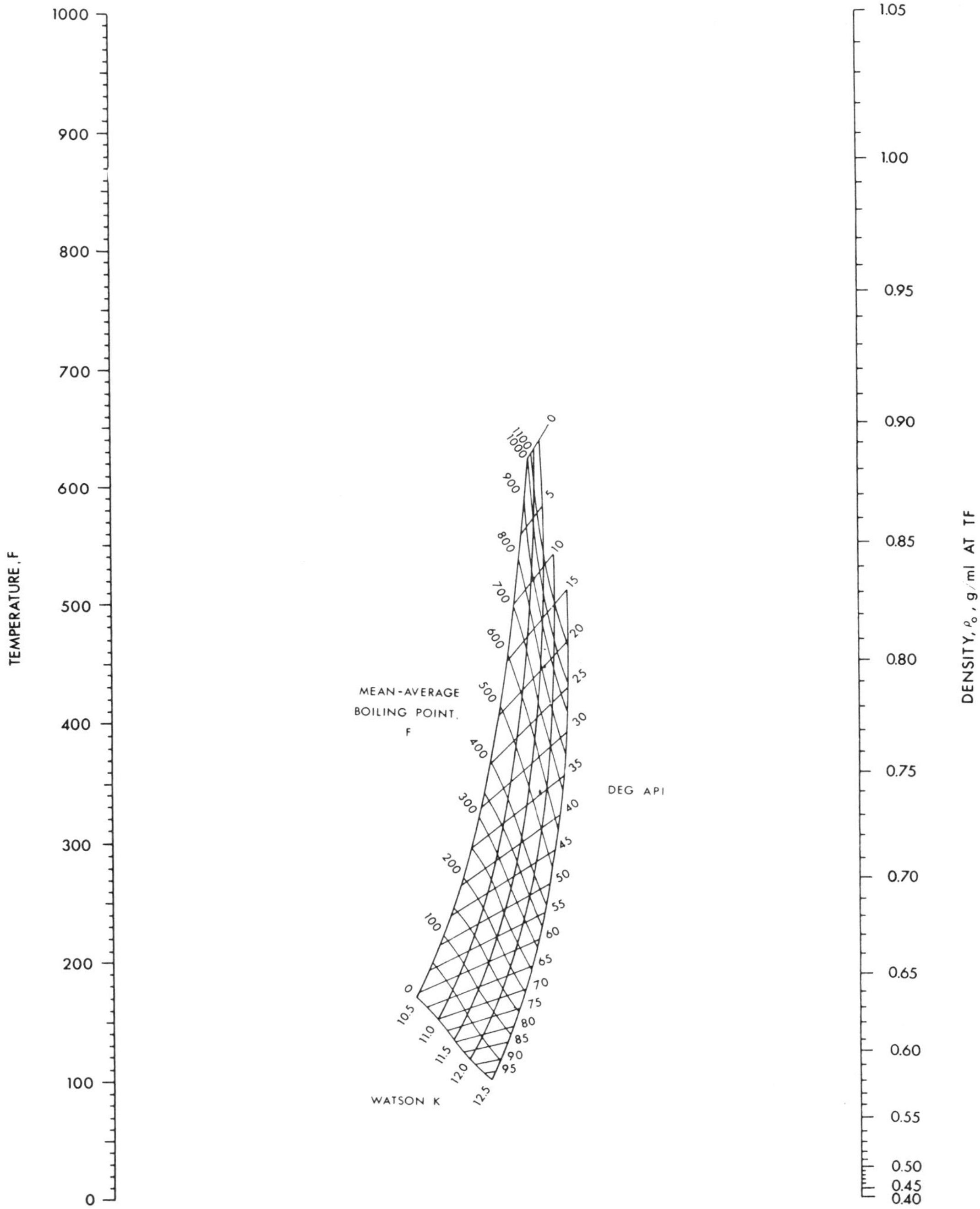

Figure A.2.1 Density of Petroleum Fractions (from API 1976)

Rocks

The following are typical values for the density of the solid rock material in reservoirs.

	kg/m^3
Sandstone	2630
Carbonates	2680

Conversion Factors

The densities in the previous sections are expressed in kilograms per cubic meters. The values can be converted to other units using the following factors.

$$1000 \text{ kg/m}^3 = 1 \text{ g/cm}^3 = 62.43 \text{ lb/ft}^3$$

BIBLIOGRAPHY

API, "Technical Data Book—Petroleum Refining," Washington, D.C., American Petroleum Institute (1976), 6.51.

ASTM, "Petroleum Measurement Tables," 7th Printing, Philadelphia, American Society for Testing Materials (1968).

BULKOWSKI, P. and PRILL, G., "Research Council of Alberta, Internal Report February 1978," results reported by D. B. Robinson in "The Thermodynamic and Transport Properties of Bitumens and Heavy Oils," AOSTRA, Edmonton (1984).

EJIOGU, G. C. and FIORI, M., "High-Pressure Saturated Steam Correlations," *JPT,* 1585–1590, December 1987.

FAROUQ ALI, S. M., "Steam Injection" in *Secondary and Tertiary Oil Recovery Processes* published by the Interstate Oil Compact Commission, Oklahoma City, Oklahoma (September 1974).

GEWERS, C., Imperial Oil Limited unpublished data from a 1965 report quoted by T. C. Boberg, "Thermal Methods of Oil Recovery," New York: John Wiley (1988).

Appendix 3

Thermal Conductivity of Oil Reservoir Materials

Unconsolidated Oil Sands

The conductivity of oil sand mixtures must fall in the range of the conductivities of the individual components—sand, water, oil and gas. These conductivities at 120°F are approximately as follows.

	Thermal Conductivity W/m °C
Sand	2.85 to 7.7
Water	0.64
Oil	0.093
Air	0.024

The thermal conductivity of the sand grains depends upon their composition. In particular, quartz has a much higher thermal conductivity than most other sand grains, and, as a result, the average conductivity is largely determined by the quartz content. An extensive listing of the conductivities of a wide range of minerals is given by Cermak (1967) and by Cermak and Ryback (1982); abstracted values are given next.

Thermal Conductivities and Specific Heats of Rock-Forming Minerals

Thermal Conductivity at Room Temperature W/m °C	
Quartz	7.69
Chert	4.53
Flint	3.71
Vitrous silica	1.36
Calcite	3.57
Dolomite	5.50
Feldspars	2.3–2.5

(Cermak and Rybach 1982)

Somerton, Keese, and Chu (1974) found that the average thermal conductivity of the sand grain material can be estimated by using the following equation (this has been converted to SI units).

$$K_{hs} = 2.86 + 4.85G \text{ W/m°C} \tag{1}$$

where G is the volume fraction of quartz in the solid.

The thermal conductivity of a porous reservoir solid is less than that calculated from (1) because the pores are filled with fluids having a lower thermal conductivity than that of the sand grains. This effect is exaggerated in an unconsolidated sand because of the poor contact between the grains in the direction of heat flow. As a result, the effective thermal conductivity of unconsolidated sands is dependent upon the thermal conductivity of the fluid in the pores. Water, because it has a much higher conductivity than oil or gas, greatly enhances the overall conductivity of an unconsolidated sand. Furthermore, in water-wet sand, a small water saturation has an exaggerated effect because the water tends to bridge the intergranular contact regions.

The effective thermal conductivity of dry sands is increased by a factor of 6 to 8 when they are saturated with water or brine; in contrast, the conductivity of consolidated rocks increases by a much smaller factor as a result of water saturation.

Somerton et al. (1974) developed the following equation for predicting thermal conductivity of oil sands as a function of the nature of the sand grains, the porosity, and the water saturation. This equation has been converted to give the result in SI units so as to be consistent with the remainder of the material in this section. Conversion factors are given on page 497.

$$K_h = 1.273 - 2.25\phi + 0.390K_{hs}S_w^{0.5} \tag{2}$$

where K_h is the thermal conductivity of the composite material in W/m °C
ϕ is the fractional porosity
K_{hs} is the average thermal conductivity of the grain material estimated from equation 1
S_w is the fractional water saturation

An important conclusion which Somerton et al. draw from their data is that, for a given water saturation, it makes little difference whether the remaining saturation is oil or gas; equation 2 can be used for either case.

As an example of the use of equation 2, the following thermal conductivities have been predicted for two oil sands, each having a porosity of 35%. In one case, the sand is assumed to be pure quartz and in the other, a mixture containing 30% quartz and 70% of other grains such as feldspars.

Calculated Thermal Conductivity for Oil Sands with $\phi = 0.35$

Water Saturation	Thermal Conductivity W/m °C	
	100% Quartz	30% Quartz
0.2	1.83	1.24
0.4	2.39	1.55
0.6	2.81	1.79
0.8	3.17	1.99
1.0	3.49	2.17

Comparison of Measured Thermal Conductivity of Tar Sand with Prediction from Somerton's Formula

Scott and Seto (1986) have reported measurements of the thermal conductivity of oil and water saturated core samples of Athabasca tar sand using both steady-state and transient heating methods.

The sample they used had the following properties:

Porosity	0.35
Quartz content of sand	0.97
Water saturation	0.267
Oil saturation	0.733

Somerton's formulas would predict

$$K_{hs} = 2.86 + 4.85 \times 0.97 = 7.56 \text{ W/m °C}$$

and

$$K_h = 1.273 - 2.25 \times 0.35 + 0.390 \times 7.56\sqrt{0.267}$$
$$= 2.01 \text{ W/m °C}$$

Setting $S_w = 0$ and to 1 gives

$$K_{h\,\text{dry}} = 0.49 \text{ W/m °C}$$

and

$$K_{h\,\text{wet}} = 3.43 \text{ W/m °C}$$

Scott and Seto's measurements are compared to these predictions in the following table.

Values of Thermal Conductivity W/m °C

	Predicted	Scott and Seto 20°C	
		Steady State	Transient
Dry	0.49	0.43	0.44
Oil saturated	2.01		2.0
Water saturated	3.43	4.03[(1)]	3.47

[(1)]Thought to be high by Scott and Seto because of some convective heat transfer.

The agreement between the predicted and experimental values is excellent.

Consolidated Porous Rocks

The thermal conductivities of consolidated porous rocks are higher than those of unconsolidated sands because of the continuous nature of the rock matrix. Also, consolidated rocks generally have lower porosities than sands, and their conductivity is less influenced by the nature of the pore fluids. The thermal conductivities of a large number of sandstone materials have been measured by Cermak (1967). His measurements of the thermal conductivities of dried sandstones are shown as a function of porosity in Figure A.3.1. Although the data are scattered, there is a trend for the thermal conductivity to decrease, as might be expected, with increasing porosity. The scatter in the data reflects the varying nature of the solid material as well as, the varying geometry of the pores. The intercept for zero porosity in Figure A.3.1 appears to be rather lower than might be expected from equation 1.

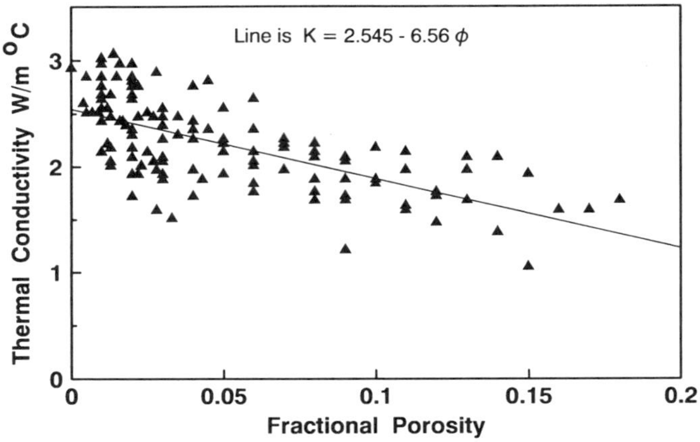

Figure A.3.1 Thermal Conductivity of Dry Sandstones (Data from Cermak 1967)

Cermak also measured the thermal conductivities of the same sandstone samples when saturated with water. The ratio of the wet to the dry thermal conductivities is plotted against porosity in Figure A.3.2. Each comparison showed an increase in thermal conductivity with water saturation, and the data show a general trend, which is represented by the exponential curve that comes from Cermak's paper or, almost equally well, by the straight line.

COMPARISON OF THERMAL CONDUCTIVITIES OF CONSOLIDATED AND UNCONSOLIDATED SANDSTONES

In Figure A.3.3, predicted curves for the thermal conductivities of dry and water saturated, unconsolidated, and consolidated sands are plotted as a function of porosity. The larger effect of water saturation on the conductivity of unconsolidated sands as compared to that on consolidated sands can be seen. Somerton's correlation predicts that the conductivity of unconsolidated sands varies linearly with the square root of the water saturation. As a result, the line for 25% water saturation falls halfway between that for the dry sand the fully water-saturated sand.

There are few data on the effect of partial water saturation on the conductivity of consolidated sands. It is reasonable to assume that this effect would be more nearly linear than for unconsolidated sands because of the absence of grain to grain thermal resistance. A reasonable assumption might be that the thermal conductivity varies linearly with water saturation.

Thermal Conductivity of Hydrocarbon Liquids

These may be estimated from the following correlation (Bland and Davidson 1967)

$$K_h = \frac{0.117}{d}(1 - 0.00054T)$$

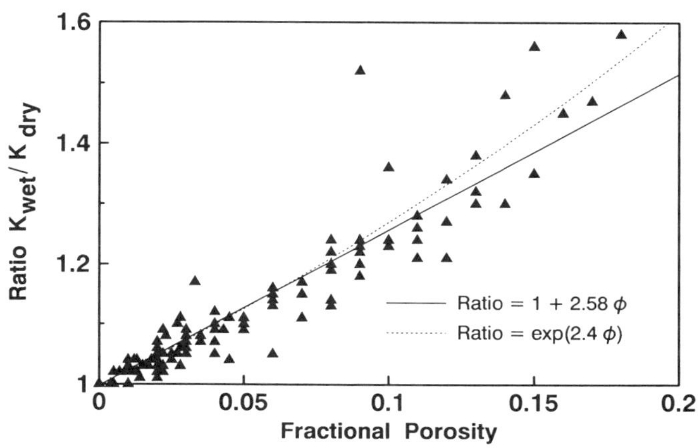

Figure A.3.2 Effect of Water Saturation on Thermal Conductivity (Data for Sandstones from Cermak 1967)

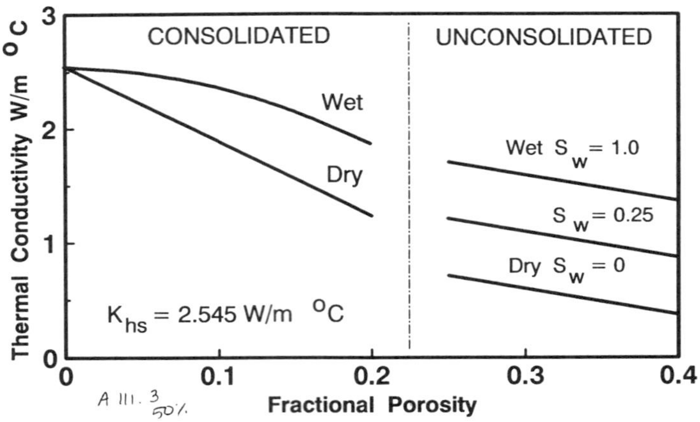

Figure A.3.3 Comparison of Correlated Thermal Conductivities for Consolidated and Unconsolidated Sandstones

where K_h is thermal conductivity, W/m °C
d is specific gravity, 60/60°F
T is temperature, °C.

Bland and Davidson list their source of data as U.S. Bur. Std. Misc. Pub. 97.

Thermal Conductivity of Liquid Water

The thermal conductivity of liquid water at its boiling point can be estimated from

$$K_h = 0.57 + 1.69 \times 10^{-3}T - 6.01 \times 10^{-6}T^2$$

where K_h is in W/m °C
T is in °C (T = 0 to 300)

For example, if $T = 150$, $K_h = 0.69$ W/m °C.

The thermal conductivity of liquid water reaches a maximum at a temperature of about 140°C.

Thermal Conductivity of Over- and Underburden Materials

The thermal conductivities of overburden materials in W/m °C may be estimated from the following:

CLAYSTONES AND CONSOLIDATED CLAYS

$$\left.\begin{aligned} K_{h\,\mathrm{dry}} &= 2.2 - 5.5\phi \\ K_{h\,\mathrm{wet}} &= 2.2 - 1.5\phi \end{aligned}\right\}$$ Correlations based on data from Cermak (1967), $\phi = 0$ to 0.2

SILTSTONES

$$\left.\begin{aligned} K_{h\,\mathrm{dry}} &= 2.0 - 4.2\phi \\ K_{h\,\mathrm{wet}} &= 2.0 - 0.5\phi \end{aligned}\right\}$$ Correlations based on data from Cermak (1967), $\phi = 0$ to 0.24

$$\left.\begin{aligned} K_{h\,\text{dry}} &= 0.7 \\ K_{h\,\text{wet}} &= 2.6 \quad \text{(est)} \end{aligned}\right\} \quad \text{Farouq Ali (1974), } \phi = 0.19$$

LIMESTONE

$$K_h = 2.6 \qquad \phi = 0 \text{ (from Cermak 1967)}$$

$$\left.\begin{aligned} K_{h\,\text{dry}} &= 1.7 \\ K_{h\,\text{wet}} &= 3.5 \end{aligned}\right\} \quad \text{Farouq Ali (1974) } \phi = 0.19$$

SHALE

$$\left.\begin{aligned} K_{h\,\text{dry}} &= 1.0 \\ K_{h\,\text{wet}} &= 1.7 \end{aligned}\right\} \quad \text{Farouq Ali (1974)}$$

Thermal Conductivities of Miscellaneous Materials

	W/m °C		
	Gas (1 atm.)	Liquid	Solid
Hydrogen	0.179		
Helium	0.143		
Nitrogen	0.024		
Water	0.024	0.60	2.22
Methane	0.033		
Propane	0.017	0.08	
Toluene		0.14	
Silver			419
Copper			389
Aluminium			201
Steel			46
Wood			0.1–0.2
Rock			1.8
Glass (Pyrex)			1.1
Insulating Materials (See Appendix 8)			0.02–0.12

CONVERSION FACTORS

$$1 \text{ W/m °C} = 0.5778 \text{ Btu/h ft °F} = 0.002388 \text{ cal/s cm °C}$$

BIBLIOGRAPHY

BLAND, W. F. and DAVIDSON, R. L., *Petroleum Processing Handbook,* New York: McGraw-Hill (1967).

CERMAK, V., "Coefficient of thermal conductivity of some sediments, its dependence on the density and on water-content of rocks," *Chemie der Erde,* 26: 271–278 (1967).

CERMAK, V. and RYBACH, L., "Thermal Conductivity and specific heat of minerals and rocks," in Landolt-Bornstein, *Numerical Data and Functional Relationships in Science and Technology, New Series,* vol. 4, New York: Springer-Verlag (1982), 305 ff.

FAROUQ ALI, S. M., "Steam Injection," in *Secondary and Tertiary Oil Recovery Processes,* Oklahoma City, Okla.: Interstate Oil Compact Commission, (1974).

HOUGEN, O. A. and WATSON, K. M., *Chemical Process Principles,* vol. 1, 2d printing, New York: John Wiley (1950), 334.

SCOTT, J. D. and SETO, A. C., "Thermal Property Measurements on Oil Sands," *JCPT,* 70–77 (November-December 1986).

SOMERTON, W. H., KEESE, J. A., and CHU, S. L., "Thermal Behaviour of Unconsolidated Oil Sands," *SPEJ,* 513–521 (October 1974).

Appendix 4

Heat Capacities and Enthalpies

Sandstones

$$C_s = 0.715 + 0.001707T - 1.908 \times 10^{-6}T^2$$

where C_s is the heat capacity of rock material at T°C (0–300) measured in kJ/kg °C.

This equation represents the average curve shown in Figure A.4.1. The data for the individual sandstones shown are taken from Cassis et al. 1985.

Carbonate

$$C_s = 0.823 + 0.001511T - 1.438 \times 10^{-6}T^2$$

This equation is based on that given by Cassis et al. (1985) for a sample of the carbonate rock from the Grosmount reservoir in Alberta. It is plotted in Figure A.4.1.

Clays

Curves for a range of clay materials are shown in Figure A.4.2 including some (the broken lines in the figure) for dehydrated clays. Also shown in the figure is a dotted line for the fine material that was separated from a sample of Athabasca tar sand.

The differences between the various clay samples is probably due largely to a variation in their water contents.

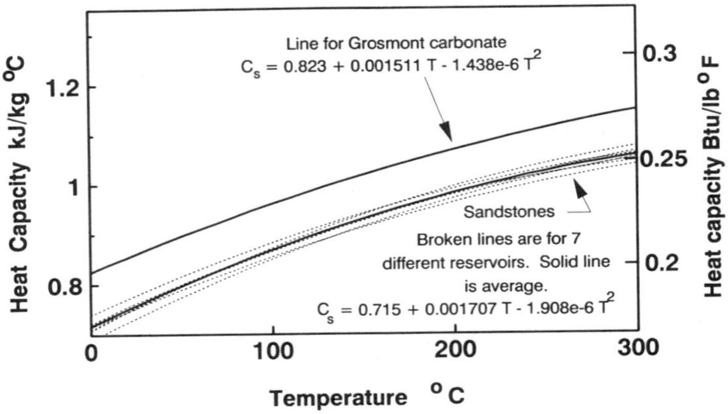

Figure A.4.1 Heat Capacities of Reservoir Solids (Data from Cassis et al. 1985)

Oils

The heat capacity of hydrocarbon oils depends upon the nature of the oil and upon temperature.

A useful correlation for predicting the heat capacity of oils is[1]

$$C_o = [1 + 0.055(K - 11.7)][(2.961 - 1.331S) + (0.00614 - 0.00231S)T]$$

where C_o is the heat capacity of the oil, kJ/kg °C
K is the Watson characterization factor of the oil
S is the specific gravity at 60/60 °F
T is the temperature, °C

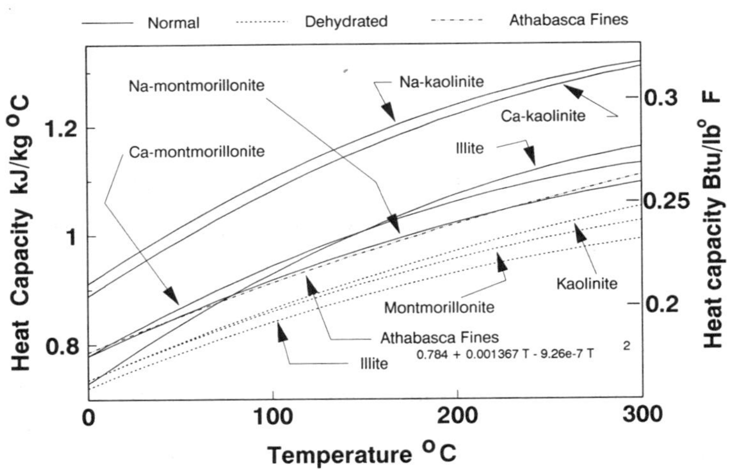

Figure A.4.2 Heat Capacities of Clays (Data from Cassis et al. 1985)

[1]This correlation is due to Hougen and Watson (1950). The algebraic representation above is based upon the equation given in British units in Bland and Davidson (1967).

For typical heavy crude oils, $K \cong 11.3$. Also, by definition,

$$S = \frac{141.5}{131.5 + °API}$$

This equation predicts the heat capacity as a linear function of temperature.

Experimental data for the heat capacities of 13 different heavy oils and bitumens have been published by Cassis et al. (1985). They are plotted as fine dotted lines in Figure A.4.3. The average of these data is given by

$$C_o = 1.605 + 0.004361T - 4.046 \times 10^{-6}T^2 \qquad 0°C < T < 300°C$$

The agreement with the Hougen and Watson correlation is good up to about 200°C. Above this, the experimental data deviate below the correlation. The experimental data given by Cassis et al. are stated to have an accuracy of about ±5% (≅ ±0.03 kJ/kg) at 300°C.

Water

The heat capacity of liquid water at saturation conditions can be represented, with excellent accuracy, by the equations

$$C_w = 4.182 - 1.5 \times 10^{-4}T + 3.44 \times 10^{-7}T^2 + 4.26 \times 10^{-8}T^3, \qquad 10°C < T \leq 240°C$$

and

$$C_w = 11.550 - 0.064518T + 1.5087 \times 10^{-4}T^2, \qquad 240°C \leq T < 300°C$$

C_w is given in kJ/kg °C.

The specific heat of water cannot be represented accurately by a quadratic equation over a wide range of temperature.

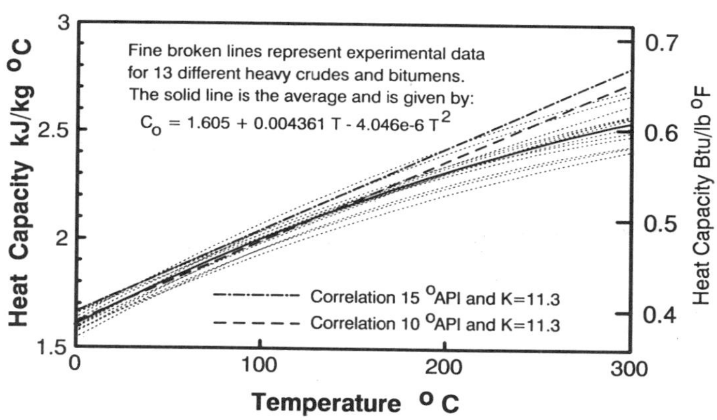

Figure A.4.3 Heat Capacities of Bitumens and Heavy Oils (Data from Cassis et al. 1985)

Heat Capacities of Common Gases

These may be predicted by the following equation:

$$C_p = a + bT + \frac{c'}{T^2}$$

where C_p is in kJ/kg K
T is in K = t°C + 273.15

and the coefficients a, b, and c' (if used) are read from the following table.

Name	Molecular Weight	a	b	c'	Applicable Temperature Range K
Hydrogen	2.0158	13.75	0.001682		273–2,500
Oxygen	31.9988	1.08	0.000034	−24,560	300–5,000
Nitrogen	28.0134	0.97	0.000149		300–3,000
Air	28.85	1.00	0.000123	−5,721	300–3,000
Carbon dioxide	44.00955	0.98	0.000261	−18,600	273–1,200
Carbon monoxide	28.01055	0.99	0.000179		273–2,500
Methane	16.04275	1.39	0.003001		273–1,200

The values of the coefficients are based upon the data summarized in Perry and Chilton (1973).

Average Heat Capacities between T_1 and T_2.

If $C = a + bT + cT^2$, then

$$\overline{C} = a + b\left(\frac{T_1 + T_2}{2}\right) + c\left(\frac{T_1^2 + T_1T_2 + T_2^2}{3}\right)$$

If $C = a + bT + c'/T^2$, then

$$\overline{C} = a + b\left(\frac{T_1 + T_2}{2}\right) + \frac{c'}{T_1T_2}$$

Change in Enthalpy between T_1 and T_2

$$\Delta H = \int_{T_1}^{T_2} C\,dT = a(T_2 - T_1) + b\left(\frac{T_2^2 - T_1^2}{2}\right) + c\left(\frac{T_2^3 - T_1^3}{3}\right)$$

or

$$\Delta H = a(T_2 - T_1) + b\left(\frac{T_2^2 - T_1^2}{2}\right) + c'\left(\frac{T_2 - T_1}{T_1T_2}\right)$$

Volumetric Heat Capacities of Reservoir Material

$$\rho C = (1 - \phi)\rho_s C_s + \phi S_w \rho_w C_w + \phi S_o \rho_o C_o + \phi S_g \rho_g S_g$$

The last term is usually negligible.

Conversion Factors

1 Btu/lb °F = 1 cal/g °C = 1 kcal/kg °C = 1 CHU/lb °C = 4.187 kJ/kg °C

BIBLIOGRAPHY

BLAND, W. F. and DAVIDSON, R. L., *Petroleum Processing Handbook,* New York: McGraw-Hill (1967).

CASSIS, R., FULLER, N., HEPLER, L. G., MCLEAN, R. J. C., SKAUGE, A., SRINIVASAN, N. S., and YAN, H., "Specific Heat Capacities of Bitumens and Heavy Oils, Reservoir Minerals, Clays, Dehydrated Clays, Asphaltenes and Cokes," *AOSTRA J. of Research,* 1, no. 3: 163–173 (1985).

HOUGEN, O. A. and WATSON, K. M., *Chemical Process Principles,* New York: John Wiley (1950).

PERRY, R. H. and CHILTON, C. H., *Chemical Engineers Handbook, 5th Edition,* McGraw-Hill, New York (1973).

Appendix 5

Viscosities

Viscosity of Crude Oil

The viscosity of crude oils can be represented as a function of temperature with accuracy by the equation[1]

$$\log_{10}[\log_{10}(\nu + 0.7)] = m \log_{10}(T + 273) + b$$

where ν is the kinematic viscosity, cs
T is the temperature, °C
m, and b, are constants which are characteristic of the oil

This equation forms the basis of the ASTM viscosity-temperature graph paper, on which viscosity-temperature curves of hydrocarbon liquids plot as straight lines (ASTM 1989). It has been used frequently to correlate, interpolate and extrapolate experimental data for the viscosity of heavy crudes and bitumens.

In order to define a line for a particular crude, it is necessary to fix two parameters, such as two points on the line or a single point together with the slope of the line.

[1]The history of this equation is described by Wright (1968). Walther (1928) described the idea of plotting viscosity versus temperature on a plot of log log(ν) against log T. In 1932, he modified this to a plot of log log(ν + 0.8) versus log T. The first ASTM chart issued in 1932 also used 0.8 in the ordinate function. Although this equation was identical to the Walther formula, the ASTM committee credited its introduction to the work of MacCoull, who introduced the idea of plotting log log(ν + const.) against log T in a Texas Company publication in 1921. Various constants and adjustable constants have been used in the subsequent versions of the ASTM paper. The work by Wright, which forms the basis for the current ASTM chart, uses 0.7 for viscosities down to $\nu = 1.3$ cs. Below that viscosity, a more complex expression is substituted for the constant 0.7.

If only one measured value is available for the viscosity of a heavy crude, then a prediction can be made if the slope of the viscosity line can be estimated.

It frequently happens that the lines for a family of oils fall as nearly parallel straight lines—i.e., have a common value of m. This allows a viscosity line to be drawn by making it parallel to lines for similar crude oils.

Although the ASTM graph paper refers to *kinematic* viscosities in centistokes, it has also been used by some authors to correlate *dynamic* viscosities in *centipoise*. For example, Svrcek and Mehrotra (1988) have shown that the viscosities of a range of bitumens can be represented for temperatures up to 130°C by the relation

$$\log_{10}[\log_{10}(\mu + 0.7)] = b' - 3.63029 \log_{10}(T + 273)$$

where T is the temperature, °C
b' is a constant for each oil

For mathematical reasons, if the density of the oil is a linear function of temperature and the dynamic viscosity of the oil follows the preceding equation, then the viscosity cannot also plot as a straight line on the ASTM paper using kinematic viscosity. However, the curve will be almost straight and, because of the precision of typical viscosity measurements, it is usually difficult to determine which graph gives the best straight line.

There are essentially no measured viscosity data for heavy crude oils above 150°C, and this makes the use of any correlation for the extrapolation of viscosities to the temperatures used in thermal recovery somewhat questionable.

The best straight line drawn on the kinematic viscosity basis will be less steep than that for the dynamic viscosity plot. The value of m for the Svrcek-Mehrotra data plotted as kinematic viscosity would be about 2% smaller, or $m = -3.5556$; i.e.,

$$\log_{10}[\log_{10}(\nu + 0.7)] = b - 3.5556 \log_{10}(T + 273)$$

Examination of the data for a wide range of asphaltic heavy crudes ranging from relatively fluid conventional heavy oils to the most viscous of bitumens indicates that rather than being constant, the absolute value of the slope tends to increase for oils of lower viscosity.

The correlation of m with b for a wide range of heavy crudes from Canada, the United States, and South America in Figure A.5.1 shows that the viscosity of these oils can be estimated from[2]

$$\log_{10}[\log_{10}(\nu + 0.7)] = m \log_{10}(T + 273) + b$$

where b is a constant for any particular oil
m is given by $m = 0.3249 - 0.4106b$

[2]The data shown in these plots were obtained by using the crude oil viscosities reported in papers describing various thermal recovery projects in Canada, the United States and South America. The papers were those that happened to be in the author's library and were not selected specifically. For each oil, two viscosity points were chosen. These were converted from centistokes to centipoise or vice versa, with an allowance being made for the effect of temperature on density. The high degree of correlation found between b and m is surprising; this presumably reflects the similar constitution of these particular heavy oils.

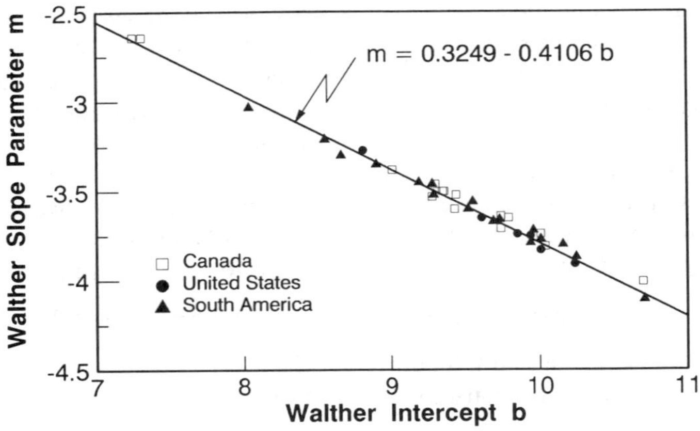

Figure A.5.1 Correlation Between Walther Parameters for Heavy Crude Oils

These relations define a family of straight lines, which pass through a common pole on an ASTM plot; see Figure A.5.2. The tendency for lines for oils from a common family to pass through a particular pole has been used in lubricating oil technology, particularly in Europe, where the position of the pole has been used as a correlating factor to represent the type of oil. This system is comparable to the viscosity index classification, which is more commonly used (see Bondi (1951) for a discussion of Walther and Ubbelohde's pole height (Ubbelohde 1940)).

The viscosity-temperature line for a given crude oil can be estimated by joining this pole (at $-1°C$, 1.5×10^6 cs) to a point representing a measured viscosity of the oil. For best accuracy for thermal recovery studies, this measured point should be at a temperature as high as possible. This will minimize the degree of extrapolation which is required in the utilization of the data. Because it is the viscosities at

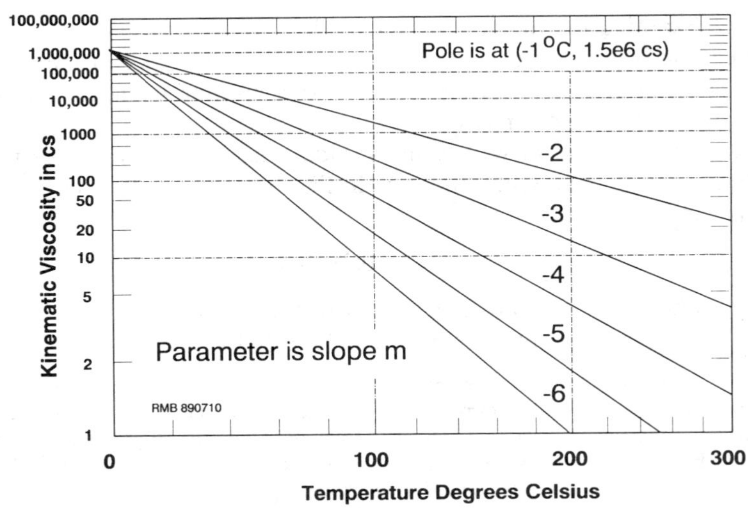

Figure A.5.2 Viscosity-Temperature Chart for Heavy Crudes Kinematic Viscosity

high temperatures that are of interest in thermal recovery calculations, the exact location of the pole is not of great importance, particularly if the measured viscosity is for a relatively high temperature.

If the measured viscosity data are given as dynamic viscosities, they may be converted to kinematic viscosity by utilizing oil densities estimated from the correlations in Appendix 2. It is possible to plot dynamic viscosity directly on Figure A.5.2 by interpreting the units of the vertical scale as centipoise. In this case, a similar plot to that of Figure A.5.1 yields a best straight line of

$$m' = 0.3464 - 0.4127b'$$

where m' and b' refer to the equation

$$\log_{10}[\log_{10}(\mu + 0.7)] = m' \log_{10}(T + 273) + b'$$

In such a plot, the pole should be at the position (-8°C, 8.1×10^6 cp). A chart suitable for this application is given in Figure A.5.3.

Effect of pressure Increasing the pressure on a liquid also increases the viscosity. While the effect is generally small, it can be significant.

Mehrotra and Svrcek (1986) show that the viscosity of a sample of Athabasca bitumen varied according to the following correlation.

$$\ln \ln(\mu) = 22.8515 - 3.5784 \ln(T) + 0.00511938P_g$$

where μ is the viscosity in centipoise at a temperature of T (K) and a gauge pressure of P_g (MPa).

This can be written using common logarithms as

$$\log \log(\mu) = 9.56204 - 3.5784 \log(T) + 0.002223P_g$$

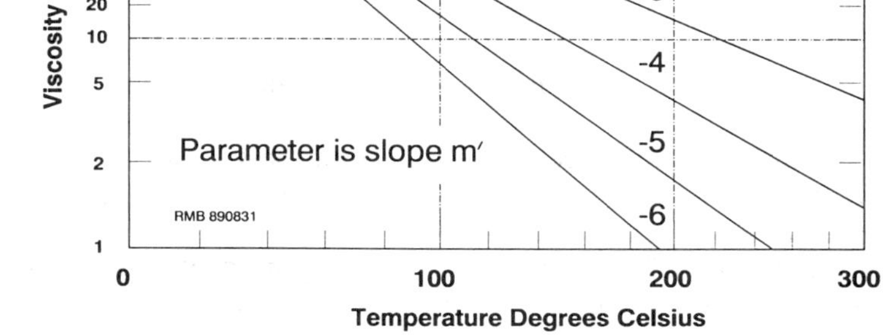

Figure A.5.3 Viscosity-Temperature Chart for Heavy Crudes Dynamic Viscosity

In the temperature range they investigate, μ would be relatively large and $\log\log(\mu)$ would be nearly equal to $\log\log(\mu + 0.7)$.

The effect of pressure is thus to raise the viscosity temperature line on a chart such as that in Figure A.5.3 by a distance of $0.002223P_g$.

The position of the pole in Figure A.5.3 is thus a function of pressure.

Pressure, MPa	0.1	5	10
Pole position, cP $\times 10^6$	8.1	12.1	28.3

The data in the following table show that the effect of pressure is larger at higher viscosities (i.e., at lower temperatures).

Viscosity of oil at atmospheric pressure (cp)	Viscosity Divided by Viscosity at 1 atmosphere	
	5 MPa	10 MPa
8.1×10^6	1.5	3.5
10^6	1.42	3.0
10^5	1.34	2.5
10^4	1.26	2.1
10^3	1.19	1.7
10^2	1.12	1.4
10	1.06	1.2

At temperatures where the oil becomes reasonably mobile—e.g., where $\mu < 100$ cp—the effect of pressure is relatively small.

Effect of dissolved methane In the reservoir, the oil is typically nearly saturated with natural gas at the reservoir pressure. Under these conditions, there are two effects upon viscosity:

1. The dissolved gas acts like a solvent and tends to reduce the viscosity.
2. The pressure tends to increase the viscosity.

The dilution effect is larger than the pressure one, and there is an overall lower viscosity.

Mehrotra and Svrcek have measured the viscosities of several bitumens that have been saturated with methane over a temperature range of about 20°C to 120°C and at pressures up to 10 MPa.

The effect of the dissolved methane upon the viscosity of three different bitumens over this range of conditions is shown in Figure A.5.4.

If, for a constant concentration of dissolved methane, the viscosity line on a plot such as that in Figure A.5.3 moves down by a constant vertical distance, then

$$\log\log(\mu_o + 0.7) - \log\log(\mu + 0.7)$$

will be constant; μ_o represents the viscosity of the gas-free (or dead oil) and μ, that of the live oil.

This is the same as saying that $\log[\log(\mu_o + 0.7)/\log(\mu + 0.7)]$ is constant, or that $\log(\mu_o + 0.7)/\log(\mu + 0.7)$ is constant. If the term 0.7 is neglected compared to μ, then it can be seen that this ratio is the same as the ordinate in Figure A.5.4.

The data in Figure A.5.4 are from measurements carried out over the whole range of temperatures. It is interesting that the data for Cold Lake crude, which are the solid circles in Figure A.5.4, correlate better than the earlier data for the other crudes. It may be that this reduced scatter for the Cold Lake crude is indicative of more consistent measurements.

That the experimental data can be correlated in the manner shown in Figure A.5.4 indicates that the viscosity of a bitumen containing a constant concentration of dissolved methane will fall below the line for the gas-free bitumen at the same pressure by a distance

$$y = 1 + 0.0021x \tag{1}$$

where x is the concentration of methane in SCF/B (89 SCF/B = 1 wt% CH_4 approx.).

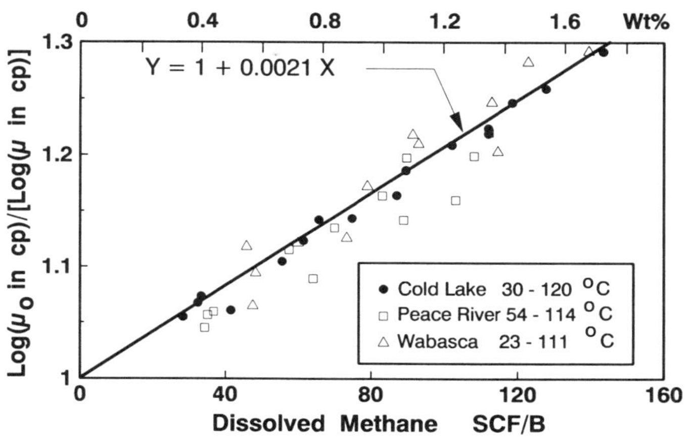

Figure A.5.4 Effect of Dissolved Methane upon the Viscosity of Bitumens. Correlation of Data from Mehrotra and Svrcek (1988, 1985a and 1985b). μ_o is the Viscosity of Bitumen in cp with No Dissolved Methane; μ is the Viscosity with Methane.

The solubility of the methane in the same bitumens has been correlated against temperature and pressure by the equation

$$\text{SCF/B} = P \exp\left(1.54 + \frac{343}{\text{T}}\right)$$

$$= 4.66\text{P} \exp\left(\frac{343}{T}\right) \tag{2}$$

where T is in K

P is the absolute pressure in MPa

The data are compared with this correlation in Figure A.5.5. Although there is considerable scatter in this diagram, the vertical scale is relatively large, and it will be seen that all but one experimental point fall within ±20% of the predicted solubilities. Again, the more recent data for the Cold Lake bitumen appear more consistent.

The methane concentration can be eliminated from equations 1 and 2 to provide a correlation that predicts the viscosity of the live oil as a function of pressure and the viscosity of the dead oil at the same temperature and pressure. This correlation is

$$\mu = \exp\left[\frac{\ln \mu_o}{1 + 0.0093P \exp(343/T)}\right]$$

$$= \exp\left\{\frac{\ln \mu_{oo} \exp[0.00511938(P - 0.1)]}{1 + 0.0093P \exp(343/T)}\right\} \tag{3}$$

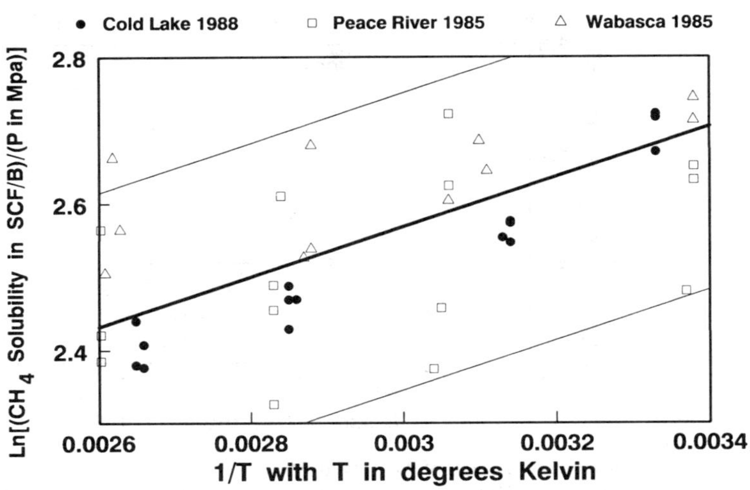

Figure A.5.5 Solubility of Methane in Bitumens. Correlation of Data from Mehrotra and Svrcek (1988, 1985a and 1985b). The Thick Line is Given By $Y = 1.54 + 343/T$. The Thinner Lines Correspond To Solubilities Which Are 20% Higher and 20% Lower Than Those Given By the Above Equation.

where μ is the live oil viscosity in cp
μ_o is the dead oil viscosity in cp
μ_{oo} is the dead oil viscosity in cp at the same temperature and atmospheric pressure
P is the pressure in MPa
T is the absolute temperature in K

Predictions were made using this equation for each of the data points, and the results are shown as ratios to the corresponding measured viscosities in Figure A.5.6. The correlation appears to be able to predict the viscosities to within about ±20%. Again it will be noted that the Cold Lake data appear more consistent.

The curves in Figure A.5.7 show the ratio of the viscosities for live and dead oils as a function of temperature and pressure calculated by means of equation 3 for a bitumen having a viscosity of 120 cp at 100°C and a dead-oil viscosity line that goes through the pole of Figure A.5.4. The effect of dissolved methane on the oil viscosity is indicated to decrease markedly as the temperature is raised and the oil becomes more fluid. It is also very dependent upon the pressure. This graph is for bitumen that is *saturated* with methane.

Viscosity of Water and Steam

The dynamic viscosities of saturated steam and of water at its boiling point are plotted in Figure A.5.8. At lower temperatures, the dynamic viscosity of water is much greater than that of steam, but, as the temperature is increased, the two viscosities approach each other and become equal at the critical point.

Figure A.5.9 shows the kinematic viscosity of saturated steam and water plotted against temperature. Because of its lower density, the kinematic viscosity of steam is much higher than that of liquid water at lower temperatures. Again, the

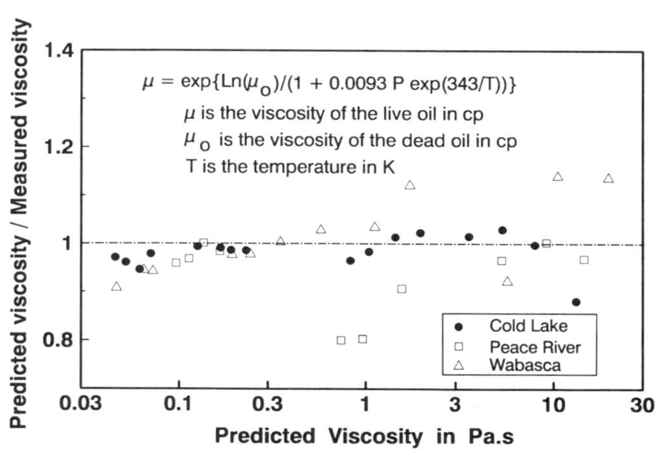

Figure A.5.6 Ratio of Predicted and Measured Viscosities of Bitumens. Data from Mehrotra and Svrcek (1988, 1985a and 1985b)

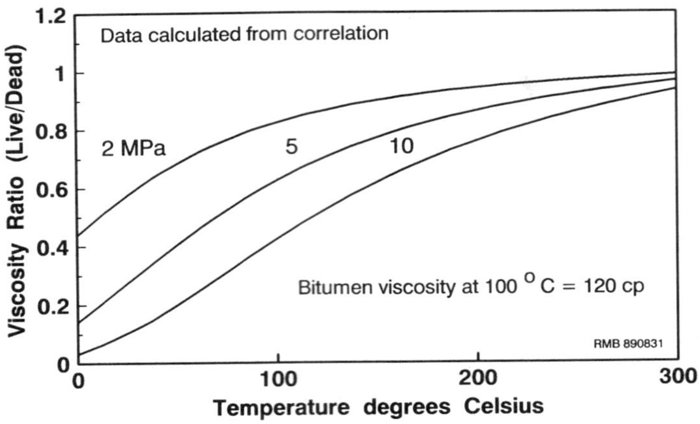

Figure A.5.7 Effect of Temperature and Pressure on the Calculated Viscosities of Methane-Saturated Bitumen

viscosities converge at the critical point. Because of the higher kinematic viscosity of steam, a much higher pressure gradient is required to achieve a given mass flux in a porous medium for steam than for liquid water.

The dynamic viscosities of saturated steam and liquid water can be estimated from the following correlation equations.

Saturated steam

$$\mu_s = 0.00879 + 0.0000354T \pm 1\%$$

where μ_s is in cp
$90°C < T < 300°C$

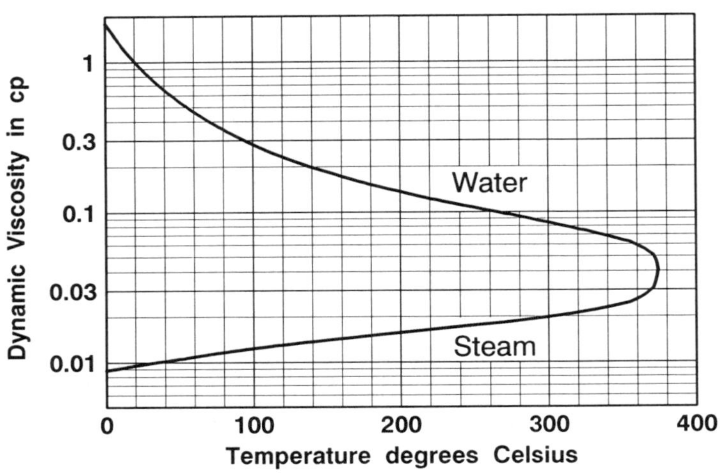

Figure A.5.8 Dynamic Viscosity of Saturated Steam and Water (Data from Schmidt and Grigull 1981)

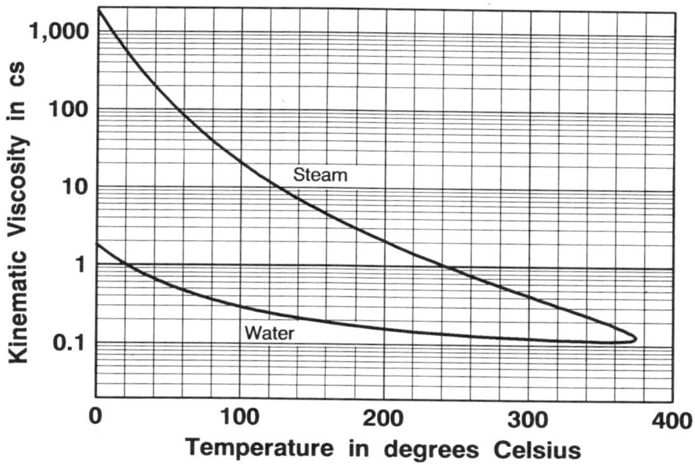

Figure A.5.9 Kinematic Viscosity of Saturated Steam and Water (Data from Schmidt and Grigull 1981)

Water at boiling point

10–100 °C

$$\frac{1}{\mu_w} = 0.5462 + 0.021192T + 0.00008934T^2 \pm 1\%$$

where μ_w is in cp
T is in °C

100–300 °C

$$\frac{1}{\mu_w} = 0.040079T - 0.4997 \pm 1.5\%$$

where μ_w is in cp
T is in °C

Conversion Factors

DYNAMIC VISCOSITY (M/LT)

The SI unit of dynamic viscosity is Pa s, or the equivalent kg m^{-1} s^{-1}.

The traditional unit is the centipoise (cp) = 0.01 p, where

$$1 \text{ poise} = 1 \text{ g cm}^{-1} \text{ s}^{-1} = 0.1 \text{ kg m}^{-1} \text{ s}^{-}$$

Thus

$$1 \text{ cp} = 0.001 \text{ Pa s}$$
$$= 1 \text{ mPa s}$$

If consistent equations are employed that use ft-lb and days, then dynamic viscosities should be expressed as lb/ft d.

$$1 \text{ cp} = \frac{0.01 \times 86400 \times 30.48}{453.6} = 58.06 \text{ lb/ft d}$$

KINEMATIC VISCOSITY (L^2/T)

By definition, $\nu = \mu/\rho$. The SI unit of kinematic viscosity is

$$\frac{\text{Pa s}}{(\text{kg/m}^3)} = \text{m}^2/\text{s}$$

The traditional unit is the centistoke (cs) = 0.01 stoke, where

$$1 \text{ stoke} = 1 \text{ cm}^2/\text{s} = 10^{-4} \text{ m}^2/\text{s}$$

Thus

$$1 \text{ cs} = 10^{-6} \text{ m}^2/\text{s}$$

BIBLIOGRAPHY

ASTM "Viscosity Temperature Charts for Liquid Petroleum Products," *ASTM* Standard D341-87 in 1989 *Annual Report of Standards,* vol. 05.01, Philadelphia: ASTM (1989).

BONDI, A., *Physical Chemistry of Lubricating Oils,* New York: Reinhold (1951).

MACCOULL, N., *Lubrication,* New York: The Texas Co. (June 1921), 65.

MEHROTRA, A. K. and SVRCEK, W.Y., "Viscosity, Density and Gas Solubility Data for Oil Sand Bitumens, Part II: Peace River Bitumen Saturated with N_2, CO, CH_4, CO_2 and C_2H_6," *AOSTRA J. of Research,* 1, no. 4: 269–279 (1985).

MEHROTRA, A. K. and SVRCEK, W.Y., "Viscosity, Density and Gas Solubility Data for Oil Sand Bitumens, Part III: Wabasca Bitumen Saturated with N_2, CO, CH_4, CO_2 and C_2H_6," *AOSTRA J. of Research,* 2, no. 2: 83–93 (1985).

MEHROTRA, A. K. and SVRCEK, W.Y., "Viscosity of Compressed Athabasca Bitumen," *Can. J. Chem. Eng.,* 64: 844–847 (October 1986).

MEHROTRA, A. K. and SVRCEK, W.Y., "Properties of Cold Lake Bitumen Saturated with Pure Gases and Gas Mixtures," *Can. J. Chem. Eng.,* 66: 656–665 (August 1988).

SCHMIDT, E. and GRIGULL, U., *Properties of Water and Steam in SI Units,* Berlin: Springer-Verlag (1981).

SVRCEK, W.Y. and MEHROTRA, A. K., "One Parameter Correlation for Bitumen Viscosity," *Chem. Eng. Res. Des.,* 66: 323–326 (July 1988).

UBBELOHDE, L., "Zur Viskosimetrie," Leipzig (1940).

WALTHER, C., ERDÖL UND TEER, 4: 510 (1928).

WALTHER, C., ERDÖL UND TEER, 7: 382 (1931).

WRIGHT, W. A., "An Improved Viscosity-Temperature Chart for Hydrocarbons," *J. of Materials,* 4, no. 1: 19–27 (1968).

Appendix 6

Heats of Combustion

Hydrocarbon Liquids

The heat of combustion of typical liquid hydrocarbon streams can be estimated from the following equations (API, 1983).

Gross heat of combustion

$$-\Delta H_g = 41{,}105 + 154.9(\text{API}) - 0.735(\text{API})^2 - 0.00326(\text{API})^3$$

Net heat of combustion

$$-\Delta H_n = 39{,}068 + 126.8(\text{API}) - 0.505(\text{API})^2 - 0.00442(\text{API})^3$$

where $-\Delta H_g$ and $-\Delta H_n$ are the gross and net heats of combustion in kJ/kg at 15.6°C and API is the API gravity of the oil.

These equations are based upon a correlation given by Maxwell (1950) for oils having the following sulphur contents.

°API	0	5	10	15	20	25	30	35
Wt% S	2.95	2.35	1.80	1.35	1.00	0.7	0.4	0.3

For oils having different sulphur contents, a quantity given by

$$[0.01(-\Delta H) - 40.5]\,\Delta S$$

should be subtracted from the preceding values.

where $-\Delta H$ is the appropriate uncorrected heat of combustion
ΔS is the sulphur content of the oil in wt% minus that interpolated from the preceding table

Example:

Estimate the heats of combustion of a 10 °API heavy crude oil containing 4.5 wt% S.

$$-\Delta H_g = 41{,}105 + 154.9 \times 10 - 0.735 \times 100 - 0.00326 \times 1000$$
$$= 42{,}577 \text{ kJ/kg}$$

$$\text{Correction} = (426 - 40.5)(4.5 - 1.8) = 1041$$

Corrected heat of combustion:

$$-\Delta H_g = 42{,}577 - 1041 = 41{,}536 \text{ kJ/kg}$$

or

$$\frac{41{,}536}{2.326} = 17{,}857 \text{ Btu/lb}$$

Similarly

$$-\Delta H_n = 39{,}068 + 126.8 \times 10 - 0.505 \times 100 - 0.00442 \times 1000$$
$$= 40{,}281 \text{ uncorrected}$$

The corrected net heat of combustion is then

$$-\Delta H_n = 40{,}281 - 1041 = 39{,}240 \text{ kJ/kg} \quad \text{or} \quad 16{,}870 \text{ Btu/lb}$$

Fuel Gases

		Heats of Combustion			
		kJ/kg		MJ/m^3	
	Molecular Weight	Gross	Net	Gross	Net
Hydrogen	2.0	142,119	120,021	12.03	10.16
Methane	16.0	55,498	50,009	37.58	33.86
Ethane	30.1	51,870	47,497	66.07	60.50
Propane	44.1	50,358	46,357	93.98	86.85

(Values based on data from Maxwell 1968.)

Solid Fuels

	Gross Heat of Combustion kJ/kg Ash-Free Basis
Coal	
Bituminous	27,800–35,400
Subbituminous	21,300–24,000
Lignite	17,300
Wood	
Green	5,500–11,000
Air Dry	12,000–14,000
Petroleum Coke (delayed)	35,450
Carbon	33,715

(Data from Perry and Chilton 1973.)

Conversion Factors

$$1 \text{ Btu/lb} = 2.326 \text{ kJ/kg}$$

For gases

$$\text{kJ/m}^3 = \text{kJ/kg}\left(\frac{\text{MW}}{23.63}\right) \qquad \text{m}^3 \text{ measured at } 15°\text{C and 1 atm}$$

$$1 \text{ Btu/ft}^3 = 37.26 \text{ kJ/m}^3$$

BIBLIOGRAPHY

API, *Technical Data Book Petroleum Refining,* American Petroleum Institute, Washington, D.C. (1983), 14-11.

MAXWELL, J. B., *Data Book on Hydrocarbons,* New York: van Nostrand (9th printing 1968 of original edition 1950).

PERRY, R. H. and CHILTON, C. H., *Chemical Engineer's Handbook,* 5th ed., New York: McGraw-Hill (1973).

Appendix 7

Air Compression Fuel Requirements

Energy Efficiency of Gas Engines[2]

HP	Engine % Efficiency[1]	Fuel Requirement SCF/HP h
85–400	28	10.0
440–800	32	8.8
880–3000	36	7.8
>3300	38	7.4

[1]Based on LHV of methane fuel (908 Btu/SCF).
[2]Based on Perry, R. H. and Chilton, C. H., *Chemical Engineer's Handbook,* 5th ed., New York: McGraw-Hill (1973), 24-14.

Approximate Energy Requirements for Air Compression

Basis: Inlet conditions 60°F and 1 atmosphere.

Discharge Pressure psia	HP h per 1000 SCF Compressed: Isothermal Compression[1]	HP h per 1000 SCF Compressed: Practical[2]
100	2.05	2.88
200	2.79	3.92
400	3.53	4.96
800	4.28	6.00
1200	4.71	6.60

[1]$HP = 1.07 \ln(P_2/P_1)$.
[2]Assuming practical HP is 40% greater than isothermal, i.e., $HP = 1.5 \ln(P_2/P_1)$.

Appendix 8

Thermal Insulation

Thermal insulation practices were reviewed in an article by Neisel and Verschoor (1981), which contains 59 references. See Table A.8.1.

TABLE A.8.1 Thermal Conductivities of Block, Board, and Pipe Insulation

Material	Temp range or maximum, °C	Thermal Conductivity, mW/(m °C)					
		0°C	24°C	100°C	200°C	400°C	540°C
Polyurethane	−73 to 110	24	25				
Polystyrene		31	35				
Cellular elastomeric		41	43				
Cork pipe insulation			48				
Cellulose fiber board			55				
Mineral fiber							
blanket	204	35	37	51			
block	204		41	52	71		
board	982				83	116	
pipe insulation	650		45	58	75		
Cellular glass	−268 to 427		60	74	100		
Calcium silicate							
block	649			66	79	106	
board	650 (577 kg/m^3 or 36 lb/ft^3)			123	126		
Diatomaceous earth	870				100	109	117
Diatomaceous earth	1040				108	121	130
Expanded perlite	816				79	105	

Neisel and Verschoor (1981)

TABLE A.8.2 Annular Insulation for Steam Injection Wells

Material	Thermal Conductivity W/m K
Evacuated jacket with layers of foil and ceramic fiber	0.07–0.0002[(1)]
Sodium silicate foam	0.029[(2)]

[(1)]Depends on gas remaining within jacket. Lowest values obtained using krypton with a getter to absorb H_2. Effective conductivity is higher than these values because of losses at hot spots. (See Meldau 1988).

[(2)]From Boberg (1988). Foamed silicate insulation is formed as a layer over the exterior of the well tubing; the layer is typically $\frac{1}{4}$ to $\frac{1}{2}$ in. thick.

(see Chapters 2 and 8)

BIBLIOGRAPHY

BOBERG, T. C., *Thermal Methods of Oil Recovery,* New York: John Wiley (1988).

MELDAU, R. F., "Reducing Well Bore Heat Loss," Reprints of papers in the Thermal Well Completions Seminar held at the 4th International Conference on Heavy Crudes and Tar Sands (UNITAR), Edmonton, Alberta (August 7–12, 1988). Presented and published by the Canadian Heavy Oil Association.

NEISEL, R. H. and VERSCHOOR, J. D., "Insulation Thermal," in Kirk-Othmer *Encyclopedia of Chemical Technology,* 3d ed., 13: 591–605, New York: John Wiley (1981).

Appendix 9

Thermal Properties of Steam

Saturation Pressure and Temperature

The steam saturation pressure corresponding to a temperature T °C can be calculated from the following correlations:

P in MPa and T in °C

$$\log_{10} P = 9.8809 - \left[2.42223 + \frac{326.57}{T + 273.15}\right]^2, \qquad 100°\text{C} < \text{T} < 275°\text{C}$$

This equation is due to Sanford Moss (1903).
Error lies in the range of $+0.6\%$ to -0.4% of P and

$$\log_{10} P = 4.4988 - \left(\frac{2043}{T + 273.15}\right), \qquad 275°\text{C} < T < 374.2 \quad \text{(critical temperature)}$$

Error lies in the range $\pm 0.6\%$.
These equations may also be rearranged to give the saturation temperature T_S °C in terms of the pressure.

$$T_S = \frac{326.57}{(\sqrt{9.8809 - \log_{10} P} - 2.42223)} - 273.15, \qquad 0.1 < P < 6 \text{ MPa}, \qquad T_S \text{ in °C}$$

Error ± 0.24°C

and

$$T_S = \frac{2043}{4.4988 - \log_{10} P} - 273.15$$

$$6 < P < 22.12 \text{ MPa} \quad \text{(critical pressure)}$$

$$\text{Error } \pm 0.4°\text{C}$$

These expressions are shown converted to British units next.

***P* in psia and *T* in °F**

$$\log_{10} P = 12.0424 - \left(2.42223 + \frac{587.83}{T + 459.67}\right)^2, \qquad 212°\text{F} < T < 527°\text{F}$$

$$\log_{10} P = 6.6602 - \frac{3677.3}{T + 459.67}, \qquad 527°\text{F} < 705.6°\text{F} \quad \text{(critical temperature)}$$

$$T_S = \frac{587.83}{(\sqrt{12.0424 - \log_{10} P} - 2.42223)} - 459.67,$$

$$14.5 \text{ psia} < P < 870 \text{ psia}$$

$$T_S = \frac{3677.3}{6.6602 - \log_{10} P} - 459.67,$$

$$870 \text{ psia} < P < 3212 \text{ psia} \quad \text{(critical pressure)}$$

Enthalpies of Saturated Liquid and Vapor:

These may be read from Table 4.6 or Figure 4.2.

The following correlations reproduce the enthalpies given in the steam tables within 0.15% and are useful in computer programs.

Temperature range 100–275°C

$$\mathbf{H}_L = -14.54 + 4.5196T - 0.002771T^2 + 0.00000922T^3$$

$$\mathbf{H}_V = 2523.43 + 1.3556T + 0.003561T^2 - 0.00001824T^3$$

$$\mathbf{H}_V - \mathbf{H}_L = 2537.97 - 3.1640T + 0.006332T^2 - 0.00002746T^3$$

Temperature range 275–357°C

$$\mathbf{H}_L = -3899.18 + 45.085T - 0.14418T^2 + 0.00017390T^3$$

$$\mathbf{H}_V = 9457.68 - 70.942T + 0.25514T^2 - 0.00031058T^3$$

$$\mathbf{H}_V - \mathbf{H}_L = 13356.86 - 116.027T + 0.39932T^2 - 0.00048448T^3$$

where T is in °C and $\mathbf{H}_V$ and $\mathbf{H}_L$ are in kJ/kg.

In order to minimize computation, it is desirable to convert polynomial expressions such as

$$y = a_0 + a_1x + a_2x^2 + a_3x^3$$

to the nested form

$$y = a_0 + [a_1 + (a_2 + a_3x)x]x$$

in computer programs.

BIBLIOGRAPHY

MOSS, SANFORD A., "General Law for Vapor Pressures," Phys. Rev. 16: 356–363 (1903).

Index